Earth's Changing Surface

Earth's Changing Surface

An Introduction to
Geomorphology

M. J. SELBY

CLARENDON PRESS · OXFORD

OXFORD
UNIVERSITY PRESS

Great Clarendon Street, Oxford OX2 6DP

Oxford University Press is a department of the University of Oxford.
It furthers the University's objective of excellence in research, scholarship,
and education by publishing worldwide in

Oxford New York

Athens Auckland Bangkok Bogotá Buenos Aires Calcutta
Cape Town Chennai Dar es Salaam Delhi Florence Hong Kong Istanbul
Karachi Kuala Lumpur Madrid Melbourne Mexico City Mumbai
Nairobi Paris São Paulo Singapore Taipei Tokyo Toronto Warsaw

with associated companies in Berlin Ibadan

Published in the United States
by Oxford University Press Inc., New York

First published 1985

British Library Cataloguing in Publication Data

Selby, M. J.
Earth's changing surface: an introduction to geomorphology.
1. Landforms
I. Title
551.4   GB401.5
ISBN 0-19-823252-7
ISBN 0-19-823251-9 (Pbk)

Library of Congress Cataloging in Publication Data

Selby, M. J. (Michael John)
Earth's changing surface.
Bibliography: p.
Includes index.
1. Geomorphology.   I. Title.
GB401.5.S45   1985   551.4   85-3062
ISBN 0-19-823252-7
ISBN 0-19-823251-9 (Pbk.)

10 9 8

Printed in Great Britain by the Alden Press
Osney Mead, Oxford

Preface

Since about 1960 the earth sciences have been undergoing revolutions in concepts which have been compared with those revolutions in biology and physics which are associated with Darwin's theory of evolution of organisms and Einstein's theory of relativity.

The most dramatic of these concepts in earth sciences is that associated with the theory of global tectonics, which provides a unifying view of how the major features of Earth's surface have evolved and taken their present form. The second revolution has developed from the still emerging ideas of how Earth's climate has changed in the last 50 million years or so, and particularly in the last 4 million years. The effects that these climatic changes have had, upon suites of weathering, erosional, depositional, and organic processes responsible for modifying the land surface, is still being clarified.

Less revolutionary, but no less fundamental, is the increasing concern of geomorphologists with the measurement and study of processes and the resistance of rock and soil to them. The themes of tectonics and structure occupy the first five chapters of this book. Processes and resistance are the concern of Chapters 6 to 15. The last part of the book is introduced in Chapters 16 and 17, where the significance of, and evidence for, climatic change and long-term evolution of landforms are discussed. Chapters 18 to 20 attempt to review the evidence for the impact of climate and climatic change upon the major bioclimatic zones of the continents. These chapters are both the summa-tion of the earlier chapters, and the place where the evidence for development of large landform units under complex environmental processes is discussed. Stress is laid upon the importance of relict forms and deposits, from Tertiary and Pleistocene times, for an understanding of the modern land-surface. These chapters are my attempt at providing that synthesis, of our art and science, which is seldom attempted but which should be the culmination of our collective efforts as geomorphologists.

In writing this book I have been greatly helped by friends and colleagues: my wife has shared with Frank Bailey and Ken Stewart the task of turning my rough drawings into understandable figures; Rex Julian and Wendy Forbes prepared all of the photographic prints. All photographs not other-wise acknowledged are my own. The onerous task of typing the text was undertaken with unfailing goodwill and cheerfulness by Christine White and Elaine Norton. I received valued advice and comment on parts of the text from Dr A. P. W. Hodder, Mr P. J. J. Kamp, Dr T. R. Healy, and unknown referees.

The fieldwork upon which many of my observations are based was carried out during leave from the University of Waikato and on expeditions of its Antarctic Research Unit. I am most grateful to the Council of the University for making this possible.

August 1983   M. J. Selby
Hamilton, New Zealand

Contents

Acknowledgements

Thanks are due to the following publishers and learned bodies for permission to reproduce copyright material.

Academic Press, Fig. 17.8.

Allen & Unwin, Figs. 6.8, 9.6, 10.12.

American Association for the Advancement of Science, Fig. 3.9.

American Association of Petroleum Geologists, Fig. 10.29.

American Geophysical Union, Figs. 1.5, 4.4, 4.6, 4.14.

American Journal of Science, Fig. 9.22.

American Society of Civil Engineers, Fig. 6.12.

Association of American Geographers, Figs. 1.6, 1.14.

Australian National University Press, Figs. 4.40, 11.8, 11.10f.

A. A. Balkema Publishers, Capetown, Fig. 19.6.

British Geological Survey, Fig. 5.16

Cambridge University Press, Fig. 3.4.

University of Chicago Press, Figs. 4.34, 9.1, 10.31, 16.9, 16.11

Clarendon Press, Fig. 7.8.

David and Charles, Fig. 16.1.

Dover, New York, Fig. 3.1.

Edward Arnold (Publishers) Ltd., Figs. 15.19c, d.

Elsevier Publishing Co., Figs. 4.23, 5.7a, b, c, 6.10, 9.6, 19.2.

Evolution, Fig. 16.8.

W. H. Freeman, Figs. A1, 5.2, 10.13, 10.17.

Gebrüder Borntraeger, Figs. 1.8, 1.11, 1.12, 4.30, 4.35b, c, d, e, f, 8.15, 8.16.

Geological Society of America, Figs. 3.7, 5.13b, 10.8a, b, 14.6, 16.9, 18.5, 4.19c, 4.35.

George Allen and Unwin, Figs. 6.8, 10.11, 10.12.

The Glaciological Society, Fig. 15.6e, f.

Her Majesty's Stationery Office, Fig. 5.16.

Institute of Arctic and Alpine Research, Fig. 14.3.

International Association of Engineering Geology, Figs. 1.13, 3.13a, b.

International Association for Scientific Hydrology, Fig. 9.20.

Interscience, London, Fig. 16.16.

Professor L. C. King, Figs. 17.4a, b.

Longman Group Ltd., Figs. 4.36, 4.39c, d, 4.41b.

Longman Paul Ltd., Figs. 5.15, 13.14a, 13.8a.

Macmillan Journals, Figs. 4.19a, b, d, e, 16.5a, b, 19.5.

Martinus Nijhoff, Fig. 4.29.

McGraw-Hill, Fig. 4.20.

Methuen, Figs. 13.26, 18.7c.

Mouton Publishers, The Hague, Figs. 16.13a, b, 16.15a, b.

National Association of Geology Teachers, Fig. 13.24.

Oliver and Boyd, Fig. 3.3.

Oliver & Boyd and Plenum Publishing Corp., Fig. 7.9a.

Oxford University Press, Figs. 6.5, 7.5, 8.4, 8.6, 8.13, 8.14.

Prentice-Hall Inc., Figs. 4.42, 13.7.

Princeton University Press, Figs. 1.8, 1.11.

Quaternary Research, Figs. 13.10b, 16.27, 16.32, 19.3, 19.9, 20.6.

A. H. and A. W. Reed, Fig. 4.11.

The Regents of the University of Colorado, Fig. 14.3.

The Royal Astronomical Society, Figs. 3.6, 4.32.

The Royal Geographical Society, Fig. 11.5.

The Royal Society, Fig. 3.5.

The Royal Society & Mrs. M. Synge, Fig. 13.10d.

W. B. Saunders, Fig. 3.15.

Science Information Division, N.Z.D.S.I.R., Figs. 5.6, 6.15.

Scientific American, Figs. 4.22, 13.11.

Seel House Press, Fig. 16.2.

Syracuse University Press, Figs. 4.43, 4.44.

United States Geological Survey, Figs. 5.3, 5.5, 5.9, 5.10, 8.10.

Van Nostrand Reinhold Co. Inc., Fig. 11.2.

John Wiley and Sons, Figs. 4.15, 8.17, 10.28, 15.16, 15.18, 15.20, 16.12, 16.16a, b, c, 16.17.
Wiley of Canada, Fig. 16.26.
Yale University Press, Fig. 16.30a.

National Aeronautics and Space Administration and the Technology Application Center, University of New Mexico, for satellite photographs.

Notation

A	area	T	transpiration
A_m	meander amplitude	U	approach velocity
BP	before present (by convention: before AD 1950)	u	pore water pressure
		V	stress induced by water in a joint, velocity
C	celerity		
C_s	suspended sediment concentration	v	velocity
		W	weight, width
c	cohesion	w	channel width
c'	effective cohesion	z	thickness, depth
c_j	cohesion along a joint	α (alpha)	angle of failure plane
D	diameter	β (beta)	angle of slope, angle of attack
d	depth, diameter	γ (gamma)	unit weight
E	energy, evaporation, Young's modulus	γ_{ice}	unit weight of ice
		γ_{sub}	submerged unit weight
e	kinetic energy	γ_w	unit weight of water
F	factor of safety, frictional drag	δ (delta)	change
F_e	Froude Number	ε (epsilon)	strain rate
G	rigidity modulus	θ (theta)	angle
g	acceleration due to gravity	λ (lambda)	meander wave length
h	thickness	μ (mu)	coefficient of dynamic friction
K	bulk modulus	v (nu)	Poisson's ratio, kinematic viscosity
L	length		
L_s	suspended sediment	ρ (rho)	density
l	length	σ (sigma)	mass density, stress
m	mass	σ_n	normal stress
n	number of particles, Manning's roughness coefficient	τ (tau)	shear stress
		τ_b	basal shear stress
P	precipitation, wetted perimeter	τ_c	plastic yield stress
p	pressure	$\bar{\tau}_o$	average boundary shear stress
Q	discharge	ϕ (phi)	angle of plane sliding friction
q_s	sediment discharge	ϕ'	effective angle of plane sliding friction
R	hydraulic radius		
R_e	Reynolds Number	$\tan\phi$	coefficient of plane sliding friction
S	storage, slope		
s	resistance to shear		

1   Basic Concepts of Geomorphology

Information on the surface of Earth is drawn from many sciences: geophysics, geochemistry, geology, climatology, hydrology, biology, pedology, glaciology, geomorphology, and others. Not one of these scientific disciplines can exist independently of the others and all earth scientists must have a broad understanding of the structures, materials, processes, and landforms which compose Earth's surface.

It is evident to anyone who has watched waves acting on a beach, or a river transporting sediment, that the surface of Earth is constantly changing and that the rate of change varies both in space across the surface and through time. There is thus a historical aspect to the study of the surface, as well as a dynamic element which is concerned with understanding the action of processes of change upon resisting rock and soil materials and geological structures. *Structures*, *materials*, *processes*, and the *history* of changing landforms, are the four essential components of a study of the nature and origin of the modern land surface, and they form the main sections of this book.

The existence of life on Earth, and the changes on its surface, are made possible by the size of this planet and its position in the solar system. If Earth had been so small as to lose its atmosphere because its force of gravity was too weak to keep the gases from escaping into space, or so large that its gravitational force had held the gases captive in the rocks, then there would have been no atmosphere, no water, no life, and no soil. The narrow range of temperatures occurring at the surface—from about $-80\ ^{\circ}\mathrm{C}$ to $+100\ ^{\circ}\mathrm{C}$—occupies only about 2 per cent of the range from absolute zero to the temperature of the Sun's surface. It is within this narrow range that all life and all surface processes occur.

Energy for landform change

The Earth's interior is a heat-engine, fuelled by radioactivity, running at a rate that is critical. Were it running more slowly all geological activity would proceed at a slower pace: volcanoes may not have spewed out the steam and other gases from which the oceans and atmosphere formed; iron might not have separated out to form a liquid core, and a magnetic field could not have developed; the resulting Earth would have had a cratered surface as lifeless as that of the Moon. An Earth with a faster heat-engine would be equally uninhabitable with widespread constant volcanic eruptions and earthquakes, and a dust-filled atmosphere of dense gases and violent electrical storms. Perhaps the Earth once had such a surface, but it is the present moderate speed of the heat-engine, the moderate temperatures, the existing size and position of the planet in the solar system which make possible the geological, atmospheric, and life processes which form our changing environment.

The internal heat drives convection currents in the mantle of plastic rock below the rigid surface of the Earth. The convection currents are thought to be the mechanism by which rigid plates of rock, composing the crust, are separated, pushed against one another, or rotated, causing great rifts in the crust where plates separate, as along the line of the Red Sea and African rift-valleys, or high mountain chains where the plates collide—as in the Himalayas. It is the motion of plates which gives rise to earthquakes, and the high heat flow towards the surface along the plate boundaries which produces volcanoes.

These internal geothermally driven processes are said to be *endogenetic* whereas the external forces operating at the ground surface, as a result of energy from solar radiation, are said to be *exogenetic* (Fig. 1.1).

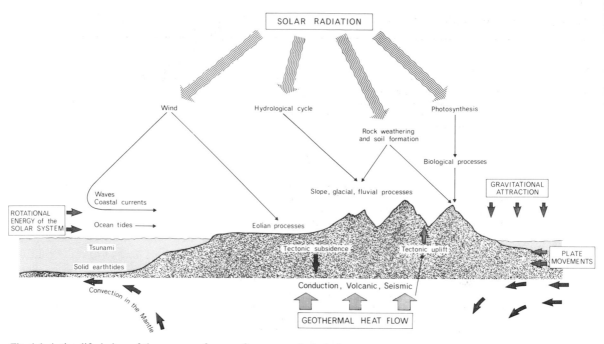

Fig. 1.1 A simplified view of the sources of energy for geomorphological processes.

Solar radiation intercepted by the Earth drives the atmosphere by setting up convection currents in it which are noticed as wind; it causes water to evaporate from the sea, rivers, and lakes, and so sets in motion the interchange of water, from the surface into the atmosphere and back to the surface, known as the *hydrological cycle*; and provides the energy for biological processes. The rotation of the planets and their positions in relation to one another, particularly to the Sun, Moon, and Earth, cause tides, which are most noticeable in water bodies and which have most obvious effect on the coasts, but which also cause motions (earth tides) in rocks of the crust.

Gravitational forces provide energy less directly but, by attracting all surface material towards the centre of the Earth, impart potential energy to rock and soil in elevated positions.

Inheritance from the past

Landforms are not created instantaneously. The smallest, such as a ripple in dune sands, may develop in a few minutes, but large landforms of continental dimensions, or even large units of continents such as mountain chains or extensive plains, have a history of tens or hundreds of millions of years. It is usually true that the larger the landform the more complex and the longer is its history; consequently the methods used to study it are related to its magnitude (Table 1.1). The environment at the surface of the Earth is not constant: climate changes partly because the amount of solar radiation trapped in the Earth's atmosphere is not constant, and partly because the atmosphere and the surfaces of land and ocean are changing. The geothermal heat flow is also not uniform through geological time and varies from one part of the crust to another. These changes, together with the gradual and slow evolution of life, and therefore change in biological processes, require that the breakdown of rock, and its removal from the ground surface, was different in the geological past from what it is today. Many processes which shape landforms are controlled or strongly influenced by organisms. This is obvious from the protective effect of vegetation upon soils. Seventy million years ago no sod-forming grasses had evolved and, although herbaceous plants and shrubs no doubt covered many hills, it seems probable that processes of erosion may have been more effective in the absence of grasses than they are where grasses form a closed ground cover. Animals, especially *Homo sapiens*, have also done

Table 1.1   Scale of Landform Units

Magnitude	Size km²	Example	Basis
1st	10^7	(i) Continent, ocean basin (ii) Mountain chain (Andes to Alaska Cordillera) (iii) Tropical zone	Geophysical. Geotectonic. Climatic.
2nd	10^6	(i) Baltic shield (ii) Mountain range (Rocky Mts.) (iii) Tropical rain forest	Subdivisions of 1st magnitude.
3rd	10^5	(i) Black Sea (ii) Anti-Atlas, Erg Chech (iii) Sahelian zone	Subdivisions of 2nd magnitude. Structural units, sub-divisions of main climatic zones.
4th	10^4	Massif Central, Taupo Volcanic Zone, Weald	Structural unit of limited regional extent. Single set of geomorphic processes.
5th	10^3	Chalk hills of Picardy–Normandy, Cotswolds, Chott Djerid	Tectonic and often lithologic individuality, lower limit of isostatic compensation.
6th	10^2	Individual anticlines and synclines in a folded range, cuestas, fault block, volcano	Lithology important, no isostatic compensation. Repetition of relief features.
7th	10	Local relief, inselberg, pediment, river gap, estuary	Lithology and exogenetic processes dominant.
8th	1	Local mudflow, gully, barchan, tor	Dominated by exogenetic processes.
9th	m²	Polygon, terracette	Repetition of forms, statistical and experimental studies relevant.
10th	mm²	Ventifact, ripple	

much to reshape the land surface, particularly in the last few thousand years.

The 'normal' condition of the Earth's surface is for there to be no permanent ice sheets or large ice caps. In the last 20 million years (20 My), and perhaps longer, there have been large volumes of ice on Antarctica and in the last four million years on Greenland also. Over the last 2 My ice has periodically built up on the northern hemisphere continents and then wasted away. In the last 1 My these periods of ice sheet formation on northern North America and north-western Europe have been rather regular in duration, with cold periods of severe ice accumulation lasting for 70 000 (70 ky) to 100 ky and intervening periods with climates similar to those of the present lasting 10–30 ky. At the same time as ice accumulated and melted, climate in other parts of the world changed, with variations in temperature, precipitation, and storminess.

The severe changes of climate which are characteristic of the last 20 My have been accompanied by major movements of the Earth's crust. Huge plates forming the rigid surface of the Earth have been moving with respect to one another at rates of up to 10 cm/y and mountain ranges have been rising at rates of up to 1 cm/y. Earthquakes, volcanic eruptions, and crustal subsidence have also added their effect to modifying the surface.

Past exogenetic and endogenetic processes have obviously not proceeded with the same intensity and distribution as they do at the present time: even the dominant processes at one locality are not constant. What has not changed are the basic physical and chemical laws which govern the development and alteration of rocks. It is an article of faith of the scientist that these 'natural laws' are constant and it is this belief which we accept as the *principle of actualism.*

Changes in intensity, kind, and distribution of processes have sometimes left their imprint upon the land surface. Thus old soils may be left buried by younger deposits at the base of a hill, or an irregular mound of rock fragments left as a deposit by a now vanished glacier. These deposits provide clues to the past processes which have acted in the landscape and also show that the modern land surface has a history which has influenced its present form.

Time-scales

The age of the Earth is thought to be about 4 600 millions years (4.6 Gy) and the oldest known rocks are about 3.8 Gy. Geologists have divided the life-span of the Earth into a time-scale (Fig. 1.2). The longest time units are *eras*: each era may be subdivided into *periods*, and each period into *epochs*. The names of the periods are taken from a geographical locality in which the *formations* (large rock units) of that period were either first studied or are well displayed. The Jurassic Period, for example, is named from the Jura Mountains of the France–Switzerland border. Correlation between rock units of similar ages, and subdivision of geological time from the earliest Paleozoic to the present, is based upon the assemblages of fossil organisms found in the rocks.

For geological purposes the time-scale must necessarily cover all of the time in which rocks have formed, but landforms of today have developed during only a small part of that time. The ocean basins are no more than 200 My old and the oldest plains on Earth have developed over much the same time-span. Most landforms are far younger that this with a history that seldom extends back before the beginning of the Cenozoic Era and more commonly has developed in the last

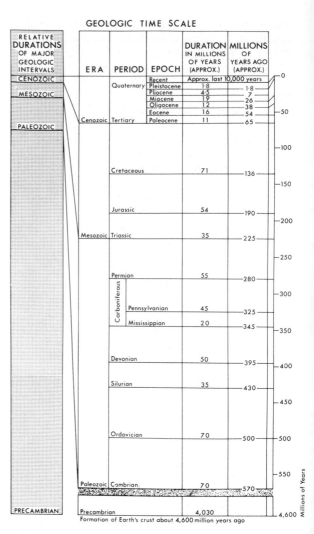

Fig. 1.2 The geological time-scale for the Phanerozoic, or time since the Precambrian. The scale is under constant review and several versions are in use.

20–30 My. Individual river terraces and small valleys may be less than 20 ky old, and many landslide scars and gullies only a few days or years old. The time-span with which the student of landforms (i.e. a geomorphologist) is usually concerned is thus the middle to late Cenozoic with particular emphasis on the Quaternary. The age of the lower boundary of the Quaternary is not agreed by all earth scientists but is commonly given as either 1.8 or 1.6 My.

The traditional method of geological dating using the evolution of fossils is one of *relative dating* in which one organism, or one rock unit, is regarded as being older than another. Relative dating is consequently based upon interpretations of sequences of events and materials and is contrasted with *absolute dating* methods which provide ages in years.

Most relative dating methods rely upon application of one or more of the three basic principles of stratigraphy. (1) *The principle of superposition* states that in any succession of layered rocks (*strata*), not severely deformed, the oldest stratum lies at the bottom, with successively younger ones above. This principle is the basis of relative ages of all strata and their contained fossils. (2) *The principle of original horizontality* states that because sedimentary particles separate from fluids under gravitational influence, stratification originally must be nearly horizontal; steeply inclined strata, therefore, have suffered subsequent disturbance. (3) *The principle of original lateral continuity* states that strata originally extended in all directions until they thinned to zero, or terminated against the edges of the original basin of deposition.

Isotopic dating

The absolute ages of rock units, and therefore of divisions of the geological time-scale, are derived from measurement of radioactive isotopes. For the age of a rock to be determined a mineral is needed that contained atoms of a radioactive isotope when it originally crystallized. Each original, that is *parent isotope*, decays to a *daughter isotope* which is retained in the same crystal. The rate of decay is usually expressed in terms of the *half-life*, which is the time required for one-half of the original number of radioactive atoms to decay. At the end of the first half-life after a radioactive element is incorporated into a new mineral, one-half is left; at the end of a second half-life, one-quarter is left; at the end of a third half-life, an eighth is left, and so on. If the data are plotted on a graph the decay rate can be seen to be exponential (Fig. 1.3).

To date an igneous rock, crystals such as zircon, muscovite, or biotite, are extracted from a sample. Such minerals contain the isotopes uranium-235 and uranium-238 which are gradually converted to lead-207 and lead-206 respectively. The crystals can be dissolved, to get the uranium and lead into a solution; and ratios of $^{235}U : ^{207}Pb$ and $^{238}U : ^{206}Pb$

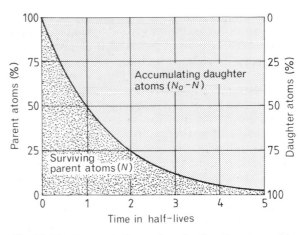

Fig. 1.3 The decrease in the number of radioactive atoms with time. On a scale with the percentages plotted logarithmically the decay curve is a straight line. If N_0 is the original number of parent atoms, and N is the number surviving after any time, the decay process can be expressed as: $N = N_0 \times (\frac{1}{2})^n$, where n is the number of half-lives of the decay system concerned.

can be measured with a mass spectrometer. Provided that all of the uranium and lead isotopes have been retained within the crystals since they first formed, and that no daughter isotopes were present originally, the age of crystallization can be calculated because the rates of decay of all the common radioactive isotopes are constant and are known from accurate laboratory measurement on pure samples.

The decay series of uranium to lead isotopes, and of rubidium to strontium is useful only for minerals older than 10 My and so is seldom of use to geomorphologists, but the decay of potassium-40 to argon-40 ($^{40}K : ^{40}Ar$) can be used to date minerals from nearly all igneous and metamorphic rocks as young as 100 ky, and is of great importance for dating both whole rocks and volcanic ashes.

Sea water contains an appreciable quantity of uranium in solution, but virtually none of its daughter isotopes. This means that carbonate organisms such as plankton, corals, and shellfish, and inorganic carbonates, will have taken up uranium, but none of its daughter isotopes. Provided that a sample of carbonate has remained a closed system, not taking up or losing daughter isotopes, then the extent to which daughter isotopes of uranium have appeared in the sample can provide a measure of age. The extent to which thorium-230 (^{230}Th) and protoactinium-231

(^{231}Pa) have appeared can be used to give ages back to 300 ky and 150 ky respectively. The ^{230}Th method has been very important in dating corals, and hence in dating changing sea-levels, and in dating flowstone and other limestone cave carbonates.

Organic materials younger than 40 ky are commonly dated by the carbon-14 (^{14}C) they contain. During growth, plants and animals steadily incorporate a small amount of ^{14}C, along with other carbon isotopes contained in the CO_2 of the atmosphere, into their tissue. At death this absorption ceases and the relative amount of ^{14}C in the tissue is the same as the ratio in the atmosphere (we assume that this ratio has remained nearly constant during the last 100 ky), but it steadily decreases after death as the ^{14}C radioactivity decays. The amount of ^{14}C remaining is measured by counting the decaying particles being emitted by a sample, and this count is used to calculate the age since death of the organism.

The method of counting decaying particles permits ages as old as 40 ky BP (Before Present) to be determined, and many thousands of dates have been measured by this method. Since 1975 improved methods, and especially the use of particle accelerators to measure directly the ^{14}C atoms in the sample, rather than count the decay particles, has permitted the dating of materials as old as 75 ky and may eventually allow dating of materials as old as 100 ky.

Carbon-14 dating is of great importance to studies of landform change and archaeology during the last 40 ky, but some dates are suspect because the samples have been contaminated by older carbon, such as lignite or coal, or younger carbon carried by percolating ground water. Soil organic materials, shells of coastal organisms, and bone are notoriously difficult to date reliably.

Other methods of dating

Fission track dating of volcanic glasses and minerals, such as apatite, zircon, sphene, and mica, is possible because they contain traces of ^{238}U. This decays by spontaneous fission (explosive division into two fragments). The rates of the fission are very slow—^{238}U fissions spontaneously with a half-life of about 8×10^{15} years. The amount of energy released in each spontaneous fission is relatively large at about 200 meV, and the resulting fragments tear into the surrounding material

for a distance of about 10 micrometres (μm) before they are stopped. The resulting damage can be revealed by etching the glass or crystal with hydrofluoric acid.

The number of fission etch pits depends only upon the accumulation period (age) and the uranium content of the sample. If the uranium content is known, the fission track age can be deduced from the number of fission tracks. The method is of increasing importance in volcanic regions such as New Zealand, Japan, and parts of USA where it has provided ages not only of volcanic events, but also of sediments, soils, and other deposits interlayered with volcanic ashes. The method has been most commonly applied to materials ranging from 1 ky to 5 My in age.

Dendrochronology, or tree-ring chronology, is a useful technique for dating living wood. Annual growth rings are composed of wood cells with large lumens (i.e. spaces in the cells) which grow in the spring, and smaller cells in summer and autumn. The number of rings of summer or winter cells is an indicator of tree age. The technique is usefully applied to dating young deposits left by glaciers or rivers where the time for colonization of the fresh deposits by trees can be estimated. Most dendrochronological ages are for the last 1–2 ky, but chronologies for 5–7.5 ky have been produced for study of past climates and archaeological remains by overlapping of patterns of rings in which years of stress, such as droughts, have produced recognizable signatures (Fig. 1.4).

Weathering rinds result from the chemical alteration of a 'skin' of rock around an unaltered core. On deposits of similar materials, but left at different times by rivers or glaciers, it is often possible to detect differences in the depth or colour of the alteration rind. This observation does not provide an 'age' for the time since deposition, but it does make possible the mapping and identification of materials of different phases of deposition.

Rhyolitic volcanic glass (obsidian) undergoes progressive alteration through the slow absorption of water. The thickness of the hydrated outer layer, or the number of hydrated and spalled layers from a pebble, have been used as indications of relative age of the deposit on which the pebble occurs. At some sites the obsidian hydration can be set into a time-scale provided by ^{40}K : ^{40}Ar dates of the volcanic rock.

Lichenometry is another relative method of dating in which the diameter of a rosette of lichen

A.

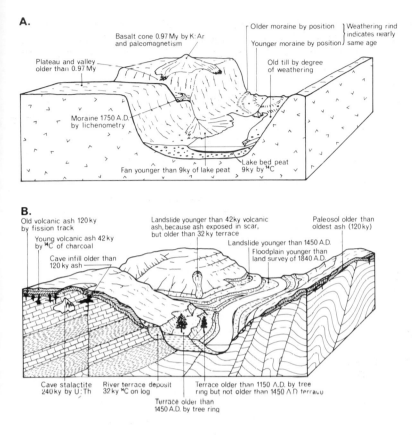

Basalt cone 0.97 My by K:Ar and paleomagnetism

Plateau and valley older than 0.97 My

Older moraine by position ⎫ Weathering rind indicates nearly same age
Younger moraine by position ⎭

Old till by degree of weathering

Moraine 1750 A.D. by lichenometry

Fan younger than 9ky of lake peat

Lake bed peat 9ky by ¹⁴C

Fig. 1.4 Sites at which various marker beds or features may be used to date deposits and, by implication, the depositional landforms composed of them. The landform surface cannot be older than the youngest material below or on it; it may be younger than the youngest marker found in the region; that marker may not have formed on the landform, or it may have been removed by erosion. Most landscapes have landforms with a range of ages.

B.

Old volcanic ash 120 ky by fission track

Young volcanic ash 42 ky by ¹⁴C of charcoal

Cave infill older than 120 ky ash

Landslide younger than 42 ky volcanic ash, because ash exposed in scar, but older than 32 ky terrace

Landslide younger than 1450 A.D.

Floodplain younger than land survey of 1840 A.D.

Paleosol older than oldest ash (120 ky)

Cave stalactite 240 ky by U:Th

River terrace deposit 32 ky ¹⁴C on log

Terrace older than 1450 A.D. by tree ring

Terrace older than 1150 A.D. by tree ring but not older than 1450 A.D. terrace

is assumed to be a direct indication of the time since deposition of the rock on which the lichen is growing. The method is simple to apply, requiring the use of a ruler only, and has been used to date many glacial moraines formed in the last 1 ky.

Paleomagnetic dating is an increasingly important method for correlating deposits. It is based on the knowledge that the Earth's magnetic field resembles that of a bar magnet located at the Earth's centre. The axis of the imaginary bar magnet (the *geomagnetic axis*) emerges from the Earth's surface at the *magnetic poles* where it presently forms an angle of about 20° with respect to the *geographic axis*. The geographic and magnetic poles do not coincide.

The Earth's magnetic field has undergone repeated changes in polarity in which the north and south magnetic poles have switched places while the axis has remained unchanged in position. This phenomenon is called *magnetic polarity reversal*. Reversal occurs by a weakening of the geomagnetic field to zero and then a rebuilding of the field in the opposite direction, i.e. reversed polar-

ity. The time-intervals between polarity reversals seem to be irregular.

The pattern of reversals is preserved in volcanic rocks because basaltic lavas contain minor amounts of oxides of iron and titanium. At the high temperatures of a molten lava the minerals have no natural magnetism but, as cooling occurs, each crystallized magnetic mineral passes a critical temperature or *Curie point* (770 °C for iron). The newly formed mineral particles assume a direction parallel with the lines of force of the Earth's magnetic field. With further cooling this *thermal remnant magnetism* becomes permanently locked within the solid rock and is a record of the direction of Earth's magnetic field at the time of cooling, that is a record of paleomagnetism. The time of reversals can be obtained from $^{40}K:^{40}Ar$ dates of basalt crystallized just before and after the reversal.

Certain minerals produced during weathering, such as hematite (Fe_2O_3), and some magnetic minerals laid down in sediments in oceans and lakes, also preserve a record of polarity at the time of their deposition. This *detrital magnetism* permits correlation of weathering products, soils, and sediments with the record derived from basalt.

The pattern of reversals during the Cenozoic is now well known and dated (Fig. 1.5). Polarity such as that of today is referred to as a *normal epoch* and opposite polarity a *reversed epoch*.

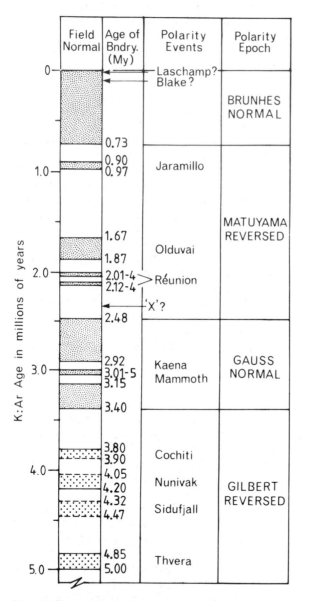

Fig. 1.5 The geomagnetic time-scale for the Cenozoic as revised in 1979. The scale differs from earlier versions and is itself liable to be revised (after Mankinen and Dalrymple, 1979).

Geomagnetic epochs last for hundreds of thousands of years but they may be interrupted by polarity *events* of tens of thousands of years. The geomagnetic time-scale is most commonly used to correlate and date marine deposits, and igneous rocks, but it is also valuable for correlating and dating some terrestrial deposits.

Objectives and history of geomorphology

Geomorphology may be defined as the science which studies the nature and history of landforms and the processes of weathering, erosion, and deposition which created them. As such it has attracted, and overlapped with, the work of geologists, geographers, soil scientists, and hydrologists.

It is possible to trace geomorphic ideas in the writings of such early scientists as Aristotle and Leonardo da Vinci, but modern earth sciences started to flourish only in the eighteenth century when a few workers began to recognize the evidence available in 'the field' and to accord it priority over the 'revealed truth' as accepted in literal interpretations of the Bible.

Amongst the most important early works is *The Theory of the Earth, with Proofs and Illustrations* by James Hutton, which was published in 1788 in volume one of the Transactions of the Royal Society of Edinburgh. In this work Hutton denied the possibility of the Earth being created during the six days of Creation, described in the book of Genesis, and the probability of catastrophic causes—such as the Flood on which Noah launched his Ark—modelling the Earth's surface. Hutton, from his extensive field-work, deduced that landforms are shaped by the slow action of running water eroding the land, and that the waste deposited by rivers in the sea is the material from which sedimentary rocks are formed. He postulated uplift of these rocks to form new land which in its turn would be eroded by water. The continuing processes of erosion, sedimentation, and uplift, acting at a steady rate in which Hutton saw 'No vestige of a beginning, no prospect of an end', are the basis of what was to become the *principle of uniformitarianism*.

Hutton's work was largely unrecognized during his lifetime; partly because the Transactions of the Royal Society of Edinburgh had a very limited circulation, and partly because his writing style is difficult and often obscure. He was, however, most fortunate in his friend, John Playfair, Professor of Mathematics at Edinburgh University, who was a believer in the validity of Hutton's work and possessor of a lucid style. Playfair edited and rewrote Hutton's work and in 1802 published his book *Illustrations of the Huttonian Theory of the Earth*. Playfair made many original contributions and provided detailed accounts of rivers, erosion, and raised beaches. His work on rivers was particularly important and he insisted that (1) rivers cut their own valleys; (2) the angle of slope of each river shows an adjustment towards equilibrium of the velocity and discharge of water with the amount of material carried; and (3) a whole river system is integrated by the mutual adjustment of the constituent parts. The first statement is still recognized as 'Playfair's Law', the second foreshadows the modern concept of grade, and the third represents the principle of accordant junctions.

Playfair's work excited many attempts at rebuttal and cannot be regarded as a widely accepted view until Charles Lyell, a brilliant talker, writer, and systematizer, started the publication of his work *The Principles of Geology* which underwent a series of revisions, in many editions, from 1833 to 1875. Lyell was unusual in that he travelled widely and was able to interpret his observations in such a manner that Ramsay, the first Director of the British Geological Survey, was able to say of him 'We collect the data and Lyell teaches us to comprehend the meaning of it.' Lyell was the high priest of uniformitarianism and so the most prominent opponent of the supporters of catastrophist ideas, especially that of the Flood. The one major defect of Lyell's work was that towards the end of his life he became convinced that streams could not erode land masses to form plains, but that marine erosion must be responsible for such bevelling and that the scarps of such areas as south-eastern England are old sea cliffs.

The theory of the Flood died hard because there was considerable evidence which could be interpreted as support for it. In many parts of Europe deposits comprising boulders and gravels, loosely known as *drift*, were found far from their probable source and sometimes in such unexpected sites as hilltops. The obvious conclusion was that they were left by the Flood. In 1836 Jean de Charpentier, Director of Mines for the Canton of Vaud, and Louis Agassiz, Professor of Natural History at Neuchâtel and a world authority on fossil fishes, together examined the glaciers around Mont Blanc. Charpentier convinced Agassiz of the power of glaciers to erode and transport rock debris. Agassiz was thus not the originator of the concept of glaciation—that honour is shared amongst a number of Swiss and French naturalists, alpine guides, and engineers—but it was not until Agassiz published his *Études sur les Glaciers* in 1840 that an alternative glacial origin for drift (which we now know to be glacial till, erratics, and outwash deposits) became acceptable.

Agassiz elaborated a theory of an ice age, even proposing that the ice formed a cap through which the Alps pushed up and caused the ice to shatter so that huge icebergs rafted blocks of rock debris away from the mountains to the surrounding lowlands, where they were left after the ice melted. It is probably fair to say that most of Agassiz's original ideas about glaciation were wrong, but he lent his prestige to the ideas of lesser-known men and spread these ideas by his writing and lectures. As a traveller, however, Agassiz did demonstrate the applicability of theories of glaciation to highland Britain and later to North America. In Britain it was Lyell who produced the most convincing evidence in favour of a glacial origin for drift deposits, and he had a powerful influence on the acceptance of glacial theory.

European ideas were spread to North America by Lyell who travelled in eastern USA and south-eastern Canada, and by Agassiz who went to lecture in Boston in 1846 and finally made his home in America.

North America

Surveys carried out by the United States Geographical and Geological Survey in western USA, during the second half of the nineteenth century, form the basis for many important contributions to geomorphological theory. Outstanding contributions came from J. W. Powell, C. E. Dutton, and, above all, G. K. Gilbert. In the semi-arid areas the relationship between landforms and processes, and landforms and geological structures, was both clearer and more readily demonstrated than in the fully vegetated and soil-covered areas of eastern USA and Europe. The great size of such features as the Grand Canyon of the Colorado testified to the power of rivers to erode, but Powell's explanation of the history of the Colorado Plateau was a pioneer study in another sense—it added greatly to geomorphological theory by developing a new classification of mountains, by structure and genesis; he produced a classification of dislocations; a classification of valleys; and a genetic classification of drainage systems. His classification of drainage involved the new ideas of superimposed drainage and antecedent drainage; associated with it was the idea that the physical history of a region might be read in part from a study of its drainage system in relation to its rock structure. Perhaps the most influential new idea he produced was that of base level.

G. K. Gilbert was, for many years, Powell's assistant and was perhaps the most outstanding student of geomorphological processes in the history of the subject. Gilbert's *Report on the Geology of the Henry Mountains* (1877) is the first major treatment by any geologist of the mechanics of fluvial erosion. He wrote more comprehensively, in English, about weathering and erosion than anyone before him, and systematized much of his own and Powell's views on erosion, transport, deposition, and equilibrium. Under the terms *law of uniform slopes, law of declivities, law of divides*, and *law of structure* he expressed the basis of modern knowledge on rock resistance, and the relationship between slope, energy available for erosion, and stream discharge: his work on rivers culminated in the monograph *The Transportation of Debris by Running Water* (1914).

Twentieth century

Unhappily the pioneer work of theory development by Powell and Gilbert became overshadowed by the vast outpouring of papers and books by W. M. Davis who, at times, seemed to confuse his theories with basic data and became more involved with concepts of evolution than with understanding the mechanisms of landform change. In the English-speaking countries, Davis's theory of a cycle of erosion eventually became more pervasive than it merited. To some extent this was due to his disciples C. A. Cotton in New Zealand and S. W. Wooldridge in England. The cycle of erosion and its associated method of denudation chronology are discussed more fully in Chapter 17.

During the twentieth century three alternative approaches to that of Davis can be identified. The German Walther Penck made crustal instability and movement the core of his approach to land-

forms, and hillslopes in particular. In contrast to Davis, Penck emphasized that crustal movements are often continuous, of varying intensity, and relatively long-lived. He attempted to use the forms of hillslopes as indicators of rates of uplift, and especially of variations in the rates. The stepped profiles of many spurs in the Black Forest, for example, he regarded as evidence for phases of uplift broken by periods of relative stability. The interest in crustal instability over Late Cenozoic time, also called the study of *neotectonics*, is still an important theme in many Russian works, especially those of Yu. A. Mescherikov and I. P. Gerasimov.

The second theme is that of climatic geomorphology, and the third that of process studies.

The influence of climate on landform creating (*morphogenetic*) processes and associated landforms had been evident to the explorers of the American West, but elsewhere most of the early contributors to this branch of study were Germans: von Richthofen in Asia, and in Africa J. Walther and S. Passarge. All three had worked in deserts and put much emphasis on the effectiveness of wind action. When they were working little was known of climatic change and they did not believe that running water could be a major contributor to desert planation.

Concepts of zonality and climatic influences in geomorphology owe much to the work of the Russian pedologist Vasilü Dokuchayev and he influenced such European workers as E. de Martonne whose *Le Climat Facteur du Relief*, published in 1913, ranks with the proceedings of the 1926 Düsseldorf Symposium *Morphologie der Klimazonen*, edited by F. Thorbecke, as a pioneer effort in this area of study. Other major works on landforms and climate include E.-F. Gautier's *Le Sahara* (1923), H. Mortensen's study of the north Chilean desert (1927), Karl Sapper's *Geomorphologie der feuchten Tropen* (1935), and more recently the major studies of J. Büdel in Germany and those of J. Tricart and A. Cailleux in France (see below), with the shorter collection of essays by P. Birot *Le Cycle d'Érosion sous les Different Climats* (1960).

Quantitative and process geomorphology has a relatively long history in Europe and is founded in the need of engineers to understand the power of fluvial erosion in order to control alpine torrents. Alexandre Surell's *Études sur les Torrents des Hautes-Alpes* (1841) ranks with the work of G. K.

Gilbert in this regard. During the first half of the twentieth century progress still depended upon the work of a limited number of outstanding people such as R. A. Bagnold whose *Physics of Blown Sand and Desert Dunes* (1941), is still the leading work in that area. In fluvial studies the contributions of F. Hjulström and A. Sundborg, in the 1930s and 1950s respectively, were the forerunners of the seminal papers of the United States Geological Survey investigators, and especially L. B. Leopold, published in the 1950s. Also at this time emerged the trend towards statistical analysis of quantitative data, and this owes much to the influence of A. N. Strahler.

Since about 1950 detailed process studies have progressed in four ways: firstly, they are more firmly based in the physical sciences and especially upon detailed measurements made in both the field and laboratory; secondly, analysis is more rigorous and makes full use of statistical techniques; thirdly, there are now considerable numbers of investigators so large volumes of data can be collected, and often processed with the aid of computers, and comparisons made between phenomena at several sites. The outstanding thinker and scientist is still essential to the development of theory and research ideas, but the leaders are now backed by a considerable body of investigators. A fourth trend has also made many of these detailed investigations possible—that is the provision of funds for research positions in universities and for field and laboratory equipment.

Until recently geomorphology, like many other branches of science, was almost exclusively a European and North American pursuit, with knowledge about the rest of the world being largely reported by people from these two regions. The need to assess the resources of newly developing lands, and the increases in university education around the world, have assisted the establishment of indigenous groups of geomorphologists in many countries. Developing countries have recruited many of the initiators of research and teaching from Europe and USA in the period since 1945, but the current generation of young scholars was born locally. In Australia and Papua New Guinea, the Land Resources and Regional Surveys of the Commonwealth Scientific and Industrial Research Organization (CSIRO) not only provided new information on landforms, soils, water resources, and vegetation, but also provided extensive field experiences for people who later set up research programmes in the local universities. Research, especially in the area of Quaternary studies, is flourishing in several parts of South America and Africa where local investigators often work with visiting scientists. In Japan geomorphology is thriving, with several large institutions producing unique work in process study, neotectonics, and volcanism: unfortunately a language barrier inhibits full appreciation of Japanese achievements. Similarly much Russian work is little known, and it often reaches English speakers at second hand via French and German reviews.

A trend since about 1960 is the recognition that geomorphology has an important contribution to make to *geotechnology* (earth sciences applied to the solution of engineering problems) and environmental management. Applied geomorphology offers not only employment for geomorphologists but a valuable testing ground for geomorphological research.

Review

The most important attempts to explain the major features of the Earth's surface are necessarily concerned with world-wide patterns and phenomena. By far the most important, convincing, and powerful concept concerned with the origin of major landforms and structural features is the theory of plate tectonics: this forms the subject-matter of Chapters 2, 3, and 4. The search for an equivalent theory relating landforms and climatically induced processes has, as yet, been less successful. The relationship between landforms and individual processes forms the subject-matter of Chapters 7 to 15, but the most important broadly based approaches to the understanding of exogenetic processes and landform development, and of the recently developed study of human influences on the landscape are discussed briefly in the five remaining sections of this chapter.

The development of a scheme of (1) *climatic geomorphology* relating suites of landforms to the climatic environment responsible for current and dominant exogenetic processes has been a major concern of both French and German geomorphologists. (2) *Climatogenetic geomorphology* is a concept developed particularly by German scientists; it seeks to explain the historical development of the landscape in terms of the processes which have shaped it over the latest part of geological time and pays attention to the interpretation of past environments from relict features. In the last 20 years, or so, a concern with the nature of exogenetic processes has been accompanied by an emphasis on segments of the evolving modern landscape in terms of the balances, or (3) *equilibrium* between landforms and processes. The study of (4) *geomorphic systems*, in which the interrelationships between the components of the landscape and the processes which modify them are explored, has become an important part of the recognition that (5) *people are major geomorphic agents*, as well as being a framework for many integrated studies.

Climatic geomorphology

Two broad and opposing views of the influence of climate on landforms may be recognized in the geomorphological literature. Opponents of a climatologically based approach may deny that differing suites of processes create recognizably different landforms: L. C. King (1957), for example, said 'all forms of hillslope occur in all geographic and climatic environments'; and A. N. Strahler (1952) wrote,

The geomorphic processes we observe are, after all, basically the various forms of shear, or failure of materials which may be classified as fluid, plastic or elastic substances, responding to stresses which are mostly gravitational but may also be molecular . . . the type of failure . . . determines the geomorphic process and form.

In contrast to the King and Strahler approach, which is often associated with English-speaking scientists, and especially with geologists and engineers, is that which stresses the effect of climatic factors upon vegetation and soils and, through their effects, the creation of distinctive rates and styles of landform development in different climatic zones. This second approach is most commonly associated with French- and German-speaking geographers: it is perhaps most sharply expressed in the contention of J. Büdel that exogenous processes, in their climatic variation, create the morphological picture of the Earth, while geological structures and tectonics merely influence and modify local landforms.

At least seven problems have to be recognized in any scheme of climatic-landform classification: (1)

the limited value of climatic statistics for process study; (2) variation in the degree to which processes are confined to specific climatic zones; (3) the scale at which processes create distinctive landforms; (4) the influence of tectonic, structural, and lithological effects on landforms; (5) selection of criteria for defining climatically controlled landform associations; (6) the influence of man; and (7) landforms relict from past processes.

Climatic statistics

A very simplistic approach to defining the zonation of geomorphic processes is illustrated in Fig. 1.6. Using mean annual temperatures and rainfalls only, L. C. Peltier attempted to recognize those conditions under which particular processes are likely to be most effective and then, by combining them, identify the dominant processes with regions which are specified partly on the basis of climate (e.g. arid, periglacial) and partly by the dominant vegetation growing in a climatic zone (e.g. selva, savanna). It has been assumed that the result identifies a *morphogenetic region*, that is a region in which a distinctive complex of erosional, transportational, and depositional processes is responsible for landform development.

The major deficiencies of this approach are: (1) it fails to recognize the importance of intensity and periodicity of geomorphic processes in landform development; by failing to include variations in seasonality of climate, and extreme events such as storms, it assumes that the average condition of climate is dominant; and (2) it fails to recognize that soil and vegetation are interposed between the climatic elements and the developing landforms.

Only in deserts and glacial regions is the climate directly effective, because only there is there a widespread absence of vegetation. Outside the biological deserts the impact of raindrops, rate of infiltration, temperature changes, velocity of surface runoff and storage of water, are all modified by the vegetation cover; furthermore vegetation directly influences the soil cover by its provision of organic matter, its effect on soil weathering, permeability, aggregation, and therefore soil erodibility.

Underlying a simple average climate approach is an assumption that processes act in the same way in all climates, but we have only to note the effects of greater air density in cold climates upon the transporting capacity of air, or the influence of

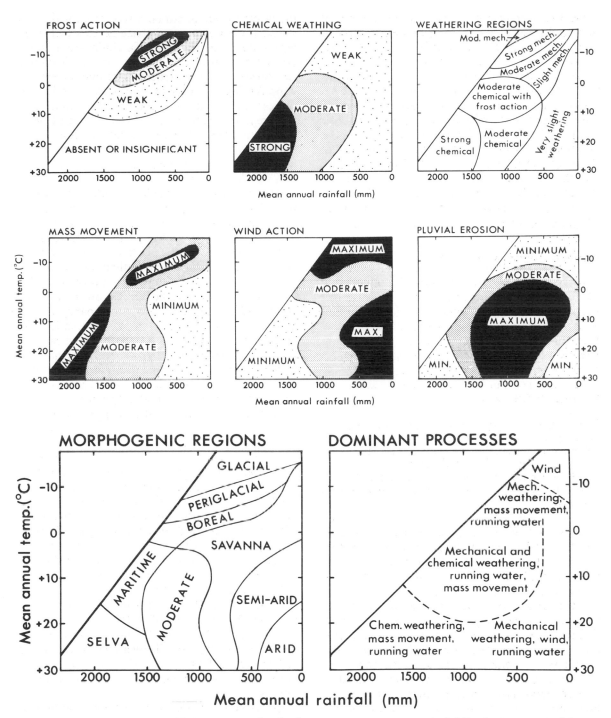

Fig. 1.6 Diagrams showing L. C. Peltier's attempt to relate dominant processes to mean annual rainfall and temperature and thus to define morphogenetic regions. The diagram of dominant processes excludes glacial processes.

snowmelt and soil ice upon runoff and fluvial processes in the periglacial zones, to recognize that this is not the case.

Processes and climatic zones

Every phenomenon or process whose global expression and extent is closely related to a specific climatic region may be said to be *zonal*. Thus tropical rainforests, coral reefs, and polar ice sheets are clearly zonal and their environments constitute *morphoclimatic zones*.

Many processes are world-wide, or occur in several zones, and are therefore said to be *azonal*. Thus wave action on beaches and all endogenetic processes are azonal.

Some phenomena may be primarily characteristic of a particular morphoclimatic zone but can occur elsewhere in a more limited way and are said to be *extrazonal*. Thus glaciers and some patterned ground phenomena may occur on tropical mountains. The same phenomena may therefore be zonal in one place and extrazonal in another, but never zonal in one and azonal in another.

Those phenomena which occur in several morphoclimatic zones without being world-wide may be called *polyzonal*, e.g. running water and wind action.

The concept of zonality is not as simple as it may at first appear, as mixed phenomena are common. Thus a newly erupted volcano has a form which is entirely dependent upon azonal endogenetic processes; but after it ceased to be active, zonal glacial processes, or polyzonal fluvial processes, may come to dominate its form. Wind action is azonal, but it is most effective in the absence of vegetation, and zonal phenomena control whether it carries snow, sand, salt, or silt, and temperature controls its viscosity. Polyzonal fluvial processes dominate many of the landforms of Ethiopia in the Nile headwaters, but once that river crosses the Egyptian desert it is clearly extrazonal.

Some landforms may be zonal in one morphoclimatic zone but extrazonal elsewhere: badlands may thus be zonal in a semi-arid climate, but extrazonal in a humid climate where a landslide has exposed a rock of limited resistance to rill erosion.

Scale

There is no single process of erosion or deposition that produces all the landforms of a morphoclimatic zone, but rather a complex of agents which are effective to a degree which is controlled or modified by geological structure and by lithology. We can visualize an hierarchy in which a broad regional climate, such as that of a hot desert, places its imprint upon all landforms, but within that region different processes or controls predominate. Thus wind may control a dune field, but periodic floods a playa, and structure an upstanding plateau. The climatic belts are consequently more uniform than the landscapes they cover, and structural influences usually increase in importance as landforms decline in area.

Structural influences

Over the great eroded plains (*planation surfaces*) of inner Asia and the southern hemisphere continents it is often difficult to discern structural, lithological, or tectonic influences, except in occasional residual hills (*inselbergs*) of resistant rocks or stepped landscapes of polycyclic origin. In mountain zones the structural influences may become dominant because altitudinal changes in climate and processes of erosion and deposition are so rapid that they modify the detail of a landscape rather than control its overall form.

Selection of criteria

In Fig. 1.6 is displayed one example of an attempt to distinguish morphogenetic zones on a climatic-vegetational basis. Such a zonation ignores not only climatic seasons but also the results of the combination of processes which result in landform-creating mechanisms. The morphoclimatic division of the Earth must, therefore, be based upon an integration of process, vegetation, soil, and landform phenomena. In Fig. 1.7 is shown a map of the zones recognized by the French geomorphologists J. Tricart and A. Cailleux. They recognize five groupings of zones and fourteen mappable morphoclimatic provinces. Less attention is paid to the precise definition of boundaries than to the characteristic features of each province.

The cold zone is divided into four provinces:

(1) glacial province;
(2) periglacial province with permafrost;
(3) periglacial province without permafrost;
(4) forests on Quaternary permafrost.

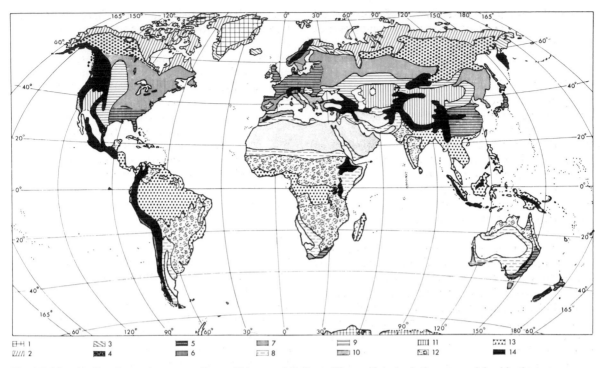

Fig. 1.7 Morphoclimatic provinces according to Tricart and Cailleux. The numbers in the key are explained in the text.

The forested middle latitude zone has three provinces:

(5) maritime province with mild winters;
(6) continental province with severe winters;
(7) Mediterranean province with dry summers.

The arid and sub-arid zone of low and middle latitudes has four provinces:

(8) steppe and semi-desert with mild winters;
(9) steppes with severe winters;
(10) deserts with mild winters;
(11) deserts with severe winters.

The intertropical zone has two provinces:

(12) savanna province;
(13) forest province.

(14) *Mountain regions* are treated separately, for in them altitudinal zonation is of prime importance.

Influence of people

The impact of people upon the landscape has been so great, and is so rapidly increasing, that humans must be regarded as the most effective geomorphic agents in many morphogenetic regions. These effects are discussed in more detail in the last section of this chapter, but it is necessary to note that the impact of people is not uniformly distributed in the morphoclimatic provinces: five broad areas of people-induced morphogenetic systems may be recognized.

(1) The areas of nineteenth-century agricultural settlement, by colonists originally from Europe, in USA, Canada, South Africa, Australia, Argentina, Brazil, and New Zealand, were regions where the use of machinery made rapid conversion from natural vegetation to cropping and grazing, for export production, very effective. This process was frequently accompanied by severe soil erosion, and destruction of a mixed vegetation cover, as agricultural monocultures and grazing lands were extended. The application of soil conservation methods has now reduced erosion to rates which are often no greater than those under natural conditions and in some cases are lower rates.

The Amazon rainforest is now being exploited

under rather similar conditions, and with equally disastrous results, to those of mid-nineteenth-century settlement mentioned above. In USSR similar erosion is resulting from attempts to extend wheat production into semi-arid steppes.

(2) The overpopulated lands of traditional agricultural methods, such as the Italian Apennines, Mexico, India, the Algerian Tell, and many small areas in Greece, Turkey, and Spain, are areas of severe erosion because of poverty. The necessity of using all land for food production, the lack of capital for construction of conservation devices, or even of fencing wire for control of grazing animals, has increased rates of erosion and made poverty and overpopulation even more serious.

(3) In many parts of the lesser developed world firewood is the sole source of fuel for cooking and domestic heating. Deforestation is thus an almost universal phenomenon in much of India and the woodlands of Africa. In areas of lower population density shifting agriculture was often common in forested and savanna areas, but the time allowed for the forest to recover from clearances declined as populations increased, so that soil deterioration is again a common phenomenon. Many governments are encouraging people to settle permanently and to improve their agricultural skills, but this can usually only be done with foreign aid.

Along the southern edge of the Sahara in the Sahel and in semi-arid margins of many deserts, overgrazing, animal trampling, scrub burning, and excessive cultivation have caused local patches of true desert to form, especially around water-holes. There is probably no general advance of deserts, but increasing pressure of human population has caused serious ecological damage and produced a rapid change in morphogenetic processes. In 1977, about 630 million people—14 per cent of the world's population—lived in arid or semi-arid lands and faced lives of increasing poverty and lessening resources.

(4) The monsoon lands of Asia suffer severe natural processes of erosion in the wet season, and where poverty, collapse of political authority, and loss of traditional conservation skills have followed overpopulation, or warfare, then again devegetation and erosion have increased.

(5) In the older settled farmlands of Europe the spread of monoculture, enlargement of fields, increasing use of heavy machinery, and change of methods have periodically caused increases in local erosion. By contrast intense land use in areas

such as Holland is sometimes accompanied by decreases in erosion.

Perhaps the major effect of people on rates and forms of morphogenetic processes has been through the spread of engineering structures and other buildings, especially in the nineteenth and twentieth centuries.

Relicts

In nearly every province some relict of past processes may be found. In some areas, such as formerly glaciated valleys or plains, the relict landforms dominate those currently developing; elsewhere, as in the tropical lowland rainforests, the relict landforms may be less obvious. The presence of relicts, however, has to be incorporated into any comprehensive account of the modern landscape and in the method of study called *climatogenetic geomorphology* the inheritance from the past becomes a key to recognizing the nature of the modern landscape.

Climatogenetic geomorphology

The major exponent of climatogenetic geomorphology as a systematic method of studying landscapes is the German J. Büdel. Büdel bases his work on the postulate that endogenetic processes are responsible for the distribution of zones of uplift and depression, and consequently for creating the possibility of selective weathering and erosion of weak rocks; he contends, however, that the influence of structure on relief is merely a passive influence and that active relief formation is brought about by exogenetic processes. He points out that if the Alps were solely the product of endogenetic uplift they would appear as a massive dome about 10 km high, but exogenetic processes have worn more than half of this away since the early Tertiary and created from this erosion the landscape we see today.

The modern landscape is not the result of presently active processes, but is the result of all processes in the past and present, and many of these former processes and process-combinations have left traces (relicts) which can be recognized, and the successive generations of relief development can be deciphered.

The lines of evidence used in reconstructing events of the past include: the form of the land surface, especially remnant uplifted planation surfaces and residual hills (inselbergs) rising above them; the forms of valleys and the benches, pediments, and terraces associated with them; the products of weathering which have been left by ancient chemical and physical alterations, especially ancient soils (*paleosols*) and chemically cemented soil horizons (*duricrusts*), for these are regarded as being the clearest indicators of distinctive climates. The age of past landscape elements can be ascertained from stratigraphic relationships, from the presence of fossiliferous deposits, by paleomagnetic determinations on iron-rich duricrusts (*ferricretes*); fossil pollen and macrofossils are key indicators of past climatic regimes. Reconstructions of relict relief generations are verified by comparing ancient landforms with those currently developing in modern morphoclimatic zones.

Büdel described five major climatomorphological regions in his early (1948) analysis and this has been further refined to ten regions in later studies (1977) (Fig. 1.8).

(1) *A glacial zone* in which ice streams have created extensive erosional and depositional land-

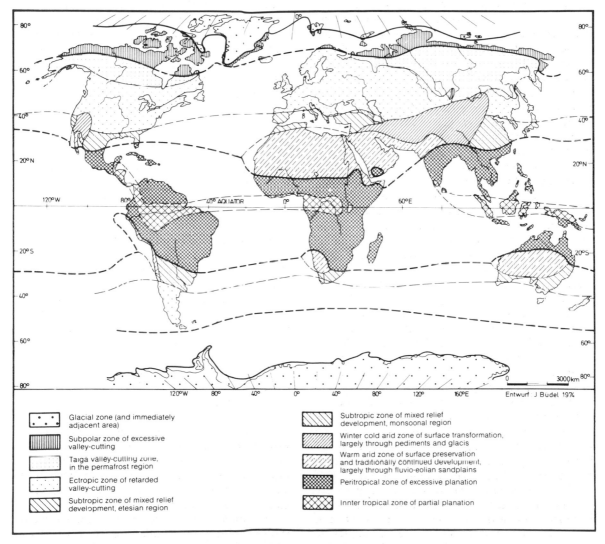

Fig. 1.8 The present-day climatomorphological zones of the Earth according to Büdel (1977).

forms and eliminated most traces of earlier land-scapes. In areas where ice cover has been thin, however, some remnants of preglacial landforms have survived.

(2) *Subpolar zone of valley cutting.* This is the modern periglacial zone of subpolar frost tundra. Extreme valley formation occurs in places like Spitzbergen as mechanical weathering shatters the surface rocks, creates patterned ground, and, most importantly, permits gelifluction, sheetwash, and rill-wash to carry mobile debris into the valley floors. During early summer snowmelt, enormous floods transport this debris and valley floors are deepened and widened by corrasion. During the autumn the thin snow cover permits the cold wave (down to $-30\,°C$) to enter the rock and extreme contraction occurs within the upper permafrost so that deep cracks open. These cracks are held open by the formation of ice-lenses. During the even colder winter the process is continued and cracks are widened further. Within a hundred years bedrock can be torn into small rock masses within a matrix of ice. This upper part of the land surface Büdel calls the *ice-rind*. The upper part of the ice-rind may thaw in summer so that the river beds are composed of shattered rock which can be transported by floods. During the Holocene alone valley-floor lowering has exceeded 10–30 metres in Spitzbergen. Büdel believes that similar extreme processes occurred in central Europe during the glacials and that this accounts for the incision of 1–3 mm/y which seems to have occurred during glacial periods. The same process accounts for valley widening in areas not occupied by valley

glaciers during glacials, and is in contrast to relatively weak Holocene river erosion in intergla-cial periods in Europe. (Note: some workers in the subpolar regions have denied the existence of the ice-rind mechanism and, by implication, the sig-nificance of pronounced valley formation in this zone; see Chapter 14.)

(3) *The ectropic zone of retarded valley forma-tion* includes most of the mid-latitude regions, and today it is characterized by moderately active processes, but often also by relict features of cold climate valley formation (zone 2).

(4) *A subtropical zone of pediment and valley formation* exists between zones 3 and 5. It is characteristically an area of strongly seasonal climate, with severe winter cold in inner Asia, and many relict tropical features in the southern hemisphere continents. The seasonal climate en-courages sheet-wash so that eroded foot-slopes (*pediments*) and rather steeper wash-slopes (called *glacis*) on weaker rocks, are the characteristic landforms (Plate 1.1). This belt is divided into four units in the 1977 classification.

(5) *The tropical zone of planation surface forma-tion* is regarded as the most widespread zone and one which has formerly dominated large areas now belonging within zones 1 to 4.

Humid tropical weathering, and especially that of the seasonally wet and dry tropics, has charac-teristics which are of great importance for land-form development. The weathering is almost exclusively chemical and relatively rapid. It is most effective upon certain types of rocks, such as granites in which both micas and feldspars are

1.1 Pediments at the foot of hills within the Cape Fold Mountains. The nature of the pediments is evident from the cut bedrock surface exposed in the road cutting which crosses the toe of the pediments.

weathered readily to clay minerals, and penetrates most deeply along open, closely spaced joints and where there is abundant soil moisture. As a consequence weathering is most effective at the base of slopes, which shed water onto the slope foot, and least effective on steep uplands, well-drained plateaux, or on rocks such as quartzites which are not weathered to clay minerals. In the lowlands the *weathering front* (i.e. the zone of contact of the regolith with the underlying rock) is irregular because of differential rock resistance and variations in soil moisture.

Weathering profiles are commonly deep (4–10 m) as the rate of weathering, in areas of low relief, exceeds the capacity of rivers to remove debris. Soils are characteristically red in colour because of the presence of iron, and chemical weathering has carried clay mineral formation to extremes so that kaolinite and iron and aluminium oxides dominate the regolith. Exposure of the subsoil, after erosion of the top soil, frequently leads to hardening of iron oxides and the formation of duricrusts.

The rivers have loads high in solutes and fine-grained clays and sands, but carry few coarse materials because of the lack of physical weathering. These rivers, therefore, have little capacity to cut through outcrops of massive bedrock which are exposed as rivers remove the weak regolith materials. The long profiles of many tropical rivers are consequently broken by rockbars, waterfalls, and rapids, and between these local base levels the rivers migrate laterally across the deeply weathered plains, opening broad basins for themselves. Thus lateral planation, surface rainsplash, sheetwash, and also solution processes, lower the ground surface in the basins and plains while the weathering front extends down into bedrock. The combination of surface lateral debris removal and regolith deepening Büdel calls *double planation*. Where ground lowering catches up with lowering of the weathering front bedrock may be exposed and eventually form a tor or an inselberg (Fig. 1.9 and Fig. 8.19).

In a seasonally dry, tropical climate slope wash processes are commonly effective and ferricretes (also called *laterites*) are exposed and they become caprocks on escarpments. Tropical planation surfaces with deep weathering profiles, which are etched into the bedrock, are called *etchplains*. Such

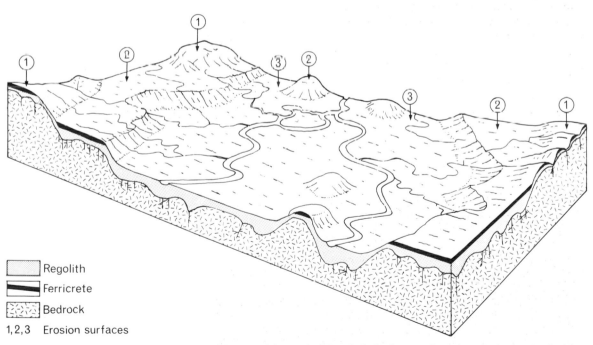

Regolith
Ferricrete
Bedrock
1,2,3 Erosion surfaces

Fig. 1.9 Double planation surface in the seasonally wet and dry tropics. Note the lack of stream channel erosion and stream incision. The plains are dominated by sheetwash. The most elevated inselbergs are older than the less elevated which are only exposed as wash removes regolith material from the ground surface. After the concept of Büdel.

plains may be dissected as a result of climatic change, vegetation removal, or lowering of local base level so that the regolith is removed and an irregular bedrock surface, or one with a thin soil cover, is developed—this is an *incised etchplain* (Fig. 1.10). An inner tropical zone is distinguished in the 1977 scheme.

Büdel and other workers have identified remnants of etchplains in many areas which no longer have tropical climates. Such etchplains sometimes have remnant kaolinized soil, red loam soils, or

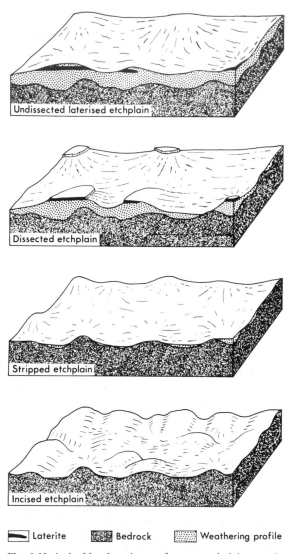

Fig. 1.10 A double planation surface, or etchplain may be stripped of its regolith, especially as a result of climatic change, and an incised etchplain with little regolith may develop.

remnants of ferricretes on them. Some of these relict etchplain surfaces can be recognized as having an essentially late Mesozoic or Tertiary age of formation. The progressive change of dominant morphogenetic processes is indicated in Fig. 1.11. In this scheme Büdel recognized the gradual cooling of the continents during the Tertiary Period and the frequent major fluctuations of climate during the Pleistocene. Recognition of the numerous fluctuations of Quaternary climates led C. A. Cotton to emphasize the importance of *alternating morphogenetic environments*.

The ultimate aim of climatogenetic geomorphology is illustrated in Fig. 1.12 showing two areas of central Europe in which Büdel recognizes (1) remnants of Tertiary etchplains, (2) broad pediments of a possible late Pliocene dry period, (3) Pleistocene deep valley incision in periglacial climates, and (4) the Holocene valley floors. This model implies that modification of plateau surfaces is a very slow process, that ancient landforms (*paleoforms*) can survive with little modification for tens of millions of years, and that active erosion is largely confined to valley sides and, more importantly, valley floors. Its triumph is that it gives a historical framework to geomorphological thinking about climatically controlled processes and landforms.

Problems of interpretation in climatic and climatogenetic geomorphology

The following two basic principles underlie attempts to relate landforms and climate: actualism and uniformitarianism.

(1) The study of modern geological phenomena provides a key for understanding similar phenomena in the past. This is the *principle of actualism*.

There is, however, considerable evidence that combinations of processes and responses to them may have been different in the past from the present. It has been shown that early Tertiary climates were, over large areas, considerably warmer than those of today; this may well have caused chemical process rates to be quicker, but it may also have made them more effective. Many ferricretes, for example, in Africa and Australia, may have formed in conditions which no longer occur and there are no *silcretes* (silica-rich duricrusts) developing anywhere at present. So little is known for sure about silcrete formation that we cannot say if the reason is climatic or due to the

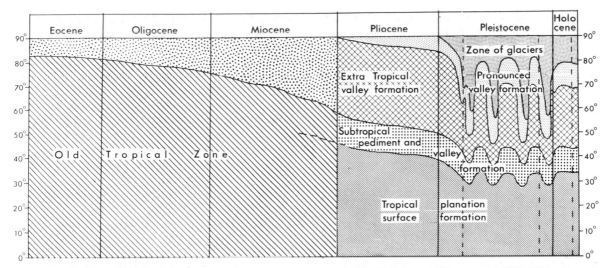

Fig. 1.11 Climatomorphological zones of the northern hemisphere in a meridional belt from the Equator to the North Pole, through Central Europe—since early Tertiary times. Note that the numerous fluctuations of late Cenozoic climate are not represented realistically (after Büdel, 1948 and 1977).

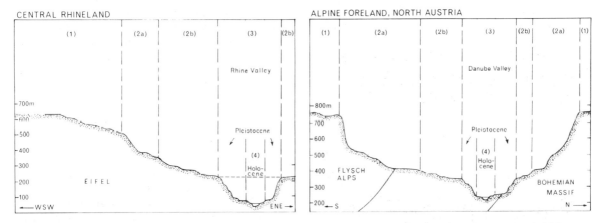

Fig. 1.12 Relief generations in central Europe according to Büdel (1979): (1) Tertiary etchplains; (2a) Pliocene to early Pleistocene surfaces of uncertain origin; (2b) broad terraces and pediments above main valleys date from the early Quaternary; (3) Pleistocene valleys; (4) Holocene floodplains and low terraces. This scheme implies that more than 95 per cent of the relief of Europe is pre-Holocene in age and throws doubt on the value of recognizing modern morphoclimatic zones.

lack of a necessary, but rare, combination of pH and oxidizing potential, with available soil and flood waters.

In humid environments soils and vegetation provide a buffer between climatic processes and rocks and regoliths. We know that the grasses, with their sod-forming characteristics, did not evolve until the early Tertiary and that many plant species and plant communities were eliminated, confined in distribution, or changed in structure,

during the late Cenozoic. With what effects upon landforming processes we do not know.

One of the key lines of evidence in paleoclimatic reconstructions is paleosols. Soils reflect a fragile equilibrium between relief, climate, and vegetation. The existence of a soil at a given place implies that mechanical erosion is slower than soil-forming processes. The recognition of normal, truncated, and buried soils with profiles, is an important aid in distinguishing the relative rates and distribution of erosion over the landscape and therefore is a key indicator of landscape stability. Buried soils within sedimentary sequences, as on old floodplains, are indicators of periodic depositional events.

Paleosols are themselves indicators of climatic regimes and deeply weathered red soils, and strongly kaolinized soils, are regarded as evidence of warm, humid climatic conditions during their formation. There is, however, some evidence that given enough time similar soils can form in cooler climates, so paleosols may indicate only long times of stability rather than particular climatic environments for soil formation.

Measurements of the rates of modern processes are sometimes used to estimate the age of a landform: the implication being that an old landform cannot survive in an area of rapid denudation. Whilst this is a useful method it can be misleading. It is evident from earlier discussions that planation surfaces can develop over periods of tens of millions of years and remnants may survive for 100–200 My. This is particularly the case where climates have been arid or semi-arid for much of the Cenozoic, or where structural conditions are favourable, as on the chalk uplands of Britain.

The fundamental problem in establishing the validity of the climatogenetic approach to landform explanation is the requirement to recognize characteristic suites of landforms for major modern climatic zones and then, by analogy, to relate relict landforms (paleoforms) to an assumed climatic regime. Many critics believe that this fundamental problem has not yet been validated, and some go so far as to believe that, because the modern landscape has the imprint on it of all past climates and their variations, this task may be impossible. As a further objection we must recall the problems inherent in the recognition of *convergence* (or *equifinality*) in which several assemblages of processes may be capable of producing one class of landforms.

(2) The view that the rates of geological processes are essentially constant is known as the *principle of uniformitarianism*. In the sense that the fundamental laws of physics or biology are the same through time this principle underlies all science but, if it is held to mean that rates of change in geological time are essentially uniform, then it is open to question. Climate, ice cover, base level, and vegetation cover have changed repeatedly,

and often rapidly, in late Cenozoic time: this is certainly different from the conditions during most of the early Cenozoic. In steep mountains very large rock-slides may catastrophically change the landscape and their effects may still be noticable 100 000 years later. By contrast the next valley may never have experienced anything more severe than an occasional small soil-slide.

The area of ground affected by an extreme event, the frequency with which large magnitude events occur, the proportion of total landform change they produce, and the time which it takes for their effect to be covered, or modified, by less catastrophic but more frequent events, can seldom be interpreted from features older than a few hundred thousand years. Such considerations are, however, at the core of many modern studies of landscape equilibrium.

Equilibrium

The geomorphic importance of an erosional event is governed by the magnitude of the energy it expends upon the landscape. The greater the magnitude of an event the lower is the probability that it will occur again in the immediate future. Recurrence intervals, or frequencies, are expressed as probabilities, so a landslide of such size that it has a 10-year return period has a 10 per cent chance of occurring in any one year, and a 100-year landslide a 1 per cent chance. Very large landslides are likely to occur, on average, only once in many thousands of years.

This simple statistical approach has the limitation that climatic change alters probabilities, so that an increase in rainfall may increase the chances of a particular size of landslide occurring.

The concept of magnitude and frequency of landform-changing events implies that the frequent, or continuous events, such as solution and soil creep, are approximately in balance, or equilibrium, with soil development, vegetation cover, and the rate of erosion. This 'normal' rate of change is broken by the rarer event, and after it the landscape gradually heals, but the larger the severe event the longer is the healing or *recovery time* (also called the *relaxation time*).

Landform change can consequently be viewed as occurring within periods of *steady state* in which the gradual processes dominate, followed by a disruptive *event* and then a recovery period in which landforms adjust to the event (e.g. landslide scars and debris are covered by vegetation and soil formation occurs) (Fig. 1.13). The disruptive event has to have sufficient energy to cross a *threshold* of available resistance. Seen over a very long time the landscape still appears to be in equilibrium with the forces acting on it, but because this is not a smooth or static equilibrium it is often called *dynamic equilibrium*.

During periods of climatic change, and hence vegetation change, there will be periods in which the long-term equilibrium is broken and rapid landform change can occur. This may be recognized by increases in the extent of gully or rill erosion, for example, or by increases in sediment yield which reach a peak during vegetation disturbance, and decline again as vegetation recovers (Fig. 1.14).

The value of the concept of equilibrium, and breaks in it, is that it focuses our thinking on the

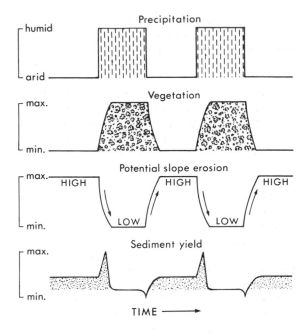

Biogeomorphic response to climatic change

Fig. 1.14 The response of vegetation and rates of erosion to a major change in precipitation. The climatic change is shown as being abrupt but the vegetation and, therefore, the erosional change is lagged but with peak sediment yield when the equilibrium is first broken and soil erosion is initiated (after Knox, 1972).

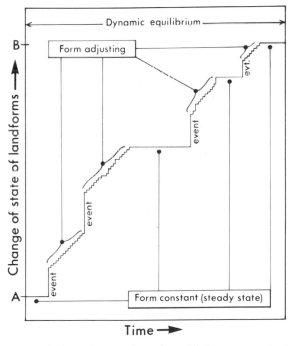

Fig. 1.13 Within the term dynamic equilibrium are contained the three states of (1) a land-forming event; (2) the adjustment of form which follows that event; and (3) a period of steady state in which there is virtually no adjustment of form. The curve which represents the change of landforms with time may, therefore, rise very steeply, gradually, or slightly, depending upon the magnitude and frequency of the dominant process (after Selby, 1974).

variability in rates of change and their effects. It also brings to our attention those parts of the landscape which are particularly sensitive to change (usually areas of high and steep relief), and forms a bridge between short-term process studies and the historical approach of denudation chronology and climatogenetic geomorphology.

The concept of a dynamic equilibrium between force and resistance in landforming events is not new: it is implicit, particularly, in the writing of G. K. Gilbert (1877). More recently Hack (1960) has suggested that whole landscapes may evolve without any change in form, where the rate of stream channel downcutting and the rate of erosion from valley-side slopes are in quasi-equilibrium. Hack suggested that the ridge and ravine topography of parts of the Appalachians may be evolving in this way. Such an explanation can, however, only apply for a limited segment of geological time, as in the longer term there must be reduction of the landscape towards base level.

Geomorphic systems

Recognition of the relationships between energy put into a unit of the landscape and its effect upon many related components of that landscape has encouraged some students of process–landform relationships to study whole systems. As an example, consider the input of energy of a storm into a catchment. It creates a landslide: landslide debris slides into the stream channel and causes channel bars to form, and this creates both deposition and some channel bank erosion where the flow is redirected towards the bank: sediment is also transported out of the valley. After the landslide, infiltration capacity on the landslide scar is reduced until soil and vegetation can cover the scar: channel bank erosion gradually diminishes as the bar is removed. Eventually most signs of the landslide are eliminated. The catchment was responding in a complex way tending to absorb the energy of the storm, transmitting storm energy as water and sediment, and then recovering with relatively little permanent modification. The catchment was acting as a system of components, or *variables*, which are interconnected so that a change in one component had an effect upon other components.

At one scale the whole solar system can be considered as a *system*, which is a structured set of objects which are related to one another and operate together. Most geomorphic systems are *open systems* which transmit energy and mass (as did the catchment mentioned above). The tendency for them to adjust the relationships between the components is the cause of steady-state conditions, and the obvious pattern of their components (e.g. the hierarchy of stream tributaries—Fig. 10.25).

At certain scales some natural systems can be considered to be closed: the Earth for example receives almost no mass and transmits none into space, and the hydrological cycle losses little mass and receives little. In both cases these may be regarded as *closed systems*, if we ignore the arrival of meteorites and the production of juvenile water.

The linkages between components of an open system cause mutual adjustments which are called *feedback*. Feedback means that as one variable affects a second variable, the second one in turn causes a change in the first variable. Feedback can be direct, influencing the two variables only, or indirect by modification of other variables. It can be negative or positive.

In *negative feedback* the change in the second variable causes a change in the first such that it tends to return to its former state and produces self-regulation, and a dynamic steady state. In the example given above, of the landslide, its debris may infill the valley floor and direct the stream flow away from the slope foot, thus helping to stabilize the slope and permitting vegetation to colonize the scar and the slope to re-establish a steady state.

In *positive feedback* the change in the second variable causes the first one to change still further in the direction of the initial change. The landslide scar had a lower infiltration rate than the initial soil. The scar consequently became a site of increased runoff, rill and gully erosion; alternatively the scar could heal but the depression could become the site of expanded channel network storm flow and so be further eroded. In both cases positive feedback enhanced the erosive effect of the landslide. Eventually, however, the depression could be so lowered that it would become a site of deposition. Self-generation of positive feedback carries the seeds of its own elimination. Glacial erosion is often associated with positive feedback as valley-floor deepening causes ice thickening and compressive flow and hence greater erosion. Another example is the way in which inselbergs shed water onto the surrounding lowlands, encouraging weathering and erosion at the slope foot and consequently an increase in the relative relief of the inselbergs.

Concepts of equilibrium and systems are the most appropriate to the study of modern erosion processes and the resulting landforms. They are of little value for studies of the long-term evolution of whole landscapes in which relicts from former environments and paleoforms may be vital clues in historical reconstructions. This is not to deny that isostatic adjustments occur after erosional retreat of scarps in Africa, nor that uplift in the Appalachians can occur at the same rate as downcutting, but that the differences in time-scale make the systems approach most appropriate where the changes can be measured and the feedback mechanisms observed. It is, therefore, an approach which relates well to study of processes and landform responses to them, and to the situation in which actualism applies. It is not relevant to the time- and space-scales considered in evolution of erosional plains or climatogenetic geomorphology.

People as geomorphic agents

The impact of the world's human population on the surface of the Earth is approximately in proportion to the numbers of people, their distribution, social organization, level of technology, and use of mineral energy. Each person in the modern industrialized world uses, on average, 2×10^7 grams (20 tonnes) of new mineral material annually. For 1000 million people, about 15 per cent of the global population by AD 2000, the annual usage (2×10^{16} g) equals in mass the most impressive geological processes of Earth—the mass of ocean crust formed, of rock eroded from the continents, and of land raised tectonically. If we add to this estimate the amount of soil moved annually in agriculture, then there is no doubt that humans have become the most important agents modifying the land surface. The brief comments permitted by available space in this chapter cannot do justice to people as geomorphic agents.

The relationship between people and the land surface is not only that of creating new landforms: human activity modifies the rate and intensity of natural processes; people study and overcome the limitations imposed by the natural environment; they are subject to natural hazards; and by evaluating resources, and managing them, they seek to control the environment.

People as creators of new landforms

Only about 15 per cent of the land surface remains in a natural state. In some areas, such as the Netherlands, the entire ground surface has been modified or created by human activity. The saying, 'God made the sea, but the Dutch made the land' is close to the truth in much of the Netherlands.

The most intensely altered surfaces are those of the urban areas where concrete, tarmac, and buildings control erosion and hydrological processes. The agricultural settlements—often with terraces, irrigation systems, and always with large-scale changes of the vegetation—which are set up to supply the urban population have effects upon landscapes which may extend for thousands of kilometres away from the urban market. Reser-

voirs, quarries, mines, spoil heaps, road, canal and rail cuttings, coastal protection works, river control works and hydroelectricity schemes are all new landforms or modifiers of processes.

The most dramatic influence on rates of natural processes is a result of vegetation change and the spread of agriculture. The erosion in the Dust Bowl States of the American West in the 1930s is only one example of erosion. The pattern is being repeated in much of Africa, parts of USSR, Brazil, and India at the present time (see this chapter, p. 16).

Problems of using the environment

The contributions of earth scientists to solutions of environmental problems fall into five classes. (1) Data must be collected about: the form of the land surface; the processes of erosion and deposition; potential hazards such as landslides, mudflows, floods, land subsidence (Table 1.2), saline soils which could cause deterioration of building materials, and weak soil or rock in areas to be used for

Table 1.2 Estimates of Average Annual Losses, and the Potential for Sudden Loss, from Geologic and Hydrologic Hazards in the USA

Hazard	Annual loss (in billion dollars)	Sudden loss potential (in billion dollars)
Earthquakes—ground-shaking, surface-faulting, earthquake-induced ground failures, tsunamis	0.6	50
Floods—flash floods, riverine floods, tidal floods	3	5
Ground failures—landslides, expansive soils, subsidence	4	6
Volcanic eruptions—tephra, lateral blasts, pyroclastic flows, mudflows, lava flows	(Note 1)	3

Notes: (1) Past data are too limited to determine average annual losses for volcanic eruptions. The 1978 eruption of Mount St Helens in the State of Washington will provide a reference for the future.
(2) Some loss estimates may be too high or too low by a factor of 2. Figures are 1980 values.
Source: US Geological Survey Professional Paper 1240-B, 1981.

foundations of roads and buildings; also about soil, water, and mineral resources. (2) The factors which influence the safety and permanence of disposal of waste materials on the land, and in waterways, must be understood. (3) Permanent water resources must be assessed for quantity and quality. (4) Mineral resources must be identified and mapped: this is especially important for large-volume, but low-value, materials such as sand, gravel, clay, and limestone for roads, cement, and concrete, as these can be made inaccessible by urban development, and alternative supplies may be expensive to transport. (5) It has become increasingly important to monitor changes in the physical environment caused by human activities so that remedies may be applied where harmful effects are recognized.

Geomorphologists have a part to play in the effective evaluation and planning of the safe use of the natural and human environment.

Further reading

Methods of dating landforms and deposits are discussed in the books edited by Michael and Ralph (1971) and by Goudie (1981). General principles of geological dating and the time-scale are described in all introductory geology texts of which those by Strahler (1981) and Dott and Batten (1981) are excellent examples. A review of the need for revision, and a revised geologic time-scale, are provided by Harland *et al.* (1982).

Two outstanding volumes describe the history of geomorphology: Chorley, Dunn, and Beckinsale (1964) deal with the subject to 1893 and the later volume, Chorley, Beckinsale, and Dunn (1973), is devoted to the life and work of W. M. Davis. A brief review was published in 1968 by Beckinsale and Chorley. The study by Davies (1969) gives a more detailed account of early British geomorphology. A number of facsimile editions of early major works are now available: they include James Hutton's papers (with an introduction by Eyles, 1970), Playfair's *Illustrations of the Huttonian Theory of the Earth* with an introduction by White (1956) and *Études sur les Glaciers* by Agassiz, translated by Carozzi (1967). The series of volumes 'Benchmark Papers in Geology' contains facsimile extracts from many key papers which form the basis of the subject, and are the primary data for an appreciation of the history of geomorphology. The volumes edited by

Schumm (1972) and Schumm and Mosley (1973) are good examples of this series. The biography of G. K. Gilbert by Pyne (1980) provides not only an insight into the development of geological exploration in USA and with it the US Geological Survey, but also the method of work of one of the greatest, some would say the greatest, of geomorphologists. It is a particularly revealing study of the contrast between the evolutionary ideas of Davis, and Gilbert's search for an understanding of the physical laws which control geomorphic processes.

French and German work on climatic geomorphology is available as a translation of Tricart and Cailleux (1972) and in the collection of essays edited by Derbyshire (1973). The latter includes a translation of important essays by J. Büdel and other German geomorphologists. Büdel's study (1977) is available in translation (1982). Geomorphic systems are examined in detail by Chorley and Kennedy (1971), and a comparison of German and British approaches to geomorphology can be obtained from the collection of papers published in *Zeitschrift für Geomorphologie, Supplement 36* (1980). A very useful study of paleoforms is that of Twidale (1976), and Finkl (1982) has emphasized the alternating stability and instability of cratonic planation surfaces. Hsü (1983) has stressed the importance of rare but major events (neo-catastrophism). The collection of essays edited by Yochelson (1980) demonstrate the contribution of G. K. Gilbert to the development of geomorphic theory and methods in experimental research, systems analysis, and equilibrium concepts. Schumm (1979) has reviewed modern ideas on equilibrium and thresholds.

Early studies, which develop the theme of human modifications of Earth's surface, include the book by Marsh (1864) and the compilation edited by Thomas (1956). The book by Goudie (1981) gives a well-organized review of the impact of Man on his environment: the volume edited by Coates (1976) and his systematic study (1981) provide many examples. The books edited by McKenzie and Utgard (1972) and Detwyler (1971) are good examples of the many review volumes discussing the relationships between people and their environment. The volume edited by Gregory and Walling (1979) and that by Cooke and Doornkamp (1974) are more explicitly directed at geomorphologists, and emphasize the impact of people on processes. Many books discuss natural hazards, but those by Scheidegger (1975), Bolt *et al.* (1975), and Whittow (1980) give valuable discussions of the mechanics of the processes. Applied geomorphology and its role in environ-mental management is the theme of Cooke and Doornkamp (1974), and the authors of chapters in the volumes edited by Hails (1977) and by Craig and Craft (1980).

2 Major Features of Earth

A glance at a map of the major physical features of the Earth's surface tells us that the continents, mountain chains, and ocean floor features have a varied relief and elevation (Plate 2.1). This implies that something must have happened to permit the mountains to stand above the continental plains, and the continents above the ocean basins, at some stage during the Earth's history.

A *hypsographic curve* (Fig. 2.1) showing the distribution by height and depth of the Earth's solid surface indicates that 29 per cent of the total surface is above sea-level: this represents a volume of rock of about 1.3×10^8 km^3. It is estimated that about 13.6 km^3 of rock material is eroded from the continents each year and deposited in the oceans. At this rate of erosion it would take only about 10 My to reduce the continents to sea-level, but we know that the Earth is about 4600 My old and the geological record indicates that deposition of rock material has been occurring in the shallow seas on the margins of continents, and on the continental slope, for much of geological time. It is clear that processes have operated, and are still operating, to maintain the relief of the continents.

Mountains, plains, and isostasy

That mountain ranges and continental plains survive, for much greater periods than 10 My, implies that there must be some process compensating for erosion from the surface. Such a process has to be within the Earth, and this implies that mountains may differ in some way from continental plains and ocean floors. The first indication of a possible cause of this difference came in 1735 from a French scientist, Pierre Bouguer, who was surveying in the Andes.

When setting up a theodolite a surveyor hangs a plumb-bob beneath the instrument to act as a vertical reference line above a survey point. New-ton's *law of gravitation* states that any two bodies attract each other with a force that is directly proportional to the product of their masses and inversely proportional to the square of the distance separating them, or

$$F = \frac{G\,M_1\,M_2}{r^2}$$

where $M_1\,M_2$ are the two masses, r is the distance separating them, and F is the gravitational force between them. G is the universal constant of gravitation ($= 6.668 \times 10^{-11}$ N m^2/kg^2).

Bouguer expected that his plumb-bob would have been attracted vertically by the mass of the Earth, but also deflected horizontally by the mass of the mountains around him. The deflection towards the mountains was not, however, as large as he calculated it should be. His calculations took the form:

$$F_{earth} = \frac{G\,M_1\,(\text{the plumb-bob})\,M_2\,(\text{the Earth})}{r[\text{the distance from the station to the centre of the Earth}]^2}$$

and

$$F_{mountain} = \frac{G\,M_1\,(\text{the plumb-bob})\,M_2\,(\text{the mass of the mountains})}{r[\text{the distance from the station to the centre of mass of the mountains}]^2}$$

Bouguer's observations were ignored but, about 100 years later, during the survey of India the same problem was recognized: the plumb-bob was deflected by only one-third of that expected from observations of the volume of the Himalayas and measurements of the density of rock samples taken from the mountains. It seemed most unlikely that the mountains are deficient in mass because they are hollow; rather it seemed probable that the

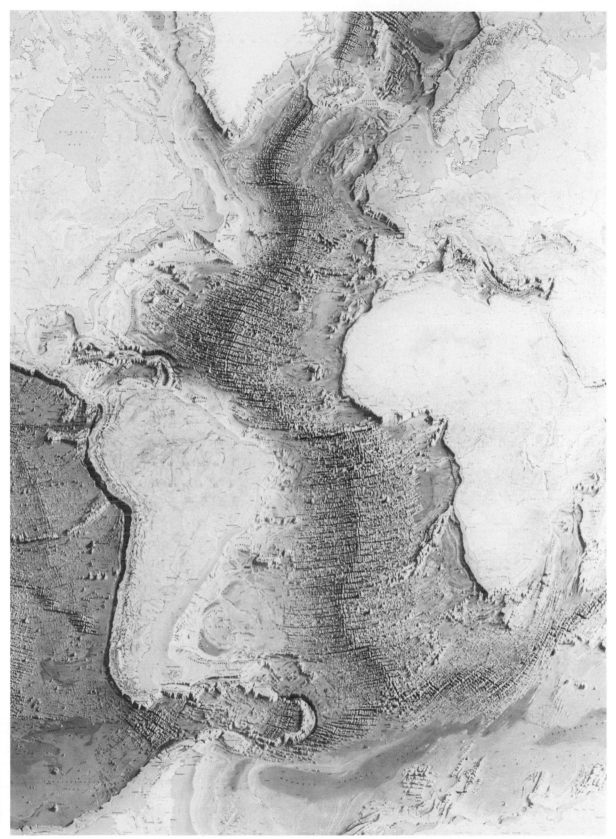

2.1 The main physical features of the continents and ocean floors. (*Copyright*: Dr Marie Tharp).

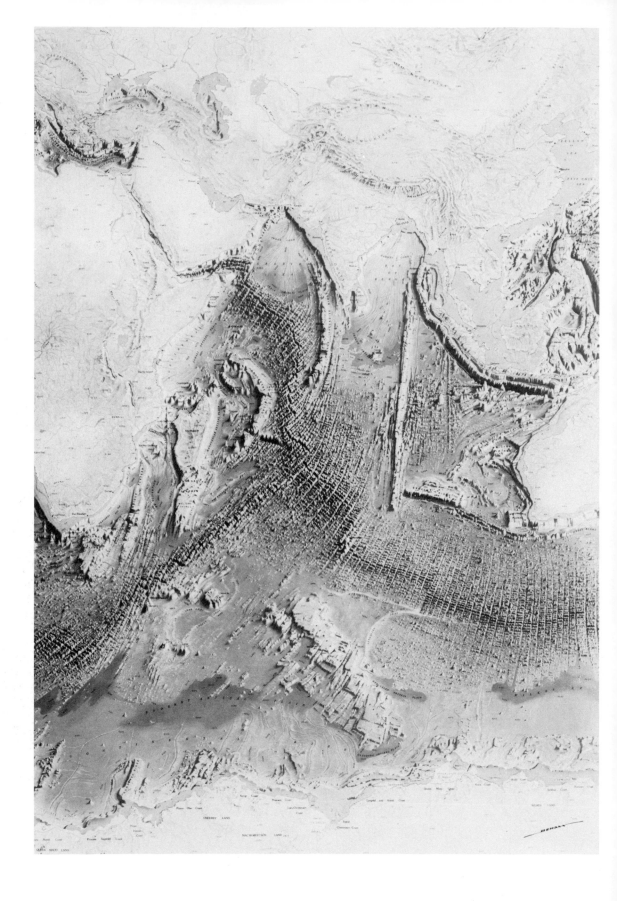

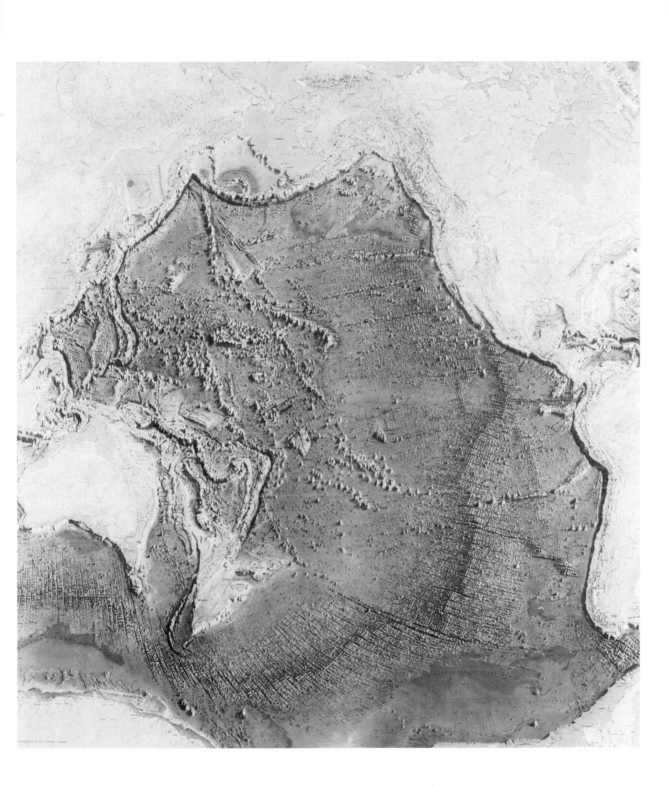

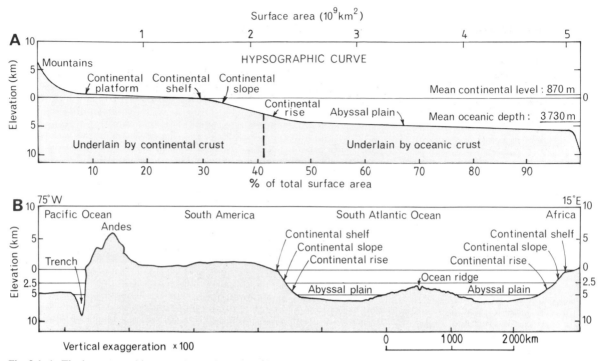

Fig. 2.1 A. The hyposographic curve shows the percentage of the Earth's surface at various elevations and the main features are identified. It can be seen that the continental platform and abyssal plain together form most of the surface.

B. A cross-section from the Andes to Africa showing the distribution of the features along it.

mountains are, on average, less dense than the rocks at depth under the plains at the mountain foot. The first plausible hypothesis to explain this problem was offered by G. B. Airy in 1855. He suggested that the crust could be likened to rafts of timber floating on water (Fig. 2.2B, C). Thick pieces of timber float higher above the water surface than thin pieces, and similarly thick sections of the Earth's crust will float on a liquid or plastic substratum of greater density. Airy was suggesting that mountains have a deep *root* of lower-density rock which the plains lack.

Four years after Airy published his work, J. H. Pratt offered an alternative hypothesis suggesting that anywhere on Earth columns of rock of equal diameter would have equal weight down to a level of compensation at the base of the crust. By this hypothesis rock columns below mountains must have a lower density, because of their greater length, than shorter rock columns beneath plains.

Both Airy's and Pratt's hypotheses imply that surface irregularities are balanced by differences in density of rock below the major features of the

crust. This state of balance is described as the concept of *isostasy* and most of the Earth's surface is thought to be in a state of *isostatic equilibrium*. Losses of rock from the surface of continents, and deposition on the margins of ocean basins, also imply that isostatic readjustments occur continually. We know from repeated surveys that parts of the crust which have been loaded by ice sheets, and are now ice free, are rising isostatically in compensation for the loss of mass as the ice melted (see Chapter 13, p. 351). Such evidence is the strongest available in favour of the concept of isostasy. Surveys of gravitational attraction over the Earth's surface, however, reveal that isostatic adjustment is not everywhere complete and that gravitational attraction in many places departs from what it is calculated to be. Such areas are said to be areas of *gravity anomalies* (see Appendix 2). Many volcanoes on the ocean floor, for example, are formed in 1–5 My and produce an excess of mass on the crust: this creates a positive anomaly until the mass sinks isostatically and equilibrium is again attained. In this example the

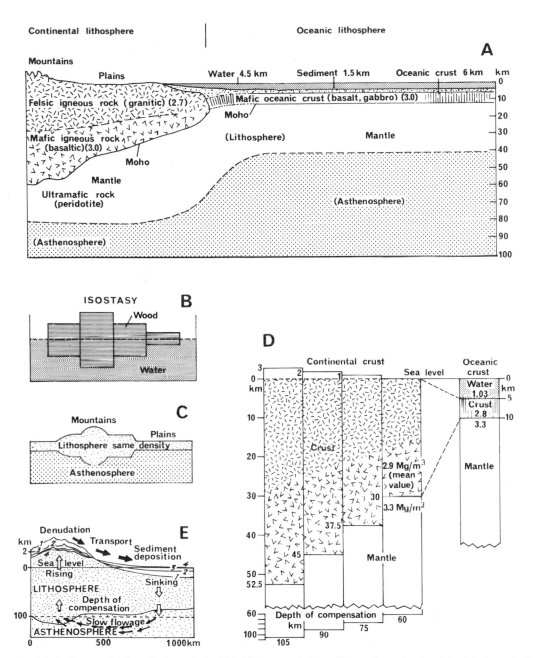

Fig. 2.2 A. A generalized profile through the *felsic* (*fel*dspar and *si*lica-rich) continental rocks of the sial, the mafic (*ma*gnesium and *fe*rric iron-rich) sima, the *ultramafic* (extremely mafic) rock of the upper mantle and into the asthenosphere. The average densities are also given.

B and C indicate Airy's theory of isostasy, and D shows columns through the crust and into the mantle to illustrate the idea of a level of compensation which is in accordance with Pratt's theory. Below the compensation level the changes in load on the Earth's surface are equalized, or compensated, by slow flowage of the asthenosphere, E. (D is based on data of Woollard, 1966.)

adjustment is slower than the rate of formation of the volcano. In many continental areas the rate of erosion from uplands and the rate of deposition along coasts, and in deltas, appears to be at a rate which is a little more rapid than can be compensated by isostatic adjustments in which the mountains rise and deltas sink, for many mountains are areas of negative anomalies and large deltas are areas of positive anomalies.

The evidence available to us is not adequate for us to decide whether Airy or Platt was closest to the truth. In Fig. 2.2A is shown a theoretical section through the crust with values for the density of the major classes of rock (in units of Mg/m^3): this figure implies an Airy-type of compensation and is regarded as a good explanation of mountains in continents. From the columns in Fig. 2.2D it can also be seen that at a depth in the mantle there is a level at which compensation is complete and this observation suggests that some elements of Pratt's hypotheses are valid, especially for the gross differences between continental and oceanic crust.

All available evidence indicates that the crustal rocks of the continents have an average composition close to that of granite with a density of about 2.7–2.85 Mg/m^3. The continental crust varies in thickness from 25 to 70 km, with an average of about 35 km. Because granitic rocks are largely *al*uminium *si*licates the upper continental crust is often termed the *sial*.

The oceanic crust is about 5–9 km thick with a density of about 3.0 Mg/m^3. Its main rock type is basalt composed of *ma*gnesium *si*licate minerals and the oceanic crust, and crust at depth beneath continents, is therefore sometimes called the *sima*.

Rocks of the mantle are thought to be mainly iron and magnesium silicates with smaller quantities of aluminium, calcium, and sodium. *Xenoliths* (i.e. rock fragments trapped in igneous rock masses) suggest that the outer mantle rock is largely composed of peridotite, a rock similar in composition to basalt and consisting of the mineral olivine with lesser amounts of magnetite and pyroxene.

The core of the Earth is thought to be largely of iron with smaller amounts of silicon, nickel, and sulphur.

Internal structure of Earth

The most advanced drilling techniques permit direct recovery of samples, and measurements down a borehole, from a depth of about 10 km. Evidence of the internal structure of the Earth consequently comes mainly from three indirect sources: from meteorites, seismic waves, and volcanoes.

Meteorites are fragments of debris from space which have become trapped in Earth's gravity field. It is thought that they were all formed at about the same time as the solar system developed. Some meteorites, known as *chondrites*, are thought to have a simple history and have not suffered metamorphism. Both $^{40}K : ^{40}Ar$ and $^{87}Rb : ^{87}Sr$ dating indicate that meteorites were formed about 4600 My ago and this is the best indication we have of the age of the Earth.

About 100 tonnes of extra-terrestrial dust enter the Earth's atmosphere each day, but this burns up with the frictional heat generated as it passes through the atmosphere. Larger fragments survive this passage and hit the surface and may be recovered (Plate 2.2). These meteorites are of two types: stony meteorites composed of rock very similar to the xenoliths of peridotite which are found in some lavas; and metallic meteorites composed of various iron–nickel alloys which may be similar to the dense material forming the core of the Earth, and may consequently be the only samples we can obtain of this type of material.

In addition to numerous small (1–50 cm diameter) meteorite fragments which have been recovered, the Earth collides with large bodies called *Apollo objects*, but only with a frequency of once in several million years. Apollo objects are interplanetary bodies whose orbits cross that of the Earth. It is estimated that there are approximately 1200 Apollos with diameters greater than 1 km. About 100 Apollo craters, each larger than 1 km in diameter, have been recognized on Earth. Impact cratering is a fundamental process in the history of the planets, but on Earth craters and other superficial features from impact are progressively destroyed by weathering and erosion, and by movements of the crust. The best-preserved craters are consequently young, with most being of Quaternary age, and have been formed in the relatively stable surfaces of ancient continental plains.

Proof that a crater-like feature is the result of a meteorite impact is derived from the survival of meteorite fragments, and study of the rocks beneath and at the margins of an impact site: these

2.2 A nickel-iron meteorite found in Antarctica. Its blue-black metallic sheen has been preserved in the very dry air. Scale is shown by the hammer.

rocks have fractures and other structures produced by shock waves and also glassy minerals caused by intense local heating at impact.

The energies involved in a major impact event are enormous. A meteorite about 200 m across would impact the Earth at a velocity of about 25 km/s and release an amount of energy equivalent to the explosion of about 100 Mtonnes of TNT. This small body would have enough energy to produce a crater 4 km in diameter. From a statistical analysis of the number and ages of known craters, it is estimated that a 10–20 km crater is formed somewhere on the land surface of Earth every few million years. Such impacts may be of considerable geological importance but they are too infrequent to be of geomorphological significance. Suggestions that very large, nearly circular features like Hudson Bay may be ancient impact sites have not been validated.

Seismic waves are shock waves produced as an *earthquake* when accumulated strain in the crust is suddenly released. The strain, that is deformation, may be associated with the movement of molten magma in the crust or the rupture of rock to produce a *fault* or fracture. The site at which seismic waves are generated is termed the *focus* of the earthquake and the point on the Earth's surface directly above the focus is the *epicentre* (Fig. 2.3). If the energy released in the earthquake is of sufficient magnitude the shock waves may pass through the Earth and be recorded at seismograph stations located in many parts of the globe. The records made at these stations are used to interpret the nature of the materials through which they have passed.

Seismic waves are of two principal types known as *P waves* and *S waves*, and a third type is the *surface wave*. The P wave is a primary or pressure wave which is received at a seismograph station before the S wave, or secondary shear wave (see Appendix 3).

Primary waves travel through both solids and liquids at a speed which is proportional to the density of the material. By determining the speed of the wave it is then possible to learn something of the density and state of the material through which the wave has travelled. Secondary waves can only travel through solids, so if they meet a liquid layer within the Earth they are stopped. Studies of seismic waves show that down to a certain depth in the Earth the waves travel relatively slowly. But once a wave reaches this depth its velocity increases markedly: the velocity of P waves, for example, jumps from about 6 or 7 km/s to about 8 km/s. The surface layer of the Earth above this depth is the *crust* and the *mantle* below it is of denser rock. The boundary itself is called the *Mohorovičić discontinuity* or *Moho*, after the Serbian seismologist who discovered it in 1909 (Fig. 2.2A and Fig. 2.4B).

From studies of the records of thousands of earthquakes, recorded at stations in many parts of the world, it has been determined that the crust is the outer part of the *lithosphere* which consists of solid, brittle, strong rock capable of transmitting shock waves at high velocity. At a depth as great as 125 km beneath continents, and about 50 km beneath the oceans, the lithosphere has a boundary with the *asthenosphere*, which is about 250 km thick, and which contains partially molten rock or a mush of crystals in liquid. This weak rock is close

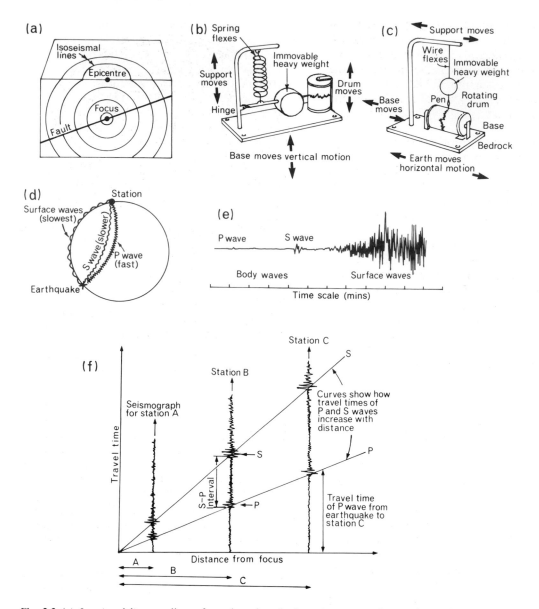

Fig. 2.3 (a) *Isoseismal lines*, or lines of equal earthquake intensity, arranged around an epicentre above a focus. In (b) the seismograph uses a heavy weight which remains stable, while the rotating drum attached to bedrock oscillates vertically as the shock wave passes. In (c) the weight remains stable while the ground and drum move horizontally. Such instruments are now being replaced by electronic devices. In (d) the different types of earthquake shock waves travel at different rates and are recorded on the seismograph (e). The distance of the recording station from the focus is indicated by the time interval between the arrival of P and S waves.

to its melting point of 1400 °C and is capable of transmitting shock waves at relatively low velocities only. The asthenosphere is part of the *upper mantle*. The boundary between the mantle and the outer core occurs at a depth of 2900 km. The core is liquid in character to a depth of 5100 km but the inner core, with a radius of 1100 km, is solid (Fig. 2.4A).

Seismic waves are to geophysicists what X-rays are to the medical practitioner. They permit investigations of an inaccessible interior. Without seismic waves we would have virtually no information about the interior and could not develop testable hypotheses of the mechanisms of deformation of the crust.

Volcanoes are useful indicators of internal structure of the crust because they are the sites at which rock from the lower part of the lithosphere is brought to the surface. Cooling magma, rapidly ascended from depth, may contain blocks of unaltered mantle rock, peridotite, as xenoliths. Studies of volcanic regions thus provide useful confirmation of conclusions drawn from meteorites and seismic wave propagation.

Conclusion

The concept of isostasy is so well established and confirmed by such convincing evidence that it is now regarded as one of the fundamental laws governing the nature of the crust. It provides, however, only part of the explanation of mountains and plains; for erosion from the surface, followed by uplift, must eventually expose the root material and then the sima from below the roots if isostatic compensation were endless. The sima is not exposed and, furthermore, young mountain systems are being created. The causes of the distribution and development of Earth's major physical features are best explained by the theory of global tectonics (Chapter 3).

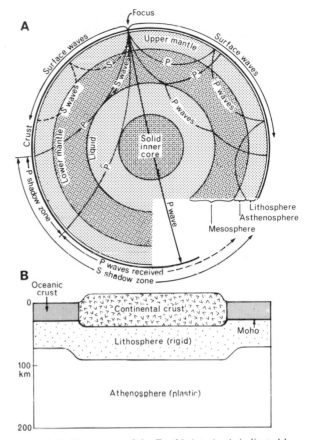

Fig. 2.4 A. The nature of the Earth's interior is indicated by earthquake wave records. Major earthquakes send out waves which can be detected around the Earth, but beyond a distance 120° from the epicentre neither P nor S waves are recorded. The S waves never reappear, indicating that they are stopped by a liquid outer core. The P waves emerge strongly beyond 143°, but with slower travel times indicating that they have passed through a zone of low seismic velocity. The zone in which no waves can be detected is the *shadow zone*.

B. The generalized nature of the lithosphere indicated by seismic records: the crust is the upper part of the lithosphere.

3 Global Tectonics

Until the early 1960s most earth scientists thought of the Earth as a rigid body with fixed continents and permanent ocean basins. Now most scientists believe that the Earth's surface is a brittle, solid crust, in constant motion and composed of six major, and many minor, plate-like blocks which interact with one another. These plates repeatedly collide, break apart, and collide again. The results of such movements are new ocean basins, mountains, earthquakes, volcanoes, and other major structural features.

The scientific theory which has transformed the earth sciences is called *global tectonics*. It will be noted that this revolutionary concept is called a 'theory' in recognition of the extreme difficulty of direct observation and proof in this branch of the earth sciences. The inaccessibility of the interior of the Earth, and the slow rate of many processes, make direct measurement impossible, so that evidence is derived from many sources. As a result we still have no entirely convincing explanations for many of the basic mechanisms of the origin of the planet and activity within it—such as motions (convection currents?) within the mantle, the Earth's magnetic field, and deep-focus earthquakes.

Continental drift

It has been recognized for several hundred years that the surface of the continents is not stable. Leonardo da Vinci, and other early naturalists, realized that the existence of sea shells in rocks on mountain tops must mean that those rocks were formed below sea-level. That land can also sink below the sea was shown by the presence of remains of submerged cities around the shores of the Mediterranean Sea. Thus it was established that the land surface can rise and fall. It took much longer for people to realize that there is evidence

that the crust of the Earth can also move in a horizontal direction.

As early as AD 1620 Sir Francis Bacon noticed that the newly discovered outlines of the continents of South America and Africa would fit together like the pieces of a jigsaw puzzle, if only they were moved. He speculated that such a fit could hardly be an accident but he did not envisage that the continents could have once been joined together. In 1858 Antonio Snider postulated that the continents of the Old and New Worlds were separated during the great flood described in the Bible and generally known as 'The Deluge', or 'Noah's Flood'. The pulling apart of the continents was accepted by several British writers of the nineteenth century, such as George Darwin, and the Revd Osmond Fisher who speculated that the Moon had been torn out of the Pacific at an early stage in the Earth's history, and that the continents had drifted apart as a result. Fisher even suggested that the drift was caused by convection currents within the fluid interior.

Ideas of a mobile crust were generally out of favour at the end of the nineteenth century because the Earth was thought to be progressively cooling and becoming increasingly rigid. Heat from internal radioactivity had yet to be discovered.

The starting-point for modern ideas was the monumental study, *The Face of the Earth*, by Eduard Suess. Suess recognized that the mountain ranges on the southern and eastern borders of Eurasia exhibit large-scale overfolding and thrusting of the strata, and hence evidence of lateral compression. He thought that this may have been produced by movement of continents towards the ocean consequent upon cooling of the crust. The American F. B. Taylor (1910) proposed the first clear theory of continental drift when he suggested that the patterns of the fold mountains, recognized by Suess, could be attributed to a slow creep of the continents towards ocean basins. Taylor, however, did not develop his ideas or seek evidence to test his theory. The first comprehensive study of the evidence for a mobile crust was made by the German meteorologist and polar explorer Alfred Wegener. Wegener was studying the evidence for changes in the Earth's climate and was particularly impressed by the evidence that the southern continents–South America, South Africa, India, and Australia—had all experienced extensive glaciation at about the same time in Permian–Carboniferous Periods. This glaciation was presumably rather like that now being experienced in Antarctica where a huge ice sheet covers the continent. When such an ice sheet diminishes, or disappears, it leaves behind debris of eroded rocks called *till*, and also deep scratches in the surface rocks where the ice has ground boulders against its bed. The directions in which the scratches point indicate the direction in which ice has moved, and the distribution of the till shows the extent of the ice sheet. It became clear to Wegener that the ice sheet of the southern continents could best be explained if those continents had once been joined together. He therefore proposed an hypothesis of *continental drift*, in which he described the course of the continents as they moved from their former positions on the surface of the Earth towards their present position (Fig. 3.1). Wegener's hypothesis was first given a public presentation in 1912, and has its best known expression in the third edition of his book, *The Origin of Continents and Oceans*, published in 1924.

Evidence from the fossil record

Both Snider and Suess had noted that there is a close similarity in the fossil plants preserved in the coal formations of the southern continents, and Suess proposed that the supercontinent on which the coal swamps formed should be called *Gondwanaland* after the key province in India where the coal is well preserved. The distinctive Permian flora which existed after glaciation is known as the *Glossopteris flora* from the readily identifiable leaves which characterize the assemblage of some twenty genera. This flora has no similarities with northern floras of that time.

It is possible to believe that similar plant communities may develop on widely separated continents because seeds and spores can be carried by wind, waves, and birds. The presence of large terrestrial vertebrates (land animals with backbones) of closely related species on each of the southern continents is far more convincing evidence of past land connections because they cannot be carried by wind or birds; they cannot fly or swim across oceans; nor can they drift on logs. Their wide distribution can only mean that they were once able to move from continent to continent over dry land.

It might be objected that animals able to walk could, by taking a long route, actually move between all of the continents, except to Antarctica, for only narrow shallow seas separate Asia and North America, and Asia and Australia, and we know that sea-level was much lower and that dry-land connections existed only a few thousand years ago. It is difficult to believe that whole groups of animals could migrate across several climatic zones, but the possibility does exist.

The evidence from the vertebrates is not as conclusive as that from the plants but one small reptile, *Mesosaurus*, which has no known close relatives, is a very useful piece of evidence. It is found only in the Dwyka formation of western Cape Province, South Africa, and in the Itarare formation of Brazil. No trace of it has been found along the land route of above 32 000 km it would have had to take to travel between these two areas, and it appears most unlikely to have been able to swim the 4800 km wide Atlantic. The most probable solution is to assume that the very similar Dwyka and Itarare formations were once one (Fig. 3.2, 3.3).

A serious gap in the fossil evidence existed because nothing was then known about Antarctica, yet that continent was of major importance in linking Australia and India with Africa. In the late 1960s, however, fossil fish, and a considerable collection of fossil amphibians and reptiles, with affinities to genera of the other southern continents, were found in the Transantarctic mountains within 100 km of the Beardmore Glacier.

Many paleontologists were content to believe that the land connections between the southern continents were land-bridges—of the kind which now links North and South America through Panama. It is to his credit that Wegener pointed out the geophysical impossibility of continental land-bridges existing and then sinking into the oceanic crust. He based his argument upon the principle of isostasy.

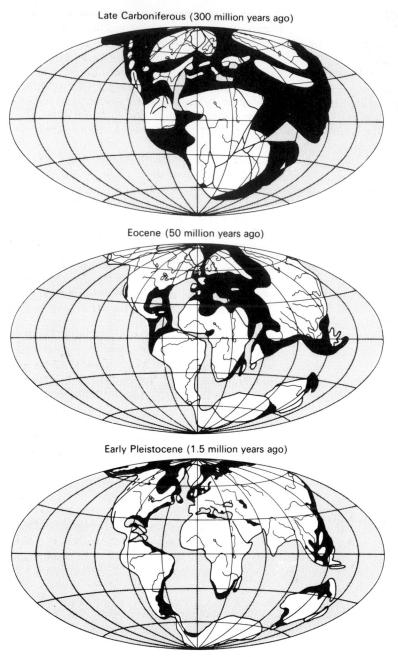

Late Carboniferous (300 million years ago)

Eocene (50 million years ago)

Early Pleistocene (1.5 million years ago)

Fig. 3.1 Reconstruction of the distribution of the continents for three periods according to Wegener's theory of continental drift. Africa is placed in its present position as a standard of reference. The heavily shaded areas represent shallow seas. Modern ideas of ages have been added (redrawn from Wegener, 1924).

There were many opponents of Wegener's ideas. Some, no doubt, objected simply because he was an outsider—a meteorologist—intruding into the realm of geology, but the fundamental problem was that there was no known mechanism which could cause the drift. Much of the opposition came from North American geologists, but a few southern hemisphere geologists did believe in continental drift—notably the South African A. L. du Toit, and the New Zealander L. C. King (who spent most of his professional life in South Africa), and S. W. Carey of Tasmania. In Britain Arthur

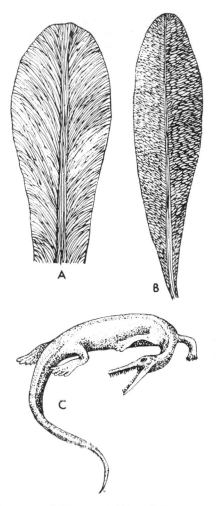

Fig. 3.2 Leaves of *Glossopteris* (A), and *Gangamopteris* (B), which are characteristic plants of Gondwanaland Permian flora. The reptile *Mesosaurus* (C) is about 60 cm long and occurs in the Permian rocks of South Africa and southern Brazil.

Holmes, through his excellent textbook *Principles of Physical Geology*, kept the theory alive through the 1930s, 40s, and 50s, and proposed that the mechanism for driving crustal movements was the generation of convection currents in the mantle by heat produced from radioactivity.

Evidence from paleomagnetism

The first important modern evidence in favour of continental drift came from paleomagnetic studies carried out by British scientists under the leadership of S. K. Runcorn and P. M. S. Blackett. Rock samples were collected from many localities in Europe and North America and the position of the magnetic pole in each geological period, from the Precambrian Era to the present, was determined. The results were plotted on a globe and they showed that the pole for each period has changed systematically. About 500 million years ago it was in the central Pacific and subsequently it has moved towards Japan and across north-eastern Asia. This phenomenon is called *polar wandering*. This may be a misleading term as it is most unlikely that the poles have wandered, rather the continents have moved while the poles were stationary. The magnetic pole path was also found to correspond closely with the paleoclimatic pole, which was found from examination of all fossil evidence for climates of each geological period. This observation implied that the various locations of the ancient paleomagnetic poles were indicative of changes in the orientation of the Earth's axis of rotation with respect to the continents.

One outstanding feature of the data plots on the globe is that although the position of the pole, as determined from studies of North American samples, is similar to that of the pole indicated by European samples, it is not identical, and the discrepancy appears to be systematic (Fig. 3.4). If scientists had been alive in, for example, the Permian Period they would have found that the magnetic poles for rocks forming at different sites all over the world would coincide—just as they do today. That the pole paths we reconstruct are different for one period for two continents suggests that since the rocks cooled below the Curie point the continents have moved apart, and the distance of separation of the poles determined for Europe and North America, at one geological period, is the distance those continents have since moved apart. It was also found by Runcorn's team that if the Atlantic were closed up, as Wegener suggested, then the magnetic poles, for the Permian Period, of each continent do coincide. Polar wandering curves indicate movement only in a relative north–south sense and do not provide information on past longitudinal positions, but they can *only* be reconciled by invoking continental drift.

Sea-floor spreading

The discoveries which have led to the modern

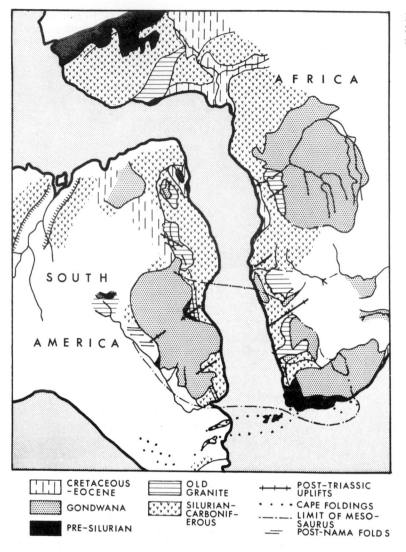

Fig. 3.3 Geological similarities between South America and Africa, as recognised by Du Toit (1937).

CRETACEOUS -EOCENE

GONDWANA

PRE-SILURIAN

OLD GRANITE

SILURIAN- CARBONIF- EROUS

POST-TRIASSIC UPLIFTS

CAPE FOLDINGS

LIMIT OF MESO- SAURUS POST-NAMA FOLD S

ideas, superseding Wegener's concept of continental drift, started in about 1950 when there began intensive studies of the features of the floors of the ocean basins. Seismic studies showed that the rigid crust of the Earth beneath the oceans is very much thinner than the rigid crust beneath the continents. The second discovery was that the long ridge of mountains which divides the floor of the Atlantic Ocean into eastern and western halves is actually part of an undersea mountain system which continues throughout all the oceans of the world. These world-encircling ridges are swellings of the ocean floor some 2000 km in width and tens of thousands of kilometres in length. The crests of the ridges are 2–3 km higher than the flat plains which make up most of the ocean floors (Plate 2.1, Fig. 2.1B).

In 1960, Harry Hess of Princeton University took up the convection current hypothesis and applied it to the ocean floor, proposing that the mid-ocean ridges are situated over the rising limbs of such currents, and that the currents push apart the rigid crust above the rising limb. Then new rock is injected into the crack to form new crust as it cools. Such a process occurring all along the mid-ocean ridges causes a spreading of the newly created crust away from the centre of the ridge. Carried to its logical conclusion the concept

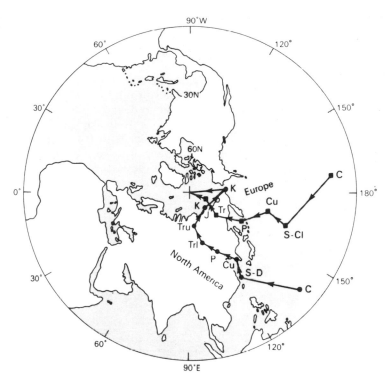

Fig. 3.4 Comparison of the apparent polar-wandering paths for North America (circles) and Europe (squares). The circles and squares themselves represent the essentially stable regions of each continent for the different geologic periods. The following letter symbols are used to designate the various periods:
K–Cretaceous; Tr–Triassic; Tru–Upper Triassic; Trl–Lower Triassic; P–Permian; Cu–Upper Carboniferous; S-D–Silurian-Devonian; S-Cl–Silurian Lower Carboniferous; C–Cambrian (after McElhinny, 1973).

implies that the formation of the whole of the Atlantic and Indian Ocean basins occurred by spreading of older crust away from the mid-ocean ridges, and its replacement by newly formed crust. From a consideration of the earlier ideas about continental drift, and the age of the rock units in Africa and South America which were thought to have been joined together at the times the rocks were formed, Hess concluded that the sea-floor might be spreading at the rate of approximately 1 cm per year on each flank. In other words he thought that the Atlantic was getting 2 cm wider each year. Such a rate of movement would mean that the Atlantic Basin is about 200 million years old.

Having postulated that the ocean basin is expanding, Hess was faced with two possibilities: either the Earth is expanding, or the old crust is being destroyed at the same rate as new crust is being created. Hess felt that it was physically impossible for the Earth to expand at the rate required to accommodate the new material and that the oceanic crust must, therefore, be constantly reabsorbed into the mantle. He suggested that deep ocean trenches around the margins of

the Pacific Ocean Basin are the sites at which old oceanic crust descends into the mantle.

In brief then, Hess was suggesting that the oceanic crust can be considered as part of a conveyor belt of rock which slowly rises at the mid-ocean ridges where it cools and solidifies, and is transported away to the edge of the ocean basins where it descends into the mantle to be heated and reabsorbed, presumably to become part of a future rising rock mass at a mid-ocean ridge. The process which provides the energy for this is a series of large convection cells. In this view continents and parts of the ocean floor are being passively drifted upon the backs of convection cells.

Supporting evidence

Confirmation of geometric and geological fits of continents on either side of the Atlantic became available in the 1960s. A Cambridge University team, under Sir Edward Bullard, showed that the geometric fit is very close if the reasonable assumption is adopted that the true edge of the continent is the 1000 m depth contour, rather than the modern coastline (Fig. 3.5). Geological studies using radiometrically dated rock units showed also that the pattern of rock ages for Africa and South America has a remarkably close correspondence.

Even before Hess produced his ideas it had been noticed that the pattern of earthquakes round the world is very distinctive. The arrays of earthquake-recording seismographs in the world were increased in number in the 1950s and 1960s because they could also be used to detect underground atomic explosions. These military uses were of direct benefit to earth scientists, for the records showed that the mid-ocean ridges are centres of shallow earthquakes, and that the trenches around the Pacific are centres of medium-depth and deep earthquakes (Fig. 3.13). Not only this, but the patterns of earthquakes around the trenches are systematic: they appear to define a plane dipping to depths as great as 700 km or more. The earthquake pattern together with the active volcanoes along the mid-ocean ridge are all readily explained by Hess's hypothesis, for shallow earthquakes and volcanic activity might be expected

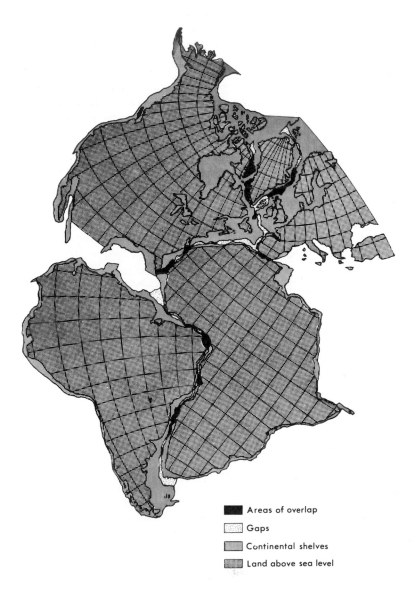

Fig. 3.5 The geometric fit of continents after closure of the Atlantic to the 1000 m depth contour (after Bullard *et al.*, 1965).

Areas of overlap

Gaps

Continental shelves

Land above sea level

where new ocean floor is being created, and deeper earthquakes expected where slabs of old ocean floor are descending into the mantle (Fig. 3.6).

Other lines of evidence in favour of sea-floor spreading are: the sediments of the ocean floor get thicker away from the mid-ocean ridges and also their lower layers are older with increasing distance from the ridges; heat flow in the crust was also found to be anomalously high near the ridges.

The evidence which most scientists regard as being overwhelmingly in support of Hess's concept developed from ideas published, in 1963, by two geologists at Cambridge University— Frederick Vine and Drummond Matthews.

The strength of the Earth's magnetic field can be measured with a magnetometer. The total field is made up of, essentially, two parts:

1 the Earth's general field which varies smoothly from about 25 000 gammas near the equator to 65 000 gammas near the magnetic poles;

2 a local or anomalous component with a range of about 1000 gammas, arising from magnetic minerals in rocks of the Earth's crust.

The difference between the measured and calculated strength at a point is a measure of the anomalous component.

A number of voyages had been made in the 1950s by ships towing magnetometers and, from the data collected, maps were made of the magnetic anomaly pattern of the ocean floor. A survey carried out off the coast of California had revealed that the anomaly pattern forms a series of stripes which are aligned with major fractures—known as the Murray, Pioneer, and Mendocino fracture zones. The existence of such a pattern was quite unexpected and, for a while, unexplained (Fig. 3.7).

Vine and Matthews accepted Hess's ideas and, believing that the new rock being created at the mid-ocean ridges is mostly basalt, suggested that the basalt, on cooling below the Curie point,

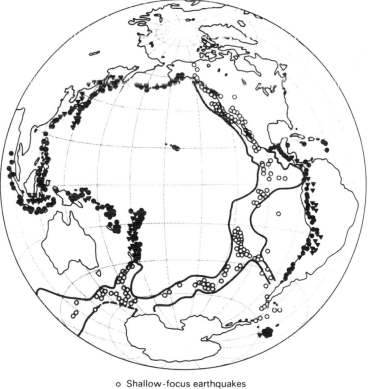

Fig. 3.6 The pattern of earthquakes in and around the Pacific basin (after Girdler, 1964).

o Shallow-focus earthquakes
• Intermediate-focus earthquakes
▼ Deep-focus earthquakes

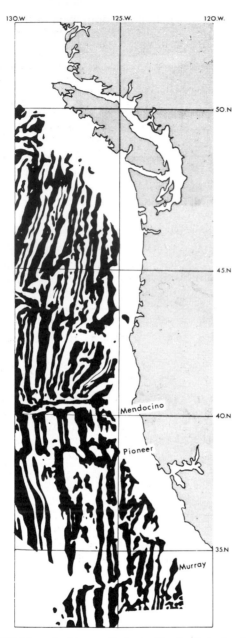

Fig. 3.7 Magnetic stripes on the floor of the Pacific basin off the west coast of the United States of America (redrawn after Raff and Mason, 1961).

direction of the field as the newly created rock moves away from the mid-ocean ridges. To complete their idea, Vine and Matthews made a further assumption: that, as new sea-floor forms and spreads away from the ridge crests, the Earth's magnetic field intermittently reverses the direction of its polarity, thus producing stripes of alternately normally and reversely magnetized rock parallel to the ridge axis. Magnetometer studies soon revealed that the striped pattern occurs not only in the eastern Pacific but in the Atlantic and Indian oceans (Figs. 3.8, 3.9). Subsequently the patterns have been found around the mid-ocean ridges in all the oceans.

At about the same time as the recognition of the magnetic anomaly pattern, improvements were made in methods of dating basalt. By 1966 so many rock samples from terrestrial lava flows had been dated that the age of the magnetic reversals during the last three and a half million years had been determined; it thus became possible to calculate the age of the rocks in each 'stripe' on either side of the mid-ocean ridges, and to calculate the rate at which spreading was taking place. These calculations revealed spreading rates, varying from 1 cm/year on each flank of the ridge in the North Atlantic, to 9 cm/year per ridge flank in the equatorial Pacific. These rates are remarkably

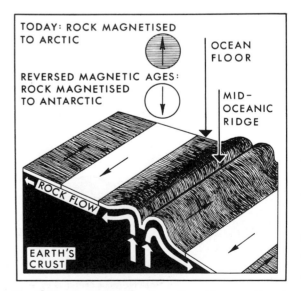

Fig. 3.8 Sea-floor spreading from the mid-ocean ridges is revealed by the striped pattern of the magnetic anomalies.

would record the direction of the magnetic field at that time. The intensity of this remnant or permanent magnetization in basalts is greater than that induced by the Earth's present magnetic field, and so a permanent record is maintained of the

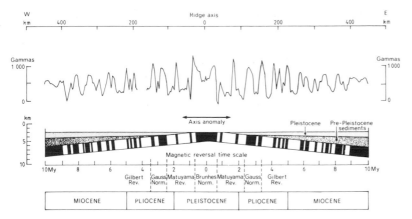

Fig. 3.9 Ocean crust magnetic anomalies plotted as profiles of measured magnetic intensity along straight-line crossings of the South Pacific–Antarctic oceanic ridge. Note the symmetry on either side of the ridge axis. It is the sharp variations in intensity which produce the striped patterns recognized in magnetic maps (Fig. 3.7). The polarity reversal episodes recorded in the crust, and the thickness of sediments is idealized in the lower half of the figure (after Pitman and Heirtzler, 1966).

good confirmations of Hess's original calculations. Further studies of remnant magnetism and the age of ocean-floor sediments have all given general confirmation of the ideas of Hess, Vine, and Matthews. In detail some parts of the ocean floor are more complicated than the original simple ideas indicated, but it is now clear that perhaps 50 per cent of the present deep sea-floor (that is, one-third of the Earth's surface) has been created during the last 65 million years, or 1.5 per cent of geological time.

It is an interesting reflection on the conservatism of the scientific community that, at the same time as Vine and Matthews were producing their explanation for the striped pattern of magnetic anomalies, L. Morley of Canada should have independently developed a very similar explanation but had it rejected for publication by several leading scientific journals.

Transform faults

An hypothesis for the origin and nature of the fracture zones which offset the magnetic anomaly pattern, was suggested, in 1965, by J. Tuzo Wilson. He proposed that they are *transform faults* (Fig. 3.10). The unique features of transform faults result from the motion of crust on a sphere. When blocks move apart they move between lines of 'latitude' relative to poles of rotation, known as *Eulerian poles* after the mathematician who described the geometry of motion on a sphere. Such poles are not related to geographical or magnetic poles, but solely to the ends of longitudinal fracture zones (Fig. 3.11). As the longitudinal fracture zone opens there will be greater motion near the equator of motion than near its poles, so the crust of the sphere is broken into latitudinal bands which spread at slightly different rates from one another. If entire blocks of the crust on either side of a fault move laterally in opposing directions the fault is said to be transcurrent (Fig. 3.12A). In transform motion the shearing takes place only between the offset ends of sections of the oceanic ridges or subduction zones (between b and c in Fig. 3.12B). Beyond the ends of the oceanic ridges (a–b and c–d) there is no opposed movement along the fracture zones: these lines are *healed transform faults* or *transform scars* left by earlier motions or produced by different rates of

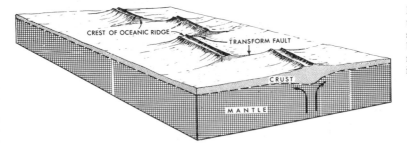

Fig. 3.10 Transform faults offsetting the oceanic ridge axis. The ridges are shown as single features for simplicity, but old ridges are usually sets of multiple ridges. The term 'crust' here includes the whole oceanic lithosphere.

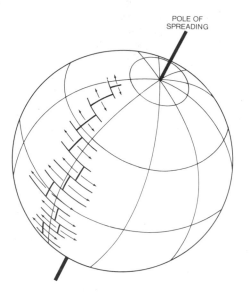

Fig. 3.11 Formation of spreading ridges with offsets along transform faults. The fracture zones are normal to the spreading axis and the rate of spreading, indicated by the length of the arrows, increases in proportion to the distance from the poles towards the equator of spreading.

movement of adjacent blocks travelling in the same direction. They are visible in bathymetric maps (Plate 2.1) because they appear as narrow ridges or scarps, and the fault trace may be filled with intrusions of basalt which has solidified. Proof of the activity along the transform fault, and

inactivity along the scars, is provided by seismic studies.

Confirmation

Confirmation of sea-floor spreading has come from a number of sources, but the Deep Sea Drilling Project (DSDP), initiated in the United States in 1968, has done most to confirm the hypothesis. A consortium of research institutions called JOIDES (Joint Oceanographic Institutions Deep Earth Sampling) was established to sample the sediment and crust beneath the oceans using the most advanced technology which could be built into a new drilling ship, the *Glomar Challenger*. Since then this ship has recovered cores from most parts of the world ocean and made possible an enormous advance in understanding of Earth structure and the record of climatic change preserved in the sediment (see Chapter 16). The work is still continuing, although since 1975 it has been an international enterprise called IPOD (the International Phase of Ocean Drilling) financed by USA, USSR, the Federal Republic of Germany, France, the United Kingdom, and Japan.

Starting in 1971 Project FAMOUS (French–American Mid-Ocean Undersea Study) has been studying the mid-ocean ridge of the North Atlantic, about 650 km south-west of the Azores, using deep diving submersibles and sampling and photo-

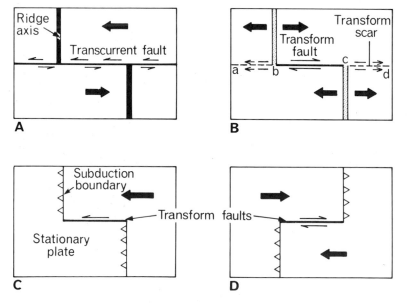

Fig. 3.12 Movements along a transcurrent fault and along three types of transform fault.

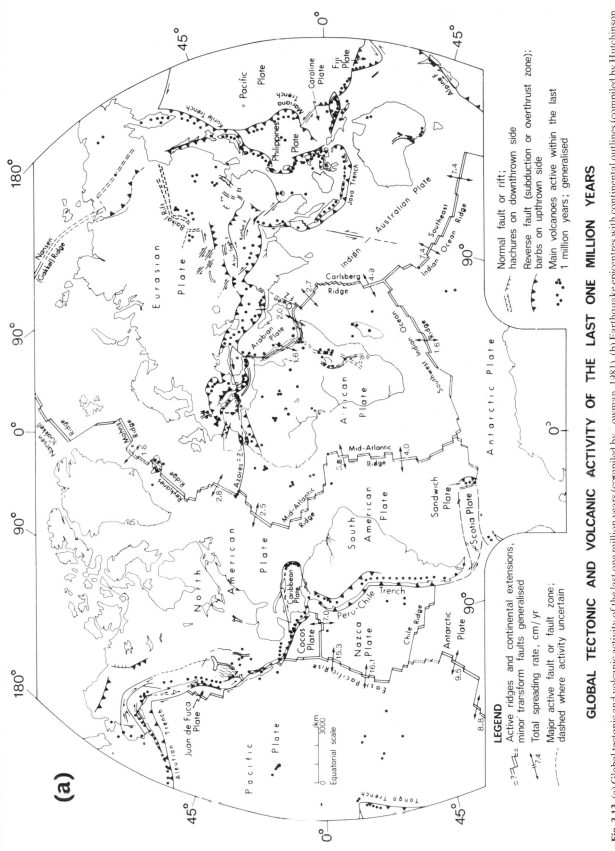

Fig 3.13 (a) Global tectonic and volcanic activity of the last one million years (compiled by Lowman, 1981). (b) Earthquake epicentres with continental outlines (compiled by Hutchinson and Lowman, from Lowman, 1981).

GLOBAL TECTONIC AND VOLCANIC ACTIVITY OF THE LAST ONE MILLION YEARS

LEGEND

Active ridges and continental extensions; minor transform faults generalised

Total spreading rate, cm/yr

Major active fault or fault zone; dashed where activity uncertain

Normal fault or rift; hachures on downthrown side

Reverse fault (subduction or overthrust zone); barbs on upthrown side

Main volcanoes active within the last 1 million years; generalised

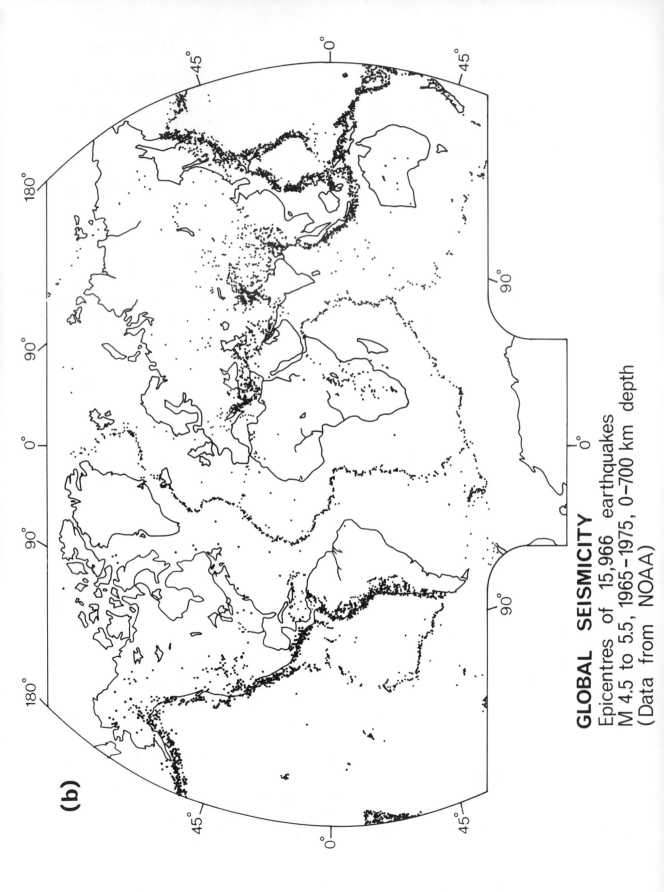

(b)

GLOBAL SEISMICITY
Epicentres of 15,966 earthquakes
M 4.5 to 5.5, 1965–1975, 0–700 km depth
(Data from NOAA)

graphing the ocean floor at a depth of 2–3 km. As expected the narrow floor of the axial rift zone proved to be underlain with very fresh basaltic lava flows, with the characteristic bulbous pillow-form produced when lava is chilled in contact with sea water. The high rate of heat flow from the ocean floor suggests that a magma chamber is present no more than 2 km below the floor. Fractures up to 3 m in width, oriented parallel with the ridge axis, confirm the active spreading associated with injection of fresh lava into the crust.

Iceland is one part of the mid-ocean ridge which rises above the surface of the ocean and has become a site of intense geological studies since the mid-1960s. Here too the active spreading of the crust and injection of new sheets of basalt into the spreading rifts has been observed directly, and traced back for the last several million years.

Plate tectonics

In 1967 yet another concept, that of plate tectonics, was added to the collection of ideas about the nature of the crust, by D. P. McKenzie and R. L. Parker and independently by W. J. Morgan. Xavier Le Pichon investigated the possibility of analysing the geometry of sea-floor spreading in terms of relative movements on the surface of a sphere. These workers realized that the distribution of earthquakes, mid-ocean ridges, transform faults, young mountain chains, and the deep ocean trenches around the rim of the Pacific Ocean floor, fit into a pattern which can best be explained if the Earth's crust is regarded as being made up of a number of nearly rigid plates which move relatively to one another. Thus was born the term *plate tectonics* which has absorbed earlier ideas of continental drift and sea-floor spreading. The noun *tectonics* means the study of structural features of the crust and their origin.

There are six major plates bounded by ridges, trenches, faults, or young fold mountains:

1 Antarctica and the surrounding ocean floor;
2 the Americas with the western Atlantic floor;
3 the Pacific floor;
4 India to Australia and New Zealand;
5 Africa with the eastern Atlantic floor;
6 Eurasia with the adjacent ocean floors.

Several minor plates have been recognized off the Pacific coast of the Americas, in the Carib-

bean, in the area of the Philippines, and in the Mediterranean to Arabia belt (Fig. 3.13a, b).

There are four major types of plate boundary. (1) Where two plates are moving apart from each other, as at mid-ocean ridges, new crust is formed and the boundary is said to be *constructive, accreting,* or *diverging.* (2) Where two plates are moving towards each other a *destructive,* or *converging,* boundary is formed. Oceanic trenches are sites at which one plate sinks below another, or is *subducted* or consumed. (3) *Collision* boundaries are also sites of convergence, but at these neither plate is consumed. An example of a collision boundary is the line of the Himalayan chains formed where the Indian plate meets the Asian plate. (4) A *conservative* boundary may form along a transform fault where the relative plate motion is parallel to the boundary. The San Andreas Fault of California is interpreted as being a boundary of this type (Fig. 3.14) with the Pacific side of the fault being moved northwards relative to a south-moving block on the continental side of the fault. The processes and landforms of collision and conservative boundaries are discussed in Chapter 4 where the features of mountain building can be treated together.

Points at which three plates meet are called *triple junctions* (Fig. 3.15). At such junctions, subduction, transform motion, or divergence may occur in any combination. It is the existence of

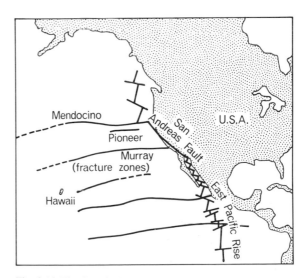

Fig. 3.14 The San Andreas Fault, and fracture zones of the East Pacific Rise.

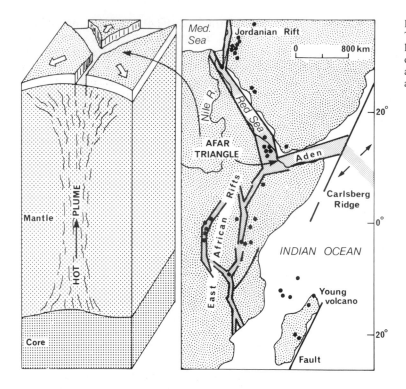

Fig. 3.15 Representation of the Afar Triangle as a triple junction above a hot mantle plume. The African rifts are displayed as spreading rifts with accompanying volcanism (after Dott and Batten, 1981).

triple junctions which permits sharp changes of direction to occur along mid-ocean ridges, which are otherwise straight or offset at right angles along transform faults.

The location of plate boundaries cannot be stable where an ocean basin is being widened: thus the southern part of the American plate is overriding the Nazca plate off the west coast of South America. Around two sides of Africa and all sides of Antarctica there are spreading ridges. It is geometrically impossible for new sea floor to be created at all these ridges and to spread towards Africa and Antarctica unless oceanic crust is being subducted below continental crust. There are, however, no trenches around these continents and the oceanic ridges must be migrating farther and farther away from the continents as more oceanic crust is produced.

The inevitability of mid-ocean ridge migration indicates that such ridges are not necessarily in the middle of oceans—consideration of the Pacific shows that *mid*-ocean ridge is a term which properly refers to Atlantic type conditions. In the central American region an oceanic ridge appears to have collided with a continent. The San

Andreas Fault is a transform fault across this ridge (Fig. 3.14), and the oceanic crust east of the ridge appears to have sunk beneath California and the Basin and Range Province of the south-west desert states of Nevada, Arizona, and Utah, where it may be responsible for stretching the crust above it and causing block-faulted mountains (Fig. 4.10).

Construction of oceanic ridges

When the sea-floor spreading theory was first proposed it was commonly accepted that the oceanic ridge is a portion of the upwelling mantle driven by convection currents and then cooled. This very simple idea has been modified. It appears from results of seismic studies, indicating that shallow earthquakes which result from fracturing or faulting are common along the ridges, from Project FAMOUS observations, and from studies of ancient oceanic ridges now preserved on continents, that a more complex situation exists (Fig. 3.16).

Ascending magma from the asthenosphere wells up below the cracking and widening ridge. As the magma approaches the ocean floor the pressure

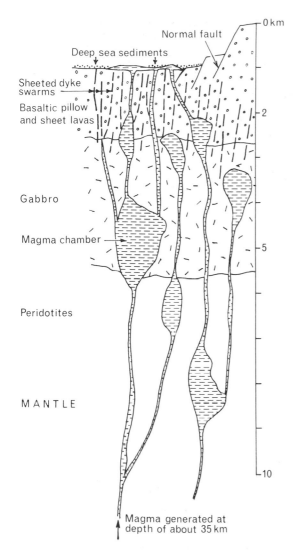

0 km

Normal fault

Deep sea sediments

Sheeted dyke
swarms

Basaltic pillow
and sheet lavas

2

Gabbro

Magma chamber

5

Peridotites

MANTLE

10

Magma generated at
depth of about 35 km

Fig. 3.16 Cross-section of oceanic crust and mantle close to a mid-ocean ridge. Magma is generated as mantle rock melts. Being less dense than the mantle it rises but is held up in magma chambers where its composition changes due to formation of new mineral crystals. Gabbro is formed by slow cooling of some of this magma, but the remainder continues to rise and forms basaltic lava sheets and pillow lavas. The boundary between the gabbro and basalt lava zones is cut by sheeted dykes which are the remains of lava conduits. The whole gabbro–basalt complex is called an ophiolite suite.

decreases and partial melting is increased. Some of the magma cools below the crust, where coarsely crystalline gabbro is formed, and some is forced into the fractures in the crust as wall-like intrusions called dykes, or flows out as basalt lava—

especially pillow lava. The accreting edge of the plate is thus composed of basalt lava flows overlying basalt dykes infilling fractures in older basalts. As the newest sea floor is the hottest, and therefore the weakest, part of the crust, the crests of ridges continue to be sites of spreading.

As the new sea floor spreads away from both sides of the ridge it gradually cools from the surface downward and thickens. The solidified portion of the plate thus becomes cooler, thicker, and denser, because of thermal contraction, away from the ridge. The ocean floor also becomes deeper away from the ridge as a result of this mechanism.

Sediment formed of the skeletons of marine animals, of wind-blown dust from the continents and, more rarely, of material carried away from the continents by *turbidity currents* is collected in the basins between fractured blocks of oceanic crust (a turbidity current is a rapid underwater flow of a mix of sediments and water, plunging downslope under gravity before settling out on the deep-sea floor). Almost invariably sediments are thickest above old crust close to the continents and thinnest near the young spreading ridges.

The revised view of plate accretion and cooling raises the important question: what force drives the plates? Are plates dragged along by the convection currents below them? Are they pushed by the weight of elevated plate edges at the ridges? Are they pulled by the subduction of cooling plates below the consuming margins (Fig. 3.17). The answer to these questions is not yet available, but it seems probable that all three mechanisms operate, with sinking oceanic crustal rock being reabsorbed into the mantle, and possibly recycled in a newly rising magma.

The result of these processes are oceanic ridges consisting of high, tilted and faulted blocks forming long mountain ranges with a total length of 64 000 km. The ridge has a total width of 2000–2400 km with a surface which rises in a series of steps from abyssal plains on both sides to the central ridges, which have elevations of up to 3700 m above the abyssal plains. An axial rift marks the spreading plate boundary.

Destructive boundaries

Much of the evidence for the nature of destructive plate boundaries comes from seismic studies. According to depth, three classes of earthquake

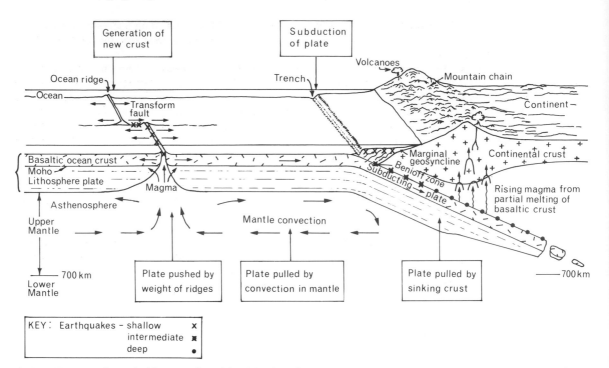

Fig. 3.17 Features of oceanic ridge spreading with subduction of oceanic lithosphere beneath the continental edge of a neighbouring plate.

are recognized: (1) *shallow*, with a focus within the upper 50 km of crust; (2) *intermediate*, with a focus from 50 to 250 km depth; and (3) *deep*, with a focus deeper than 250 km and a maximum depth of about 650–700 km. Below about 650 km the mantle material behaves as a plastic substance, rather than a brittle one, and so does not fracture and produce a shock wave.

Where the foci of earthquakes are mapped, it is usual to find that close to a marginal oceanic trench all earthquakes are shallow, and closer to the continent they are of intermediate and then deep focus (Figure 3.17). This is clearly true in the case of the Pacific coast of South America and in the area of Japan. The significance of this pattern was explained independently by the American Hugo Benioff, and the Japanese K. Wadati: dipping zones of earthquake foci, which are now recognized as evidence for subduction, are called *Benioff zones*, although it would be more appropriate to call them *Wadati–Benioff zones*.

The second line of evidence had been discovered earlier by the Dutch geophysicist, Vening Meinesz, who found that the trenches around the East and West Indies are belts of negative gravity

anomalies. The third line of evidence is that trenches are areas of low heat flow in the crust, compared with the high heat flow zones of the axial rifts of the oceanic ridges.

What happens at the upper surface of the crust during a collision depends upon the nature of the crust. (1) Where oceanic crust collides with oceanic crust an *island arc* may be formed above a subduction zone, as in the Aleutians or the Marianas. (2) Where oceanic crust collides with continental crust the oceanic crust may be subducted below the continent and a mountain chain formed above the subduction zone, as along the Andean coast of South America. (3) In a few localities the collision of oceanic and continental crust may result in a flake of oceanic crust being *obducted* over continental crust while the majority of the oceanic crust is subducted, as in Cyprus. (4) Where two continental blocks collide the sediments between them may be piled up into a mountain, as in the Himalayas. Only the first two types of collision will be discussed in this chapter and the other two will be described more fully in Chapter 4.

(1) Most of the *island arcs* of the world occur

around the western and northern edges of the Pacific Ocean, where the Pacific plate is being subducted below the oceanic edges of the Eurasian and American plates. As the down-bending lithospheric plate descends into the mantle it is heated, and it is thought that the crustal basalt melts and forms new magma bodies. Some of the magma differentiates as the crystallization temperatures of the constituent minerals are passed, so that those minerals which melt first become concentrated in the rising magma bodies. As a result of differentiation the mafic materials from basalt stay behind but the felsic components rise so that the newly

created magma is of intermediate composition which produces andesite volcanoes and lava flows (Fig. 3.18).

Andesite and basalt outpouring first forms a chain of submarine volcanoes and then, as they grow above sea-level, a chain of volcanic islands. Eventually larger islands may form with many eruptive centres. All of these stages can be recognized in the island arcs of the western Pacific (Fig. 3.19). So common are andesitic volcanoes around the margin of the Pacific that the line of the island arcs, and the continental volcanoes of the Pacific coasts of the Americas, is called the *Andesite Line*.

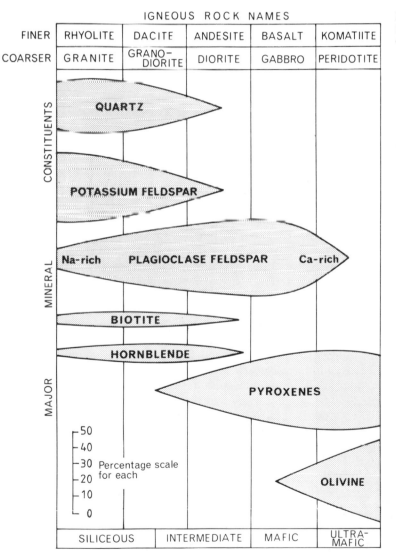

Fig. 3.18 Classification of extrusive and intrusive igneous rocks and their major compositions.

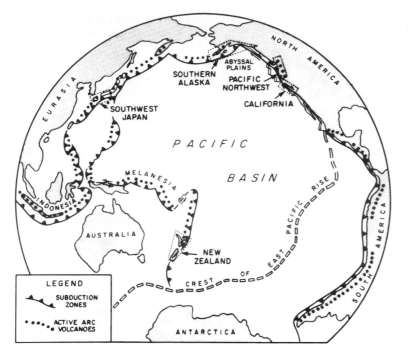

Fig. 3.19 Island arcs and the andesite line of Pacific Basin volcanoes.

Chains of volcanic islands like the Aleutians and Marianas form a limited source of sediment for the deep trenches along their margin: such trenches are regarded as barren, and the *backarc basin* between the islands and the continent may also be an area of thin oceanic crust and of limited sedimentation on its side nearest the island arc.

In the region of Japan the situation is more complex, for Japan is underlain by thick continental crust yielding large volumes of sediment which are infilling the Ryukyu Trench and much of the Sea of Japan basin. The Sea of Japan has thin oceanic crust with up to 2 km of sediment lying on it. Seismic evidence of an active Wadati–Benioff zone beneath Japan indicates that there is a subducting plate.

Two hypotheses have been offered to explain how oceanic crust came to be located in backarc basins. (1) The *entrapment hypothesis* suggests that a subduction boundary was formed within the oceanic crust and thus isolated the backarc crust behind younger arc volcanoes. This explains the situation of the Marianas (Fig. 3.20), and it appears to explain the Aleutian Islands and the Bering Sea, as the Aleutians are relatively young Cenozoic volcanoes and the Bering oceanic crust is

of Mesozoic age, as indicated by paleomagnetism and $^{40}K : {}^{40}Ar$ dates. (2) In the case of the Japan arc, the Sea of Japan crust is younger than the older rocks of Japan, which are Paleozoic. In this case an hypothesis of *lateral drift* appears to be appropriate: it is proposed that subduction occurs beneath a continental margin, as in Fig. 3.17, and that a slice of continental crust separates from the main continental crust opening up a rift which will progressively widen, and be filled with new oceanic crust and become a backarc basin. The curvature of such arcs away from the Eurasian Plate, the high heat flow beneath the Sea of Japan, and the age of the crust, all support this hypothesis. As spreading continues a separate convection current system may develop. The Sea of Japan basin may have been formed by multiple rifting as no clear magnetic stripe pattern has been recognized in its oceanic crust. In the Lau Basin west of Tonga, and the Shikoku Basin of the eastern Philippine Sea, anomaly stripes are recognized and these are thought to be active backarc basins of the type shown in Fig. 3.21A.

Some subduction arcs are known to have very steeply dipping subducting slabs as well as actively spreading backarc basins. Because the spreading is

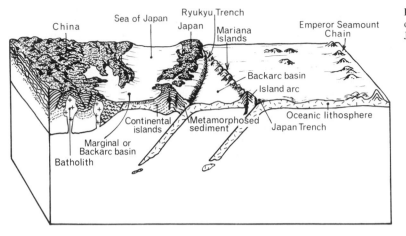

Fig. 3.20 Plate tectonic reconstruction of the major crustal features in the Japan region.

accomplished by a retreating motion of the continental plate, the new oceanic lithosphere is continually subjected to tension, and so relieved of compression that very intense earthquakes do not occur. The arc is not greatly uplifted and so the trench is barren; furthermore the arc may be subjected to tectonic erosion if fragments of the upper slab are carried down with the subducting plate. This type of subduction arc is sometimes called the *Mariana type*.

(2) *Subduction arcs of continental margins*, such as the Andes of Chile and Peru, are usually characterized by a low dip angle of the subducting slab, which presses against the overlying plate. The oceanic plate may have a low upward bulge oceanward of the trench, and the entire subduction zone is subjected to high compression, severe scraping between the two plates, and hence intense earthquakes. This type of subduction has been called *Chilean type*. It has been speculated that the Chilean type is associated with relatively thin, warm, low-density oceanic crust which consequently sinks less rapidly into the mantle than thicker, colder, denser crust of the Mariana type. This working hypothesis appears to be in conformity with evidence from several sources—seismic, paleomagnetism, rock age, plate geometry, stresses, and heat flow.

Along continental margins, and to a lesser extent island volcanic arcs which stand high above sea-level, large quantities of sediment are carried from the land and swept across the continental shelves by currents. At the outer edges of shelves turbidity currents carry the sediment down the continental slope and along submarine canyons into the trenches. At a site of active subduction some of this sediment is trapped between the upper and lower plates and carried down into the mantle, but most is scraped into crumpled underthrust wedges which pile up, one against the other (Fig. 13.21B). The wedges progressively steepen in dip and are pushed upwards into the newly deposited sediments overlying them, which in their turn form folds draped over the wedge crests.

In many places groups of wedges may be pushed up to form *structural highs* which trap younger sediments in *forearc basins*. A complex of wedges is known as an *accretionary prism*: such features, with forearc deposits, may continue to grow for periods of up to 100 My. Forearc basins may, alternatively, be formed above a fragment of oceanic crust on the continental side of a subducting slab and then this oceanic crust may fracture and be incorporated in the accretionary prism. During a later period of uplift of the prism to form mountains, old oceanic crust may then be found associated with highly faulted, steeply folded sediments, ranging from mudstones to conglomerates in particle size. In addition blocks of oceanic crust may be torn off the upper surface of the subducting slab. The resulting complex of highly contorted continental sediments, volcanic debris, deep ocean floor pelagic sediments and fragments of oceanic pillow lavas, basalt dykes and blocks of gabbro with, perhaps, fragments of peridotite is called a *mélange*. The various rocks of oceanic crust together are called an *ophiolite suite* and they, together with mélange sediments, are clear

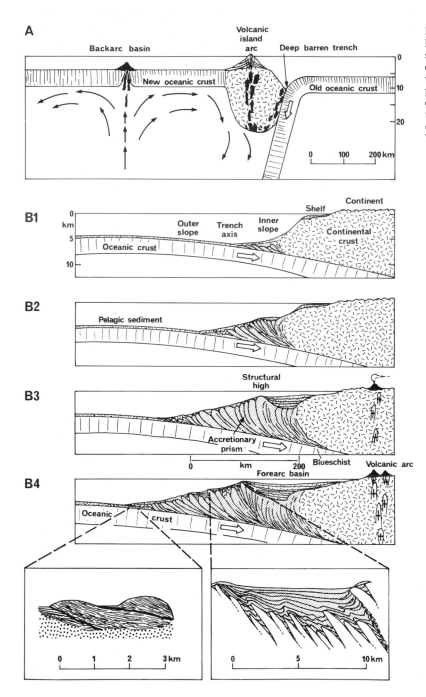

Fig. 3.21 A, An island arc with high-angle subduction and backarc spreading. B1–4, Evolution of a continental margin subduction zone with formation of an accretionary prism and forearc basin. The enlargements at the bottom indicate the thrusting and folding in weak sediments of the prism (based on reports in Talwani and Pitman, 1977).

evidence of the processes of accretion accompanied by subduction, even at sites where subduction ceased long ago. (see Fig. 3.16).

Global tectonics: the great synthesis

The value of plate tectonic theory is that it has, for the first time, provided a framework within which we can understand the mechanism and relationship of the Earth's major physical features—continents, ocean basins, mountain systems, volcanoes, earthquakes, ocean trenches, mid-ocean ridges, and rock suites. It has been suggested that this unifying concept should be called *global tectonics*.

There has already been some attempt to apply plate tectonic theory to the period before about 200 million years ago when the single continent, which Wegener called Pangaea, first started to break up. By using the presence of eroded mélanges with ophiolite suites, the existence of old plate boundaries is inferred, and the origin of old mountain systems, such as the Appalachians, the Caledonian mountain systems of Scotland and Scandinavia, and the Urals of Asia may be explained by pre-Pangaean plate tectonics.

Using such an approach it may be possible to reconstruct Earth history in a dynamic form. But before we can fully accept such reconstructions we need to understand more about the mechanics of tectonics, the driving forces moving the plates, and the chemistry of crustal changes.

Reconstructions of Earth history for the time before 200 million years ago are based upon polar-wandering curves and the direct geological evidence provided by the distribution of desert sandstones, coal-beds, coral reefs, glacial deposits, and desert salt deposits. All of this evidence can be used to reconstruct a climatic pattern which prevailed at the time such rock formations were being produced, and the climatic pattern can be used to recreate the past geography, and to check on the evidence from polar-wandering analyses and ophiolite suites.

For the time since the beginning of the Triassic period, over 200 million years ago, it is possible to reconstruct Earth history using both polar wandering and sea-floor spreading data. In the Triassic the one continental mass, Pangaea, was surrounded by a universal ocean, called Panthalassa, which was the predecessor of the Pacific Ocean. A deeply penetrating arm of this ocean, the Tethys

Sea, penetrated between Africa and Eurasia and this was the forerunner of the Mediterranean Sea (Fig. 3.22). By the end of the Triassic Period, 180 million years ago, the land mass had become two supercontinents, Laurasia and Gondwana, as a result of the opening of new ocean basins south of North America and east of Africa. The movement of South America towards the Pacific began the formation of the Andes. From the end of the Jurassic Period, about 135 million years ago, these trends continued and India separated from the other southern continents. By the beginning of the Tertiary Period, about 65 million years ago, the South Atlantic was open and the North Atlantic beginning to form. Of great importance also was the separation of South America and Australia from Antarctica. This permitted the Southern Ocean to develop a full westerly circulation, to bring moist air into Antarctica, which was by then in a full polar position, and to permit the development of snow- and ice-fields there. This development was the precursor to the present ice age.

During the Tertiary the oceans continued to widen: the collision of India with Eurasia pinched up Tethyan sediments to form the Himalayan and Alpine chains, and the movement of the Americas towards the Pacific continued the formation of the Andes and fold mountain ranges to the north. Island arc systems developed in the western Pacific, and Africa split from Arabia along the line of the Red Sea where a new ocean floor has been initiated.

If these trends are projected into the future it has been suggested that East Africa will split along the line of the great rift-valleys, Australia will continue to drive northwards, and the Atlantic will open further. Continued spreading of South America towards the Pacific Basin will cause overriding of the southern East Pacific Rise in much the same way as California has overriden its northern end.

Objections to global tectonics

In spite of the very wide acceptance of global tectonic theory, there remain many problems and questions which the simple forms of the theory have not yet accommodated. (1) The length of spreading oceanic ridges is considerably greater than the length of subduction zones. This implies that there must be zones of convergence of oceanic plate and increased rates of subduction there, but

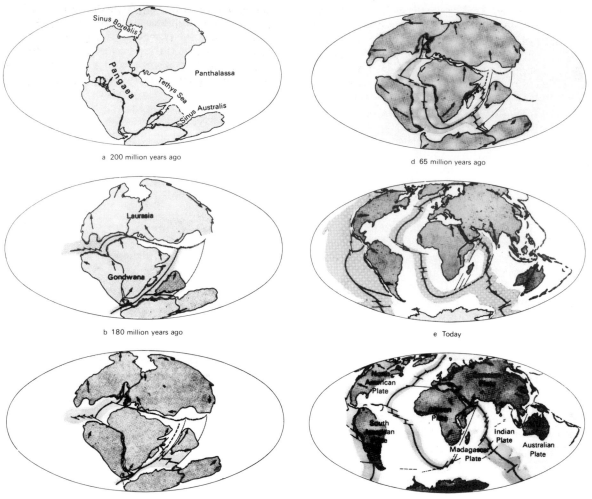

a 200 million years ago

b 180 million years ago

c 135 million years ago

d 65 million years ago

e Today

f 50 million years from today

Fig. 3.22 Stages in the break-up of Pangaea to form the modern continents (redrawn and simplified after Dietz and Holden, 1970).

no such zones have been detected. (2) Spreading occurs at all oceanic ridges but subduction is virtually confined to the margins of the Pacific. Why? (3) Spreading is symmetrical around Africa and Antarctica which have increased the area of their plates since the Mesozoic by 50 per cent, and yet there is no evidence of subduction at the margins of these plates. The cause of the resulting migration of the spreading ridges is not known. (4) Many of the features of the western edge of North America are not understood. (5) The presence of numerous small plate fragments, in areas like the Mediterranean, makes a simple theory appear inadequate. (5) Broad uparching of areas like the

Eastern Highlands of Australia does not appear to fit into known patterns of global tectonics.

That there should be many unanswered questions is not surprising, but their existence is a warning that our present ideas may be too simplistic to provide an adequate model for all tectonic patterns.

Further reading

There are many texts giving reviews of plate tectonic theory. A well-illustrated introduction is the collection of papers from *Scientific American* introduced by Wilson (1976). Collections of papers edited by Bird and Isacks (1972) and by Bird (1980) also provide ready access to major contributions to the subject. The books by Uyeda (1978) and Takeuchi, Uyeda, and Kanamori (1970) provide a more integrated treatment, and the collection of research papers edited by Cox (1973) provides a valuable insight into the way in which many individuals have made their contributions to modern theory. Reviews of the theory with emphasis on the development of ideas are offered by Wyllie (1971) and Hallam (1973). Important newer works include those of Watkins *et al.* (1979), and Talwani and Pitman (1977).

4 Tectonic and Structural Landforms

Tectonics is concerned with the form, pattern, and evolution of the Earth's major features such as mountain ranges, plateaux, and island arcs. *Structure* is a term used with reference to small features such as folds, faults, and joints. The discussion in this chapter will begin with the largest, tectonic, features of the continents and progress towards smaller features.

Cratons

All of the continents have a core of stable continental crust composed wholly, or largely, of Precambrian rocks with complex structures: such segments of continental crust are termed *cratons*. Two main types of surface occur on cratons. (1) *Shields* are areas over which ancient igneous and metamorphic rocks are exposed, and the relief is generally subdued and a result of selective erosion along lines of weakness related to joints, folds, or weaker rock which underlies depressions, and harder rock which underlies elevated areas. (2) *Platforms* are parts of cratons on which, largely undeformed, sedimentary rocks lie upon basement igneous and metamorphic rocks (Fig. 4.1).

The oldest parts of cratons have nuclei of rocks with ages of 2.5 Gy or greater. These nuclei are surrounded by belts of younger rocks of middle to late Precambrian age. This pattern suggests that cratons began as relatively small units and grew by the welding on of younger units. The oldest rocks are metamorphosed lavas and coarse-grained sedimentary rocks, which are interpreted as the remnants of ancient island arc volcanoes and trench sediments, to which further belts have been added as a result of ocean–continent and continent–continent collisions. The zones of contact are marked by ophiolite belts with their ultramafic rocks: such belts of contact are called *sutures*. Elements of a history of repeated collisions and accretion of lithospheric plates are interpreted from the recognition of sutures and the dating of the rocks.

Progressive erosion over much of late Precambrian and early *Phanerozoic* time (i.e. post-Precambrian time) has cut away the upper rock units raised in the collisions, and left those rocks formed, or altered, at depth in regimes of high temperature and pressure. Most shield surfaces, like those of Canada, now have limited surface relief. Cratons are said to be stable features of the crust, but this stability is a relative feature, for broad up- and down-warping, localized arching and doming, and large-scale rifting, as well as intrusion of young granite masses and dolerites, are features of the margins of cratons.

Broad vertical movements of continental crust are said to be *epeirogenic* as they do not involve severe deformation of the rocks but only localized tilting, and this is in contrast to the severe *orogenic* deformation associated with folding of sediments, faulting, and severe metamorphism. Epeirogenic movements may cause submergence of parts of cratons to depths of 4000 m, as is indicated by the thickness of the sedimentary strata upon basement rocks. Such depositional periods may last for up to 150 My. Depositional basins within a craton are sometimes the result of stretching of cratonic crust, with thinning and downwarp sagging. The stretching may cause local rifting of a craton margin, or interior, and it is usually assumed that this involves either an upwelling hot plume in the mantle producing elevation and tension, or a downwelling below a growing basin. Up- and down-welling may be associated either with convection currents, or with *phase-changes* as temperature and, or, pressure changes in the asthenosphere cause gabbro to change into the denser form of ecologite, or the reverse. Density increases or decreases cause volume changes, and hence

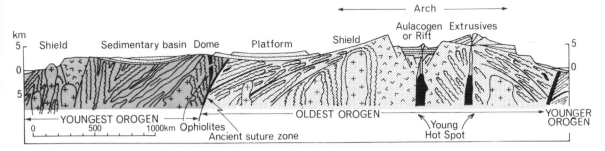

Fig. 4.1 Features of cratons.

subsidence or uplift respectively. Such temperature changes may be associated with heat flow variations without convection.

Another cause of marine sediment deposition on cratons is a world-wide (i.e. *eustatic*) sea-level rise. Such a *marine transgression* may result from growth of oceanic ridges during periods of rapid sea-floor spreading. One such period of world-wide transgression on a large scale occurred in the middle Cretaceous Period (Fig. 4.2) and extended into Tertiary times. Marine, shallow-water Cretaceous rocks which have suffered little deformation are, consequently, common components of platform and continental shelf sediments (Fig. 4.3). Transgressions associated with increases in the volume of oceanic ridges are followed by *regressions* as the oceanic ridges become cooler, subside, and subduction effects increase.

Two well-known intra-cratonic sedimentary basins are the Michigan basin of USA which, forming in early Paleozoic time, was filled with nearly 4000 m of sediment, including coral limestones and salt deposits of an arid climate and shallow lagoonal sea, and the Pannonian Basin, of central Europe, contained within the sweep of the Carpathians. This basin's surface is the Hungarian Plain, with its infill of late Tertiary continental sediments overlying a faulted, thin (20–25 km) continental crust.

Because of the lack of deformation the landforms of platforms are essentially controlled by the details of jointing and lithology, but the dip of the beds, with resulting scarps controlled by the relative resistance of their rocks, are the result of craton warping (Plate 4.1).

While it must be recognized that some of the belts of similar-aged rocks in cratons may owe their origin to periods of deep crustal metamorphism without collisions and folding, the concept of craton accretion and growth appears to be a very plausible hypothesis. It is probable that during the Cambrian there existed five continental cratons—forming the core of North American, European, Siberian, and Chinese continents—and the huge southern hemisphere continent of Gondwanaland. By the Ordovician Period the North American and European cratons had collided to form the Caledonide orogen which produced the chain of mountains running from western Scandinavia, through eastern Greenland, north-western Britain, Newfoundland, the Maritime Provinces of Canada, and most of New England. The belt of these Caledonian Mountains is consequently in structural continuity through three land masses which are now widely separated. Also during the Silurian Period the European and Siberian continental plates had collided to form the Uralide chain in an orogeny which continued until the Permian. These collisions thus produced Laurasia.

Towards the end of the Carboniferous Period Gondwanaland, which was moving northwards, collided with Laurasia to create the Hercynide belt which includes the main part of the Appalachians of North America, parts of North Africa, and many blocks in Europe—such as the Massif Central, Vosges, Bohemia, and south-western Britain. These systems are shown in Figure 4.21 and will be discussed below.

Geosynclines and geoclines

Long, narrow sedimentary basins attracted much interest from early geologists, and the name *geosyncline* was given to them as a result of the work of two Americans, James Hall and James Dana, who studied the rocks of the Appalachians and concluded that they had been formed in an

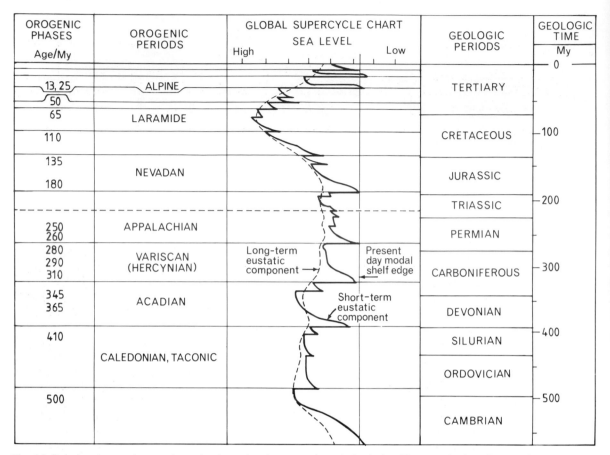

OROGENIC PHASES Age/My	OROGENIC PERIODS	GLOBAL SUPERCYCLE CHART SEA LEVEL	GEOLOGIC PERIODS	GEOLOGIC TIME My
		High ⎯ Low		0
13, 25 50	ALPINE		TERTIARY	
65	LARAMIDE			
110			CRETACEOUS	100
135 180	NEVADAN		JURASSIC	
			TRIASSIC	200
250 260	APPALACHIAN		PERMIAN	
280 290 310	VARISCAN (HERCYNIAN)	Long-term eustatic component → ← Present day modal shelf edge	CARBONIFEROUS	300
345 365	ACADIAN	Short-term eustatic component	DEVONIAN	
410	CALEDONIAN, TACONIC		SILURIAN	400
			ORDOVICIAN	
500			CAMBRIAN	500

Fig. 4.2 Relative changes in eustatic sea-levels, and major orogenic periods, during Phanerozoic time. Some writers propose a correlation between sea-level rises and the development of an orogeny, and rapid falls of sea-level mark the end of most orogenies. The sea-level curves are those of Vail (1977).

elongated subsiding trough, underlain by older basement rocks of a craton. The sediment to fill the trough was presumed to come from bordering uplands. Where parallel troughs were separated by an uparched belt, this belt was called a *geanticline*. A fundamental tenet of this classical model was that periods of sediment deposition always ended in a mountain-building period in which the strata of the geosyncline were folded and severely faulted. It was implied that orogenies occur only where geosynclinal deposition has taken place.

The discoveries of the second half of the twentieth century indicate that forearc basins, and similar features of subduction margins, may be regarded as having some characteristics in common with those of the classical geosyncline, but

that the concept of an intra-cratonic trough bordered by mountains contributing sediment has to be abandoned. A more useful modern definition is that a geosyncline is a thick, rapidly accumulating body of sediment formed within a long, narrow belt which is usually parallel to the margin of a continent. The sediment may accumulate in a trough or trench, or it may accumulate on the continental shelf and slope, and at the foot of the continental slope. Sediments of a geosyncline may thus be deposited in shallow or deep seas or on continents. The sediment may be underlain by continental or oceanic crust, and deposited on either an active or a passive continental margin. Geosynclines may be controlled by tectonic activity, but they are never the cause of that activity.

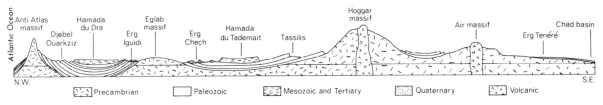

Fig. 4.3 A generalized diagrammatic cross-section of the Saharan craton, from Morocco to Chad, showing the major basins and sedimentary platforms and also the exposures of the shield.

4.1 Gently dipping sedimentary rocks of a platform on the African craton, Algeria. Erosion has cut the dip slopes into triangular facets called chevrons. The basin structure is revealed by the plan form of the strata (photo by USAAF).

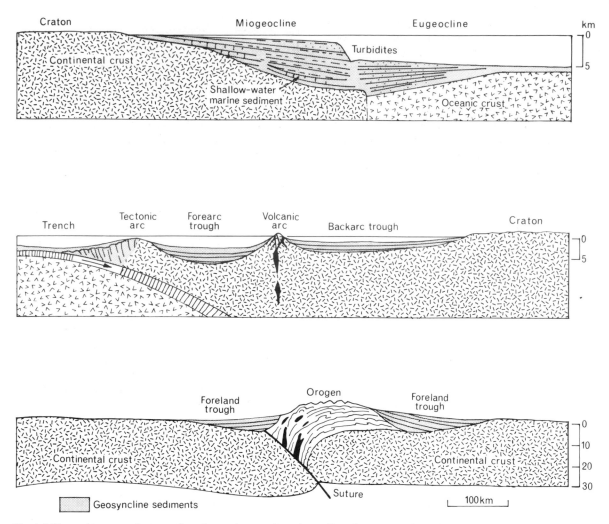

Fig. 4.4 The major types of geosyncline. At passive margins miogeoclines form on continental crust and eugeoclines on oceanic crust. At active margins of subduction forearc and backarc troughs are the characteristic geosynclines: both are on continental crust. In continental collisions foreland troughs form on the continental crust on both sides of the orogen (based on reports in Talwani and Pitman, 1977; and on Dewey and Bird, 1970).

The sediment may form gently tilted strata of uniform dip, in which case the term *geocline* is more appropriate; it may have a broadly down-warped or synclinal form, or it may be involved in the intense deformation of accretionary wedges (Fig. 4.4). Geosynclines and geoclines are, consequently, stratigraphic terms not necessarily related to tectonism. Because of the implications of the classical definition of a geosyncline, some modern geologists prefer to avoid the term and instead use a neutral phrase such as 'basin of deposition', even though this may also be misleading.

Geoclines of passive continental margins

Geoclines of passive continental margins are of two types: *miogeoclines* are the wedges of shallow-water, marine sediments which form on continental shelves; and *eugeoclines* are the wedges of deep-water sediment formed at the foot of the continental slope and lying on oceanic crust. Both types of geocline wedges are formed during prolonged slow subsidence of the lithosphere so the sediment may reach considerable thicknesses.

Along many Atlantic margins, shelf sediments

have a greatest thickness exceeding 10 km. In the Gulf of Mexico the miogeocline sediments have a thickness of 20 km at the outer edge of the shelf. They are composed of sandstones, mudstones, shelly limestones, coral limestones, and calcareous mudstones. The formations thicken towards the edge of the shelf. The basal rocks overlying Paleozoic basement are Triassic salt-beds, and above them is a nearly conformable sequence of later Mesozoic and Cenozoic sediment. In the deeper water, seaward of the continental slope, the eugeosyncline sediments have accumulated to maximum thicknesses of 10 km. These rocks are *turbidites* which have been carried by turbidity currents from the edge of the shelf and have settled out in alternating beds of sand and mud. Such graded beds are characteristic of deep-water deposition at the margins of continents. Eugeocline sediments contain volcanic materials only where submarine volcanoes have been formed in oceanic crust above a hotspot, within the eugeocline basin.

The very thick sediment accumulations of all Atlantic passive margins have formed since the opening of that ocean basin. The miogeoclinal units are all of shallow-water form, even though alternating transgressions and regressions have affected the depth of water on the shelves, and this implies that the continental margins have progressively sunk. The sinking is an isostatic response to crustal thinning and the effect of sediment loading. Crustal thinning was probably caused by a combination of cooling (and hence contraction), extension of the upper brittle crust along deep curved *listric* faults, and ductile flow and metamorphism in the lower continental crust. It is this sinking which has permitted uninterrupted sediment accumulation in the miogeoclines for nearly 200 My. The presence of oil, and salt domes acting as oil traps, has made the miogeoclines areas of great economic importance. The remarkably uniform water depths on continental shelves around the world imply that there is an approximate world-wide equilibrium between sediment supply from the land on the one hand, and the coastal currents distributing the sediment, the sinking of the geoclinal wedges, and the transgressions and regressions of late Cenozoic time, on the other hand. Only the shelves around Antarctica are notably deeper than the average and this may be due to the load of ice on that continent.

Geosynclines of subduction boundaries

Because the major subduction zones are around the Pacific Ocean the most extensive examples of geosynclinal basins are found there too. The only examples in the Atlantic Basin are the Caribbean and Scotia arcs. Seismic reflection profiling has revealed details of the submarine geology of the subduction zone formed where the Indian–Australian Plate is sinking beneath the Eurasian Plate and forming the Indonesian tectonic arc, the only such arc in the Indian Ocean. This area provides examples of the major types of geosyncline.

Sections through the Indonesian arc (Fig. 4.5) show a trench with a floor at a depth of about 6 km: seawards of the axis of the trench, a wedge of sediments is little deformed, but towards the accretionary prism the turbidites are increasingly incorporated into thrust wedges with mélange characteristics. Those sediments which are not subducted are forced upwards into an elevated ridge called a *tectonic arc*: this has the characteristics of a geanticline. Off the Sumatra coast the tectonic arc emerges above sea-level and forms the Mentawai Islands, but to the east the arc lies at 1–2 km below sea-level for most of its length, except in Timor where it emerges again.

On the continental side of the tectonic arc is a geosynclinal *forearc trough* which receives sediment from the Indonesian islands. These sands and clays are transported across the trough by turbidity currents and their uniform bedding indicates a lack of deformation. A forearc trough may capture all of the sediment eroding off the land and, in the absence of a tectonic arc standing above sea-level, the trench may be deprived of new sediment.

The high axial mountain ranges of Sumatra and Java form a *volcanic arc*. This arc is another type of geanticline because it has a basement of continental crust of Mesozoic rocks which have been uparched. North of the volcanic arc is a *backarc trough*, 100–200 km wide, representing a third geosyncline. In Sumatra and Java this is a low coastal plain, underlain by Cenozoic sediments, but east of Java this trough deepens and is entirely submarine.

North of the backarc trough lies a *continental foreland* of stable continental crust with rocks of Paleozoic and early Mesozoic age in the Malay Peninsula and Borneo (Kalimantan).

The curvature of the Indonesian arcs is con-

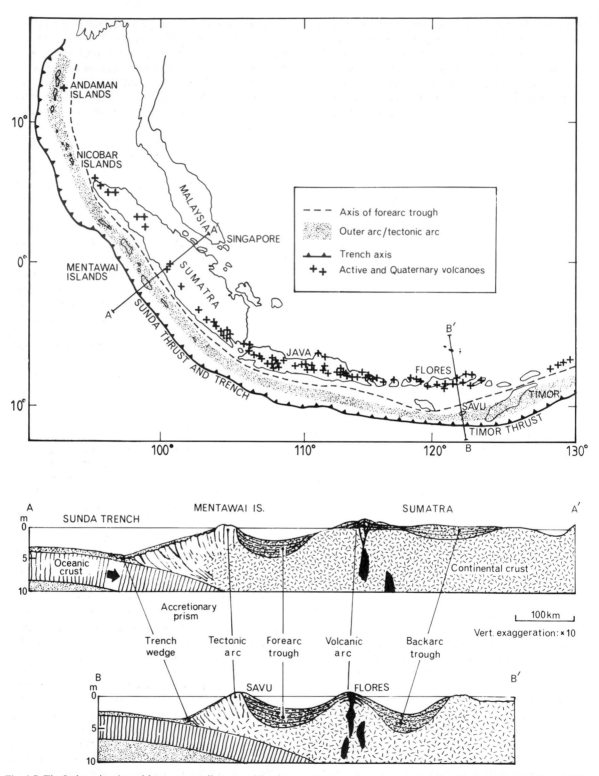

Fig. 4.5 The Indonesian Arc with an outer sedimentary island arc and an inner volcanic arc (based on Karig, 1977; Hamilton, 1977; and on Association of American Petroleum Geologists, 1982).

trolled by the angle at which the subducting plate descends into the asthenosphere. This situation can be understood by considering the angle of a cut made in an orange. If the cut forms a radius of the spherical orange the cut (and hence the arc) is a straight line drawn across the flattened peel, but if the cut removes a thin slice the arc of the cut is part of the circumference of a small circle.

The pattern of tectonic and volcanic arcs with continental forelands separated by geosynclines is different from the classical concept of a geosyncline formed between two continental forelands. The classical concept has little or no application to our present understanding of global tectonics, but the pattern associated with tectonic and volcanic arcs can be recognized in the rocks of many ancient orogenic arcs.

This simplified account does, however, ignore a number of 'awkward' facts and alternative interpretations. The most difficult observation to accommodate in a simple subduction model of island arc tectonics is the clear evidence of tensional faulting in the backarc regions. This is clearly seen in areas of the western Pacific such as the Philippine Sea. Alternative views of rising blocks causing crustal extension have been put forward. It seems possible that localized convection cells form above the subducting plate (as in Fig. 3.12A) and cause extension. There is, however, not always direct evidence of such currents from heat flow data. The reservations expressed here may be expressions of the lack of data; they may be fundamentally in conflict with plate tectonic reconstructions; or they may turn out to be misleading. They remain, however, a reminder that not all plate tectonic theories should be accepted uncritically.

Orogens

An *orogen* is the total mass of rock deformed during an orogeny. There is still much debate about the meaning of the term 'orogeny' because of its original association with folding of geosynclinal deposits, which were thought to be squeezed upwards to form mountains composed of many compressional folds. A simple folding mechanism is now known to be inadequate to form mountains, but the term is still useful if it is regarded as including compressional folding, severe overthrusting, intrusion of magma on a massive scale, large-volume sliding of masses of sediment under the influence of gravity, broad uparching, and very large-scale faulting and uplift of crustal blocks. Such processes in various combinations produce orogenic belts, or orogens, but not necessarily mountains. Two major types of orogeny are recognized: (1) a cordilleran type which results from an episode of subduction in which geosynclinal sediments are severely deformed and intruded by large volumes of magma; and (2) a continental collision orogeny which results from the trapping of oceanic crust and sediment between two masses of continental crust, with consequent severe deformation and usually with uplift.

An orogeny was formerly thought to be a rapid and relatively short-lived event of a few million years. It is now recognized that many orogens have a very complex history extending over many tens of millions of years in which there were periods of rapid and slower deformation (also called *diastrophism*). It is sometimes debatable whether a particular orogen or mountain system should be regarded as the result of several orogenies, or of one complex and long-duration orogeny which terminated in uplift of a mountain mass.

Cordilleran orogens

A schematic hypothesis of a cordilleran orogeny was published by Dewey and Bird in 1970. Initially, they assume that a passive continental margin bears mio- and eu-geoclines (Fig. 4.6), and that this margin is then converted into an active one by the formation of a subduction zone. The oceanic crust breaks seawards of the contact between oceanic and continental lithosphere and the eugeoclinal turbidite wedge remains attached to the continental plate. The attached slice of oceanic crust is bowed upwards, then broken into a number of overthrust slices. Melting of the upper surface of the sinking plate, and of the trench sediments it has carried down, forms basaltic magmas that rise behind the uparched oceanic crust and they form submarine volcanoes, or rise above sea-level as an island arc. As the volcanic arc grows and widens, the nature of the rising magma and the volcanoes at the surface changes. The parent is basalt of oceanic crust but, as this is subducted, it is heated and subjected to high pressure and changes to *eclogite* (a metamorphic rock containing pyroxene rich in sodium and

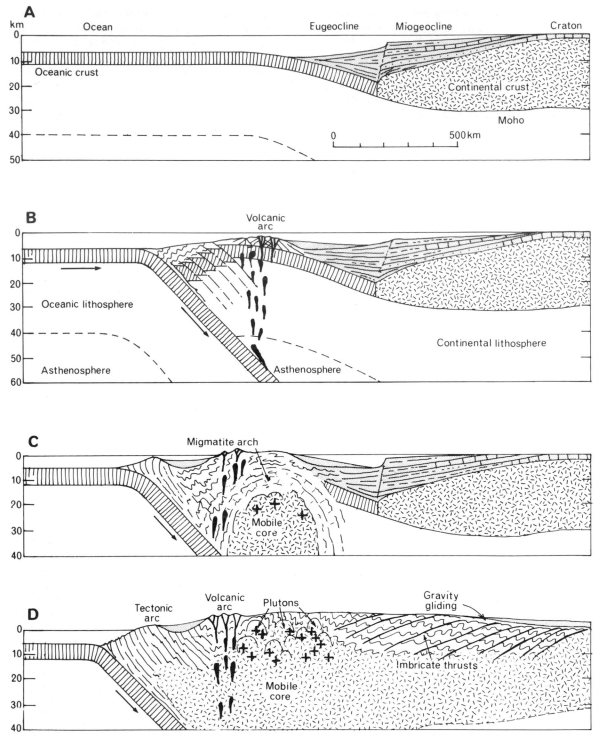

Fig. 4.6 Possible stages of a cordilleran orogeny. A stable passive margin suffers subduction, then arching of a mobile core, and finally emplacement of granite batholiths and large-scale overthrusting and gravity sliding onto the continental foreland (based on Dewey and Bird, 1970).

calcium and a variety of garnet rich in magnesium). Three processes operate to generate new magmas: (1) partial melting of eclogite at depth; (2) fractional crystallization of basaltic magma and (3) contamination of magma with subducted trench sediment and water. The efficacy of these processes and their intensity varies with depth. Consequently, although all of the new magmas are of andesitic composition, those rising from greatest depth produce alkaline andesites rich in feldspars with high sodium or potassium: such magmas form the volcanoes on the continental side of the developing orogen. Seawards of this the magmas are calc-alkaline with andesites rich in feldspars with a high calcium content and lower potassium content.

As subduction continues the continental plate edge suffers transformation and metamorphism, with the development of a mobile core of mafic (gabbro) and intermediate (diorite) composition together with more felsic granodiorite. This mobile core expands and causes thermal uparching of the entire volcanic arc and much of the eugeocline. As the arch rises above sea-level erosion sets in and it becomes a source of sediment for the deep-water basins (a type of geosyncline) on either side of it. Mild heating, but strong compression of the deepest sediment above the core, causes the sediment to be metamorphosed to schist, and the upper zones of the core are converted to migmatites —metamorphic gneisses in which there are alternating bands of granitic form and of altered sedimentary rock (Plate 4.2a, b, c).

The final stage of the orogeny is characterized by intense thrusting and overthrusting of the miogeosynclinal sediments, with large-scale gravity sliding and folding of young sediments above the thrust blocks. At the same time granite batholiths on a huge scale may form and reach the surface to form andesitic and rhyolitic volcanoes together with huge sheets of ignimbrite. The processes of granite batholith formation and the different types of rhyolitic eruptions are discussed in Chapter 5. Debris from the rising mountain chains is carried both inland by streams where it may form large alluvial fans of coarse gravels and sands, and seawards into the forearc trough. The terrestrial deposits of the fans may be buckled into open folds and the forearc sediments suffer deformation by gravity sliding. The orogeny draws to a close as subduction ceases and the supply of

magma declines. Cooling of the mobile core makes the whole orogen more rigid and any subsequent deformation can only be by large-scale faulting and not by folding. Later erosion removes the upper rocks of the orogen and exposes the plutons, the thrust structures, and metamorphic rocks. The landforms of the orogen cease to be tectonic but are erosional with individual structures and rock units being expressed in the landscape as a result of preferential erosion of the weakest units.

The Andes are regarded as a type area of a cordilleran orogen, but their history is complicated and our knowledge of events in the early Paleozoic is very limited. The presence of mica-rich sediment, from granitic crust, along the western margin of South America in Devonian times is regarded as evidence that a continent then existed in what is now ocean, and this continent provided sediment to geosynclines in the area of what is now Peru and Ecuador. It has been speculated that this continent, called 'Pacifica', included parts of north-eastern Australia, part of the New Zealand region and many fragments which are now embedded in the western part of North America, and eastern USSR, where they are known as *suspect terranes* because of their unknown or uncertain origin. During Permian and Triassic times the Andean geosyncline became unstable: the sedimentary strata were warped and buckled into large folds, and magma migrated upwards as large batholiths and formed rhyolitic volcanoes at the surface. The roots of these

4.2 (a) A sample of gabbro showing its coarse texture which is evidence of slow crystallization at depth in the crust.

4.2 (b) Flow folds of a gneiss. The scale is 40 cm long.

4.2 (c) The structure of a schist (same scale as in (b)).

volcanoes are still found in the Paleozoic sediments forming the eastern ranges of the Andes (Plate 4.3), where they may be prominent because of the superior resistance of their rocks.

It is postulated that Pacifica broke up and an ocean formed. Then, about 190 million years ago in early Jurassic times, the Nazca plate began spreading towards South America and was subducted several hundred km off the coast. Partially differentiated magma from the plate formed andesitic volcanoes as an island arc. The feeder pipes of some of these Jurassic volcanoes are still found in the western Andes where they are frequently the source of valuable minerals such as silver.

By the end of Jurassic times, 135 million years ago, the Atlantic had started to open and the orogeny of the Andean belt increased in vigour as South America migrated towards the Pacific: a new igneous belt was formed on the continental side of the Jurassic belt with large batholiths at depth and volcanoes at the surface. This phase of intrusion and extrusion began at the beginning of Cretaceous time and reached its peak at the beginning of the Cenozoic, that is during the period 100–50 million years ago. Many of these batholiths have been raised during a Cenozoic period of uplift, have then been severely eroded by intense stream action, and are new exposed in the eastern and western ranges of the Andes (Plate

4.3 The Eastern cordillera of the Andes, east of La Paz. The high peak Illimani (6440 m) is on intrusive rocks. The dark rocks in the foreground are Paleozoic slates. Note the gently sloping surfaces dipping to the west (to the right) which are partly erosional across old rock and partly remnants of Quaternary fans dipping towards the Altiplano.

4.4). During this compressive phase of Nazca plate and American plate collision, the geosynclinal folded sediments of the eastern Andes were thrust-faulted on a huge scale so that slices of rock from the Precambrian crust were pushed into contorted Paleozoic sediments and schists. The rising ranges were eroded at the same time as they were uplifted and sheets of sediments were laid down in the headwaters of the rivers flowing towards the Amazon, and in the broad basin formed between the eastern and western cordillera. These basin sediments accumulated to a thickness of several kilometres and formed the Altiplano. The Altiplano is now a relatively flat-floored basin between the ranges and contains the rather shallow, but wide and long, Lake Titicaca (Fig. 4.7).

Uplift of the Andes continued throughout the Cenozoic and is still continuing. The rate of uplift has not been constant but varies so that it has the form of pulses of accelerated epeirogenesis. At the same time huge outpourings of ignimbrite (highly silicic lava and ash flows) spread over much of the ranges of northern Chile and southern Peru to form the relatively level plateaux of the Puna. In the last 4–5 million years these plateaux have been gently warped upwards at rates of 1–4 mm/year and large conical volcanoes, formed of andesite, built along their crests (Plate 4.5).

The Andean belt, like many island arcs, is an area of severe earthquake activity with some of these earthquakes being located offshore in the trenches. Some earthquakes may be generated by thrusting and phase changes of basalt to eclogite, but others are produced by subduction. As oceanic crust descends into the mantle, at a rate of say 5 cm/y, it drags part of the crust on the land side down with it. When the deformation caused by this dragging reaches a certain strain, the rock ceases behaving elastically and fractures as a brittle solid, releasing a seismic shock wave with

4.4 Headwaters of the Urubamba River, in Peru, where it cuts through a granite batholith.

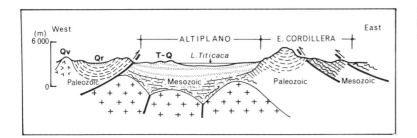

Fig. 4.7 A schematic section through the Altiplano of Peru. The key is the same as that of Fig. 4–8a.

high intensity. Nearly all of the world's high seismic energy release occurs in the circum-Pacific belt at shallow depths. The fracture occurs as a fault and the landward side of the crust rebounds and is uplifted. If a severe earthquake occurs once every 100 years in a zone where subduction is occurring at 5 cm/y the rebound would be 5 metres. This is about the amount of displacement which commonly occurs along major faults during such earthquakes.

The history as outlined here is, of course, greatly simplified. It has been reconstructed from detailed mapping and dating of rocks throughout the Andes, by many geologists, and by the careful relating of the history of one part of the orogenic belt with that of other belts. It should be noted that along most of the Andes since Jurassic times the volcanic belts have been formed on the continental side of older volcanic belts. This is the opposite situation to that existing in oceanic island arcs

4.5 The Puna east of Salar de Atacama with andesite cones rising above an ignimbrite sheet, dated at 4.5 million years, which has been raised and slightly domed during the Quaternary.

such as those of south-east Asia where new arcs form seawards and older arcs accrete onto the continental foreland.

The intense seismic and volcanic activity, and the great width of the Andean orogenic belt is probably due to the very fast rate of spreading at the East Pacific rise (15–18 cm/y). This produces a young, warm, and hence buoyant, subducting slab sinking at a low angle and interacting directly with the base of the overlying continental lithosphere across a zone 700–800 km wide. The low-angle subduction creates heat, and melts vast volumes of crust, so that huge granite plutons form at depth and, where they rise to the surface, rhyolitic and ignimbritic eruptions.

The main geological features in a transect across the Andes are shown in Fig. 4.8a and the evolution of the central Andes of Peru in Fig. 4.8b, c, d, e. It will be noted that the trench is nearly barren of sediment and that there is no accretionary wedge, so the coastal ranges are of rocks all older than the Cenozoic. This may be because the coast is a desert with no through-flowing rivers to deliver sediment to the coastal zone.

The Cordilleran Orogeny of North America began near the close of the Cretaceous Period and continued into the Cenozoic. A feature of this orogeny is the massive intrusion of granite batholiths, which began in the late Cretaceous with the emplacement of the Sierra Nevada, and continued with the Idaho batholith in the northern Rocky Mountains. The eastern ranges of the Rocky

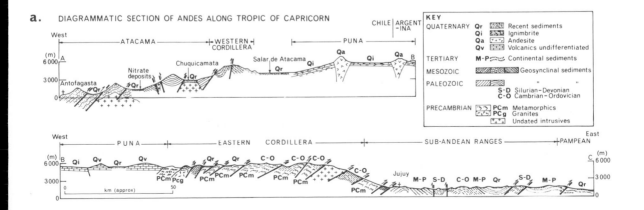

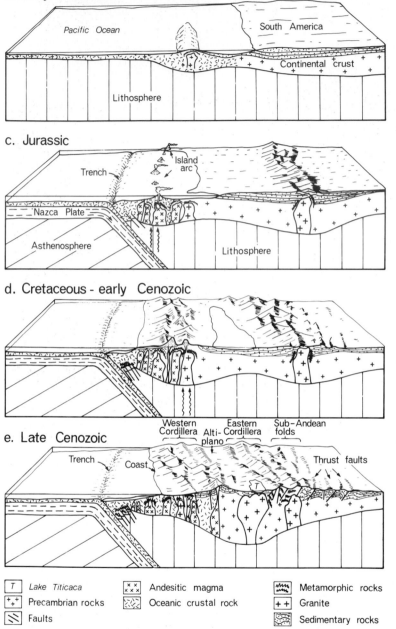

b. Early Paleozoic

c. Jurassic

d. Cretaceous - early Cenozoic

e. Late Cenozoic

Fig. 4.8 (a) A section through the Andes along the southern tropic from Antofagasta in northern Chile to northern Argentina (compiled from numerous published sources). (b, c, d, e) Block diagrams indicating how the Andes of Peru may have evolved. This section runs approximately from the coast at Tacna in a north-east direction (based in part upon James, 1973).

T	Lake Titicaca
+,+	Precambrian rocks
⩘	Faults

x x	Andesitic magma
⟋⟍	Oceanic crustal rock

⌇⌇	Metamorphic rocks
+ +	Granite
⌇	Sedimentary rocks

Mountains developed from a geosyncline of Mesozoic age by massive low-angle thrusting (Plate 4.6) and extrusion of rhyolite and andesite.

These low-angle thrusts carried immense slabs of rock eastwards over one another along a line extending from Mexico to north-western Canada (Fig. 4.9). As in other orogens, the thrust fault zone, which marks the former cratonic margin and the zone of greatest geosynclinal subsidence, became the zone of maximum uplift. Thrusting ceased in Miocene time, about 12 My ago, and this may be said to mark the end of the Cordilleran Orogeny. Uplift and tectonic processes, however, did not cease.

4.6 The Keystone thrust at the south-east end of Spring Mountain Range, west of Las Vegas, Nevada. The dark-coloured Cambrian dolomite is thrust over the pale sandstone, of probable Jurassic age (photo by John S. Shelton).

In a broad belt stretching from Montana to New Mexico many structural domes were uplifted. Domes have dimensions of 50–75 km width and lengths of 200–400 km. The vertical movement was several thousand metres along the crests. Denudation has stripped the sedimentary rocks from the crests of the uplifts, revealing Precambrian basement igneous and metamorphic rocks surrounded by the inward-facing scarps of the Mesozoic sedimentary rocks.

Beginning near the end of Miocene time, and continuing to the present, almost the entire Rocky Mountains and adjacent regions were warped up as a large unit. This caused rapid downcutting by the streams, as their gradients increased, and the formation of spectacular canyons, such as that of the Grand Canyon of the Colorado cut in the uplifted Colorado Plateau (Plate 4.7). The cause of this uplift is uncertain, but one hypothesis is that a postulated low-angle subduction of oceanic crust caused heating of the lithosphere beneath the Colorado Plateau and surrounding ranges, with resulting upward expansion. The same process may have been responsible for the widespread granitic magmatism and/or the andesite eruptions of the Cascade Range.

The Basin and Range Province of USA started to develop in late Oligocene times by lateral extension of the crust and the formation of major faults elongated in north–south directions. Individual fault-bounded blocks, many kilometres wide and tens of kilometres long, were raised or dropped

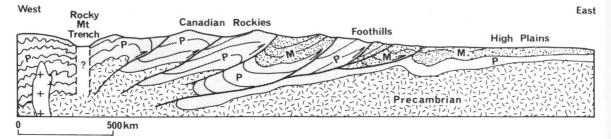

Fig. 4.9 A generalized structural section through the Canadian Rocky Mountains showing the complex thrusts over the Precambrian foreland. The Rocky Mountain Trench may be a tensional rift or the upper part of another thrust sheet (greatly simplified from Price and Doeglas, 1972).

4.7 Structural benches formed by resistant horizontal strata forming the upper walls of the Grand Canyon of the Colorado. Note the angular unconformity between the underlying Precambrian rocks and the Paleozoic strata above (photo by John S. Shelton).

through as much as 3 km. The down-dropped blocks formed *graben*, which received large volumes of sediment from the uplifted *horsts*. The cause of this crustal extension is uncertain, but two main hypotheses have been put forward: (1) high heat flow above the subducted plate may have caused doming and extension of the crust; or (2) the subducting plate may have broken, causing very high heat flow and extension of the crust for a limited period (Fig. 4.10). The high heat flow still

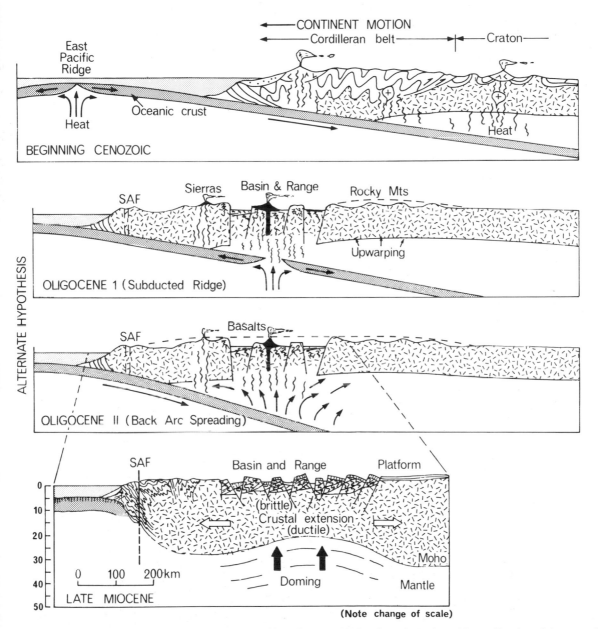

Fig. 4.10 Alternative hypotheses for the Cenozoic stretching of western USA to form the Basin and Range Province. It is assumed that a low-angle subduction caused widespread magmatism with formation of granite batholiths and andesite volcanoes. Rifting and basalt volcanism may have resulted from either (I) subduction of the East Pacific ridge with high heat flow above its central rift, or (II) high heat flow followed by mantle doming and crustal extension after subduction had ceased. SAF—San Andreas Fault (based in part on Dott and Batten, 1981).

detectable in the Province is regarded as support
for the first hypothesis, as is the continued activity
along many of the faults. If this hypothesis is
correct, then the crustal extension may be the early
stages of spreading, with the Basin and Range
Province being the early stage of development of a
marginal ocean basin like the Japan Sea.

The San Andreas Fault (see Fig. 3.14) extends
from the head of the Gulf of California to a point
on the California coast near Cape Mendocino. It is
thought by most geologists to form the boundary
between the Pacific plate and the American plate
in this sector. During the late Cretaceous and early
Cenozoic the oceanic crust was being subducted
below the American plate, from Mexico to British
Columbia, and this brought the oceanic spreading
ridge closer to North America until one segment of
the ridge met southern California, but the ridge
farther north remained out of contact. The line
joining the two ridge segments was a transform
fault which we know as the San Andreas. The
transform segment is a very wide zone of distributed faulting and aseismic strain. The meeting of
the spreading boundary with the continent caused
subduction to cease to the south, and in the north
the remnant of the unsubducted oceanic crust
(now called the Juan de Fuca plate) forms a
microplate which continued to be subducted for a
time.

Subduction of the Juan de Fuca plate remnant is
now thought to be limited or non-existent. The
evidence is, however, conflicting. The Cascade

Range andesitic volcanoes are still partly active,
especially Mount St Helens (see Chapter 5),
suggesting that subduction continues, and the
Cascades are still an active volcanic arc. There is,
however, no trench offshore. This could imply that
subduction is inactive or slight, or it may be that
there is a trench which is filled with sediment from
severe erosion of the cordilleras and Cascades.
This situation is in contrast to that of the Atacama
Desert coast of Chile.

The New Zealand region provides another
example of a transform boundary, in continental
crust, along the line of the Alpine Fault (Figs. 4.11,
4.12). New Zealand is the exposed part of a
micro-continent which broke away from Gondwanaland as the Tasman Sea basin opened. Some
of the rocks of the south-western tip, known as
Fiordland, are Gondwana remnants. The rocks
which now form the main axial ranges are geosynclinal sediments (greywacke) of late Paleozoic to
early Cretaceous age. It is probable that these
sediments were derived from a land mass to the
east of Gondwanaland and it has been proposed
that this continent was Pacifica (see the section on
the Andes). The New Zealand geosyncline was
severely folded in several phases of an early
Jurassic to early Cretaceous Rangitata Orogeny.
Many sedimentary rock units were metamorphosed to schists, and uplift and erosion occurred.

Separation of New Zealand from Gondwanaland may not have been complete until the late
Cretaceous, about 80 My ago. The Rangitata

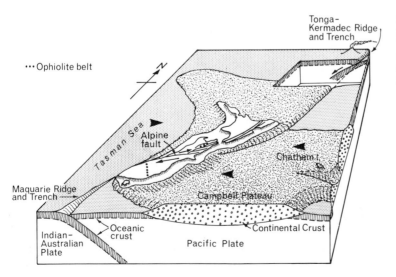

Fig. 4.11 The New Zealand microcontinent showing its ophiolite belt, the large strike-slip faults, and the subduction zones, to the south-west and north-east, with opposite dips (modified from Stevens, 1980).

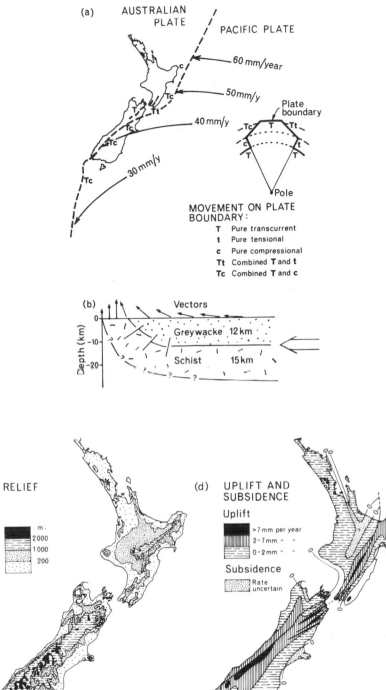

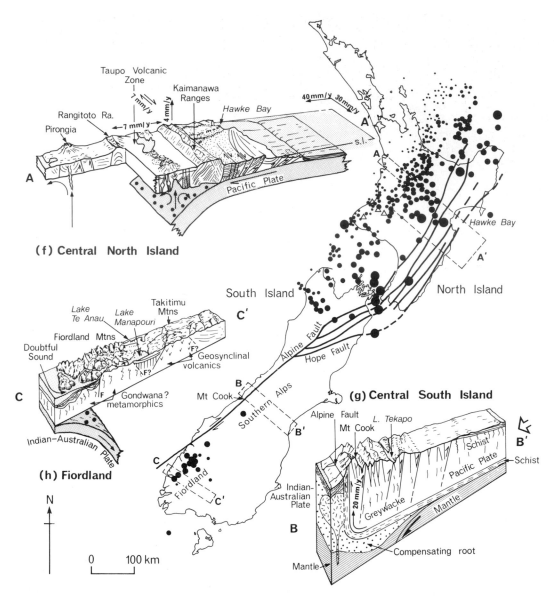

(e) INTERMEDIATE AND DEEP EARTHQUAKES. MAJOR FAULTS

Depth (km)

● 33–50   ● 33–100   • 101–150   • 151–200   • 201–250   · >250   △ 600

Fig. 4.12 (a) Rates of convergence of the Pacific with the Australian crustal plate showing the angle of convergence and resulting motion. (b) Resulting vectors in the Mt Cook region (the geology is extremely simplified). (c) Relief and (d) rates of vertical motion. (e) Pattern of earthquakes defining the dip of the subduction plate in the northern and extreme southern regions. (f) Central North Island with directions of motion, the Taupo Volcanic Zone rift with rhyolitic and andesitic volcanoes, and the basalt cone of Pirongia. Note the resulting pattern of faulting. (g) Central South Island. (h) Fiordland area of ancient rocks. (Figures based on data in Walcott and Cresswell (eds), 1979, and data provided by H. W. Wellman. Fig. (g) is partly based on Kamp, in Soons and Selby, 1982).

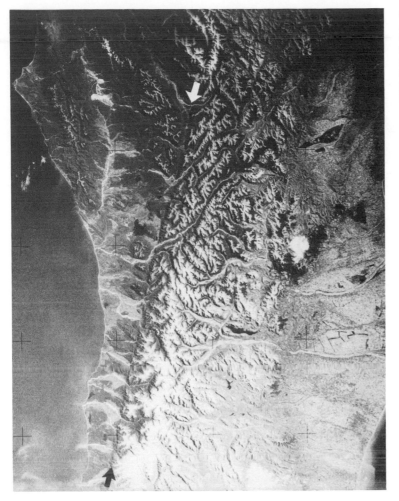

4.8 The central New Zealand Southern Alps. They terminate to the west (left) along the Alpine Fault where there is evidence of offsetting of some stream channels. Mount Cook is in the snow-field at the bottom left of the photo (satellite photo by courtesy of NASA).

Orogen has been curved into an arc—as is shown by the shape of its ophiolite belt (Fig. 4.11), but the age of this event is unclear.

In Cenozoic times the plate boundary between the Pacific and Indian–Australian plates has developed through the New Zealand micro-continent. This boundary is marked by a subduction zone and trench which extends from Tonga to central New Zealand, and a separate zone extending southwards from Fiordland towards Antarctica, but in most of South Island by the transform fault linking the two subduction zones—the Alpine Fault. Subduction directions beneath North Island are from the west as the Pacific oceanic crust sinks beneath the tectonic arc of the eastern ranges, which are still suffering underthrusting, faulting, and folding. In the south the Indian–Aus-

tralian plate is sinking eastwards below the continent. The pattern of subduction and the dip of the plates is revealed by the pattern of earthquakes and their depth (Fig. 4.12e). It can be seen also that the Taupo Volcanic Zone with its andesitic and rhyolitic volcanism is one of intra-arc rifting which continues northwards in the Havre Trough towards the Kermadec Islands.

In post-Miocene times, motion on the Alpine Fault has changed from being wholly transform, in which the west coast of South Island moved northwards relative to the east coast. Some transform motion still occurs and is indicated by offsets in river terraces, stream channels, and other features (Plate 4.8). The dominant direction of movement along the Fault is now vertical and, at its maximum in the Mount Cook region, reaches

4.9 The fault plane, showing a throw of about 1 m, produced during the Inangahua Earthquake of 1968, in Westland, New Zealand (photo by D. L. Homer, New Zealand Geological Survey).

an average of 20 mm/y. This uplift which has created the Southern Alps since the early Pliocene results from oblique convergence of the plate along the New Zealand sector of the boundary. Because the pole of rotation of the Pacific plate is to the south-east of New Zealand, that plate converges with North Island nearly at right angles (Fig. 4.12a) but increasingly obliquely and more slowly southwards. The stresses set up are, consequently, compressional in the far north, and increasingly transcurrent with compression southwards. The transcurrent motion is evident from the pattern of faults and offsets in landforms, with lateral movements of 3–6 mm/y on average along some transcurrent faults in eastern North Island, and vertical uplifts of 1–3 m during earthquakes

which may occur once in 500 years (Plate 4.9). In South Island about half of the convergence is taken up in aseismic slip within the rocks at depth in the crust, but large-scale underthrusting has produced reverse-faulted blocks suffering tilt and uplift, which takes up the other half of the convergence. Schists are thought to be shearing at depth below the greywacke and being sheared and flow-folded up the western side of the Alps (Fig. 4.12g). The rates of uplift are derived from the heights of raised beaches, tilted lake shorelines, and river terraces of known ages, by resurvey of triangulation points (Fig. 4.13), and by radiometric dating of the alpine schists.

About 70 km of crustal shortening has occurred in the last 5–6 My by subduction, crustal thicken-

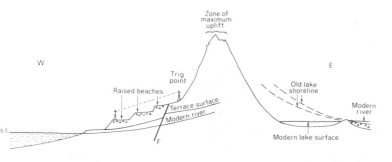

Fig. 4.13 A highly schematic representation of the types of evidence used to determine the rates of uplift across the New Zealand Southern Alps.

ing, and compressive folding and faulting. The Alps appear to be in a state of isostatic equilibrium, with the present rate of uplift, decreasing from the main drainage divide towards the east, being about equal to the present rate of erosion. This convergence of continental with continental crust is regenerating, in a young uplift and faulting phase named the Kaikoura Orogeny, an older Rangitata fold belt. The rates of uplift are so high along the main axis of the ranges that even large landforms are very young; hill blocks up to 1000 m high may be as little as 0.25 My old.

The central South Island plate boundary is a good example of a conservative plate boundary, mentioned as a minor type in the classification given in Chapter 3.

Continental collision orogens

Continental collision during the Cenozoic Era has occurred along the great tectonic line forming the southern edge of the Eurasian plate. This line of collision extends from the Himalayan Ranges in the east, through the Zagros mountains of Iran in the centre, through Turkey, the Aegean Sea, the European Alps, and the Atlas Mountains in the west. Continuity of the line is broken at either end of the Zagros Mountains where transform faults following the Gulf of Aqaba–Sea of Galilee rift (Plate 4.10) and the Owen Fracture Zone (see Fig. 3.13a) form the boundaries of the Arabian plate.

The Himalayan sector results from collision of the Indian–Australian plate with the Eurasian plate; the Zagros from collision of the Arabian plate with a fragment of the Eurasian plate; and the Atlas–Turkey sector from complex collisions of many microplates of the Mediterranean region between the African and Eurasian plates.

The general features of a continental collision orogen are shown in Fig. 4.14. As the ocean basin between converging continental blocks narrows, the sediments of marginal geosynclines and accretionary prisms are tightly compressed and deformed as the oceanic basaltic crust is broken into thrust-faulted blocks which are pushed up into the sediments. The thrust blocks ride up one over another in an imbricate pattern. The upper part of each thrust sheet of sediment bends over, under the force of gravity, into a horizontal position and the strata form very elongated flat-lying folds called *nappes*. Segments of oceanic metamorphosed crust incorporated into the nappes, or between them, as ophiolites mark the collision boundary, or suture.

Erosion of the rising thrust sheets produces fine-grained sediment which may accumulate in narrow ocean troughs, during the early stages of collision, as flysch. *Flysch* consists of shale and fine-grained sandstone which may become incorporated with the late stages of over-thrusting, and into the younger nappes. At the late stages of formation of the orogen the high mountains release coarse sands and gravel which are spread across the marginal continental forelands, or are deposited in shallow lakes and the shallow remnants of marginal seas, as sheets of thick *molasse*.

The European Alpine Orogen is a belt of mountains extending from the French Riviera to Austria. The evolution of the Alpine belt is complicated since it involves continuous motion of a large number of microplates, or fragments of continental crust, over 200 My. The outline given here is thus a gross simplification of what is known about the area and obscures many of the areas of ignorance which still exist.

The collision history began with the break-up of Pangaea in earliest Jurassic times 200 My ago. The Eurasian and African plates were separated by the Tethys Sea which began to close with a rotational movement in which the location of Gibraltar is the pivot (see Fig. 3.22 a,b).

Alpine deformation, as indicated by the radio-

4.10 The rifts forming the Gulf of Aqaba and its extension to the Sea of Galilee, and the Gulf of Suez north of the Red Sea (satellite photo by courtesy of NASA).

metric age of metamorphism, started in late Cretaceous times when compression cut off slices of continental crust under the southern shore of the Tethys Sea. These crustal slices formed a wedge that continued to advance northwards. During the early Tertiary, the wedge began to thrust over the central (or Pennine) geosyncline (Fig. 4.15). Buried under this load the Pennine rocks were broken up into thrust slices and subjected to high-pressure and high-temperature metamorphism. A crustal wedge with the more rigid Pennine nappes, formed of slabs of basement in a mélange, on its back arrived at the edge of the Helvetic geosyncline about 40 My ago. The continued advance during the Oligocene and Miocene caused the detachment of the Helvetic sedimentary cover from its basement and turned it into the Helvetic nappes (or Helvetides). Erosion of the northern alpine ranges by rivers delivered molasse into the fans, lakes, and basins to the north.

During the late Miocene and early Pliocene a crustal thrust developed, under the Aar Massif, which progressed northwards with the Helvetic, Pennine, and Austroalpine nappes on its back

(Fig. 4.16). This push detached the Molasse and Jura sedimentary sequences from their basement. The Jura was thrown into folds and the Aar Massif was uplifted. The alpine belt in Miocene times suffered severe erosion and was probably not very elevated.

The history as outlined above represents a crustal shortening of 200 km in 40 My at an average rate of 2.5 mm/y. As the movements may have been episodic the rate of deformation may have been at several centimetres per year at times. The elevation of the Alps results mostly from vertical movement during the Pliocene and Quaternary, with average uplift rates of 1–2 mm/y (see Fig. 4.35). The present relief is a result of Quaternary erosion by glaciers and rivers.

The absence of subduction under the modern Alps clearly distinguishes this type of orogeny from island arc and oceanic–continental plate convergence, but it does not preclude the possibility of subduction in the earlier stages of the orogeny. In fact the ophiolite suites now found in the Pennine nappes represent such a small portion of the oceanic crust, which must have existed

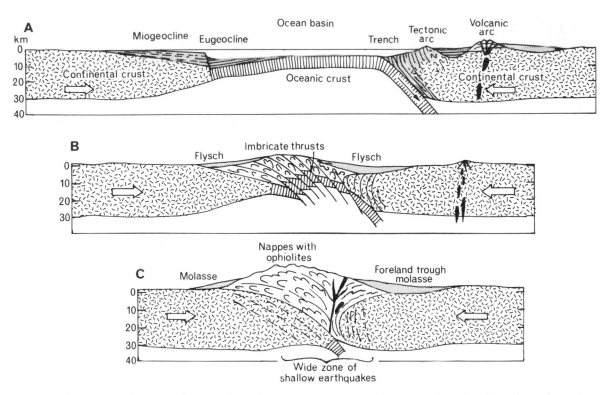

Fig. 4.14 The sequence of events leading to the formation of a continent-to-continent orogen. Note how slices of oceanic crust are thrust into sedimentary rock sequences and exposed as ophiolite complexes (based on Dewey and Bird, 1970).

beneath the Tethys, that consumption of oceanic lithosphere must have occurred in the earlier part of the Tertiary.

A feature of the modern Alpine Orogen is that all earthquakes are shallow, with foci commonly at depths of 2–8 km, and a maximum of about 20 km. Studies of seismic waves indicate also that the crust is relatively thin, being about 25 km under the Rhine basin to the north, and about 50 km thick under the central Alps. This crust is layered, with a number of zones of low seismic wave velocities.

This geophysical data is interpreted as indicating that there is now no well-defined Wadati–Benioff zone below the Alps and that there has been a history of shearing of small plate fragments and wedges under and into the Alpine crust—so thickening it. The thickening resulted in vertical uplift and hence erosion of the rising mountains.

North of the Swiss alpine ranges are the Jura mountains in which deformation is relatively less severe. The folds are more regular and thrusting is limited. Crustal shortening in this zone has been accomplished by detachment of the folded sediments from the underlying basement rocks in a *decollement* horizon formed in weak Triassic salt-rich (evaporite) sediments (see Fig. 4.29).

South of the European Alps is the Po Basin, in which terrestrial molasse deposits derived from the Alps are accumulating, and the Apennines. The Apennines have formed since the main ranges of the Alps, from the advance of the Corsican microplate towards Italy and the Adriatic. This movement started with the large-scale thrusting of Tethyan oceanic crust and its sedimentary cover over the Italian continental margin. Nappes were translated over large distances. At present the advancing Apennines are thought to be under-thrusting the Po basin along a deep fault which reaches into the basement (Fig. 4.15).

The crust underlying the Mediterranean Sea is partly continental and partly oceanic with many microplates reacting with one another, especially in the eastern area beyond the submarine ridge which links Sicily and Africa. In Cyprus, for example, the crust in mid-Tertiary times ruptured

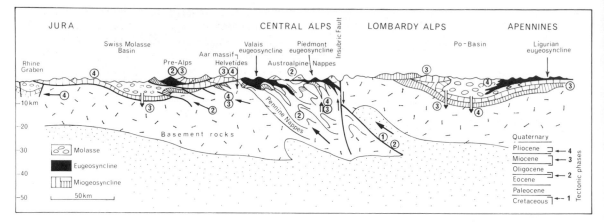

Fig. 4.15 A schematic section through the European Alps showing the main structural features and the order of their development (modified after De Jong, 1973).

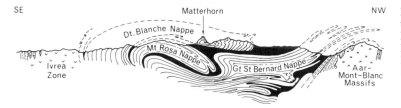

Fig. 4.16 A generalized section across the Pennine nappes near Zermatt. The Dent Blanche nappe, of which the Matterhorn is part, lies upon the Monte Rosa nappe (modified after Argand, 1934).

and ultrabasic rock from the mantle was pushed over a slice of sialic crust—a process known as obduction. The ultrabasic rocks, with pillow lavas and dykes of oceanic crust, now form the Troodos Massif. The Massif is an area of high gravity anomalies (Fig. 4.17).

The Zagros Mountains of Iran are remarkable for their numerous broad, open folds in sedimentary strata. The usual interpretation of the origin of these mountains is that a miogeocline lies upon the margin of the Arabian plate. Movement of this plate towards the Eurasian plate has caused underthrusting of the Eurasian margin and formation of imbricate thrusts of the oceanic crust along the suture, with complex folding and thrusting in the most north-easterly strata between the converging plates (Fig. 4.18).

The open folds (Plate 4.11) are thought to be produced not by compression alone, but also by décollement and by the intrusion of salt diapirs into overlying strata, and the creation of salt domes which form the core of many of the folds. The salt was probably precipitated in lagoons formed on the shallow shelf of a desert coast. As

the shelf miogeocline sank, the salt was buried by clays and limestones, but the pressure of earth movements probably caused the salt to rise as 'stalks' through the overlying strata and bow them upwards. The salt also formed a mobile plastic layer between the basement of the Arabian plate and the miogeoclinal sediments so they could be uncoupled readily from the basement and deformed into folds.

The Zagros Mountains are still seismically active suggesting that the plates are still converging and that faulting may still be occurring at depth, and that thrust blocks are partly responsible for bowing up strata in actively growing folds. Knowledge of the Zagros fold belt is unusually good because the dome structures created by salt diapirs are excellent traps for oil, and much geological information has come from oil exploration work.

The Himalayan Ranges form the highest mountains on Earth and the Tibetan Plateau, north of them, is a huge block of continental crust of 2 million km² over 5 km high, and between 50 and 70 km thick. The origin and development of these

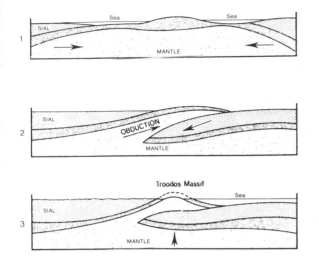

Fig. 4.17 Development of the Troodos Massif, Cyprus, by obduction and then isostatic rise of the underlying sial to heave up the mantle rocks (after Gass and Masson-Smith, 1963).

features has been recognized, since the work of Wegener, as being due to the closure of the Tethys Ocean, as India and Eurasia collided. A detailed understanding of the collision has, however, become possible only since 1980, when collabora-

tive work between French and Chinese geologists has included studies of the deep structure of Tibet using seismic traverses; reconstruction of the positions of land masses through the Mesozoic and Cenozoic Eras using paleomagnetism; isotopic dating of basement rocks, ophiolites, and volcanic units; recognition of major faults from satellite photographs; and geological mapping in hitherto inaccessible areas. The major results of this work are summarized in Fig. 4.19, and the wider pattern of major tectonic features is displayed in Plate 2.1 and Fig. 3.13a.

The Plateau, south of the Tarim Basin, consists of blocks of continental crust in contact along major sutures, of which the youngest is the Indus–Tsangpo suture (ITS) in the south, separating the Indian continent and the Lhasa block of Tibet, and the ages of the other sutures (S2, S3, S4) increase, in sequence, northwards (see Fig. 4.19b). Seismic studies indicate that each block underthrusts the one north of it, so accounting for the great thickness of the crust and the great altitude of the isostatically uplifted Plateau. Paleomagnetism indicates that Lhasa had a paleolatitude of 12°N as late as 80 My ago; at the same time, the northern margin of India may have been between

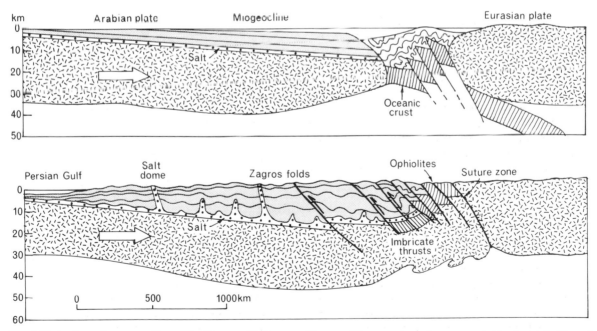

Fig. 4.18 A schematic cross-section of the Zagros collision zone. Closure of the ocean basin is accompanied by thrust-faulting of oceanic wedges into sediments and the exposure of ophiolite complexes. Folding is largely due to thrusting and to décollement from the basement along the salt beds (based upon data in Haynes and McQuillan, 1974).

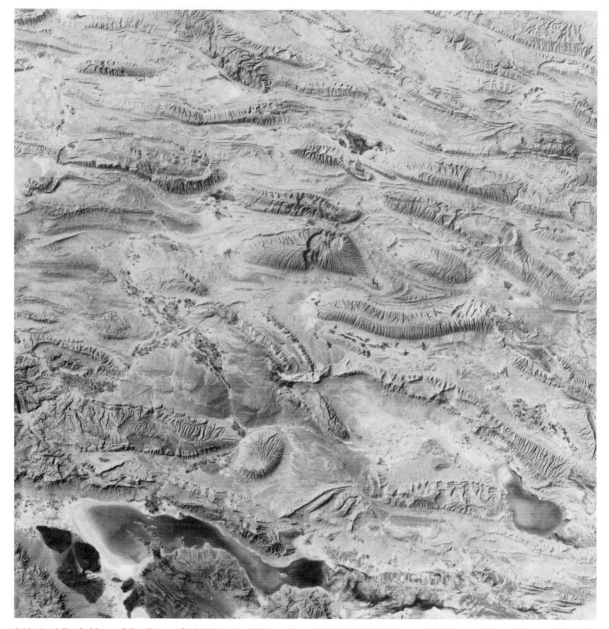

4.11 Anticlinal ridges of the Zagros fold belt east of Shiraz, Iran. Note that the flanks are deeply scored by gullies, and that the breached anticlines have inward-facing scarps (satellite photo by courtesy of NASA).

20 and 30°S, thus indicating that the Lhasa block was a separate unit from India; geological evidence from fossil floras indicates that it was separate from Gondwanaland as early as the Permian (250 My). The ITS ophiolites, the sedimentary rocks overlying them, and the Precambrian crystalline rocks of the basement have been thrust southwards over the advancing margin of India in at least four compressive stages, with the earliest stage occurring in the Upper Cretaceous,

The Himalaya Ranges along much of their length now consist of three E–W aligned units. (1)

The southern Sub-Himalayan frontal range, which includes the Siwaliks, consists of folded, molasse-like fresh-water sedimentary rocks, probably laid down by E–W trending rivers in Tertiary times. Southwards, these folded and faulted sediments are covered by the Quaternary alluvium of the Indus–Ganges plain which is up to 5000 m thick. Northwards, they are separated from the overlying rocks of the Lower Himalaya by the northwards-dipping Main Boundary Thrust (MBT) (see Fig. 4.19e). (2) The eastern Lower Himalaya are formed of Precambrian schists and gneisses derived from the Indian shield; westwards, these are partly overlain by Paleozoic sedimentary rocks which have Gondwanian affinities. These rocks form thrust sheets, rather than nappes, which have been eroded by rivers into steep-sided ridges rising to about 3000 m. (3) The Higher Himalaya are formed of Precambrian gneisses, intruded by many granite bodies and dykes of Tertiary age, which have been thrust over the crystalline rocks of the Lower Himalaya along the Main Central Thrust (MCT). In Kashmir, and along the northern edge of the High Himalaya, including Mt Everest, the crystalline rocks are partly covered by Paleozoic and Mesozoic sedimentary rocks which formed on a shallow continental shelf of the advancing Indian continent before the first uplifts of early Tertiary time. The main Himalayan uplifts did not occur until Miocene and later times.

The Himalayan Ranges have not developed from folding of geosynclinal rocks; rather they consist of great thrust sheets formed of the basement of the advancing edge of the Indian continent. This edge has not been subducted and the overthrusting southwards, and underthrusting within Tibet, have caused crustal shortening of up to 400 km. It is evident, however, that the continental collision has had even more far-reaching effects for, still active, strike-slip faults (see Fig. 4.25), especially the Altyn Tagh and Kunlun Faults acting along old sutures, form the boundaries of huge blocks of China which are being wedged eastwards and rotated in a clockwise direction. Similar movements may have extruded Indo–China 800 km to the south-east before the late Miocene and, at present, propagation of the stress across the Tien Shan may be responsible for normal faulting in the Baikal Rift and in northern China (Fig. 3.13a).

Continental convergence at average rates of 5 cm/y producing huge thrusts and wedges during 100 My, or more, is inconceivable if the mechanism is only slab pull at subduction zones (which only existed in the earliest phase of convergence of the Lhasa block with Eurasia) or ridge push (see Fig. 3.17). The drive of a long-lasting large convection cell is a more probable force.

Ancient orogens

The discussion of orogenic belts in this chapter so far has centred on those which are still developing and are essentially of Cenozoic age. There are also uplands on the continents which are no longer growing, but are being reduced by erosion: most of these developed during the Paleozoic and have a history which is related to the movement of lithospheric plates before the formation of the supercontinent, Pangaea, in Permian–Triassic times. The Paleozoic mountain systems which formed around what is now the North Atlantic will be discussed briefly, as examples of ancient orogens which still have substantial relief. Precambrian orogens are now all eroded so deeply that most form relatively level lowlands of shields, unless they have been substantially raised epeirogenically in Phanerozoic time.

It must be emphasized that what is now seen of Paleozoic orogens is not a relief formed by tectonism, but is only the eroded remnant of ancient mountain ranges. The relief which survives is due to differential erosion in which resistant rock units stand in high relief compared with weaker units. Many tectonic features from Paleozoic orogens, such as nappes, have disappeared, except for their roots, and the size and extent of an original nappe, or even its existence, must be conjectural (Fig. 4.20).

The Caledonian and Hercynian belts

The Caledonian Orogen is an early Paleozoic tectonic belt which was recognized by early geologists, such as James Hutton, in Scotland and Norway, and later geologists recognized similar rock units and fossil assemblages in the northern Appalachians and East Greenland. Modern reconstructions of the tectonic history are based on the evidence that rock units and their structures were involved in a continent-to-continent collision orogeny. Structurally the two sides of the Atlantic are mirror images. In East Greenland, and the N.

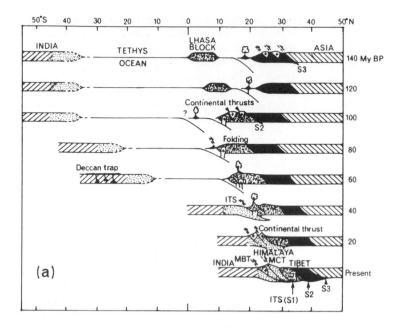

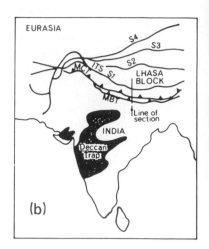

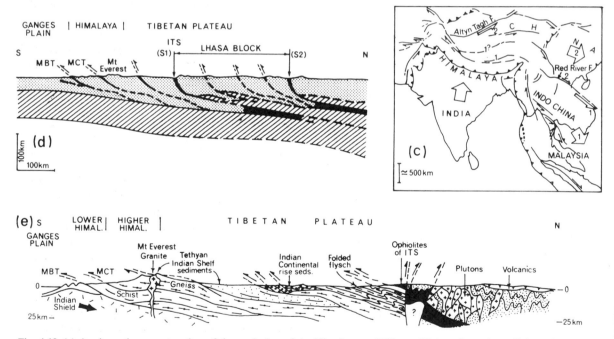

Fig. 4.19 (a) A schematic reconstruction of the evolution of the Himalaya and Tibetan Plateau from the collision of India and Eurasia. Note that the closing of the Tethys involved subduction; but continental collision, crustal shortening by underthrusting of blocks along sutures (modified from Allègre *et al.*, 1984). (b) The Himalayan thrusts and Tibetan sutures S1. . .S4 (modified from Allègre *et al.*, 1984). (c) Schematic reconstruction of Cenozoic extrusion tectonics and large faults in eastern Asia. Numbers refer to extrusion phases: 1 = 50 to 20 My BP; 2 = 20 to 0 My BP (modified from Tapponier *et al.*, 1982). (d) A simplified model of the structure of the lithosphere in Himalaya–Tibet. Oceanic crust is shown in solid black. The ITS and S2 are obducted; older sutures in the Lhasa block are rotated into the vertical and reactivated as strike-slip faults (modified from Allègre *et al.*, 1984). (e) Schematic cross-section from the Ganges Plain to the ITS (modified from Allègre *et al.*, 1984).

LEVELS OF EROSION IN MOBILE BELTS

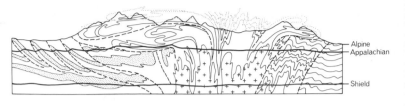

Fig. 4.20 The relationship between landforms and structure in orogens. Even in the relatively young ranges of the European Alps the erosional pattern of mountains and valleys is largely unrelated to original structure. As erosion cuts away the uplands relative resistance of harder rock units controls the form of the uplands, as in the Appalachians. In shields of Precambrian rocks most of the rock units are hard and selective erosion is largely along major joints and faults (modified from an original figure by Dott and Batten, 1981).

Appalachians, thrust-faulting carried thick wedges of rock westwards towards the North American craton. In Norway thrusting was eastward against the craton. In both cases large granite batholiths developed in the interior of the orogen and metamorphic rocks were thrust over the sedimentary rocks along the margins. The Old Red Sandstone beds of Europe and the American (Catskill) red beds are the molasse transported towards the flanks of the mountain ranges by rivers.

The duration of the orogeny was long: volcanism and diastrophism began as early as late Cambrian time and reached its peak in North America as the Taconian Orogeny of Ordovician time, and in Europe as the Caledonian which lasted into the Devonian and thus overlapped with an Acadian tectonic pulse in North America (see Fig. 4.2 and Fig. 4.21).

The most commonly accepted plate tectonic reconstruction postulates a proto-Atlantic ocean, called the *Iapetus Ocean*, being open in late Precambrian time and closing in the early Paleozoic (Fig. 4.22). Collision of the continents started to the north of this ocean and gradually extended southwards to form the Caledonian Orogen. Closure of the Iapetus Ocean created most of Laurasia, and it was the collision of Laurasia and Gondwanaland which produced the Hercynian–Appalachian Orogeny.

The Hercynian (= Variscan) Orogen is remarkably extensive. At the end of the Paleozoic several continental blocks, which were later to form Pangaea, were involved in collisions: North Africa collided with Europe, and the Urals resulted from collisions between European and Siberian plates.

The term 'Hercynian' is widely used to refer to orogens of late Paleozoic age, but it is also used to refer to a structural trend in rocks such as those of the Harz Mountains of Germany (from where the word is derived), and to certain rock units of Devonian age; as a result many European geologists refer to the orogenic belt as the 'Variscan'.

Both the Variscan and Ural belts experienced some deformation in Devonian times, but the major orogeny with widespread metamorphism and granite intrusion began in late Carboniferous time. Erosion of the rising mountains caused widespread deposition in marginal lowlands and deltas, where swamps formed and became the source materials for the major coal-beds of Europe and North America.

In contrast to the Caledonian belt, whose structural trend runs nearly parallel with the modern North Atlantic coast, the Variscan trends in Europe are mostly normal to it, as a result of opening of the Atlantic as Pangaea broke up (Fig. 4.23), and end in ridges and deep inlets (called rias) in SW Ireland and Brittany.

The Variscan uplands of Europe are now isolated blocks, but their composition is not everywhere the same. In western Britain, from the Scilly Isles to Dartmoor, the blocks are the upper parts of a continuous batholith; in the northern part of western Europe, between the Ardennes and Bohemia, they are of folded Paleozoic rocks with little metamorphism; in the central belt, from Brittany, to the Massif Central, the Vosges, and Black Forest, the rocks are highly metamorphosed; and the southern belt, from the Iberian and Moroccan Mesetas through Corsica and Sardinia, is a belt of folded rocks which are not metamorphosed. The massifs now form resistant uplands (Plate 4.12) which have suffered no later

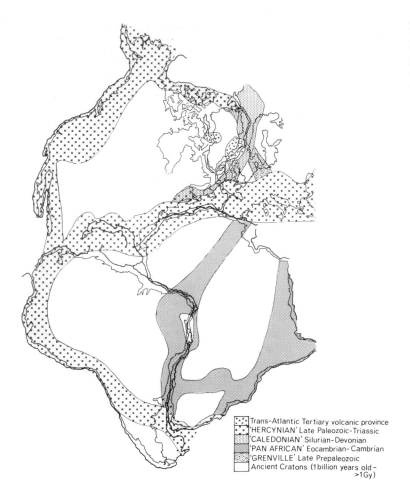

Fig. 4.21 The major Paleozoic orogenic belts of Pangaea produced by closing of the craton units (based on a figure by Dott and Batten, 1981, with the continental outline used by Bullard *et al.*, 1965).

Trans-Atlantic Tertiary volcanic province
'HERCYNIAN' Late Paleozoic-Triassic
'CALEDONIAN' Silurian-Devonian
'PAN AFRICAN' Eocambrian-Cambrian
'GRENVILLE' Late Prepaleozoic
Ancient Cratons (1 billion years old – >1 Gy)

deformation, even where they have been on the margins of the Cenozoic Alpine Orogeny.

The Appalachian Mountains have traditionally been interpreted as a compression zone involving the entire crust. The Piedmont region was regarded as being the root zone of large nappes and, to the west, foreland folds were thought to have been thrust onto the margin of the American craton. These folds have resistant rock units, such as quartzite, which are now eroded into sharp-crested ridges that zigzag across the landscape (Plates 4.13, 4.14). In the Blue Ridge Mountains, metamorphic rocks are exposed, and these are broken by numerous low-angle overthrust faults.

Modern deep-drilling and seismic studies have contradicted the theory of involvement of the whole crust in thrust-folding to the west. As is indicated in Fig. 4.24, it is now believed that a thin

thrust sheet, 6–15 km deep, including Precambrian basement, was displaced westwards at least 260 km. This sheet is internally very complex with many subsidiary thrusts which flatten downwards. This massive thrusting was the result of collisions of Gondwanaland with Laurasia and is another example of the huge scale of thrusting (like that of the Himalayas) which results from continent-to-continent collisions.

Behaviour of rock under stress

The following section is an introduction to the processes, and results, of folding and faulting. Such processes are the result of deformation, or strain, of the rock under applied stresses (see Chapter 6).

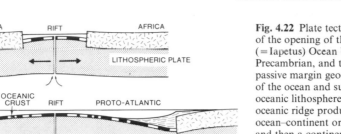

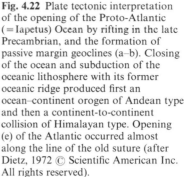

Fig. 4.22 Plate tectonic interpretation of the opening of the Proto-Atlantic (=Iapetus) Ocean by rifting in the late Precambrian, and the formation of passive margin geoclines (a–b). Closing of the ocean and subduction of the oceanic lithosphere with its former oceanic ridge produced first an ocean–continent orogen of Andean type and then a continent-to-continent collision of Himalayan type. Opening (e) of the Atlantic occurred almost along the line of the old suture (after Dietz, 1972 © Scientific American Inc. All rights reserved).

Elastic strain and plastic flow

Adding a kilogram mass to a spring balance can be seen to produce a strain in the elastic spring which is exactly proportional to the applied stress. This is the essential characteristic of *elastic strain*, but it has a limit. If too many masses are added to the balance, the spring may fail by *brittle fracture* as the *elastic limit* is exceeded. Alternatively the spring may be permanently stretched, or strained, as a result of ductility in the metal.

Rock is an elastic solid, but it is ductile at high confining stresses when *plastic flow* can occur within it. Plastic flow is a result of the loss of atoms from one mineral grain and their addition to another, or is the result of shearing between individual atoms, molecules, or mineral grains. The shearing is thus uniformly distributed through the material. For plastic deformation of a

solid to occur, a certain shearing stress must build up before the deformation can start, as the bonds holding the atoms, molecules, ions, or grains together must be broken first.

A molten rock behaves as a *viscous fluid* in which shearing motion begins as soon as a stress is applied and in which the rate of shear increases in direct proportion to the increase in shear stress. *Viscosity* is the property of fluids and solids which causes them to resist instantaneous change of shape and to produce strain that is dependent on time and the magnitude of the stress. It is a measure of the strength of the bond between the particles of the substance and is measured in units of N s/m². Some values of dynamic viscosity for materials are given in Table 4.1

Rock at depth in the crust behaves as a very viscous fluid and flows very slowly. The rates of

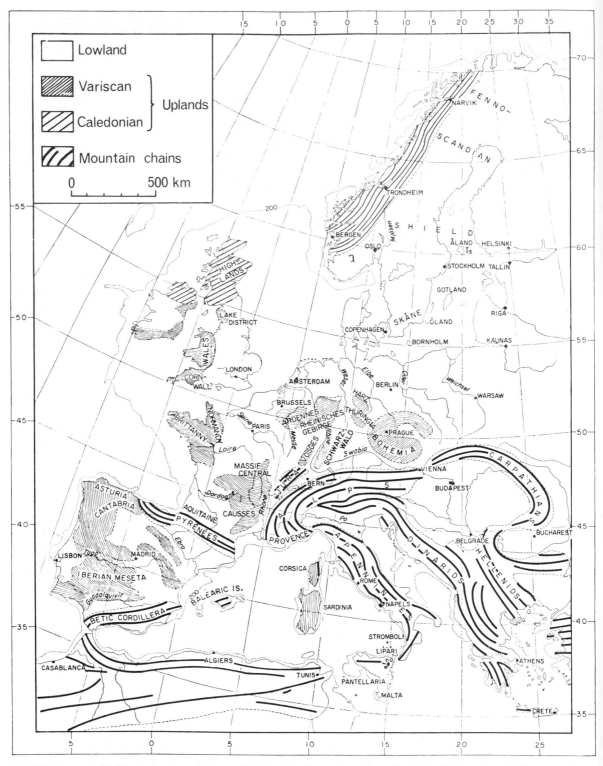

Fig. 4.23 A generalized tectonic map of western Europe, showing structural trends of the Caledonian and Variscan upland remnants and the fold mountain chains of the Alpine Orogeny (after Rutter, 1969).

4.12 Rounded uplands of the Variscan Vosges, NE France.

Table 4.1   Dynamic Viscosities (N s/m^2)

Earth's mantle	10^{20}
Shale	10^{16}
Anhydrite (evaporites)	10^{13}–10^{16}
Glacier ice	10^{12}–10^{13}
Magma	10^2–10^3
Flowing lava	10^1–10^5
Debris flows	7.5×10^2
Mudflows	2×10^2–6×10^2
Solifluction	10^2
Water at 20 °C	10^{-3}

sea-floor spreading of 2–10 cm/y are indicative of this, but it is in a solid state: we know this because it transmits seismic waves. The simultaneous behaviour as a solid and a fluid is said to be *elasticoviscous*. A well-known example is pitch (or tar), which at temperatures of 10–20 °C is a black, glassy solid which fractures when struck, yet at slightly higher temperatures, and over longer time-periods, slowly flows. Much of the deformation of rock which we see expressed in flow structures and folds is consequently controlled by the time available for strain: rapidly applied strains, as in earthquakes, produce brittle failure, but slowly applied strains can cause elasticoviscous behaviour.

Saturated granular materials

In a stratum of sedimentary rock, which has been compressed beneath an overburden, the loss of void space between grains creates high frictional strength, and the growth of crystals of silica or calcite in the voids forms a cement so that the rock also has cohesive strength. The availability of frictional and cohesive strength, and the rock characteristics of a brittle solid, may be thought to operate against the possibility of a rock being readily folded, except under very high compressive stresses and at high temperatures. It must be remembered, however, that much folding takes place before *lithifying* (i.e. rock-making) processes of compaction and cementation are complete.

The sediments of a geosyncline have a high water content, and this water may be trapped by layers of low permeability clay and silt, so that if compressive and shearing forces occur in an accretionary wedge, for example, the sediment cannot drain during deformation. In this situation the stresses are transferred to the trapped water and the sediment grains virtually float in the water and have little frictional contact with one another. As water cannot withstand a shear stress the whole sediment mass is a readily deformed semi-fluid.

Brittle, elastic, viscous, and fluid behaviour explain part of the deformation of rock and the development of folds, faults, and joints. The geological conditions associated with tectonism explain the rest of the formation of structures.

Fractures and folds

Fractures

Rock fractures may be divided into two categories: *joints* and *faults*. A joint is a crack in rock along

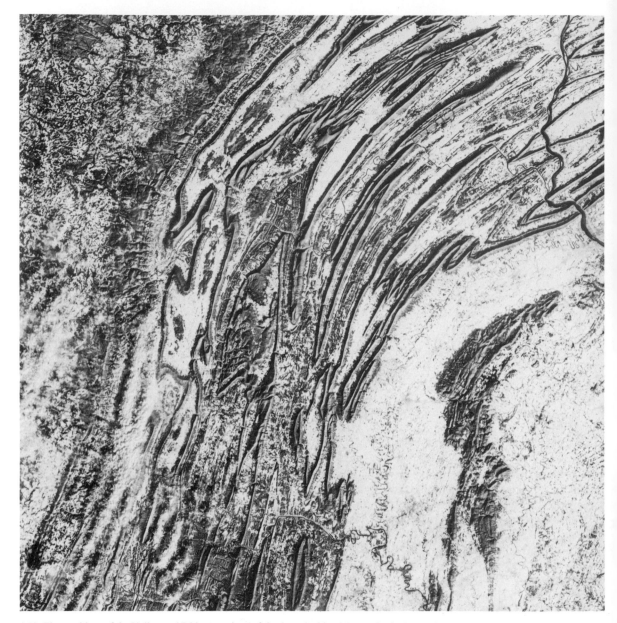

4.13 Zigzag ridges of the Valley and Ridge province of the Appalachian Mountains in Pennsylvania. Most of the wider ridges in the north-central (top centre) part of this image are breached anticlines. Harrisburg on the Susquehanna River is in the bottom centre just south of the major fold belt (satellite photo by courtesy of NASA).

which no appreciable displacement (i.e. sideways or vertical movement) has occurred; a fault is a crack along which displacement has occurred.

Joints may be highly irregular in their inclination and spacing, or they may have a discernible pattern. Patterns of joints may develop either because the original rock has natural features within it which are lines of weakness, or because of stresses resulting from crustal movement. Bedding joints are features of sedimentary rocks and are

4.14 A zigzag ridge on Tuscarora quartzite near Hollidaysburg, Pennsylvania, looking south-east (photo by John S. Shelton).

commonly aligned along boundaries at which the materials change their nature, as for example at the junction of a silty and sandy bed. The opening of the joint may occur because in the period of sediment deposition the rocks were compressed by the weight of the overburden and later, when some of this overburden was removed by erosion, the imposed stresses were released so that the rock could expand and, in so doing, fracture. Joints may also open in sedimentary beds at right angles to the bedding because the beds shrink as they dry out, or joints may be inherited from desiccation mud cracks formed during deposition.

In igneous rocks joints develop during cooling as the rock contracts, or as igneous rock is intruded into surface country rock and tension cracks are opened up in the surface of the intrusion. The regular columnar joint systems formed in cooling, and the layered joints produced by unloading, will be discussed in Chapters 6 and 8, as they can have an important influence on landforms.

A form of jointing which affects all rock types—igneous, metamorphic, and sedimentary—is that produced by compression, tension, and shearing in the crust. These stresses may cause joints to develop parallel to, or obliquely to the direction of the stress, but they impose this pattern on rocks over a large area and so provide evidence for the stress even when it is no longer being applied.

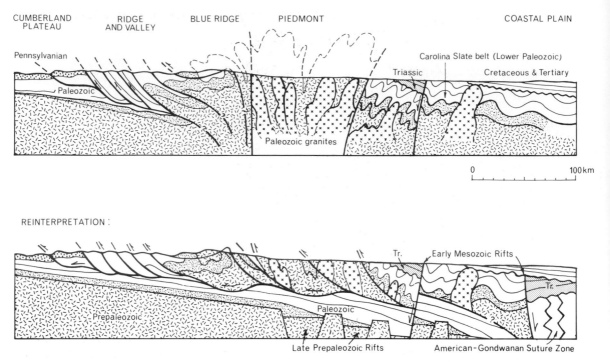

Fig. 4.24 Alternative interpretations of the structure of the central Appalachians. Older interpretations imply that there is a bilateral symmetry and deep origin of the thrusts. Newer interpretations (bottom) show much shallower thrusts becoming less steep at depth, but linking up into one major fault zone, with basement rocks being incorporated into the thrust sheet (top figure based on King, 1959, and the bottom one on Cook *et al.*, 1979, and Harris and Bayer, 1979).

Structural geologists use joint patterns and directions as one of the major lines of evidence for tectonic stress directions.

Joints are of major importance to the development of landforms because they form lines of weakness along which water, air, and plant roots can penetrate the rock and initiate its breakdown, that is *weathering*, by physical and chemical processes.

Faults are fractures along which displacement may be of any magnitude from a few millimetres to hundreds of kilometres. The very large displacements along transform faults, like the San Andreas, have been mentioned already. Very small displacements are common in sedimentary rocks which settle and shrink differentially. Other faults are produced by shear, compressive, and tensional movements in the crust.

The different categories of faulting are distinguished by the direction of motion along the fracture plane (Fig. 4.25). A *dip-slip* fault is one in which motion is up or down the dip of the fault plane and may be either *normal* where the rocks above the fault plane move down relative to the rocks below, or a *reverse* fault where the rocks above the fault move up relative to the rocks below. A reverse fault in which the dip is so small that the displacement of the overlying block is nearly horizontal is a *thrust* fault (Plate 4.6).

A *strike-slip* fault is one in which the movement is horizontal parallel to the strike of the fault. If, as an observer faces a strike-slip fault, the displacement of the block on the other side is to the right, then the fault is a *right-lateral* or *dextral* fault, and if it is to the left then it is a *left-lateral* or *sinistral* fault (Fig. 4.26). A fault with both a vertical and horizontal displacement is an *oblique-slip* fault.

Dislocation along a fault may be recognized in the field by the disruption of the rock formations on either side of it and the amount of relative displacement may be measured. But where displacements are very large, recognizing the displaced rock units may be difficult.

In the process of displacement, rock forming the

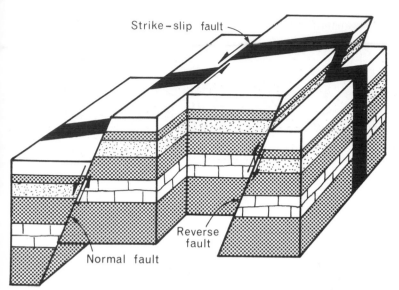

Fig. 4.25 Main types of fault.

Strike-slip fault

Reverse fault

Normal fault

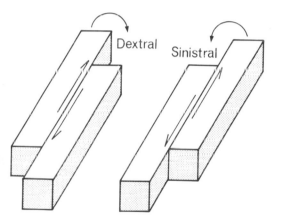

Dextral   Sinistral

Fig. 4.26 Right- and left-lateral strike-slip faults.

walls of the fault may be crushed, polished, and striated (i.e. scratched) to produce *slickensided* surfaces which may also indicate the direction of displacement. The amount of displacement in the vertical plane is called the *throw*.

Faults are most extensive in brittle rocks such as quartzites and granitic rocks, and least extensive in plastic rocks, such as clay and salt-beds, which are likely to fail by folding or flowing unless the displacement along the fault is large. Many faults, showing clear displacement in brittle rocks, die out as they penetrate more ductile rock units. Because strain energy in a rock mass is relieved by displacement along a fault, major faults usually have a spacing of at least 250–500 m unless they are

splinter fractures developed along one major plane of separation. Where strike-slip faults are associated with earthquakes an approximate rule is that a magnitude 4.0 earthquake is associated with movement along a 2–5 km rupture, a magnitude 5.0 movement along 5–10 km, and a magnitude 6.0 along 20–40 km.

The energy required to shear intact brittle rocks is large, but is considerably less in rocks which have been faulted earlier, so old faults tend to be reactivated in preference to the development of new ones.

Folds

Folding, or bending, is the most common form of deformation of layered, sedimentary rocks and may occur at any scale, from crumpling of thin beds into folds only a few centimetres long and high, to huge waves in the crust many kilometres long and hundreds of metres high. Folds may result from compression in the crust, from uplift of a block beneath a sedimentary rock cover so that the cover becomes draped over the rising block, or from gravitational sliding and folding where layered rocks slide down the flanks of a rising block and crumple.

If deformation is slight the nature of the folding is easily described, but it may be very complex and a geologist may have to employ careful mapping and measurement of structures in order to work

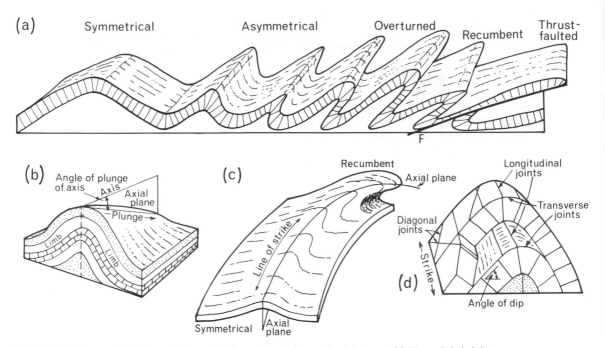

Fig. 4.27 (a) Types of fold. (b, c, d) The axis, plunge, dip, strike, and axial plane of folds, and their joints.

out the nature, extent, and age of the folds. The three basic principles of stratigraphy which are followed in such research are those of superposition, original horizontality, and of lateral continuity (Chapter 1, p. 5).

In broad, open folding the beds will retain their natural order of deposition, in both the upfolds, called *anticlines*, and in the downfolds, called *synclines*. If, however, the folds are *overturned* with the beds tilted more than 90° from the original horizontal attitude, or *recumbent* with both *limbs* of the fold rotated through nearly 180°, or more, the beds will locally have an inverted sequence with older over younger units (Fig. 4.27). In order to understand the pattern of deposition and folding the geologist then has to have a good knowledge of the ages of the units, derived from fossils or radiometric dating, or be able to trace the rock unit laterally into a sequence in which the normal bedding order is preserved. In broad zones of upfolding, called *anticlinoria*, and in *synclinoria* a complex pattern of folds may be very difficult to interpret. General up- and down-folded areas in which the stratigraphy is unknown may be called *antiforms* and *synforms*.

Folding may involve the development of a single steplike bend in otherwise gently dipping or horizontal beds so that a *monocline* is produced but, far more commonly, folds occur in elongated groups or belts.

The process of folding may involve flowage by plastic deformation of the rock, and also small-scale shearing along numerous small fractures. Plastic flow occurs readily in rocks such as limestone in which the calcite crystals deform by the internal gliding of layers of ions along cleavage planes within the mineral crystals (Plate 4.15a). Shales commonly appear to have suffered deformation in which the clay minerals have been reoriented parallel to one another by internal shearing; they also exhibit thinning in the limb of the fold and thickening in the apices of the fold. Quartzites and sandstones, in contrast, show little thickening or thinning but have much internal fracturing and small-scale faulting along which the shear is distributed by brittle failures. Under extreme conditions of pressure, sedimentary rocks may be metamorphosed so that folding involves the development of schistosity, that is, alteration, recrystallization, and parallel alignment of mineral grains in the rock.

The mechanics of folding is still a subject of

4.15 (a) Folds in Precambrian marble. Note the thinning in the limbs of the folds and thickening in the apices, which are evidence of plastic flow.

4.15 (b) Folds in quartzite showing evidence of fracturing and thrusting in the apices of the folds. Note that the folds do not control the form of the slope.

controversy. Many earlier writers believed that simple compression in the crust caused folding, and that the compression was caused by shrinking and crumpling of the crust of a cooling and contracting Earth. It is doubtful if any geologists still believe in a shrinking Earth, but not all have forsaken ideas of simple compression. It is evident from studies of accretionary wedges that compression does produce folding in saturated sediments, but such folding is usually associated with severe thrusting and deformation and cannot explain simple open folds. Fuller explanations may involve processes of décollement, thrusting, draping, gravity sliding, and nappe formation (Fig. 4.28).

Décollement was originally proposed as a mechanism to explain the broad, open folds of parts of the Jura Mountains (Fig. 4.29). The classical theory was that the limestones of the Jura resulted from the push transmitted from the Alpine nappes and that the limestones became detached from the basement of crystalline rocks because of the plastic flow occurring in the Triassic anhydrite beds. The resulting folds were sometimes faulted and sometimes formed box-like anticlines because of the ductility of the limestones. Simple push mechanisms are, however, inadequate because weak rocks which can deform plastically cannot transmit a large compressive stress over a great distance. Two alternative mechanisms are suggested by theorists.

The simplest alternative is that as the Alps were uplifted and the nappes moved downslope, the Jura were affected in the same way, and décollement was a response to large-scale gravity sliding rather than compression. The second view is that the surface folds are merely expressions of blocks of basement upfaulted into the overlying limestones which have been draped over the uplifts. In this view décollement had no part to play in the folding, but as the anhydrites are up to 700 m thick

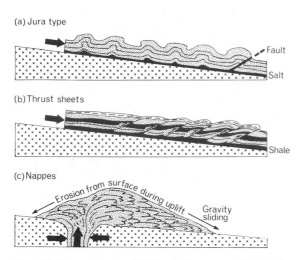

(a) Jura type

Fault

Salt

(b) Thrust sheets

Shale

(c) Nappes

Erosion from surface during uplift

Gravity sliding

Fig. 4.28 The mechanics of folding. In Jura-type folds décollement is a major result of gravity sliding above a lubricating bed. Thrusting may develop at depth as a result of gravity sliding and cause draping of upper strata over the thrust blocks. Nappes may form over a long time as a result of large-scale gravity sliding.

it seems more probable that they have had some part to play in the folding process, as they would otherwise have tended to flow and absorb the effects of thrust-faulting below them.

Large-scale gravity sliding is a well-known phenomenon: it is clearly recognizable at the scale of landslides which move along well-defined shear planes, and on a large-scale whole hill masses can be recognized as moving at rates of a few mm/y. Plate 4.16 shows a 3 km long hill of mudstone and limestone, in the eastern ranges of the North Island of New Zealand, which has moved 1.5 km. A feature of large-scale sliding is that it occurs on very low angles of slope. It must do so because very steep slopes are necessarily short, and sliding on them leads to chaotic failures or, in water, turbidites. The problem that faced early theorists is that the frictional properties of most rocks implies that a slope of about 30° is the minimum required for sliding. This suggests that very large blocks could not travel more than a few kilometres, yet some blocks and nappes have travelled many tens of kilometres on slopes of less than 2° (Plate 4.17). Suppose, however, a plastic material such as salt, clay, or calcite lies between the block and the basement: then the law of viscous fluid friction will come into play, instead of the friction of solids. Hence any force, no matter how small will succeed in moving the block. The velocity of movement will be small if the viscosity is high but

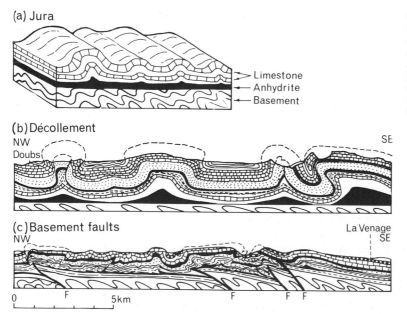

(a) Jura

Limestone
Anhydrite
Basement

(b) Décollement

NW
Doubs

SE

(c) Basement faults

NW

La Venage
SE

0   5km

F   F   F F

Fig. 4.29 Folds in the Jura Mountains. (a) A simplified section showing the main rock units and the anhydrite beds along which décollement can occur. (b) A section showing the box folds and kink folds in the French Jura. (c) Contorted folds which may be caused by thrusting of basement blocks (based, in part, on Umbgrove, 1950).

4.16 The hill indicated by the arrow has slipped downslope from the range to the left. Note the swamp around its foot, indicating a lack of adjustment in the drainage. Hawkes Bay, New Zealand (photo by J. R. Pettinga).

the sliding can take place over geological spans of time.

Where a sliding block rests on a layer of lubricant, and there is no barrier to sliding, the rate of sliding is a function of the weight of the block, the angle of slope, the viscosity of the lubricant, and its thickness. If all the dimensions and properties remain constant, there will be no deforming stresses within the block and it will slide without folding or faulting. If, however, there is an obstacle to sliding such as a reversal of slope, then the forces opposed to sliding create a stress field within the block such that lateral compression within the block occurs in the direction of dip. If this stress exceeds the strength of the rock a thrust fault will develop and the upper part of the block will continue to slide. Thrust-faulting is consequently a feature of many large folds and nappes.

The problem of gravity sliding then becomes not 'How can blocks slide?' but rather 'What stops them?'. Brief answers include: slide surfaces may become rough so that energy is consumed; part of a sliding block may come to move over an unlubricated surface; or a block may move into the sea where buoyancy nearly halves the driving force and halves the velocity (the unit weight of many

saturated rocks is close to 25 kN/m³ above sea-level, but only 15 kN/m³ below it).

Sliding of blocks can only create folds where there is a variation in the rate of movement through the block, or along its base, so that the tail tends to overtake the head of the slide, thus creating buckling and thrusting. In ductile rocks the buckling is accommodated by internal plastic flow, and in brittle rocks by internal shearing if rates of movement are fast, but even in rocks of low ductility deformation can occur over long time-spans of tens of thousands of years. Variation in sliding rates occurs mainly because of changes of slope and changes in lubrication.

If gravity sliding is the driving force of thrusting, folding, décollement, and draping then, clearly, the creation of folds cannot produce mountains. Some uplift must come before the fold. This question is considered at the end of the chapter.

Landforms controlled by faults

Faults may be distinguished as having three types of influence on landforms. (1) They form sharp boundaries of uplands, or valleys, and so give rise to *tectonic scarps*. (2) They control, or strongly influence, the alignment of drainage patterns by forming either lines of shattered, and hence weakened, rocks which are then more readily eroded so that the fault line or zone is followed by a stream channel, or a down-faulted valley becomes a drainage line which may be perpetuated in the landscape, long after the original valley form has been infilled. (3) By juxtaposing rocks of different degrees of resistance to erosion, faults permit the development of a more varied relief than would otherwise have occurred, and the formation of *fault-line scarps*. The influence of faults on the landscape may thus be both direct and readily observed, and also indirect and inheritable from past geological events.

Rift-valleys

Amongst the largest structurally controlled landforms are *rift-valleys* formed as down-faulted *graben* bounded on either flank by fault scarps along the edges of uplands (Fig. 4.30). If they were exposed, the most impressive rift-valleys would be the mid-ocean rifts, but their more limited continental equivalent gives an indication of their size and extent. The 3000 km long East African rift-valley system (Fig. 4.31) extends from Malawi in the south and diverges into two arms encompassing the 1 km high plateau containing Lake Victoria, then extends through Ethiopia to the Red Sea (Fig. 4.32).

Along much of its length the African rift system is marked by clearly defined escarpments 400–2000 m high, which are the result of normal faulting in simple scarp or steplike sets. The western branch of the rift-valley, with a floor occupied by the chain of lakes from Lake Malawi to Lake Mobutu (formerly Lake Albert) is thought to represent a mid-ocean rift in the process of formation, and in about 50 million years it may have the width and size of the present Red Sea. The extension of the rift system through Ethiopia is regarded as a transform fault linking the western rift with the Red Sea (see Fig. 3.22f).

The African rift-valleys are uniformly 30–90 km wide but the floors of the valleys vary in depth because of transverse buckling and obstruction by late Cenozoic volcanism. The deepest parts of the valleys are consequently occupied by lakes. Many of the lakes in the eastern, or Gregory, rift have no drainage outlet and are highly saline. The lakes of the northern part of the western rift drain to the Nile, those in the central section drain to the Zaïre River, and Lake Malawi in the south drains to the Indian Ocean via the Zambezi River. Lake Tanganyika, which drains to the Zaïre, is 1400 m deep and its floor is 650 m below sea-level.

The Ruwenzori massif stands in the western rift-valley. It is the highest (5100 m) non-volcanic mountain in Africa, rising nearly 4 km above the plateaux on each side of the rift, and forms a *horst* which has been raised above all of the surrounding land.

Nappe

Décollement along 10m thick dolomite beds

Basement rocks undeformed

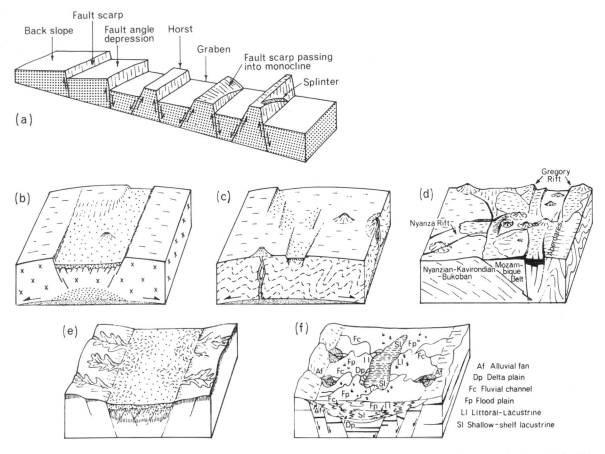

Fig. 4.30 (a) Landforms resulting from normal faulting. (b) Rifting in a brittle rock mass creates graben formation, but in (c) within a ductile shale the result is crustal stretching and the formation of shallow discontinuous basins (after Illies and Baumann, 1982). (d) Features of the Gregory Rift of East Africa (after Pickford, 1982). (e) Erosion of the graben shoulders and sedimentation in the rift may lead to further plateau uplift and graben subsidence. (f) Rift-valleys have many types of depositional environment and have been classed as a form of geosyncline (based on Burggraf and Vondra, 1982).

4.17 Surviving limbs of a nappe in the Naukluft Mountains, Namibia. The folded units slid over a décollement layer of dolomite, only 10 m thick, on slopes of less than 1°.

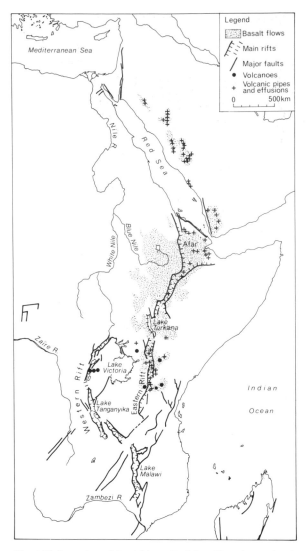

Fig. 4.31 Location of the African–Red Sea rift and associated volcanic rocks.

In plate tectonic theory it is assumed that large rift-valleys are the result of broad-scale doming above mantle plumes, followed by fracturing along the crest of the dome (see Fig. 3.15). It is evident from the drainage patterns that the African rift system has been domed along both rift-valley axes and the extensive plateau between them has sagged so that the former drainage of the plateau, which went westward towards the Zaïre, was beheaded along the line of the western rift and ponded in the central plateau to form the large,

but shallow, Lakes Victoria and Kyoga (Fig. 4.33). The present Victoria Nile drains from Lake Victoria into Kyoga, which is a drowned river-valley system, and then out over a fault scarp at Murchison Falls.

Initial uplift of the Kenyan and Ethiopian domes occurred during the late Cretaceous and early Tertiary and was accompanied by crustal thinning and volcanic action, but the formation of graben did not take place until the Miocene. Depression of the central graben allowed up to 1000 m of sediment to accumulate in parts of the rifts. Major uplift of the Kenyan dome formed large escarpments during the Pleistocene.

It is thought that epeirogenic uplift of the central plateau is still continuing at an estimated rate of 3 mm/y. Active faulting and volcanism are also continuing in the Afar region of Ethiopia, near the southern end of the Red Sea.

The Dead Sea rift is about 600 km long and 10–20 km wide, but the Sea itself occupies only that small segment where the rift valley is deepest. The eastern boundary scarp is straight and continuous, but the western side is more complex with many stepped and splintered segments resulting in a series of blocks rather than a single scarp. The Dead Sea itself occupies a complex site in which movement along a zigzag left-lateral strike-slip fault has left a gap of a type known as a *rhombochasm* (see Plate 4.10).

Many rifts have the Y-shaped pattern of triple junctions. Many triple junctions are still evident from the break-up of Pangaea. The opening of the Atlantic formed two arms and the third, or failed arm, at one site formed the Benue structural trough in Nigeria and at another formed the Amazon trough (Fig. 4.34). Some forty-five triple junctions have been recognized throughout the world: amongst the oldest is that which has as arms the Midland Valley of Scotland with its Carboniferous basaltic volcanoes, the Oslo graben of Norway, and the North Sea trough. These rifts represent the first attempt at large-scale fracturing of Pangaea.

The Rhine graben is one of the most commonly described rift-valleys in the world. It was originally thought to have a rather simple history resulting from fracturing along the crest of a rising dome. The rift is associated with high heat flow, active faults, and earthquakes; it has undergone about 5 km of widening in the last 48 million years. The initial down-faulting began in the south and

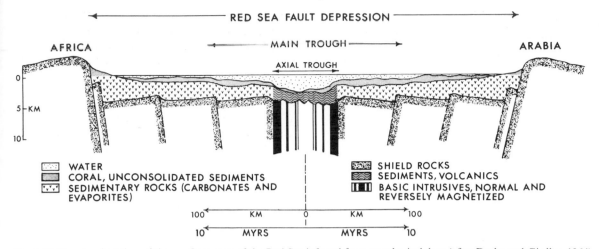

Fig. 4.32 Structural section of the northern part of the Red Sea inferred from geophysical data (after Drake and Girdler, 1964).

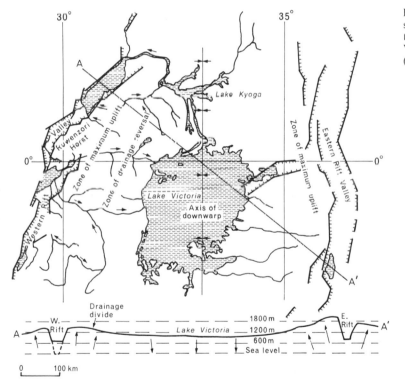

Fig. 4.33 The East African plateau showing the arching around the rifts, the broad downwarp occupied by Lake Victoria, and the drainage diversions (after Bloom, 1978).

movement along the normal faults was accompanied by volcanic intrusions which date the event as being of Eocene age. In Pliocene times the uplift of the Alps was associated with compressive movements from the south and the normal faults along the eastern boundary of the rift valley were subjected to left-lateral strike-slip motion with accompanying extensional and compressional features.

Interpretations of the Rhine graben are not in agreement. Some workers regard it as a triple junction rift with one arm being the present valley south of the junction centred on Frankfurt, a second arm stretching through the Rhine gorge to

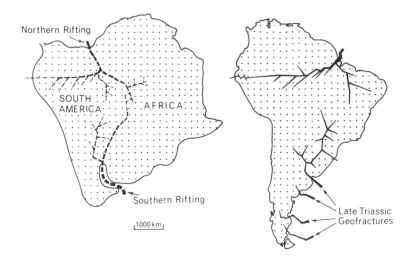

Fig. 4.34 Triple junctions formed during the opening of the southern Atlantic and the failed arms which have strongly influenced many of the major drainage lines (after Potter, 1976 © Univ. of Chicago).

Cologne and the delta, and the third arm diverging at about 120° from the others along the Hessen depression towards Kassel. An alternative hypothesis is that it was an original fracture of the crest of a dome forced up by mantle swelling, caused by subduction along an Alpine collision front. The ascending mantle body provoked the rifting, the uplift of the Black Forest and Vosges, and then as subduction ceased the movement became compressive (Fig. 4.35).

The Rhine graben illustrates two fundamental features of faulting. (1) Widespread propagation of large faults with formation of scarps can occur only in brittle, hard rocks (usually of igneous and metamorphic type). Where the upper crust is composed of schists or shales, ductile or plastic deformation will occur with stretching, thinning, and monoclinal warping. In the southern and northern Rhine rifts the faults are in brittle rocks above a mantle rise in which the Moho is uparched by nearly 24 km, but the Rhenish Shield is of shales and slates which have yielded to tension by stretching. The recent volcanism of the Shield and the continued uplift of about 1.6 mm/y, which has caused the Mosel and other rivers to incise and leave river terraces 200–300 m above present river-level, are both evidence of continuing tension in the crust. To the north the rift is buried by sediments of the delta and North Sea. (2) The Rhine rift follows Variscan trends of the Vosges and Black Forest (see Fig. 4.23) in its southern sector, and merges in the delta region with the Jurassic rift of the North Sea. Only where it crosses the Rhenish Shield does it not follow ancient lines of weakness.

It is evident from the discussion above that large-scale rifting is a feature of plate divergence and the doming is associated with upwelling of mantle material and volcanism, which is usually basaltic. It is doubtful, however, whether all rifts are associated with plate divergence and some areas, like Baikal north of Mongolia, may be the result of more localized upwelling of the mantle to form *hotspots*. The cause of hotspots is not known and even their existence is doubted by some. An alternative explanation of Lake Baikal is that it occupies a rhombochasm along the initial fracture of a diverging plate which will eventually separate the major part of Asia from China.

Features of fault scarps

Fault scarps are the exposed slope which results from vertical displacement along a fault. It is rare, however, for the actual hillside to be the faulted surface. More commonly weathering and erosion cut channels into the hillslope and this becomes covered with a soil and vegetation. The whole hillside may eventually be cut back until it is no longer located above the fault. Even faults which are still active may now have minor active scarps some distance from the foot of the main hillslope. In spite of the results of erosion the hillslope resulting from faulting is still known as a fault scarp (Fig. 4.36).

Movements along fault scarps are usually episo-

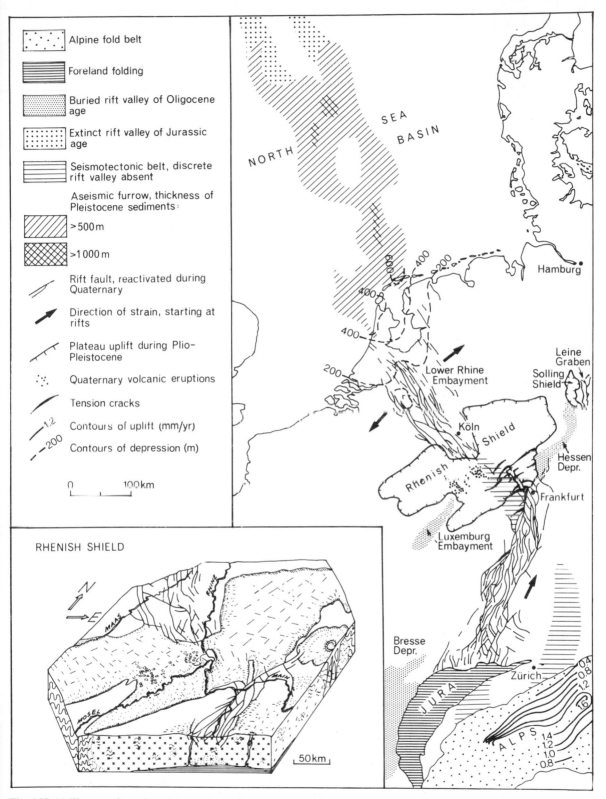

Fig. 4.35 (a) The tectonic setting of the Rhine graben. It forms a tectonically active zone joining the Alpine orogen to the central graben of the North Sea. In response to the stresses of the Alpine area, the upper Rhine zone fails by sinistral shearing along many strike-slip faults, and the lower Rhine between Köln (Cologne) and Arnhem is undergoing extensional rifting. Uplift rates are given for the central Alps. The inset block of the Rhenish Shield is a zone of active plateau uplift and recent basaltic volcanism, with ductile behaviour which has impeded extension of the faults (after Illies and Greiner, 1978, and Illies and Baumann, 1982).

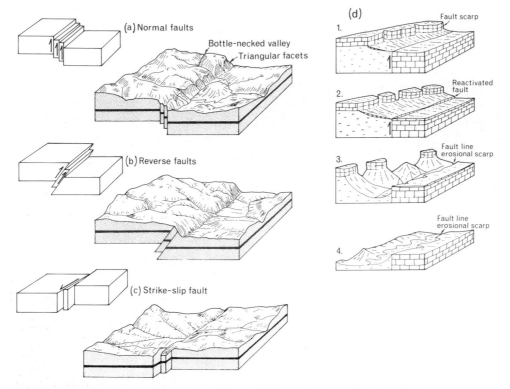

Fig. 4.36 (a) Fault scarp formed along normal faults. (b) Reverse faults produce a less distinctive scarp. (c) The strike-slip fault has produced a crush zone which is exploited by the streams. Drainage which once crossed the fault is now offset. (d) The formation of a fault-line scarp by differential erosion. The scarp ultimately (4) does not reflect the original sense of displacement (in part, after Tricart, 1968).

dic because strain in the crust builds up continually for some years, perhaps causing bulging of the ground surface, until the elastic strength of the rock is exceeded and a fracture occurs. Differential movement may then vary from a few millimetres to about 5 metres. At the known rates of strain it is rare for more than one 5 m movement to occur in less than 100 years or so.

Fault scarps, like most high, steep hillslopes, are subject to vigorous erosion. Streams flowing down the uplifted slope will usually cut deep, narrow gorge-like channels in it which extend headwards into the upland. The resulting valleys are often narrower at their lower end (i.e. bottle-necked) and the hillslope on the scarp between neighbouring valleys becomes triangular-faceted.

Over a long time the scarp may be so eroded that its original nature is no longer recognizable. Debris from the upland may then be spread as a

fan or river deposit in the lowland and the line of outcrops of the fault is buried. Erosion may later remove the river deposits and *exhume* the fault line. Because faults may cause large displacements, the rocks on either side of them may have different resistance to erosion. An exhumed fault may then become the site of a new scarp caused not by the original displacement but by varying resistance: such a scarp is distinguished as a *fault-line scarp*. The distinction may be of practical importance because a fault scarp may still be prone to movement and earthquakes, and hence a dangerous site for construction, but a fault-line scarp is not a site of active displacement.

The amount of displacement along a fault is not uniform. Consequently a scarp may pass into a simple monoclinal fold, it may merely diminish in throw, or the displacement may be shared amongst several splinters or steps. Each type of

4.18 The Pacroa Range, New Zealand: a tilted fault block formed of ignimbrite (photo by courtesy of NZ Geological Survey).

lateral termination may produce a distinctive hillslope which will eventually become less distinct as a result of erosion (Fig. 4.36).

Features associated with strike-slip faulting

Strike-slip faulting, whether it is simple lateral movement or of transform type, may produce no scarps in a landscape of low relief, but less obvious displacement may occur. It was recognized in the late 1960s and early 1970s that there are inconsistencies in data from surveying traverses, of different dates, for an area of southern California inland of Los Angeles. It became evident that a large area astride the San Andreas Fault is being gradually domed and between 1959 and 1978 the crest of the dome near Palmdale rose 25 cm. Short-term changes of elevation, both up and down, occurred in different parts of the bulge in a few months (Figure 4.37).

Behaviour along the Fault may vary. There appear to be some sections, including the part near Palmdale and a segment south of San Francisco, that are locked. In such segments strain energy builds up until a violent earthquake, and fault movement of up to 6 m lateral displacement, occurs perhaps once every 50–200 years. Along other sections slow, slight, creeping movements, averaging 3–4 cm/y, occur frequently, with the gradual release of strain energy and very minor earthquakes. These areas of slight movement may occur where either, or both, water and serpentinite lubricate the Fault. Serpentinite is a rock which behaves in a plastic manner under stress and can be squeezed into a fault.

The significance of this finding is that where

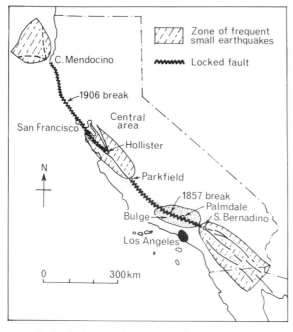

Fig. 4.37 Locked and active zones of the San Andreas Fault, California (compiled from several published sources).

sections of the fault are locked but strain energy is building up, as is shown by lack of minor earthquakes and by bulging, a large and possibly devastating earthquake may be expected. Where nearly continuous movement is possible the strain is steadily dissipated and large earthquakes are improbable. Attempts are now being made to identify locked and mobile sections of many large faults.

Bulging and minor rents are common features along strike-slip faults: less common but more obvious are the offsets of drainage channels and ridges which once crossed the fault (Plate 4.19). Half-displaced ridges which now block some valleys have been called *shutter-ridges*. Continued movement along the main fault will eventually produce a zone of shattered and weakened rock which may then become a major drainage line collecting the streams which once crossed it (Plate 4.20).

Block-faulted landscapes

Many of the pioneer accounts of block-faulted landscapes are of the desert Basin and Range areas

of south-western USA. These landscapes consist of highly folded rocks which have later been cut by steeply dipping, mostly normal, faults. Many large blocks have been uplifted to form mountains with strong backward-tilted dip slopes and eroded scarp slopes with angles which are unrelated to the dip of the faults. Because the blocks are strongly tilted they are not true horsts and the tectonically depressed basins are *fault-angle depressions* rather than true graben. A feature of the Basin and Range province is the lack of external drainage. Consequently the fault-angle depressions become filled with fan deposits, salt lakes which evaporate to form playas, and extensive sheets of river gravels. As the bases of the fault scarps are progressively buried in the deposits, derived by erosion from the tilted block mountains, the relationship between the faults and the landscape is often obscure. Where the faults are still active, however, small scarps are formed in the fan gravels, showing clearly how far the scarp in the bedrock has been cut back, by erosion, from the fault line.

In many tectonically active areas with high-angle faulting producing block-faulted mountains and wide-bottomed depressions there is external drainage. Consequently the streams may cut through the mountains in gorges, if they have been able to maintain their courses as the blocks rose, or they follow the depressions. In either case fans will be small or absent from the bases of the uplifted blocks, and basin-floors will usually be smoothed over by extensive gravels deposited by the rivers. In areas like central New Zealand, where the controlling faults are mostly high angle, thrust, dip-slip faults the scarps are deeply dissected by streams, and hillslopes have angles quite unrelated to the dip of the faults. It is the straightness of the break between upland and lowland which is the obvious indicator of the presence of active faulting. Unless the faulting is sufficiently active to repeatedly sharpen this break the relationship will become increasingly obscure, scarps will be indented as they retreat, and active faults will be indicated only by fresh scarplets across the river gravels covering the structural boundary between uplifted and depressed blocks.

It is probably true of most faulted topography that horsts with boundary faults delineating relatively untilted uplands are rare. Graben are more common, and certainly evident as rift-valleys. Tilted blocks and fault-angle depressions are the

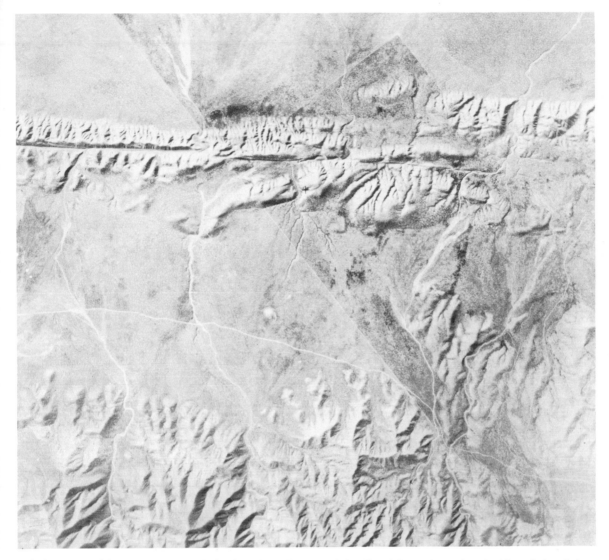

4.19 A segment of the San Andreas Fault showing the crush zone and the offset drainage lines (photo by Litton Industries).

more usual result of normal or high-angle thrust-faulting.

Landforms controlled by folds

Whether or not a distinct landform is created by a fold, or by a fault, depends upon the size of the fold in relation to the size of the main features of the landscape. Small folds, and faults with a small throw, may have little or no influence upon the detail of the relief (Plate 4.15b) which will be more strongly influenced by the spacing and dip of the joints, by the thickness of the soil cover, and by the pattern of erosion.

Large folds, as with faults of large displacement, may, however, have a strong influence upon the shape of landforms, especially while the fold is growing. After a period, which may be hundreds of thousands of years, a fold which is no longer growing will be transformed into a set of landforms which increasingly reflect the influence of erosion rather than the original shape of the fold.

4.20 Wellington, Port Nicholson (the enclosed harbour) and the straight Hutt Valley beyond. The sharp boundary scarp along the western (left) side of the Hutt Valley is the fault scarp of the still-active Wellington fault. The Hutt Valley consists of basins, now filled with river gravels, formed in a fault angle depression. Valleys in the ranges east (right) of Port Nicholson follow sets of older faults which cut the line of the Wellington fault at angles of about 15°. The ridges between these valleys extend into the Hutt Valley as shutter ridges (photo by courtesy of RNZAF).

Because the structural control of the landform gradually merges into a structural influence upon erosional forms, a clear distinction between the two is seldom possible, and both will be discussed together.

Rising anticlines and monoclines

In areas of active orogenesis, or uplift, folds can grow at measurable rates. The rate varies along the crest of the anticline, being a minimum at either end of the fold where it plunges out, and a maximum at the highest point of the crest. Rates may be measured directly by repeated surveys which have revealed vertical increases of elevation of up to 10 mm/y in Japan, and rates of up to 6 mm/y in California and central New Zealand. More commonly rates are estimated from deformed structures which are thought to have been horizontal or had a constant dip. In the south-east of North Island, New Zealand, for example, marine benches and beaches have been cut across plunging anticlines by wave action. Where the age of the beach can be determined, perhaps by ^{14}C-dating of marine shells in beach sands and gravels, and its height above sea-level measured, an estimate of the increase of elevation in that number of years can be made. If the bench cuts across a wide fold, or along the axis of a fold, it may be possible to distinguish rates of folding along different parts of the fold. In Japan many rivers cut through rising anticlines and the profiles of the river terraces are usually humped over the anticlines and depressed over the synclines (Fig. 4.38).

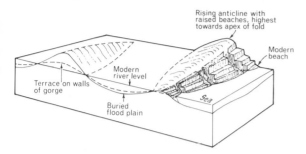

Fig. **4.38** Rates of uplift along anticlines and sinking in synclines may be assessed from the elevation of terraces and from buried floodplain deposits where these can be dated. The terrace reaches maximum altitude near the apex of the anticline, and floodplain deposits are downwarped to greatest depth near the axis of the syncline.

Because all folds grow at rates which are slow enough for erosion to have at least some effect upon the surface rocks during the folding process, it is unlikely that the surface features of the fold will remain intact for long. Gullies, stream channels, and landslides are likely to scar the upper surface of the fold. Whatever the origin of the fold, erosion will exploit the weaknesses of the structure as uplift provides the vertical relief which makes erosion by running water and other slope processes possible.

Landforms of eroded folds

Whether they are part of rising folds composed of the uppermost and youngest strata in a landscape, or older folds which have once been deeply buried by overlying rocks (Plate 4.21) and are now being exposed, as epeirogenesis gives erosional energy to the rivers, sedimentary strata are likely to have varying resistance to erosion. Such differences are

4.21 Anticlinal and synclinal structures in limestones of the Swiss Alps. These are not primary tectonic controls on landforms but the result of erosional stripping of thick overburdens to reveal resistant beds and structural influences (photo by Swissair).

emphasized by erosional processes and the structure expressed in distinctive landforms. The differential erodibility of shale, sandstone, and limestone is the dominant control on the landforms of many fold mountains of the alpine system and also of extensive basins, broad domes, and plateaux of all the continents.

Flat-lying sedimentary rocks of contrasting erodibility show the influence of varying resistance in the walls of valleys which have been incised into the sedimentary pile. The valley walls consist of steep slopes, or cliffs, capped by outcrops of the most resistant strata which form *structural* benches. The benches may owe their resistance to the greater hardness or cementation of their rocks, to the wide spacing of their joints, or to the weakness of the underlying bed which is eroded so rapidly that it undercuts the bench-forming rock and makes it appear more resistant than it really is. It is the *differential* resistance rather than an *absolute* resistance which controls the valley wall profile (see Plate 4.7).

Outcrops of horizontal strata follow the contours along the valley walls and form a relatively simple and easily recognized pattern of slopes.

Monoclinally tilted or folded strata provide more complex structural controls on landforms. The outcropping edge of a resistant stratum will usually form a cliff, or *escarpment* or *scarp*, with details of the scarp form being controlled by the joints. The *dip* slope away from the crest of the scarp may be parallel to the bedding planes and tend to resist dissection. When it is breached, and streams cut through it into underlying weaker rocks, the more resistant stratum will form a caprock (Plate 4.22a,b,c,d).

Stream incision into the flanks of a fold may produce only a series of equally spaced gullies around a dome or anticline if the uppermost stratum is thick, but most folds vary in size, shape, thickness of their strata, and hence, in the character of the landforms developed upon them. Where strata are thin the incising channels may cut into the flanks of the fold to produce gullies with dipping benches along their walls. Erosion may eventually cut so deeply into the fold that individual strata will form scarp slopes facing the apex of the fold and dip slopes forming its flanks.

As all folds, which are not symmetrical domes, plunge along the axis, a range of distinctive

4.22 (a) A hogback in sandstone with the dipslope exactly following the bedding plane, Groot Swartberg, Cape Mountains.

4.22 (b) A transverse gorge through an anticline whose outer surface is formed of Asmari limestone, Zagros Mountains. The central portion of the gorge has expanded into a small axial basin where sapping has caused undercutting of the limestone at its contact with the underlying marl. In the distance the arch has been unroofed by both transverse and local streams (photo and copyright: Aerofilms).

landforms may be produced by stream incision into their flanks. *Mesas* are large residual platforms, and *buttes* small residuals, in horizontal strata; *cuestas* are asymmetrical ridges with distinctive scarp and dip slopes; *hogback* ridges are nearly symmetrical with scarp and dip slopes having similar angles; *razorbacks* have spines of nearly vertical strata (Plate 4.22d).

Where folds consist of many strata, of varying resistance, series of scarp and dip slopes may be formed with each ridge being on a resistant stratum and each lowland on weaker rocks. Strata with uniform thicknesses and dips, over large areas, produce very regular hills called *homoclines*. The differences in resistance between strata are

exploited by the stream network and the entire system of ridges and intervening vales migrates laterally, as well as downward, as streams cut down the dip slope into the base of the scarps. This process of *homoclinal shifting* gradually causes the structurally controlled ridges to migrate a distance many times their own width as a region of dipping strata is eroded (Fig. 4.39).

As an anticline is eroded the uppermost strata may be cut through, first on the most steeply dipping part of the flanks of the fold but later, as the drainage pattern develops, the roof of the fold may be removed. If the older rock forming the core of the fold is resistant then a central massif of basement rock, which may be metamorphic,

4.22 (c) Kuh-i-Mafarun, Zagros Mountains, an unroofed anticline. The hogback of the outer flank of the fold and the inner domed core are both in limestone. The annular subsequent valley is cut along a narrow bed of flysch (photo and copyright: Aerofilms)

igneous, or sedimentary, will be exposed as at Sheep Mountain, Wyoming. But if the older rock is of low resistance it may be eroded and the result is a series of inward-facing scarps rising above a central lowland, as in the Weald of southern England (Fig. 4.40). Conversely the rim of a syncline will possess outward facing scarps, as in the Paris Basin and London Basin.

Complex folds, especially where they have been stripped of overlying strata, have a great diversity of relief and frequently display *inversion of relief*.

Instead of anticlines forming uplands, and synclines the troughs, as happens in a landscape of rising open folds, the uplands are formed of synclinal structure and the lowlands formed over anticlines. Such inversions are common in old mountain systems like the Appalachians from which several kilometres of overlying rocks have been removed. It has been suggested that the crests of anticlines are particularly weak, and therefore easily eroded, because strata are stretched across the top of a fold and joints there are opened. This

4.22 (d) A razorback with a core of sandstone. Note the vertical bedding of the peak, and the lower angle of dip on the shales which form the flank of the hill, Groot Swartberg.

seems an unlikely cause of topographic inversion and has never been demonstrated from field evidence. It is more probable that during the growth of a fold gullies will incise into the crest and extend down the flanks as they emerge: incision thus begins in the crest. During uplift of a block of folded rocks which have already been planated, resistant strata will form uplands as a result of differential erosion (Fig. 4.41 and see Plates 4.13, 4.14). The elevated areas may then be limbs of either anticlines or synclines.

Drainage patterns on folds

Two classes of terms have been coined for describing the relationship between stream channels and the structures in which they are incised. (1) The most useful class relates the channel directly to the type of geological structure so that we can speak of

strike, dip, or antidip streams for those channels which follow that trend, or attach the terms anticlinal and synclinal, as in 'synclinal strike stream', to give a more complete description. Alternatively channels parallel to the structural trend can be described as *longitudinal* streams, and those cutting across the trend as *transverse* streams. (2) Much effort in the earlier part of this century was attached to relating the stream pattern to the assumed development of drainage upon an evolving or uplifted fold belt. This genetic terminology is now little used but, because a student is likely to come across it in wider reading, its main elements are described here (Fig. 4.42).

Rivers that develop upon an original surface, such as a lava sheet or newly uplifted coastal plain, and follow its regional slope are said to be *consequent*. A less common term for small valleys which are *in*itial con*sequent* channels is *insequent*

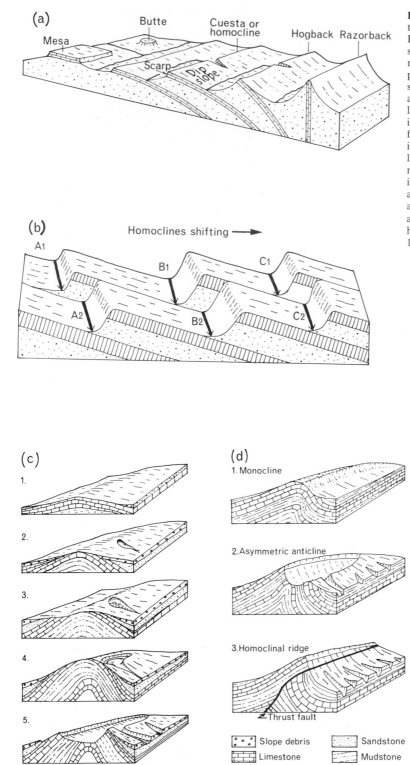

Fig. 4.39 (a) Landforms controlled by the dip of a resistant bed. (b) Homoclinal shifting resulting from stream migration down the dip. (c) A rising anticline is subject to erosional planation of the upper limestone beds so that by stage 4 the roof of the anticline is on weak mudstones or thin limestones. By stage 5 streams have cut into the weak older rock units and formed a valley, along the strike, with inward facing boundary scarps. The limestone in the core of the anticline may eventually form a resistant ridge inside this valley. (d) Erosion of an asymmetric growing fold first produces an asymmetric central valley and then, as the thrust fault develops, a homoclinal ridge (c,d after Tricart, 1968).

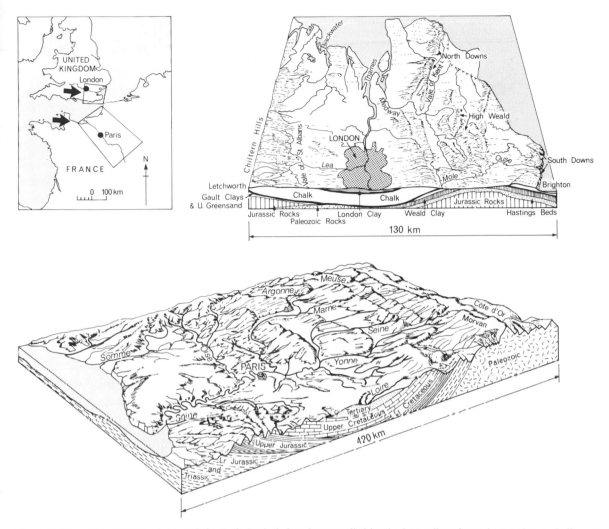

Fig. 4.40 The relief of SE England and the Paris Basin is largely controlled by the homoclines formed on resistant chalk and limestone. The structure has been simplified (the lower diagram after Twidale, 1971, and Naval Intelligence Division, Geographical Handbook, France, Vol. 1, 1942).

stream. At first all drainage will tend to be more readily incised into zones of weaker rocks leaving the more resistant rocks elevated. Thus structurally controlled *subsequent* valleys and channels are formed. Two special types of consequent channel are also recognized: a *re*newed con*sequent*, or *resequent*, stream flows approximately parallel to and in the same direction as the trunk stream; and *op*posite to the direction of the trunk consequent flows an *obsequent*. These terms should not be used with reference to the dip of the controlling strata as these classes of stream may be found on

either scarp or dip slopes. This genetic classification has the considerable disadvantage that in order to use it one has to have a full understanding of structural and drainage history of the study site.

Antecedent, superimposed, and captured drainage

Within an orogenic belt, or in an incompletely stabilized craton, a drainage pattern may be established before the folding and faulting processes have ended. Consequently a hill block may rise across a river course. The river may be ponded

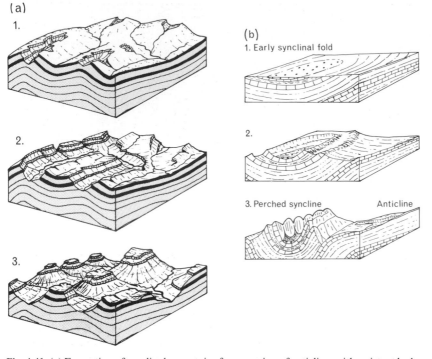

Fig. 4.41 (a) Formation of synclinal mountains from erosion of anticlines with resistant beds. (b) A planated fold has no relief initially, but continued compression and differential erosion cause the development of outward-facing scarps of a perched syncline and unroofing of a growing anticline (b, after Tricart, 1968).

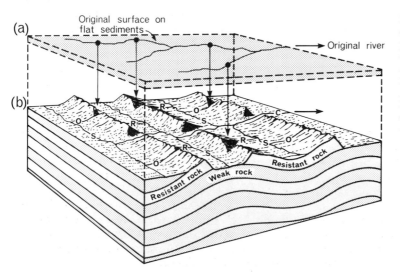

Fig. 4.42 (a) Formation of a consequent river system upon an initial plane surface. (b) The dendritic pattern of (a) is converted into a trellis pattern as the drainage is superimposed upon folded rocks. River capture leaves wind gaps at places indicated by arrows. The letters C, S, R, and O signify consequent, subsequent, resequent, and obsequent rivers and valleys (after Bloom, 1978).

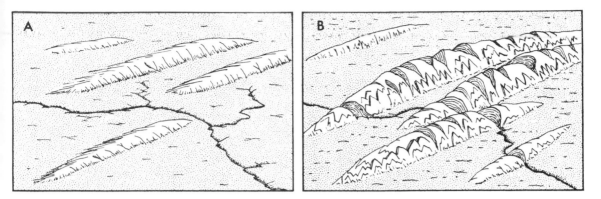

Fig. 4.43 The development of antecedent reaches along a river maintaining its course across rising anticlines (redrawn after Oberlander, 1965).

back and forced to change its direction if it cannot erode downwards as fast as the hill rises, but if the rate of erosion is as great as the rate of uplift then the channel will be maintained transverse to the structure. Such a channel is said to be *antecedent* because it preceded the rising structure (Fig. 4.43).

Superimposed (or superposed) rivers are those which have been let down onto an underlying structural system from an overlying one. An often quoted example is that of a drainage system formed on an alluvial plain which is epeirogenically uplifted. The drainage gradually incises itself until it is superimposed upon the underlying structures (Fig. 4.42).

Antecedence and superimposition are common features of orogenic belts. It is, however, unlikely that all streams in one part of a belt will be either antecedent or superimposed. More probably there will exist a detailed local relationship with structure in which some segments of channels are antecedent and others are superimposed. This type of condition has been well demonstrated for the central Zagros Mountains of Iran by T. Oberlander. Determining the history of the drainage pattern may be very difficult or even impossible. The existence of superimposition can be demonstrated only if remnants of the cover beds, from which a channel was let down, still exist: if they have been eroded away an origin can only be surmised (Fig. 4.44).

Either by chance, because it is cutting along a line of weak rocks, or because of tectonic tilting of the landscape, one river may erode into the drainage basin of another. Water may then be diverted gradually by seepage, or more suddenly

by channel wall breaching, into the capturing river. The stream which loses its headwaters is then said to be *beheaded* and the process of drainage diversion is called *stream capture* or *stream piracy*.

Longitudinal streams flowing along weaker rocks in a fold belt are particularly likely to capture segments of transverse streams which have to maintain their courses across resistant strata of a fold. Similarly streams following a fault crush belt may capture transverse streams crossing a horst or uplifted tilt block. Valleys still occupied by transverse streams are called *water gaps*, but those valley sections which lose their streams are known as *wind* or *air gaps*. The sharp changes of direction between longitudinal and transverse reaches of the stream are known as *elbows of capture*, and the resulting pattern of drainage with straight channel segments and near right-angle bends is called a *trellis drainage pattern* (Fig. 4.42).

Detailed studies of trellis drainage patterns and their history were carried out in the open fold systems of the London Basin, Weald, Paris Basin, and the Appalachians in the first half of this century. Such studies depend for their success upon detailed tracing of remnants of river terrace, alluvial fan, floodplain, and other deposits across and along valleys. As these deposits become more dissected the reliability of the reconstruction is reduced. Where the deposits cannot be correlated by reliable dating of their materials, or by included fossils or trace minerals, the study is subject to uncertainty. It is also made very difficult by even slight tectonic warping as altitudinal similarities, and differences, between deposits have been used as key lines of evidence.

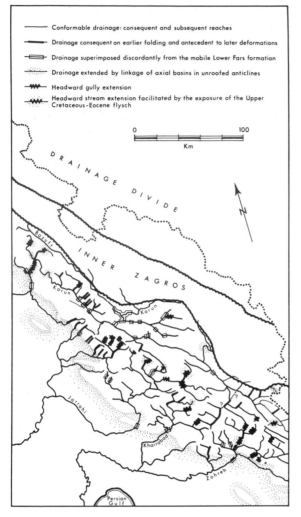

Fig. 4.44 Part of the Zagros Mountain drainage system showing the causes of transverse drainage along sections of each river course (redrawn after Oberlander, 1965).

Orogeny and mountain uplift

There is still debate amongst geologists about whether orogeny takes place at distinct times or not. Many earlier writers considered that, over the Earth as a whole, there are clear episodes of metamorphism and magmatism separated by periods of tectonic quiet; others believe that orogeny goes on almost continuously but at different sites; an alternative view is that orogenies are concentrated episodes of tectonic activity that are localized, and that groupings of such episodes are the phases to which we give names such as Caledonian, Variscan, Alpine, and so forth. The rough correlation of world sea-level rises with orogenies (Fig. 4.2) generally supports the idea of pulses of tectonic activity. The recognition that most of the alpine relief of Europe and New Zealand, and much of that of the Himalayas, is essentially of Pliocene and Pleistocene age has caused many old correlations between orogeny and uplift to be questioned.

The recognition that compressive stresses cannot be transmitted through unconsolidated sediments to form fold belts, and that indurated rocks usually fail by brittle fracture and hence faulting, has led to the conclusion that folding itself does not create mountains. An orogeny is consequently not necessarily a period of mountain building. Dating of orogenies (i.e. periods of folding, metamorphism, and magmatic intrusion) radiometrically, and of mountain uplift from the age of the sediment being deposited along the margins of a rising orogen, has led to the further conclusion that uplift may occur at the same time as orogeny, intermittently during and after orogeny, or at periods so far removed in time and space from orogenies that the two types of event are not necessarily connected.

Uplift of mountains is always essentially epeirogenic, whether it is forming broad plateaus like the Drakensberg and Lesotho mountains, domed uplands like the Eastern Highlands of Australia, faulted block mountains like the Basin and Range Province of USA, or the alpine relief of Switzerland.

The cause of mountain uplift is essentially unknown. Many writers associate it with subduction, but if we recognize that plateau uplift is not related to subduction—which it certainly is not in Africa or Tibet—and also that mountains may be only plateaux so deeply dissected that isolated peaks have been formed, then subduction cannot be a universal cause of uplift. There is ample evidence that many mountains are uplifted orogens that had been eroded down to plains (i.e. *planated*) at low altitudes while the orogen was forming and subduction was active. Uplift has occurred in the European Alps, for example, only after subduction ceased.

There may be no single universal cause of uplift of mountains. Possible causes which may operate for limited periods at limited sites are listed below.

(1) Heating of crust and upper mantle by a mantle plume which causes expansion of the

heated rock, lowers its density, and thus causes it to rise and push up the crust above it (Pratt-type isostatic compensation).

(2) Uplift due to thickening of the lithosphere by underthrusting or by floating upwards of a previously subducted block.

(3) Crustal thickening, during continent-to-continent collisions, by thrusting of wedges of one continent into the other.

(4) Overthrusting of sedimentary slices during collision.

(5) Subduction in which low-density sediments are carried down below continental crust and so force it to rise isostatically.

(6) Addition of low-density material to the base of the crust from the mantle.

(7) Chemical phase changes, with volume increases and density reduction, in the lower crust or upper mantle.

(8) Tectonic thickening of the crust and, or, upper mantle by plastic flow in the mantle.

(9) Oblique compression at a convergent–transform plate boundary, with the transportation of suspect terranes along and against a continental margin.

(10) Uplift may be renewed or prolonged as a result of surface erosion (Airy-type isostatic compensation).

The rates at which these processes take place are reflected in the rates of vertical and horizontal movement of the continental surfaces (Table 4.2).

Landforms of major structural units

On a world map of main structural units (Fig. 4.45) it is possible to recognize five main classes of structure with features which form distinctive suites of landforms: (1) orogenic belts in their first phase of development; (2) old orogens which have been through several phases of uplift (or *resurgent tectonics*) and still form uplands; (3) marginal fold belts on the continental side of the orogenic belt; (4) platforms which are blocks of rigid shield rocks with a cover of sedimentary units; and (5) shields without a sediment cover.

1. Orogenic fold belts in a first phase of development

When geosynclinal deposits are uplifted, erosion may be rapid in the monotonously uniform sediments in which structural controls on landforms are often weak. The sediments may be folded and faulted, but in the absence of lithification it is only the contrasts between limestones, clastic rocks, and igneous intrusions which are etched out by erosion. Individual folds may not influence the relief, either because they are recumbent or because the rock is uniform in its resistance. This type of situation can be seen at present in southeastern France where large areas of the Maritime Alps are composed of massive units of mudstone, and in the tectonic arc south of Sumatra.

The contrast in styles of landform developed in

Table 4.2   Maximum Mean Rates of Recent Crustal Movements

Type of crust	Vertical movements	Horizontal movements
Shield and platforms (i.e. cratons remote from plate boundaries)	1mm/1000y	undetectable
Cenozoic orogenic belts	20–50mm/y	1mm/y
Paleozoic and Mesozoic orogenic belts (often block faulted)	10mm/y maximum more commonly 5mm/100y	1mm/y
Intra-cratonic rift-zones	5mm/y	1mm/y
Rifted plate margins	10mm/y	catastrophic gravity slides

Notes: (1) Maximum rates occur for short periods only; average rates for the late Cenozoic are commonly one or two orders of magnitude lower than maximum rates; directions of uplift or sinking can be reversed.
(2) Local rates of up to 6mm/y can occur in seismically quiet areas for short periods; if maintained for 1 My they would produce a Himalayan-type chain 6000 m high.

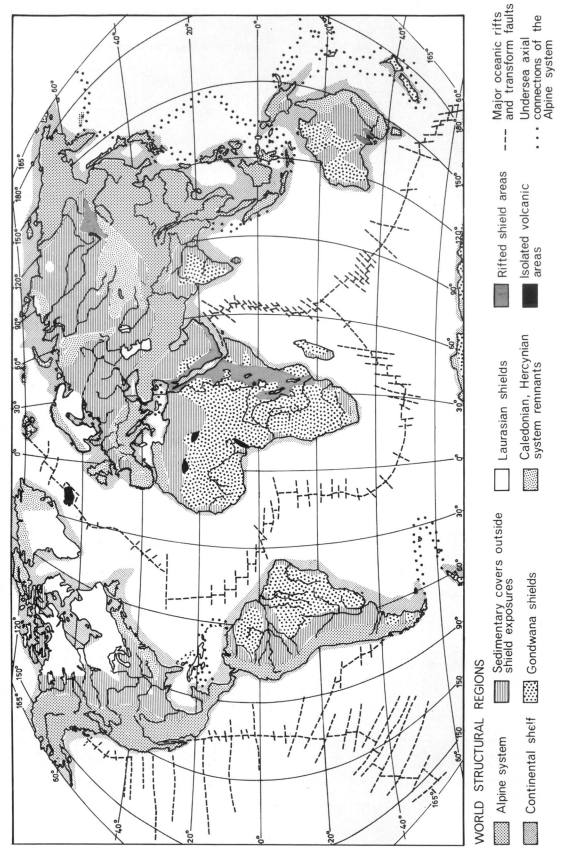

WORLD STRUCTURAL REGIONS

Alpine system

Continental shelf

Sedimentary covers outside
shield exposures

Gondwana shields

Laurasian shields

Caledonian, Hercynian
system remnants

Rifted shield areas

Isolated volcanic
areas

Major oceanic rifts
and transform faults

Undersea axial
connections of the
Alpine system

Fig. 4.45 The main structural regions of the world.

different parts of an orogen is well displayed in the Zagros Mountains of Iran, where in a broad belt running NW–SE to the east of the Persian Gulf and Mesopotomia is a fold belt with broad open anticlines and synclines formed in clastic rocks and limestones. Thick limestone beds now form the outer units of many anticlines and the synclines between them are filled with molasse derived from erosion of the growing folds (Plate 4.22b). The drainage pattern has been referred to already.

In the thrust zone to the north-east igneous and metamorphic rocks have been brought to the surface. Here there is a complex drainage pattern picking out lines of structural weakness along faults and joints, in some places traversing rising thrust blocks, and in other places being diverted around them.

2. Old orogens

The feature of old orogenic belts which makes them different from those in a first phase of folding and uplift is the greater rigidity of their rocks. An old orogen is usually composed of Paleozoic rocks which were folded (e.g. in Caledonian or Hercynian phases), injected with igneous intrusive and extrusive rocks, and metamorphosed to schists and gneisses at depth in the trenches and migma-

tite arch. During and after uplift the cordillera were eroded so that the less resistant strata were removed, and the whole belt may have been planed by erosion to a relatively even erosion surface. This happened for example in New Zealand where the Paleozoic rocks deformed in the Rangitata Orogeny were planated during the Cretaceous period. Terrestrial deposits and shallow sea deposits may be formed upon the planation surface if the sea advances over the land in a marine transgression. Subsequent earth movements will then usually give rise to large-scale faulting of the first-phase rocks which yield by brittle fracture with thrusting occurring along deep-seated reverse faults, as in the present New Zealand Kaikoura Orogeny. The igneous and metamorphic rocks of the basement may thus be brought to the surface and emplaced alongside, or through, the overlying terrestrial sediments or those of shallow (*epicontinental*) seas. The more recent sediments will be folded as a result of thrusting, and monoclines and flexures formed where the sediment is draped over rising fault blocks.

Rising blocks of Paleozoic or Precambrian rocks are not readily eroded so the planation surfaces cut across them may be preserved or resurrected as the sediment is stripped off them (Plate 4.23). Gorges may be cut around the

4.23 Quartz Hill, Wellington, New Zealand: the largest relatively uneroded, uplifted erosion surface remnant in the area (photo by D. W. McKenzie).

margins of such blocks and lithological and structural differences emphasized by the drainage, as in the western Scandinavian highlands. Where old fold belts are uplifted and stripped of their cover beds, as in the Appalachians, the drainage pattern will adjust to the lithology so that resistant beds form the uplands and weaker rocks the lowlands. Synclines may be perched to form highlands and anticlines underlie the lowlands. Some young sedimentary rocks may fill basins left as graben or fault angle depressions between horsts or tilt blocks. These sediments may be relatively undeformed and may be buried by recent sediments being carried into the basin. Thus lake silts, swamp deposits, fan and river deposits can accumulate. In arid areas salt-beds and wind-carried sand may be important. Thus the Altiplano of the Andes is still being infilled, and the central Danube basin of the Hungarian Plains is still being downwarped and infilled. A detailed study of the infill of such basins frequently provides the best evidence for reconstructing the geological history of the surrounding uplands.

One orogenic belt may have both an early folding and late faulting phase of development. The Zagros for example has a western fold belt of young growing folds in Mesozoic and Cenozoic rocks, but its thrust zone is in Paleozoic rocks which are undergoing faulting and uplift of an earlier belt. It is common, therefore, for a first-phase belt to be uplifted and the next phase of sedimentation to occur parallel to the first: over several hundred million years one orogenic belt will widen with both first and late phases of cordillera development occurring at the same time.

3. *Marginal fold belts*

Parallel with major orogenic belts, and on the continental side of them, sediment derived from the rising cordillera, or from epicontinental marine deposition, may be involved in predominantly open folds which lie upon rigid basement rocks which are usually of Precambrian or early Paleozoic age. Examples of such belts are the Jura Mountains of north-west Switzerland and central eastern France, marginal to the European Alps; the Saharan Atlas of northern Algeria lying south of the Plateau of the Chotts; the Sub-Andean fold belt of north-western Argentina, lying around the

margin of the Eastern Cordillera; and the foothills east of the Canadian Rocky Mountains.

The Sub-Andean folds are actively growing examples: they have been thrust-faulted as well as folded. Many anticlines have been eroded as they were folded so that strata outcropping on their flanks have steeper bedding dips than the topographic slope of the ground (Plate 4.24a), but where resistant beds form the outer stratum of the anticline characteristic triangular *chevrons* form the facets of the slopes between gullies. Valley floors are still being infilled with fan and stream deposits, from the rising ranges, and these are also being tilted in the continuing phase of uplift (Plate 4.24b).

The typical features of all marginal fold belts are: axes roughly parallel to the cordillera; trellis drainage with both superimposed and antecedent reaches along the transverse sections of river channels; homoclines forming the uplands facing inwards across breached anticlines; anticlinal strike valleys developed where weaker beds underlie the caprocks forming the outer strata of the anticlines; and, where there has been sufficient uplift, perched synclines form upland ridges because resistant strata are preserved in the downfold. As the severity of thrusting and uparching increases anticline flanks become steeper and more readily eroded.

4. *Platforms*

As a result of geosynclinal belt accretion, mostly in Precambrian times, platforms with sediment covers make up about three-quarters of the Earth's dry land surface. Sediment covers may be several thousand metres thick as in the inland delta of the Niger River, around Timbuktu, and the now emerged parts of the Paris Basin and south-eastern England. Folds in these sediments are different from those of the marginal belts because dips are generally of low angle—frequently being below 5°. Higher dips, up to 10°, are usually associated with faults, although locally steep dips can occur, as in the Isle of Wight.

Broad upwarping to form large, low-angle domes, or downwarping to form broad basins, is the characteristic of platform tectodynamics. Even slight changes of dip can be responsible for major changes in relief as erosion emphasizes lithological and structural features. Thus the chalk and limestone cuestas of the Paris and London Basins, and

4.24 (a) Steeply dipping sedimentary rocks forming the eroded flank of a young anticline in the Sub-Andean marginal fold belt of northern Argentina.

the quartzite cuestas of the Congo (= Zaïre) basin result from scarp and dip slope contrasts in very gently dipping strata. Beneath the sediments the basement rocks are subject to faulting and very broad warping rather than folding. Faults may not penetrate to the surface but broad draping of sediment across them influences landforms.

5. Shields

Shields are composed of brittle, rigid, granitic, gneissic, and associated rocks. They lack a sediment cover and consequently the faults and major joints becomes zones of selective erosion by the streams. Shields are not immobile, and upward or downward warping can occur over periods of several million or tens of millions of years to form broad sags or very large domes. Downwarping in the Kalahari and in eastern central Australia has led to accumulation of a relatively thick sediment

cover, and in Scandinavia rapid uplift is occurring as a result of isostatic recovery after depression under the load of ice in the last glacial period.

Because shields are ancient features which have suffered extreme bevelling of their rocks by erosion they are characterized by extensive erosional plains. These plains are involved in the broad warping which may occur, and so one surface may have a range of elevations which could lead to misinterpretation as several distinct surfaces (see Chapter 17). Cover beds, like those deposited in the Cretaceous Sea which transgressed across much of what is now the Sahara, may bury such erosion surfaces for a while and then lead to their exhumation as the sediment is removed in later erosional phases.

Elevations on shields may result from broad upwarpings, upfaulting of blocks, or more recent extrusions as at the rifted margins of the African and Indian Shields where plateau basalts have

4.24 (b) Chevrons on the flanks of an anticline in northern Argentina. Fans of debris from the rising fold are now being eroded.

flooded the ancient rocks. Large-scale upwarping may lead to deep erosion of the margin of a shield, as in Namibia and Angola, where Precambrian basalts, schists, gneiss, and metasediments are now being eroded to form a coastal plain backed by a large escarpment forming the edge of the African Plateau. Such an erosional belt has many features in common with an old orogen as erosion selectively erodes weaker rock units and stronger rocks stand as local uplands.

Conclusion

It is the internal forces in the Earth which create the major inequalities in its surface, and permit the forces of gravity to be exercised. Without these internal forces the Earth would have a uniform cover of ocean about 2600 deep and no relief. Tectonism and volcanism thus cause all major relief forms, but whether a landform is primarily

of structural origin depends on the scale at which features are considered (see Table 1.1).

It may be said that the formation of volcanic arcs and orogenic belts is the origin of the continents, and the subsequent development of cratons with platforms and shields is one of progressive stabilization, with the oldest orogens being now most stable. This is the fundamental characteristic of the evolution of continents.

It is evident, therefore, that at any site there is never a return to an initial state; there is only an evolution of landforms towards increasing exposure of the roots of orogenic belts, with periodic resurgence of belts by thrust-faulting, block-faulting, and uplift, and periods of sediment deposition in epicontinental seas or on continental platforms.

Further reading

The only modern text on tectonics and landforms

is that by Ollier (1981). Principles of structural geology are very well treated by Hills (1972) and in many other texts. The best treatment of structural geomorphology is that of Tricart (1974): this is a translation from a French text originally published in 1968 and so is essentially pre-plate tectonics in conception, but it is still worth close study because of its insights into the relationships between types of landform assemblages found on different structural units. Landforms associated with structural influences are also discussed by Twidale (1971). The study of Zagros drainage by Oberlander (1965) provides excellent examples of fold belt erosion patterns. The volumes referred to at the end of Chapter 3 are also relevant to this chapter. The 'Plate-Tectonic Map of the Circum-Pacific Region, Pacific Basin Sheet', published by the American Association of Petroleum Geologists (1982), is worthy of study for its display of relationships between tectonic processes and their results. The volume edited by Hsü (1982) has a useful collection of essays on many aspects of orogeny and mountain uplift.

5 Volcanic Landforms and Intrusions

There are on Earth about 60 000 volcanoes, of which at least 50 000 occur as seamounts on the floor of the Pacific Ocean, where they constitute up to 25 per cent of the oceanic lithosphere. There are only about 500 *active* volcanoes on Earth. By 'active' is meant that at least one eruption has been witnessed and recorded by people, although it is recognized that this is a very uncertain definition: Vesuvius, for example, was not known by Roman and Greek settlers to be a volcano until it erupted and destroyed Pompeii, Herculaneum, and Stabiae in AD 79. Most volcanoes have long periods between eruptions in which they are said to be *dormant*; those which have not erupted in historical times are said to be *extinct*; but this also is a very uncertain term because a volcano which has been inactive for 5000 years may still erupt. It is probably wise to regard all volcanoes which have erupted in the last 25 000 years as potentially active.

The term *volcano* includes both the vent through which gas and lava are erupted and the accumulation, usually in a cone form, of volcanic material around the vent. Large areas of the Earth's surface are constructed of volcanic material, notably the ocean floors and the great basaltic lava plateaus such as the Deccan Plateau of India, and the island of Iceland. The largest lava sheets were erupted during the break-up of Gondwanaland during Jurassic, Cretaceous, and early Cenozoic time (Fig. 5.1). Associated with these basalts are very thick sequences of dolerite intrusions, as sills and dykes. The Jurassic lavas and intrusions extend across central Antarctica and may represent an abortive splitting of Gondwanaland. The Cretaceous lavas are located along or near the margins of continental blocks which separated successfully. The volume of the lava expulsion of this time is remarkable. The Parana basalts of Brazil cover an area of 1 200 000 km². In southern Africa the basalts and dolerites have a thickness of up to 9 km and the lava field an area of 200 000 km². The Quaternary basalts of the Columbia and Snake plateaux (USA) cover 400 000 km².

About 62 per cent of the active volcanoes are on the rim of the Pacific Ocean in the circum-Pacific 'ring of fire' at convergent plate margins. Another 22 per cent are in Indonesia, 10 per cent in the Atlantic Ocean, and a few are in each of the Mediterranean-Middle East, the African rift-valleys, Hawaii, and other oceanic islands. The Pacific margin volcanoes are mostly in the island arcs of the Aleutians, Kuriles, Kamchatka, Japan, the arc of Melanesia, and the New Zealand-to-Tonga belt. Alaska, Central America, and Mexico have many volcanoes and a chain of them crowns the high Andes, but only seven volcanoes have been active in historical times in Canada and continental USA.

The interest of earth scientists in volcanoes is not only becuse of their importance as landforms or sites of igneous extrusion, but also because they form piles of readily dated rock and the origins of extensive blankets of fine-grained particles, known as *tephra*, which have been used to date many landforms and sedimentary deposits.

Energy of volcanic eruptions

An atomic bomb explosion releases roughly 10^{14} joules; a one megatonne hydrogen bomb 10^{16} J; the annual world electricity production is around 10^{18} J; the thermal energy in 1 km³ of lava is also about 10^{18} J, as is the world annual lightning flux. The largest known eruption—that of Laki in Iceland in 1783—had an energy of 10^{20} J (one million times the energy released by an atomic bomb), and a typical violent eruption of a single conical volcano may have an energy of 10^{12}–10^{15} J. Very large eruptions which produce great sheets of glowing

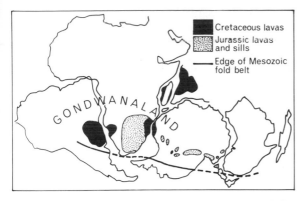

Fig. 5.1 The distribution of Mesozoic lava sheets and dolerite intrusions in Gondwanaland. The Jurassic lavas are approximately parallel to the margin of the Mesozoic fold belt, and the Cretaceous lavas are near separated continental margins (data from Cox, 1970; Smith and Hallam, 1970).

ash clouds may have energies far greater than that released at Laki.

The energy for volcanic activity is derived from the heat of magma deep in the crust. As the average temperature increase with depth into the crust is only 1 °C every 33 metres, and molten lava has a temperature of 1000 ° ± 200 °C, the magma must rise from a considerable depth—even in areas of thin crust above a hot-spot.

Whether energy is released as a quiet discharge of heat from a lava flow, or explosively, depends upon the viscosity and gas content of the magma. The viscosity of magma depends upon its composition (Fig. 5.2), pressure, and temperature. An increase in silica content produces a marked increase in the viscosity of a liquid. Basalts have a silica content of approximately 50 per cent and a viscosity of about 10 N s/m²; andesites have values of ~ 60 per cent SiO_2 and 10^2 N s/m²; granites and rhyolites 70 per cent SiO_2 and 10^4 to 10^5 N s/m². Dissolved water content of a magma also has a very great effect upon viscosity: the greater the dissolved water content the more fluid is the lava. The greatest single control, however, is the temperature of the lava. Basalts are erupted at temperatures of 1000–1200 °C, but rhyolites at about 800 °C, and as a result basalts have viscosities at least 1000 times lower than those of rhyolites.

Gas within a magma is mainly composed of water vapour, carbon dioxide, and various minor components such as hydrogen sulphide, chlorine, and hydrogen chloride. The gas is held in the magma by the confining pressure in a magma chamber. When dissolved water in the magma changes from the liquid to the gaseous state because of a decrease in pressure, its volume increases instantly one thousand times—it is this instantaneous expansion which causes explosive eruptions. If magma rises quietly to the surface, the gas may be released gradually and escape, either from large bubbles breaking on the surface of a red-hot lava pool, or by frothing the lava into a fountain of molten rock which falls back into the pool or builds up a cone of vesicular cooling lava blocks around the pool. In other volcanoes the gas may be released explosively as lava rises into zones of lower pressure. Explosive eruptions may be dominated by the gas release, which fragments the rock forming the *pipe*, or *conduit*, up which lava rises, frothing the lava and hurling it high into the atmosphere, or blasting it as a glowing cloud of molten rock, ash, and gas into the air and down the flanks of the volcano. In the 1906 eruption of Vesuvius, for example, gas was blasted from

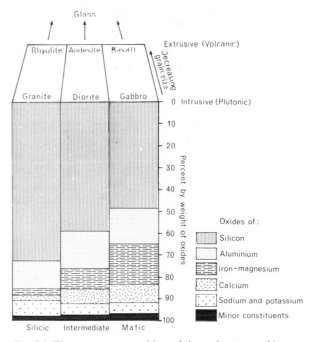

Fig. 5.2 The average composition of the major types of lava (basalt, andesite, and rhyolite) and their intrusive counterparts (gabbro, diorite, and granite). There is a systematic increase in silica and a decrease in iron and magnesium oxides in the transition from basalt to andesite to rhyolite (after Press and Siever, 1974 © W. H. Freeman & Co. All rights reserved).

a 500 m wide crater to a height of 13 km in an explosive phase which lasted for 18 hours. The eruption enlarged the crater to a diameter of 600 m.

The eruption of Agung, on Bali, in 1965, provides a good idea of the power of a rather large eruption: 5000 million tonnes of material was erupted; of that, 10 Mt of fine ash, 120 Mt of water vapour, 9 Mt of sulphuric acid, and 1 Mt of hydrochloric acid were blasted up into the stratosphere.

Some eruptions are made more violent because molten lava comes into contact with water under low pressure, as in a lake or shallow sea. The tremendous explosion of Krakatau, in the Strait between Java and Sumatra, in May 1883, may have resulted from contact between sea water and lava, or it may have been an ignimbrite eruption (see below). The explosions were heard 3200 km away in central Australia and 4800 km away in Mauritius. Ash from the eruption stayed in the atmosphere and passed around the world for many months. Most devastation was caused by the enormous sea wave, called a *tsunami*, which was generated by the explosion (see Chapter 13 for a discussion of hazards). When eruptions take place beneath deep water, as along mid-ocean ridges, the high pressures induced by the overlying ocean mass may prevent eruptions from being explosive.

Volcanic products

Lava is different from its parent magma because of its loss of gas and because it is chilled on contact with the ground and atmosphere. Very rapid chilling produces glass in the form of obsidian, and as the rate of cooling decreases, first fine and then larger mineral crystals are formed in the rock (Plate 5.1).

Lava flows

Lavas flow down hill at velocities as high as 100 km/hour, but more commonly of a few km/hour for basalts and a metre of so per hour for viscous rhyolite flows. The distance they travel is controlled by lava viscosity and slope of the ground: lava streams extending more than 30 km have been witnessed in historical times. On flat terrain basalt issuing from extensive fissures forms thin sheets which may accumulate from successive flows to form lava plateaus, but more commonly a flow is a relatively narrow and thin feature which cools most rapidly along its margins, which consequently solidify, and leaves a central zone of the flow which is most active. Rhyolite flows are so viscous that they form thick, bulbous deposits and domed volcanoes, rather than long thin flows.

Basaltic lavas have surface features named from Hawaiian examples. *Pahoehoe* (pronounced 'pahoyhoy') is a highly fluid lava that spreads in sheets. A thin, glassy skin forms on the surface of the hot flow and is dragged into ropy folds by the glowing lava beneath. *Aa* (pronounced 'ah-ah', after the agonized noise made by someone walking barefoot on its rough surface) is slower moving with a thick skin broken into jagged, clinkery blocks forming on the surface and steep front of the flow. Single flows frequently grade from *pahoehoe* near their source to *aa* downstream, as the flow loses gases and so becomes more viscous. On some flows in Iceland, *aa* blocks are the size of a house, but elsewhere more like the size of a football. Local relief on a flow is usually several metres.

As gas is released from lava, *vesicles*, or gas cavities, may be formed in the cooling rock to form sponge-like *scoria*. Extreme vesicle formation creates pumice of such low density that it floats in water; many lake and coastal, and some submarine eruptions, result in great rafts of pumice on the water surface and the pumice may float for several months. *Xenoliths*, foreign inclusions of rock, from the eruption fissure, or greater depths, may be carried along be a flow.

As a thick lava flow cools and contracts, shrinkage joints form. This is not a rapid process. It has been calculated that a flow 1 metre thick will cool from 1100 °C to 750 °C in 12 days; if it is 10 m thick this cooling will take 3 years, and if 100 m thick 30 years. To cool to air temperature a very thick flow will take thousands of years. In very thick flows the cooling joints tend to be more widely spaced, regular, and columnar, than in thinner flows, while at the margins of flows joints are more closely spaced and the rock is glassy or fine grained. Joints tend to form at right angles to the flow surface, so irregularly shaped flows have irregular joint patterns (Plate 5.2). *Lava caves* or *tubes* form within flows where more fluid lava continues to drain from within a solidifying flow. Tubes in thick flows may later become important reservoirs for ground water.

5.1 Volcanic materials. (a) Basalt with large, round gas vesicles. (b) Rhyolitic pumice with small vesicles and elongated glassy crystals. (c) Volcanic glass from a rhyolite flow. (d) Volcanic breccia with angular basalt fragments in a matrix of fine-grained tuff. (e) Pahoehoe from Mauna Loa. (f) Aa from Mauna Loa (photos (e) and (f) by R. M. Briggs).

5.2 (a) Regular columnar jointing in dolerite, Transantarctic Mountains.

Pyroclasts

Explosive volcanism, especially from rhyolitic and andesitic eruptions, produces large volumes of shattered, fragmented rocks which are collectively known as *pyroclasts*. The fragments are classed according to size: material with diameters smaller than 4 mm is termed *ash*, and in its compacted form *volcanic tuff*; fragments of 4–32 mm are termed *lapilli* and larger than 32 mm are *blocks*. A collective term for an air-fall deposit containing a mixture of fragments of these sizes is *tephra*. Blocks which originated as lava and have been thrown into the air while still molten become rounded or decorated with spiral patterns as they travel; such rounded and patterned forms are called *volcanic bombs*. Consolidated blocks are known as *volcanic breccia*.

Tephra deposits fall around and downwind from an eruptive centre. Close to their source the tephra mantle is relatively thick and it thins away from source (Fig. 5.3). The deposit from a single eruption is usually well bedded with the largest pyroclasts at the base of the bed and a progressively finer grain size towards the top. This sorting results from the projection of the smallest fragments to greatest altitudes where they stay suspended for much longer than coarse material, which falls first and closest to the source. Because it mantles the landscape a tephra dates all of the landforms it falls on as being 'pre-that eruption'. All deposits formed later overlie the tephra and are known as 'post-that eruption'. In areas like northern New Zealand and Japan where many tephra units are recognized they are used for precise dating of the landscape and of inter-bedded deposits. The dating by tephra is known as *tephrochronology*.

Eruptions which produce very large quantities of gas may give rise to glowing clouds of ash, pumice, and larger pyroclasts which move downslope at velocities in excess of 100 km/hour. Glowing avalanches are known by their French name as *nuées ardentes* (Plate 5.4). The processes of emplacement of pyroclastic flows and ignimbrites are discussed below.

Volcano types

Fissure eruptions are probably the most widespread on Earth because they occur along mid-ocean ridges. Each fissure becomes filled with a

5.2 (b) Rhyolite flows of Mayor Island, New Zealand. The dark band below the middle flows is obsidian, chilled by contact with the cold lower flow. Joints are irregular (photo by M. Buck).

dyke of lava and fluid basalt. On land lava and ash flow material spread out and may cover the fissure, which is then marked only by a row of small scoria cones (Plate 5.5).

Central vent, or *pipe*, eruptions form the most distinctive volcanic features—*cones* and *domes*. Cones may be built of lava alone, scoria, or a combination of lava flows and scoria (Fig. 5.4).

Lava cones are built by successive lava flows. Free-flowing basalt in large quantities can build up a broad *shield volcano* many tens of kilometres in circumference and more than one kilometre

high. Mauna Loa on Hawaii is the classic example of a shield volcano: it rises 4 km above sea-level, or 10 km above the ocean floor, and has a basal diameter of 100 km. The slopes of such volcanoes are gentle at 6–12° and formed of thousands of individual lava flows, each only a few metres thick and of limited extent. The island of Hawaii is formed by the overlapping of several shield volcanoes. Where the lava supply is limited small lava cones form. On the flanks of a large volcano it is common for small cones to develop where subsidiary vents have reached the surface. Rhyolite

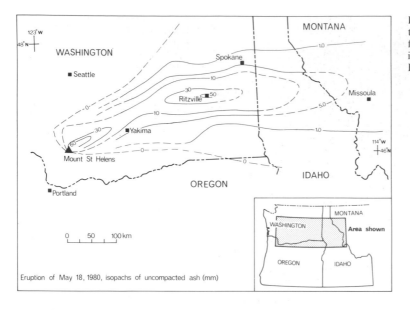

Fig. 5.3 Isopachs (i.e. lines of equal thickness of tephra) of the deposits from the eruption of Mount St Helens in 1980 (redrawn after McKnight, Feder, and Stiles, 1981).

Eruption of May 18, 1980, isopachs of uncompacted ash (mm)

Fig. 5.4 Types of volcanic cones and domes.

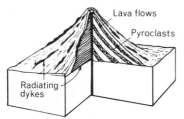

(a) Strato-volcano

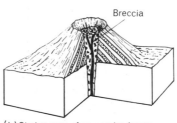

(b) Cinder cone of successive layers

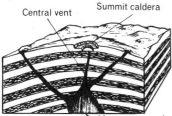

(c) Shield volcano

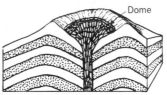

(d) Lava dome within cone of pyroclasts

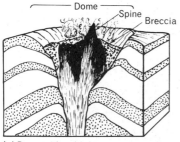

(e) Dome with spine

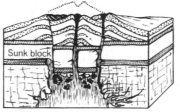

(f) Caldera

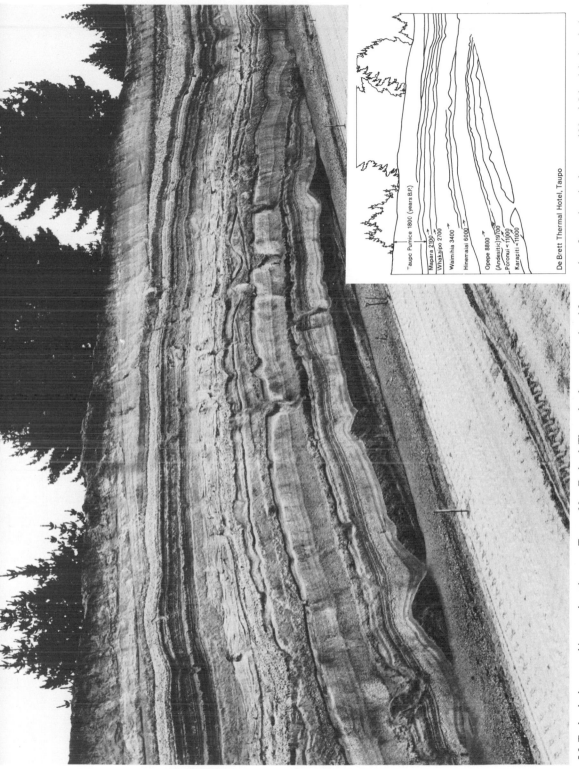

5.3 Tephra beds exposed in a road cutting near Taupo, New Zealand. The tephra have accumulated from a series of eruptions in the central North Island during the last 11 000 years. Each bed is named from its source and the ages since each eruption are given in the inset diagram. The uppermost bed resulted from the Taupo eruption 1800 years ago (photo by H. S. Gibbs).

Taupo Pumice 1800 (years B.P.)
Mapara 2100
Whakaipo 2700
Waimihia 3400
Hinemaiai 6000
Opepe 8800
(Andesitic) 9700
Poronui < 10,000
Karapiti <11000

De Brett Thermal Hotel, Taupo

(a)   (b)   (c)

5.4 Pyroclastic eruptions of Ngauruhoe volcano on 19 February, 1975. A period of gas-streaming lasting 1.5 hours was followed by a series of cannon-like eruptions which threw blocks at least 2.8 km from the vent. Initial ejecta velocities of 400 m/s were caused by steam pressures as magma heated confined meteoric water. In (a) can be seen the rising cloud of ash and gas above the vent, and dust rising on the flanks as blocks fall on them. (b) About 10 seconds later pyroclastic avalanches form as the eruptive column collapses and debris streams down the flanks at velocities of 30 m/s. (c) After about ten minutes the eruptive column had risen to an altitude of 10 000 m above the crater and pyroclastic avalanches had reached the foot of the cone. At least 3.4 Mm^3 of pyroclastic material was erupted. This is the equivalent of about 2.0 Mm^3 of dense rock. The largest block erupted had a mass of 3000 tonnes and was thrown 100 m from the rim of the vent (photos by R. B. Pillans).

5.5 Fissure eruption of basalt, Erta'Ale Range, Ethiopia (photo by H. Tazieff).

domes have steep sides and limited lateral extent because of the high viscosity of their lava (Figs. 5.5, 5.6; Plates 5.6, 5.7).

Scoria or *cinder* cones are built of fragmental material explosively ejected from the central vent. The profile of the cone is determined by the maximum angle at which the scoria blocks come to rest—known as the *angle of repose*. This angle is often in the range of 30–40°, and is steepest near the rim of the cone where the larger blocks fall (Plates 5.8, 5.9). Fine particles are carried further from the vent and may form a gently sloping apron at about 10° around the scoria cone. Scoria cones may build rapidly: Parícutin, in Mexico, erupted in 1943 and built a cinder cone 150 m high in its first six days of activity. It is now 400 m high and dormant.

Strato-volcanoes or *composite cones* are amongst the most common forms for large volcanoes, especially those of andesitic composition.

They are built of interbedded lava flows and beds of pyroclasts. Fujiyama, Vesuvius, Etna, and Stromboli are of this type. The slope of the flanks is largely controlled by the angle of repose of the pyroclasts, but intrusions of dykes, the formation of flank fissures and *parasitic cones* may create irregularities in the straight or concave slopes. Many large volcanoes have a long history of eruptions and may have several vents and craters as well as collapse structures. Consequently they have a complex form (Plates 5.10, 5.11).

Craters form the summits of most volcanoes. They are partly the upper end of the vent up which the lava rises and overflows, and partly the result of collapse of the vent walls as the magma sinks in the vent after an eruption. In explosive eruptions the upper part of the vent may be blasted away and the resulting pyroclasts fall back into the widened crater so that it has inward facing slopes of 30–40°. Because crater walls are steep, they may collapse inwards or be eroded so that the crater has a diameter several times that of the vent. Volcanic breccias may be formed in the upper vent from pyroclasts and cooling lava.

Calderas are huge collapse structures formed near the summits of very large volcanoes when a magma chamber drains and leaves an unsupported roof (Plate 5.13). Large semi-circular faults develop at the edges of the unsupported roof and the central rock mass collapses into the chamber. The subsidence may be catastrophic or gradual, with step formation of faulted blocks and a flat floor to the caldera. Small cones and dyke intrusions may be formed within the basin after the collapse. The smallest calderas have a diameter of 2 km (by definition) and the largest known on Earth is 70 km long and 45 km wide and largely buried by the ignimbrite sheets of the Yellowstone Plateau. Very large calderas may be expected from ignimbrite eruptions because they produce 1–100 km³ of ejecta in a few hours or days.

Maars are rare, shallow, circular pit-like craters, often about 1 km or less in diameter, surrounded by a thin sheet of tephra or a low tuff ring. They are believed to have been formed as gas-charged magma melts and explodes its way from the upper mantle to the surface where it ejects debris from the vent with enormous force, perhaps at supersonic velocities. The term 'maar' is the German word for lake; in the Eifel district of the German Rhineland many maars are filled with water (Plate 5.14).

A *diatreme* is the consolidated rock formed in a vent after explosive activity. Much of the rock is a breccia formed from the vent walls and fragmented lava. Diatremes formed below maars, however, are especially interesting because they may contain upper mantle rock, including diamonds. The famous Kimberley mines are in such rock.

Domes of several kinds may be formed by viscous degassed lava, usually of rhyolite or andesite. Slow-moving bulbous flows can build up large domes several kilometres in diameter and several hundred metres high (Plate 5.7). Much smaller domes may form within craters as rounded *tholoids* (Plate 5.15) or more irregular *cumulodomes* or *plug domes*. The plug domes are often cylindrical masses of lava which may have irregular surfaces and from which spine-like extrusions may project. Spines are produced as gas pressure below a cooling plug of lava forces it upwards out of the vent. The largest known is the 300 m high spine of Mont Pelée.

The surface of a growing dome is constantly stretched by intrusions of lava within it, so that its surface may be ruptured and fissures formed, from which more lava may flow or pumiceous pyroclas-

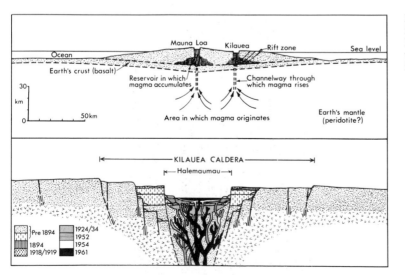

Fig. 5.5 The basalt shield volcanoes forming Hawaii, and a section through Kilauea Caldera, the central crater of which is shown in Plate 5.12 (modified from Peterson, 1967).

5.6 Rangitoto, a small basalt shield volcano in Auckland harbour, New Zealand. A cinder cone forms its steep summit (photo and copyright: Whites Aviation Photo).

5.7 Mount Tarawera, New Zealand: a rhyolite volcano of three domes formed about 900 years ago. The domes are created by expansion of a bulbous lava mass. The rift along the crest is a line of craters formed in the eruption of 1886, which ejected basaltic scoria which forms a thick cover over the domes (photo by D. L. Homer, NZ Geological Survey).

5.8 A section through the flank of a small scoria cone, showing the bedding.

5.9 The crater of Ngauruhoe showing pyroclasts and emissions of water vapour.

5.10 Mount Egmont, showing the lava flows forming resistant caprocks along spurs and the weaker beds of pyroclasts below them. The parasitic cone of Fanthams Peak is visible on the left flank: it was formed about 3500 BP. The flanks of Egmont are older than this. A tholoid occupies the summit crater (photo by D. L. Homer, NZ Geological Survey).

5.11 A chain of andesitic strato-volcanoes in the central North Island, New Zealand. In the foreground are old craters of the extinct Tongariro volcano. The symmetrical cone of active Ngauruhoe is in the centre, and the cone of quiescent Ruapehu is in the distance. Ruapehu has a hot lake just below its summit. An extensive ring plain of lahar deposits surrounds Ruapehu (photo and copyright: Aerofilms).

tics be ejected. Fragmental material from such eruptions accumulates around the dome and may nearly bury it (Fig. 5.6).

Water and volcanic activity

Magma may contain up to 4 per cent water by weight. Much of this water is lost during crystallization to form igneous rocks, and it has been calculated that the *juvenile water* from an igneous crust, 40 km thick, around the Earth, could produce all of the water now in the oceans. Most water associated with volcanic phenomena is not juvenile but *meteoric*, from the atmosphere. In humid climates it frequently fills craters, calderas, and maars: it is associated with *phreatic*, steam and mud, explosions and as groundwater it is heated and carried to the surface to form mud pools, and boiling-water pools which may periodically erupt as *geysers*.

Hot springs, geysers, steam vents, called *solfataras*, and gas vents, called *fumaroles*, may all deposit precipitated minerals in layered deposits around the pool or vent. These deposits, called *sinter* or *travertine*, have varied compositions but may include silica, sulphates, sulphur, and other

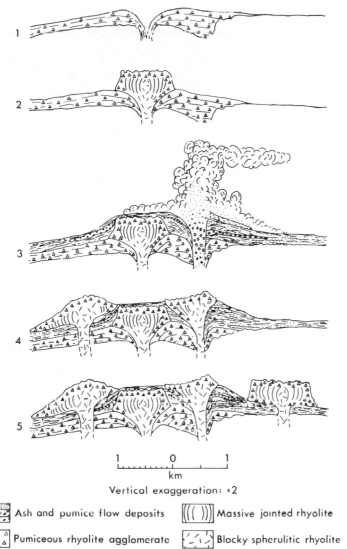

Fig. 5.6 A suggested sequence of events which were responsible for the formation of Mount Tarawera, New Zealand, see Plate 5.7 (after Cole, 1970).

minerals. They often form low mounds and, in a few places, flights of terraces down which the overflowing water discharges. Of greater general importance are the mudflows of volcanic debris called *lahars*.

Lahars most commonly occur when a crater lake or a lava-dammed lake suddenly overflows. This may be associated with an eruption or collapse of a dam, heavy rain, snowmelt, or mixing of a nuée ardente with lake water. The result is a rapid flow of mud and volcanic debris: velocities of flow may be up to 90 km/hour, and distances of travel are commonly 10–15 km and up to 180 km. The Osceloa mudflow of Puget Sound (USA) is a particularly large lahar deposit; it has a thickness varying from 1 m to 105 m and extends over an area of 320 km²; it is about 5000 years old. Lahar deposits with tephra beds form extensive hummocky *ring plains* around many strato-volcanoes. The hummocks are formed of large irregular blocks left as the more mobile mud thins and drains away (Plate 5.16a,b).

The largest lahars are the *jökulhlaups* of Iceland. These notorious flows are produced by eruptions

5.12 An aerial view of the second phase of an eruption on 15 November 1967, in Halemaumau Crater, Kilauea Caldera, Hawaii. Depth to the crater floor is about 120 m. Note the cliffs, of an older and larger caldera, at the top right (cf. fig. 5.5) (photo by A. T. Abbot, Hawaii Institute of Geophysics).

beneath shield volcanoes such as Vatna Jökull and Katla, and also by bursting of ice-dammed lakes. Recorded jökulhlaups often last for a week and discharge as much as 7 km³ of water—much more than the Amazon—in that time.

Types of eruption

The characteristics of an eruption have a major influence on the size, form, and nature of a volcano and the material it produces. In addition to the volume of solids ejected, the volume of gas and the rate of gas release largely control the degree of fragmentation of the magma and the area over which tephra is distributed. Styles of volcanic explosion have been distinguished by the volcanoes from which they were recognized, or in one case (Plinian type, after Pliny the Younger) by the name of the original describer.

Volcanic explosions

Hawaiian eruptions form where basic basaltic lavas are exuded regularly, with very little explosive activity, and produce flows or lava sheets.

5.13 The collapse caldera of Mayor Island, New Zealand (photo and copyright: Aerofilms).

5.14 A maar: Pulvermaar, Eifel, Germany (photo by courtesy of C. D. Ollier).

5.15 (a) The volcanic neck of Iharen in the Hoggar Mountains, central Sahara, with well-developed columnar jointing in its basalt. The columns are about 2 m in diameter, and the neck stands about 150 m above the desert plain. The neck is possibly the remains of a feeder pipe, which has been forced upward as a plug before the surrounding rocks were eroded.

Limited explosive release of gas may create cinder cones, seldom exceeding 300 m high at central or parasitic vents, or spatter cones of lava globules up to 100 m high as a result of *fire fountains* developed during lava release. Tephra are thrown only a few hundred metres into the air in some eruptions, and a few kilometres in more violent ones; consequently tephra are limited in their distribution to the immediate flanks and margin of the volcano by the fluidity of the lava and consequent ready escape of gas.

Strombolian eruptions, named after the 'lighthouse of the Mediterranean', are remarkable for the regularity and frequency of basaltic eruptions of mild to moderate explosiveness. Being considerably less fluid than Hawaiian lavas, the Strombolian lavas congeal in the vent, forming a crust which is fragmented during the next eruption, when debris and bombs of fresh lava are thrown into the air and fall onto the rim, or back into the crater. Fire fountains and white luminous eruption clouds are characteristic. Continuous subsidence of the volcano base, with inward seepage of sea water along fissures, may explain the regularity of the eruptions of Stromboli. The mildness of the eruption may be explained by magma rise rates of only a few metres per second and slow gas release from bursting bubbles in the lava.

Vulcanian, or *Vesuvian*, eruptions are named after the island of Vulcano in the Lipari Islands. Vulcano is presently quiet but Vesuvius is active. Periods of quiescence are terminated by the emission of gases and then the formation of small cinder cones on the crater floor. Sporadic emissions of basaltic, or andesitic, lavas take place over

5.15 (b) A section through a symmetrical tholoid above a parasitic vent on the flanks of Mount Erebus, Antarctica. Note the radial pattern of joints.

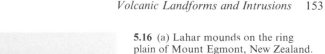

5.16 (a) Lahar mounds on the ring plain of Mount Egmont, New Zealand.

5.16 (b) Sections through lahar and tephra deposits forming the ring plain of Ruapehu, New Zealand.

several decades until the sizeable crater is filled, then a culminating blast again clears the crater. The resulting volcano has a cone which is largely formed of tephra and large blocks, with some lava flows which have escaped from the crater after explosions have shattered its walls. Vulcanian eruptions produce large volumes of fine dust and rock fragments which form many nuées ardentes on the flanks of strato-volcanoes. A typical eruption functions as a rock-crusher and dust-making machine. The eruptions of Ngauruhoe in 1975, and Mount St Helens in 1980, are typical examples of a common type which occurs somewhere around the rim of the Pacific each year. The period of quiescence is one in which gas saturates the magma in the conduit, until sufficient pressure builds up and the lava cap is disrupted by numerous discrete explosions, as the gas comes out of solution or groundwater is vapourized.

Plinian eruptions are highly explosive, with huge billowing, dense clouds of gas and tephra being propelled upwards to 5–60 km above sea-level. The volume of the explosions is usually great enough to disrupt the structure of a strato-volcano, so flank fissure eruptions, crater wall destruction, and fragmentation of solidified lava masses are common. The power of the eruptions is provided by the disruption, at depth in the conduit, of a considerable volume of juvenile magma by released gases, and the exit of gas and fragments from the crater with velocities of several hundred metres per second.

Plinian eruptions have a range of magnitudes so that it is useful to distinguish Subplinian, Plinian,

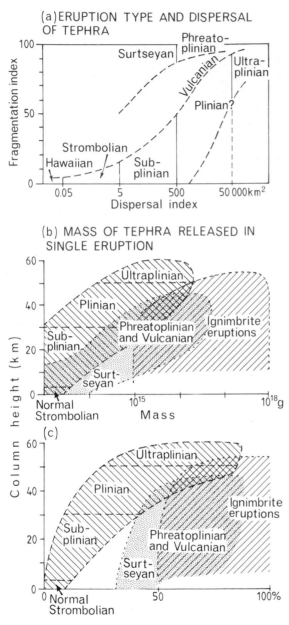

(a) ERUPTION TYPE AND DISPERSAL OF TEPHRA

(b) MASS OF TEPHRA RELEASED IN SINGLE ERUPTION

(c)

Fig. 5.7 (a) A plot of the fragmentation index (= weight per cent of tephra, finer than 1 mm, enclosed by the isopach which delimits a thickness of pyroclasts which is 1 per cent of the maximum thickness of the unit) against a dispersal index (= the area enclosed by the same isopach) (after Walker, 1980). (b, c) Plots of fine ash and dust release against height of the eruptive column for different types of eruption: b, mass of fine ash and dust released in individual eruptions; c, ash and dust as a percentage of the total mass erupted in individual eruptions (after Walker, 1981).

and Ultraplinian types in an ascending order of power, mass of material ejected, altitude to which ejecta are carried, and area on which they fall (Fig. 5.7). Subplinian eruptions overlap in power with many which are classed as Vulcanian. Ultraplinian types are, fortunately, extremely rare and that of Taupo (central North Island, New Zealand) in about AD 130, has been recognized as a characteristic example.

Complex eruptions are very common with several kinds of explosion occurring in sequence. On Vesuvius, for example, the pattern commonly has three stages. First there is the typical Vulcanian stage with fracturing of the cone and the pouring out of lava; then there is a Plinian stage with violent explosions blasting the lava into a cloud of fine ash, and the scattering of bombs; finally there is another Vulcanian stage as the crater walls collapse into the vent, resulting in blockage of the pipe and more violent explosions.

Phreatomagmatic eruptions are characterized by the entry of water in copious quantities into a vent which is submerged in the sea, a lake, under a glacier, or low-lying country with large rivers. Basaltic phreatomagmatic (called *Surtseyan*) eruptions are particularly common and produce far larger quantities of pyroclasts than other basaltic eruptions, but the ejecta are not driven up into the troposphere and they fall out of the eruption column fairly close to the source.

Phreatoplinian explosions are far more powerful and are characteristic of andesitic and rhyolitic magmas in contact with large volumes of water. They eject large volumes of pyroclasts but these are commonly washed out of the sky by rain produced from the erupted water.

Pyroclastic flow eruptions and ignimbrites

Pyroclastic flows are of several types: block flows, nuées ardentes, pumice flows, and ash flows. All are of high temperature, but they range over about eight orders of magnitude in volume, from the nuées ardentes of strato-volcano flanks with angles of rest of 15–25°, to huge pyroclastic flows deposited to form ignimbrites, with volumes of tens of cubic kilometres and nearly horizontal attitudes, discharged from massive calderas (Plate 5.17). The term 'pyroclastic flow' is now usually reserved for the actively moving flow and the term 'ignimbrite' for the resulting rock body, whether it is welded or not.

5.17 The surface of an ignimbrite sheet forming the Kaingaroa Plateau, New Zealand. Gully dissection is most evident in areas cleared of pine forest (NZ Forest Service photo by J. H. G. Johns, Crown copyright reserved).

The smaller ash flows usually develop during Vulcanian and Plinian eruptions as a result of a very high velocity (60–200 m/s) of gas flow and ejection through the vent, which carries tephra and gas several kilometres into the atmosphere. The high temperature of the gas cloud is maintained by transfer of heat from entrapped pumice and dust, so that thermal expansion of entrapped air causes the cloud to inflate and rise hundreds to thousands of metres by convective buoyancy. Parts of the cloud that are denser than air, either because of included rock fragments or because of loss of heat, collapse and flow at high velocities down the flanks of the volcano, where they are guided along valleys. Continuing loss of gas and dust from the pyroclastic flow may cause a blast, or surge, to precede the ground-hugging and valley-confined flow, and its dust may settle out as the first and bedded unit of a flow. The ground-hugging unit moves largely as a non-turbulent (i.e. laminar) plug flow, because of its density, and settles as a layer which may be a few centimetres to 100 metres thick. A highly gas-charged upper part of a flow may move as a turbulent flow over the laminar plug flow-mass below it. The fine dust that is carried to high altitudes by expanding gas will travel great distances and settle out as the last unit of a pyroclastic flow deposit. The resulting ignimbrite, consequently, has three recognizable units (Fig. 5.8) which may be underlain by an earlier

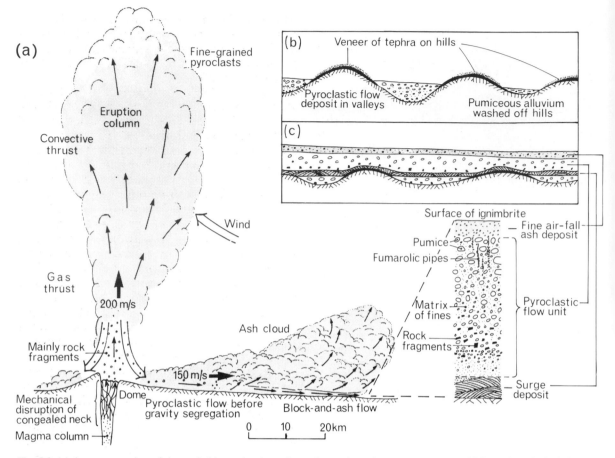

Fig. 5.8 (a) A representation of the probable mechanism of eruption and emplacement processes which produce ignimbrites. (b) Pyroclastic flow deposits of an indented relief. (c) Formation of plateau ignimbrites.

blocky Plinian unit formed as the vent was cleared. With increasing distance from the vent, first the Plinian deposit, then the surge deposit, and then the main flow unit decrease in volume and die out, so that at greater distances only a fine ash is detectable.

The distance travelled by each unit depends upon the explosive power of the parent eruption and the volume of ejecta. Very powerful Plinian and Ultraplinian eruptions drive material to such great altitudes that, as pyroclasts fall back to the ground, they reach terminal velocities and may develop energies capable of carrying ash flows considerable distances from the vent. The potential energy from the collapse of an eruption column, or the kinetic energy from the explosion

of a lava dome, is adequate to account for the distance travelled by small ash flows and avalanches. Intermediate to large flows have a greater mobility which is probably a result of a fluidized condition in which tephra are supported by an upward stream of gas which may be trapped within the flow mass, or may be derived from exsolution of gas from glass fragments in the flow. The Taupo pumice eruption of AD 130 was so powerful that an estimated 100 km³ of pumice and ash was produced from the vesiculation and fragmentation of about 20 km³ of magma. The eruption column probably reached an altitude of 50–60 km, with a volume of material equivalent to 1 Mm³ of magma being ejected per second. The great height of the column caused the thickest

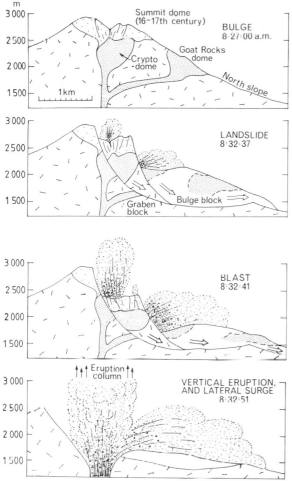

Fig. 5.9 Changes in the profile of Mount St Helens during the morning of 18 May 1980 (drawn from figures and photographs in Lipman and Mullineaux, 1981).

deposits to be formed 20 km downwind from the vent, and the energy of the ash flow was great enough for the flow to rise over 1500 m up slopes over 40 km from the vent. The capacity of flows to cross rough terrain and to leave only thin veneers of debris on the tops of hills, but thick infills in valley floors, has created very irregular distributions of some ignimbrites.

Welding of ignimbrites occurs rapidly after emplacement while the temperature is still high. The minimum welding temperature in a rhyolitic pumice unit 10–40 m thick is $600 \pm 25\,°C$. Non-welded units of comparable thickness are presumed to have been emplaced at cooler temperatures.

Much smaller than the Taupo eruption is that of

Mount St Helens which was so well observed, compared with most eruptions, that it provides much evidence for the complexity of events and products of one eruption (Figs. 5.3, 5.9, 5.10; Plates 5.18, 5.19).

Mount St Helens is a volcano in the Cascade Range of western USA. Its eruptive history began about 40 ky ago with dacitic volcanism, and continued with numerous explosive periods separated by dormant periods hundreds to thousands of years in duration. About 25 ky ago the rock type had changed to andesite and, on at least two occasions, basalt. During most eruptive periods, pyroclastic flows and lahars built fans of fragmental material around its base, and partly filled the valleys around it. Most pyroclastic flows terminated within 20 km of the cone, but lahars extended as far as 75 km down valleys.

Mount St Helens had been dormant since 1857, but became active again in March 1980 with steam blasts and earthquakes. Major swelling of the north flank indicated magmatic intrusion into the volcano. On 18 May, an earthquake of magnitude 5 caused the north flank to bulge and collapse as a 2.3 km³ landslide. Hydrothermal steam eruptions then formed a northward directed blast that devastated an area of 600 km². These events triggered a 9 hour dacitic magmatic eruption that drove a Plinian column more than 20 km high, producing ash fall-out for more than 1500 km to the east, and ash flows down the flank of the cone. Catastrophic lahars, mudflows, and floods were generated by rapid melting of snow and ice.

Small eruptions occurred during the following months and successive domes formed in the crater as leaky plugs to gas emissions, only to be fragmented by the next eruption. In 1983 a dome still caps the vent.

Volcanic centres

The life-span of a zone of volcanism may be 1–50 My, with great ages presumably being related to areas overlying mantle plumes as in the African Rift zone. Many major centres of volcanism may have shorter histories. The Taupo Volcanic Zone of New Zealand, for example (see Fig. 4.12f) probably developed about 2 My ago as a result of backarc crustal extension. Spreading rates are now 7 mm/y, and if this is an average rate, and if this is compensated for by dyke intrusion into the upper 10 km of crust, the intrusion of 8400 km³ of

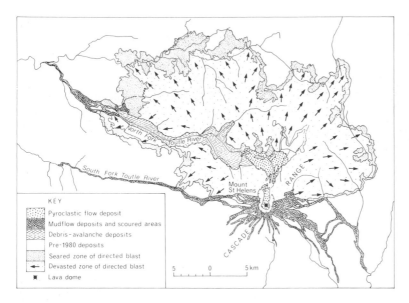

Fig. 5.10 Effects of Mount St Helens eruption of 18 May 1980, on surrounding areas. Note the shape of the directed blast zone, resulting from the landslide, and the spread of lahars along old valleys (based on Lipman and Mullineaux, 1981).

5.18 Mount St Helens, after the 1980 eruption, with a dome in the centre of the breached crater (photo by R. M. Briggs).

5.19 Terraces of pyroclastics and lahar deposits, in the valley of the Toutle River, after the 1980 eruption of Mount St Helens (photo by R. M. Briggs).

magma is required. Much of the volcanism in the Zone is rhyolitic, implying that basic basaltic magmas have partially melted granitic crust and wet subducted sediment and replaced it with rhyolite and andesite: This mechanism may have produced a further 4200 km^3 of magma—the volume of rock occupying the 2.1 km depth to subvolcanic basement. In the last 50 ky about 360 km^3 of tephra, 350 km^3 of ignimbrite, and 55 km^3 of lava have been produced. The massive pyroclastic flow eruptions have built up extensive flat plateaux of welded ignimbrite located around collapse calderas, which presumably form above the large magma chambers emptied in Ultraplinian and large Plinian ignimbritic eruptions. The huge pyroclastic, flow-generating explosive events have not formed strato-volcanoes because the power of the blasts, and of the flows, has carried ejecta away from the vents. The most powerful of all volcanic eruptions have the least impressive-looking results—a nearly horizontal plateau of ignimbrite around a broad central depression (Fig. 5.11).

Many strato-volcanoes, as in Japan, have histories of eruption of 10–100 ky. In their lives they usually pass through a dacitic to andesitic phase with pyroclastic eruptions dominating, then, as the ejecta become increasingly rich in SiO_2, lava flows become more nearly dominant. Further enrichment in SiO_2 may lead to more rhyolitic and pyroclastic eruption forms and the development of calderas.

A limit to the duration of a volcanic centre is imposed by the constancy of the originating mantle plume, hot-spot, or subducting plate. The altitude which can be reached by an individual volcanic cone is controlled by the strength of the crust on which it stands. Segments of crust bearing oceanic volcanoes become isostatically depressed and the volcanic mass sinks. Continental crust

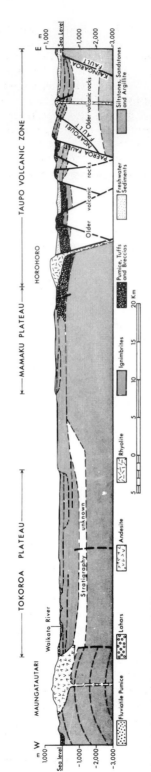

Fig. 5.11 A section through part of the North Island, New Zealand. In parts of the Taupo Volcanic Zone ignimbrites and pumice breccias are severely faulted and underlie the surface to depths exceeding 3000 m. Ignimbrite sheets are piled up to form the nearly horizontal surfaces of the Mamaku and Tokoroa plateaux. In the same volcanic region Maungatautari is formed of andesite and Horohoro of rhyolite (reproduced with permission of NZ Geological Survey, Crown copyright reserved).

more commonly develops large ring faults around a cone, which consequently sinks and leaves inward-facing fault scarps around its margin, or large-scale subsidence and caldera formation occurs.

The initial maximum height of a cone is limited by the pressure due to its magma column beneath the vent, compared with the pressure of a similar column of lithosphere, down to the level of molten rock in the asthenosphere where isostatic compensation is, presumably, complete. To calculate the maximum height of the volcano assume, for example, that the density of lithospheric plate is 3.2 Mg/m^3 and the volcano is of basalt with a liquid density of 2.7 Mg/m^3. If the depth to the asthenosphere is 45 km, then a vertical column of lithospheric rock, one metre in cross-sectional area, has a mass of $1 \times 45\,000 \times 3.2\,Mg = 144 \times 10^3$ Mg. A metre-square column of basalt with the same mass has a height of $(144 \times 10^3)/(2.7) = 53.3$ km above the asthenosphere $= 8.3$ km above the ground surface (i.e. lithospheric surface). Once the basalt column reaches this balancing elevation, no more magma can reach the apex of the volcano: the volcano can only grow further by flank eruptions.

Submarine volcanoes

Submarine volcanism associated with mid-ocean ridges has been described already; many other volcanoes, called *guyots*, are shown by bathymetric surveys to be flat-topped features with summits hundreds of metres below water-level, and no longer directly associated with spreading ridges. Chains of high islands like the Hawaiian Group are quite unrelated to any mid-ocean ridge.

Guyots were once thought to have been produced when sea-level fell as a result of the incorporation of ocean waters into ice-sheets. Cone-shaped volcanoes would then have been eroded by wave action to form a flat-topped reef and, as sea-level rose again, would either have been drowned or would have become a site on which coral reefs could grow. Fringing reefs around islands would then mark sites at which coral growth was upon shore platforms cut into the flanks of islands, too resistant to have been planed flat by the waves in the time available, and *atolls* would be formed above the rim of a fully eroded island on which coral growth had kept pace with the rise of sea-level. This theory, first propounded by Charles Darwin, seemed an adequate explanation until boreholes into atolls showed that coral limestones on the stumps of the volcanoes are sometimes more than 1000 m thick. It was recognized that sea-level could not have fallen and risen through this height, and that the glacial sea-level theory is inadequate to explain guyots and coral reefs.

The sea-floor spreading theory, however, does offer an explanation for most guyots. It is recognized that away from a mid-ocean ridge the average sea-floor depth increases and guyot crests occur at greater depths. It is suggested, therefore, that volcanoes are built above the sea surface along ridges or above hotspots in the crust, but as they are carried away from ridge crests by sea-floor spreading they are carried into deeper water. The guyots are carried away from the source of their feeder dykes and so become extinct, wave action planes the top of the volcano descending the gently sloping flanks of the mid-ocean ridges and, in some cases, first fringing reefs and later atolls are formed (Fig. 5.12). Many guyots, but no coral reefs, lie in cold waters.

An alternative theory for the origin of guyots is that they are built up by layer upon layer of submarine volcanic ejecta, and that they are flat-topped by construction rather than erosion. This theory was developed after study of Mount Asmara, in the Afar Triangle of Ethiopia, which appears to have been formed as a guyot (Plate 5.20).

The presence of the Hawaiian Group of islands in the middle of the largest ocean plate, and well away from spreading ridges appeared to be an anomaly in plate tectonic theory. Hawaii itself is the youngest island in the group, is still active, and lies at the south-eastern end of the chain. The islands become progressively older to the north-west, and all are volcanic but extinct. The chain appears to have developed from its north-west end and volcanoes to have progressively formed along a straight line, with just one bend in it, towards the south-east.

One hypothesis for the Hawaiian chain is that the volcanoes developed as a tensional fracture was propagated from the north-west to south-east. A more generally accepted hypothesis is that there are, at a number of sites within crustal plates, stationary hotspots above convective plumes rising from the lower mantle, and that magma is extruded above such plumes. As the plates moves, so extrusion is continued at the one site, but the

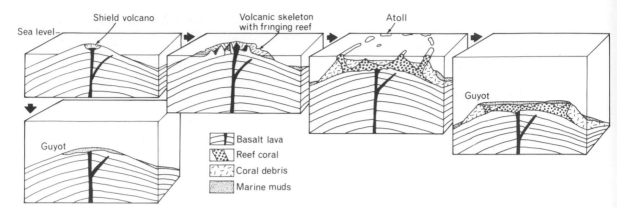

Fig. 5.12 The development of coral reefs around and on a subsiding oceanic volcano. If the rate of subsidence exceeds the rate of reef growth then the volcano becomes a guyot.

5.20 Mount Asmara standing above salt flats, Ethiopia. The mount is composed of pyroclasts which are thought to have been erupted beneath the sea, thus making the mount a guyot in origin (photo by H. Tazieff).

previously extruded volcanoes are carried away on the moving plate (Fig. 5.13). The bend in the chain may be explained by the assumption that about 40 million years ago the Pacific plate changed from a more northerly movement to the north-westerly drift of today, hence the older islands were formed during a different phase of plate motion. Support for this hypothesis is given by the fact that the chain of islands and guyots extending from the active volcanoes of Pitcairn and Macdonald Seamount have nearly parallel configurations with the Hawaiian Group.

Hot-spots below the stationary African plate are thought to give rise to the Tristan da Cunha volcanoes which have erupted on the same site for the last 18 million years. Before this the plate had been moving over the plume and so produced the volcanoes of Walvis Ridge.

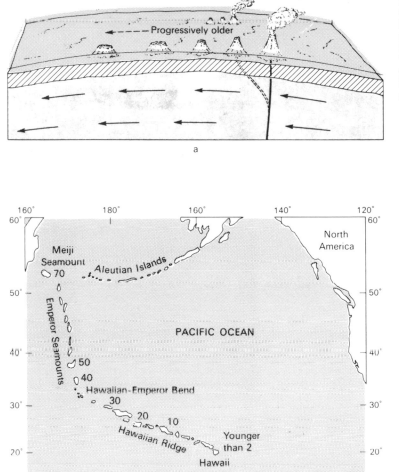

Fig. 5.13 (a) Model demonstrating the way in which the Hawaiian volcanic island chain may have formed by successive migrations away from the hotspot magma source (based on Wilson, 1963). (b) Ages, in millions of years, of volcanoes in the Hawaiian-Emperor chain (after Clague et al., 1975).

Erosion of volcanic features

The general processes and results of erosion apply to volcanic rocks as they do to all rocks, and these processes are discussed in Chapters 7–15. A few erosional features are, however, unique to volcanic regions and hence are discussed here.

Scoria, ash, and pumice deposits which have not been welded together by heat during emplacement, and not consolidated or weathered, may be extremely permeable. Consequently small scoria cones or extensive sheets of air-fall, ash flow, and nuée ardente pumice may retain surfaces with little erosion on them until water can concentrate in depressions or channels. When water has concentrated, however, erosion may be rapid and severe, with large gullies being formed in a few hours because the low-density pumice fragments float in water and are easily carried away. Gullies 400 m long and 20 m wide and deep have formed in 24 hours in pumice in the North Island of New Zealand during severe storms.

Many volcanic features are approximately con-

5.21 Lyttelton Harbour: an erosion caldera, occupied by the sea, in the basaltic shield volcano of Banks Peninsula, New Zealand. The Canterbury Plains lie beyond the volcano (photo and copyright: New Zealand Aerial Mapping Ltd., Hastings, N.Z.).

ical in form. When drainage is first initiated on such a feature there is little erosion near the crater rim because there is no catchment for water collection and, in places, because the rim is of porous scoria. The drainage channels start somewhere on the midslopes and reach their greatest depth there, but at the base channels again become smaller, as slopes decrease and streams lose their power to transport and hence drop their loads. Some channels develop from sites of landslides in hot ash, and some become preferred routes followed by lava flows so, even on a very symmetrical cone, the drainage pattern is seldom very regular. Irregularity is increased where lava flows are small in volume so that a channel cut into a slope may cut through alternate layers of resistant lava and weaker pyroclasts. A single valley may enlarge by cutting through weak rock and may capture the drainage of surrounding channels. Eventually a single channel may encroach headwards and cap-

ture the drainage from a central crater. On large volcanoes central craters may be so enlarged that they have the dimensions of calderas (although they should not be mistaken for collapse features). Banks Peninsula, New Zealand, provides excellent examples of deeply eroded and breached craters, now occupied by the sea to form Akaroa and Lyttelton Harbours (Plate 5.21).

Radial drainage requires that channel heads are close together near the crater rim and, like the spokes of a wheel, are more widely separated at the circumference of the volcano. The original flanks of the cone are consequently removed most readily on the mid and upper slopes, and on the lower slopes triangular facets, called *planezes*, may be preserved between incised channels. Later stages in erosion of the volcano involve much adjustment of the drainage, as the streams cut through the irregular layers of lava and pyroclasts of a stratovolcano until the cone form may be difficult to

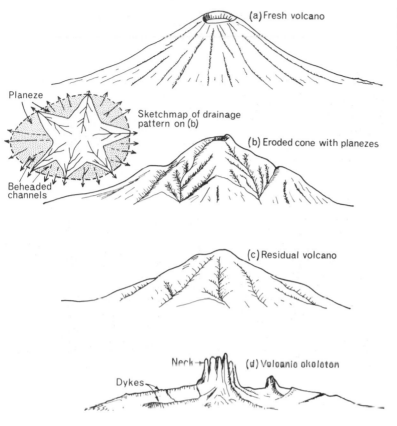

recognize—especially where the central crater has been breached. Only the exposure of cliffs at the edges of lava flows and dyke intrusions indicate the volcanic origins. Where the central vent was filled with cooled magma or recemented volcanic breccia it may form a resistant *erosional neck* standing as a steep tower, perhaps linked to feeder dykes, as the last remnant of the former volcanic pile (Fig. 5.14; Plate 5.22).

Plateaux of basalt and ignimbrite are extremely variable in their landforms, but commonly have deep gorges cut into the most resistant beds and more open valleys cut in weaker units. In many valleys the alternating hard lava and weakly welded units, or boundaries between successive lava flows, are visible and give a steplike appearance to the valley wall. Outcrops of welded and crystalline rock frequently display strongly developed, and often vertical joints. In basalts the columns so formed may be hexagonal in cross-section but in ignimbrites are more nearly cubic. The

erosion of weakly welded ash-flow deposits may leave a surface above which rise pinnacles of stronger rock which are the sites from which hot gas escaped from the cooling flow, and hence welded the rock. Such pinnacles should not be confused with the buttes and mesas left as residuals of erosion of lava sheets (Fig. 5.15).

In some sites lava flows which once filled valleys or depressions may become elevated as the surrounding weaker valley rocks are removed to leave hills capped by the lava.

Plutons

Masses of igneous rock which have solidified underground from intrusions of magma are called *plutons*. They have variable shapes, sizes, and relationships with the country rock (the invaded rock) surrounding them. *Sills*, *laccoliths*, *dykes*, *lopoliths*, *stocks*, and *batholiths* are the main forms (Fig. 5.16; Plate 5.23).

5.22 Ship Rock, New Mexico: a volcanic pipe, or diatreme, exposed by erosion of its enclosing rock. It rises 500 m above the plain which is crossed by radiating dykes (photo by John S. Shelton).

Sills, *laccoliths*, and *dykes* are emplaced by wedging open and penetrating cracks, joints, or planes of weakness. A sill is a tabular intrusion which is concordant with the main structures of the country rock, sills frequently follow bedding planes. They range in thickness from a few centimetres to hundreds of metres. The Whin Sill of northern England averages about 25 m in thickness and has an area of about 4000 km^2; the Palisades Sill of the Hudson Valley, New York, is up to 300 m thick. Where a concordant intrusion domes up the country rock it is called a laccolith. Wall-like plutons cutting discordantly across country rock are called dykes. Individual dykes can be traced across country for tens of kilometres and may be very thin or some tens of metres wide. Dykes often form as swarms with hundreds or thousands of intrusions running nearly parallel with one another. They are often exposed as a result of erosional removal of a volcano. *Ring dykes* and *cone sheets* are formed in circles, ellipses, or cones where subsidence of a large cylindrical or cone-shaped mass of country rock forms fractures which can be filled with magma.

Lopoliths and *batholiths* are massive intrusions, usually formed deeper in the crust than sills or dykes. A lopolith is a large, floored intrusion whose centre has sagged downward to form a bowl-shaped body. The Duluth lopolith of gabbro underlies Lake Superior and is estimated to be 250 km across, 15 km thick, and to have a volume of 200 000 km^3.

Batholiths are huge intrusions at least 100 km^2

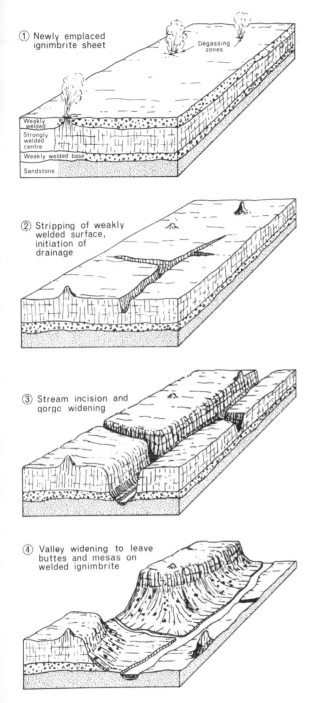

① Newly emplaced ignimbrite sheet

Degassing zones

Weakly welded

Strongly welded centre

Weakly welded base

Sandstone

② Stripping of weakly welded surface, initiation of drainage

③ Stream incision and gorge widening

④ Valley widening to leave buttes and mesas on welded ignimbrite

Fig. 5.15 Successive stages in the erosion of an ignimbrite sheet to leave buttes and mesas. Note the small resistant pinnacles left on the upper surface as weakly welded ignimbrite is removed by erosion (after Selby, in Soons and Selby, 1982).

in area. Smaller discordant intrusions are called stocks. Batholiths probably extend at least 20–30 km down into the crust and form the core of orogenic belts. Erosion of uplifted mountain chains may expose the upper surfaces of batholiths and many granite and granodiorite outcrops of fold belts and ancient shields have such an origin.

Plutons only influence landforms in detail where erosion has removed much of the overlying country rock. Then exposed sills may produce plateau surfaces similar to lava plateaux, especially where the rock is dolerite (the intruded equivalent of basalt). Dykes may form low wall-like features and *laccoliths* low domes. On the surfaces of the largest plutons, complex landforms develop in which the main joints are etched out, and large domed outcrops may form where sheet joints, parallel to the ground surface, control the landforms (Plate 5.24).

Residual masses of plutonic and extrusive rocks may form uplands because these rocks are relatively resistant. The rock masses, however, have lost most of their original form and are not recognizable by the landforms they support, but only by the mineralogy of their rocks.

The 300 km long and 45 km wide granite batholith of south-western England, for example, reaches from Dartmoor to beyond the Scilly Isles. It was formed when the Hercynian mountains were developing, at about the same time as there were volcanoes near Exeter and the Whin Sill was being intruded—some 290 million years ago. Dartmoor now has a broad rolling and indented surface in which jointing controls details of outcrops on the ridges. The Cairngorm granite of Scotland has been so deeply incised by glaciers that it is hardly recognizable as a pluton. It formed the core of the Caledonian Mountains and was emplaced about 400 million years ago. Ashes and lavas from volcanoes of this period form mountainous terrain like Ben Nevis, Glen Coe, the Ochils, and Cheviots. Around Edinburgh such features as Arthur's Seat are the remnants of a Carboniferous period volcano of which only the vent plug and some sills remain as prominent features (Fig. 5.17; Plate 5.25).

Mechanisms of intrusion

Magma rising towards the surface cannot be observed and the mechanisms have to be inferred

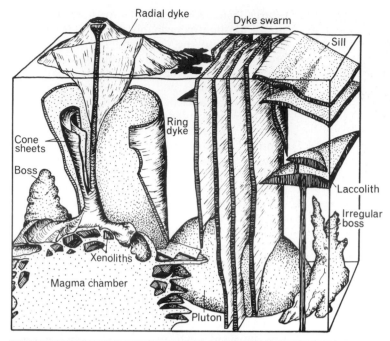

Fig. 5.16 The various forms taken by igneous intrusions (redrawn after Institute of Geological Sciences, 1974, Crown copyright reserved).

5.23 An exposure of sandstone penetrated by numerous sills of dolerite along bedding planes, and cross-cutting short dykes, Transantarctic Mountains.

5.24 Gross Spitzkoppe, a large granite dome, standing 1100 m above the Namib Desert plain.

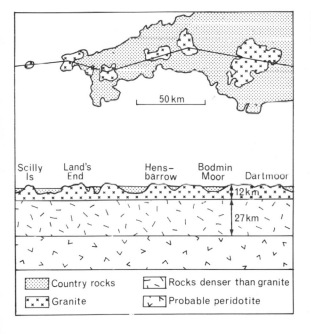

Fig. 5.17 The location of outcrops of granite forming the summits of a pluton underlying south-western England (based on Bott *et al.*, 1970).

5.25 Salisbury Crag, Edinburgh, is the remains of a dolerite sill of Carboniferous age. It was intruded into marine sediments and is associated with the large basaltic volcano whose eroded stump forms the hills of Arthur's Seat on the edge of the city. The volcanic rocks have been exposed by eroding Quaternary glaciers (photo by Institute of Geological Sciences, Crown copyright reserved). (*Copyright*: British Geological Survey).

from seismic, petrological, and eruption evidence, and from geochemical and simulation experiments. Diapirs of basalt or andesite may rise at rates of 1–5 m/y by displacing hot plastic rock and by disintegrating invaded rock which becomes xenoliths or is assimilated, at depth, into the melt (a process called *stoping*). Some magmas, such as kimberlite, may rise rapidly through the crust by *fluidization* as magma drills its way upward as a turbulent mixture of hot gas and solid fragments. Close to the ground surface, many magmas penetrate joints and force the country rock aside and upwards.

The origin of granite is, perhaps, the most controversial topic discussed by students of intrusion mechanisms. At least two mechanisms are widely recognized. (1) Primary granites may be formed at depths of 20–30 km within continental crust; they cannot be derived from mantle or subducted oceanic crust, and must be composed largely of sialic rock, which has either been differentiated from sima in earlier processes, or has been weathered and deposited as sedimentary rock. Some granites are thought to have risen as large diapirs from depth by wedging apart the country rocks: this may be the mechanism of emplacement of the 40 km wide Vredefort Dome of South Africa. Other granites may be emplaced by stoping. (2) Granitized bodies of rock have foliated and gneissic structures, but with granite composition. They gradually merge into the surrounding rocks and a 'ghost stratigraphy' of their origins may be preserved. Granitization, or *metasomatism* by soaking in granitizing fluids, takes place at great depths and, if diapiric rise follows, the diapirs may cut through older plutons and be indistinguishable from primary granites.

In all cases the rise of granite is a slow process, involving time-spans of millions of years, and the driving force is gravity. It is the relatively low density of heated rock, rich in the light elements aluminium, potassium, and sodium, which allows it to force a passage through, and into, overlying cooler and denser rock. Granite magma should not be thought of as a liquid, but as a viscous solid which deforms slowly and plastically at high temperatures and pressures.

Further reading

There are many available books concerned with volcanoes. Those by Cotton (1944), Ollier (1969b), and Green and Short (1971) are particularly concerned with volcanic landforms; the latter is well illustrated and a good reference text. Williams and McBirney (1979) discuss volcanic processes, and the books by Searle (1964) and Stearns (1966) give useful studies of volcanoes in specific areas. The paper by Nairn and Self (1978) provides a detailed study of the eruptions shown in Plate 5.4.

General texts which are useful background studies are Wilcoxson (1966) and Francis (1976). The book by Rittman (1976) is particularly good general reading.

The characteristics and emplacement of pyroclastic flows are discussed by Sparks, Self, and Walker (1973) and Sheridan (1979). Walker (1980) has given an account of the power and products of Ultraplinian explosions, and methods of classifying all eruptions. The eruption of Krakatau has been discussed by Self and Rampino (1981), and that of Mount St Helens in the volume edited by Lipman and Mullineaux (1981). The collection of papers edited by Self and Sparks (1981) contains many discussions of tephra. Descriptions of volcanic centres and their landforms are given for Japan in Yoshikawa, Kaizuka, and Ota (1981) and for New Zealand in Soons and Selby (1982).

6 Behaviour, Strength, and Resistance of Rock, Soil, and Water

The surface of the Earth is a zone within which rock and soil are acted upon by processes of weathering and erosion, and landforms are created as a consequence. The rate and manner of landform development and alteration is controlled partly by the nature, power, and frequency of operation of the processes (such as frost action, water and wind erosion), and partly by the strength, or resistance, of the rock and soil on which they act. This chapter is concerned with the nature of rock and soil strength, and the resistance to flow of water.

Definitions of *rock* and *soil* are not as universally accepted and unambiguous as is commonly assumed. To an engineer rock is a hard, elastic substance not significantly affected by immersion in water. To a geologist rock is that material which is below the depth of modern weathering, irrespective both of the degree of hardness or elasticity of the material, and of whether the material is exposed at the ground surface or is covered by an overburden. To an engineer soils are naturally occurring loose or soft deposits which either disintegrate or soften on immersion in water. To a soil scientist, soils are a complex of mineral and organic material formed at the surface of the Earth in response to physical, chemical, and biological processes acting on both hard and soft materials.

Practical definitions for geomorphological purposes incorporate aspects of all of these definitions:

Soil is a naturally occurring loose or soft deposit formed at the surface of the Earth; it is weakened or softened by immersion in water; it may be the result of physical, chemical, and biological processes acting to produce an organic-rich material; it may be formed from weathering of harder rock or older soils; it may be formed on the site in which it currently occurs; it may be of transported material; it may be deposited as a weak geological formation—but whatever its origin *a soil mass is essentially a continuous material* with few large-scale joints or fissures.

Rock is an intact material of mineral grains cemented together; it is a hard, elastic substance which does not significantly soften on immersion in water; in a mass *it is a discontinuous material with joints and fractures* which separate the mass into discrete blocks; it is the defects which largely control the resistance of a rock mass to the forces acting on it.

The term *strength* is used in three senses in geomorphology: (1) it may be used for the ability of material to resist deformation by tensile, shear, or compressive stresses; (2) it may be used to denote the ability of rock or soil to resist abrasion—a property which is controlled by the hardness of constituent minerals and the cement holding them together; and (3) it may also be used to indicate the resistance of loose or unconsolidated mineral grains, such as sand particles, to being transported by a fluid.

A *stress* is a force applied to a unit area of a substance and is measured in newtons per square metre or pascals ($1 \text{ N/m}^2 = 1 \text{ Pa}$). A *tensile stress* is a stretching force which tends to pull a rock or soil mass apart, and hence changes the shape of the material and, in a soil, may also change the volume if it causes fractures and pores to open. A *shear stress* is a force acting so that a body deforms by one part sliding over the other (Fig. 6.1a): thus a shear stress changes the shape of the body but has little effect on its volume. *Compressive stress* acts to crush a body and causes failure in a rock or soil by internal collapse of the structure due to compression of the void space, and may involve

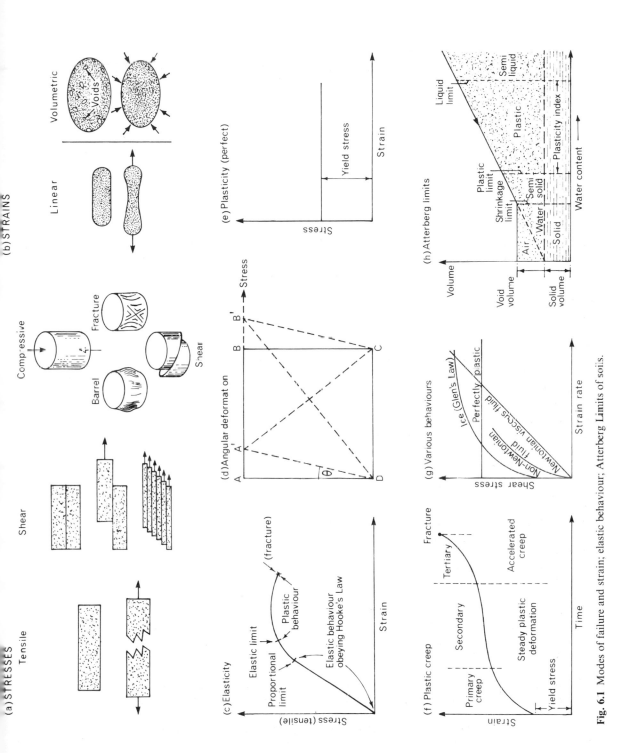

Fig. 6.1 Modes of failure and strain; elastic behaviour; Atterberg Limits of soils.

fracture of mineral grains and shearing (i.e. lateral displacement) along grain boundaries.

The change of shape or volume produced by a stress is termed a *strain* (see p. 95 and Fig. 6.1b).

Behaviour of rock and soil

The behaviour of bodies subjected to deforming forces constitutes the study of rheology. If the body entirely recovers its original size and shape after being subjected to stress, it is said to be *perfectly elastic*; if it entirely retains its altered size and shape, it is said to be *perfectly plastic*. *Viscosity* is a characteristic of solids and fluids which defines their resistance to flow. Few natural materials are purely elastic or plastic and exhibit combinations, such as elastoplastic, elasticoviscous, and plasticoviscous behaviours.

Elasticity

A spring is a mechanical model of an elastic material. The criteria of perfect elasticity are:

  (i) a given stress always produces the same strain;
 (ii) maintenance of a given stress results in constant strain; and
(iii) removal of a stress results in complete recovery of strain.

It was found experimentally by Hooke (in 1679) that, over a considerable range, the strain is proportional to the stress applied. This relation, which is termed *Hooke's Law*, holds up to the point called the *proportional limit* of the material. Beyond the *elastic limit* (Fig. 6.1c) the deformation is no longer recoverable and behaviour is said to be plastic. (Note: for rocks the practical definition of the proportional and the elastic limit is the same point.) For brittle materials, such as fine-grained igneous rocks, plastic yielding is very limited, and they fail by fracturing just beyond the elastic limit. Other rocks, such as evaporites and clay-rich moist soils, exhibit predominantly plastic behaviour with little elasticity.

Hooke's Law can be expressed in the general form:

$$\frac{\text{stress}}{\text{strain}} = \text{a constant factor for that material.}$$

A constant factor is termed a *modulus*. There are four elastic moduli for *unconfined* conditions according to the nature of the stress and strain:

(1) *Young's modulus (E)* =
$$\frac{\text{longitudinal applied force/cross-sectional area}}{\text{change in length/original length}}$$

(2) *Rigidity modulus (G)* =
$$\frac{\text{shear force/unit area}}{\text{angular deformation } \theta} \text{ (see Fig. 6.1d)}$$

(3) *Bulk modulus (K)* =
$$\frac{\text{compressive (or tensile) force/unit area}}{\text{change in volume/unit volume}}.$$

The rigidity modulus is also known as the *shear modulus* or *torsion modulus*. The bulk modulus is also known as the *modulus of incompressibility*, and its reciprocal, $1/K$, is called the compressibility.

When a specimen is stretched, not only does its length increase but its width is decreased. The lateral change in dimensions is proportional to the longitudinal strain and is a constant, known as Poisson's ratio, for that material:

(4) *Poisson's ratio (v)* =
$$\frac{\text{change in width/original width}}{\text{change in length/original length}}.$$

For both Young's modulus and Poisson's ratio the forces may be either compressive or tensile.

The four moduli of elasticity are used by seismologists studying the propagation of seismic waves, and by engineers studying the behaviour of materials being used in construction projects. They have, as yet, been largely ignored by geomorphologists, but there is evidence from a limited range of studies that there is a high statistical correlation between relief, drainage density, and some of the elastic responses to stress. Further work is needed before the mechanics of this relationship are understood.

Plasticity and creep

Where stress is applied to a perfectly plastic material, no deformation can occur until the bonds holding together the lattices of its mineral crystals are broken. Once this threshold value, or yield stress, is exceeded, deformation occurs at a constant rate as long as the stress is uniform (Fig. 6.1e). Most soils and rocks are mixtures of various materials and do not behave as pure plastic substances. A moist clay soil, for example, may initially deform rapidly as voids are reduced, then at a nearly constant rate until the deformation produces permanent weakening of the soil fabric by aligning the particles so that they offer reduced resistance to further movement. When the stress is removed the strain is permanent.

Time-dependent behaviour in earth materials is commonly referred to as *creep*. It is usually promoted by such factors as variations in temperature and water content, and by loads induced by overburdens, in addition to time. After an initial component of instantaneous elastic response to stress, the strain-time graph is curvilinear with a decreasing strain rate during a period of primary creep (Fig. 6.1f). If stress is removed at any instant during primary creep, the elastic component of strain is recovered immediately in cohesive rock—but not in a non-cohesive soil; the creep component of strain is recovered in part by a time-dependent process at a decreasing strain rate.

The primary creep phase is followed by a secondary stage of constant strain rate and permanent deformation. At high stress levels a tertiary phase of accelerating deformation may lead to rupture.

Creep is an important process in cohesive soils on slopes, and may be of large magnitude in very plastic rocks such as evaporites and porous sedimentary rocks with a high clay content. Evidence for such behaviour is the closure of tunnels and shafts and swelling of walls in cuttings. Fine-grained igneous and metamorphic rocks have such nearly pure elastic behaviour that tunnels through them do not have to be braced or lined to resist closure.

Viscosity

A liquid, like water, which deforms (or flows) at a rate which is proportional to the applied stress is said to be a Newtonian fluid. On a stress–strain rate diagram the gradient of the curve is the *dynamic viscosity* of the fluid and is related to its molecular composition (Fig. 6.1g). Viscosity is a measure of the internal frictional resistance of the fluid and is measured in units of $N s/m^2$. The coefficient of dynamic viscosity is defined as the ratio of shear stress to the rate of shear strain. Most fluids which are suspensions of solids in a liquid, such as mud- and earth-flows, are said to be non-Newtonian in behaviour because their viscosity changes as the applied stress changes.

Over a geological time-scale, and at high temperatures, rocks can flow: pitch, salt, and lead are well-known solids which flow in human time-scales and viscous behaviour of such solids is comparable to that described as secondary creep. In the shorter time-scale of normal investigation, only molten rock and saturated soils are capable of viscous behaviour. Viscosity, however, is a concept that is applied to all fluids and solids and links the behaviour of the two types of material.

In conditions in which measures of viscosity not related to applied stresses are required, the term *kinematic viscosity* is used and measured in units of m^2/s.

Mixed behaviours

Under applied loads only the hardest and densest rocks, such as basalt, behave purely elastically. Most rocks initially strain nearly plastically as their voids and fractures are closed, then respond elastically up to the proportional limit, then plastically as internal fracturing commences and shearing occurs within crystals, and finally reach a fracture point. Such mixed behaviour is characteristic of nearly all natural materials: ice for example has elements of non-Newtonian viscosity, plasticity, and brittle fracture in its response to stress.

Soil composed of clay and silt (but not larger particles) will change its behaviour with changing water content. At very low water contents soil behaves as an elastic solid and will fail by brittle fracture: at a certain water content, known as the *plastic limit*, it will deform plastically and at a higher moisture content, known as the liquid limit, it will behave as a viscous fluid. The plastic and liquid limits of any soil are characteristics controlled largely by the clay minerals of the soil, and the range of water content between these limits is the *plasticity index*. For most soils, the shear strength at the plastic limit is about $110 kN/m^2$, and at the liquid limit it is about $1.6 kN/m^2$. A pure liquid has no resistance to shear. The three limits of liquid, plastic, and shrinkage behaviour (i.e. the moisture content, of a drying soil, below which no further shrinkage will occur) are collectively known as the *Atterberg limits* (Fig. 6.1h).

Rock and soil resistance to shear stresses

Rock and soil are aggregations of solids with substantial voids which are filled, or partly filled, with water and air. Their strength depends both upon the minerals composing the aggregates and upon the forces holding the aggregates together.

In land-forming processes shear stresses are imposed upon rock and soil far more commonly than tensile or compressive stresses. The resistance to shear of the material depends upon many factors but three may be identified as being of greatest significance:

(1) the *frictional property* of the material;
(2) the *normal load* (i.e. the weight of material forcing the mineral grains against each other;
(3) the natural cementation, or *cohesion*, binding the mineral grains together.

Friction

There are two types of friction. (1) *Rolling friction* is the force resisting the rolling of a body across a surface and it is appropriate to debris rolling down a slope or along the bed of a channel. (2) *Sliding friction* is of two kinds—*static friction* which is the resistance to initial motion, and *dynamic friction* which is the resistance to continued sliding of one surface over another.

Friction between the surfaces of two mineral grains depends upon the hardness of the mineral surface, the roughness of the surface, and the number and area of the points of contact between grains. Loosely packed grains with smooth surfaces can slide past one another easily, but densely packed grains, with rough surfaces, offer high frictional resistance to such sliding and consequently have a higher resistance to shear (i.e. they have a higher shear strength) (Fig. 6.2). Hardness of grains in contact is important where a rough hard surface is in contact with a softer surface. The hard surface will have rough edges which can

plough into the soft surface, so creating a greater area of contact and greater frictional strength.

The importance of a normal load in controlling frictional resistance can be appreciated if a simple shear box is made by placing a wooden block on a table. The block is easily moved by a horizontal force imposed by the hand but, if a pile of bricks (i.e. a normal load) is placed upon the block, a much greater force is required to move the wooden block because the frictional resistance between the block and the table has been increased. The importance of surface roughness can be appreciated by putting first packed sand and then ball bearings between the block and the table. The least resistance will occur without the bricks but with the ball bearings. Frictional resistance is thus the product of normal load and a value which characterizes the packing, shape, surface roughness, and hardness of the grains. This value is called the *coefficient of plane sliding friction* (symbolized by $\tan\phi$). For purely frictional materials resistance to shear is therefore given by: $\sigma_n.\tan\phi$ (where σ_n is the normal stress). The word 'normal' is used because the stress applied by the load is at right angles (i.e. normal) to the shear plane between the block of wood and the table surface.

The derivation of the coefficient of friction is explained by reference to Fig. 6.3. The weight of the block (N) generates an equal and opposite reaction (R). If a small stress (H) is applied to the block the reaction (R) will no longer be normal to the plane of contact. It adjusts in magnitude and direction to equal the resultant N and H. The triangle of stresses represents, in magnitude and direction, the relationships between N, H, R and the angle θ. If the shear stress H is increased until the block is just about to slide, R will increase and so will θ. At the moment when sliding begins, θ will have attained its maximum value possible on that surface. That maximum value is ϕ, the angle of plane static friction, or angle of shearing resistance, and $\tan\phi = H/N$ and is the coefficient of plane sliding friction.

Cohesion

Cohesion is caused by chemical cementation of rock and soil particles and by electrostatic and electromagnetic forces between clay-size particles: it is not controlled by compressive forces holding particles together. In a sandstone with calcium carbonate, silica, or iron oxide cement, much of

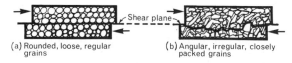

(a) Rounded, loose, regular grains

(b) Angular, irregular, closely packed grains

Fig. 6.2 Resistance to shear is relatively low between the rounded grains in box (a), but much higher frictional resistance occurs in (b).

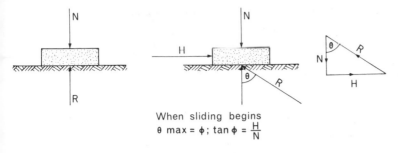

Fig. 6.3 Derivation of the coefficient of plane sliding friction.

the strength of the rock is derived from the cement. If the cementing material were removed in solution then the pile of sand grains would have no cohesion but would still have some resistance to shear derived from frictional resistance. Total resistance to shear (s) can thus be expressed as cohesion (c) plus frictional resistance ($\sigma_n.\tan\phi$), or:

$$s = c + \sigma_n.\tan\phi. \tag{6.1}$$

This equation was first expressed by a French engineer and is universally called, after him, the *Coulomb equation*.

On a graphical plot of the results of three or more measurements of the shear stress required to cause failure in soil or rock samples under various normal stresses, the values of c and ϕ may be determined from the position and slope of the line drawn through the data points. The line is the Coulomb failure line and it must always be drawn on arithmetic plots with intervals of equal magnitude (Fig. 6.4).

The process of deforming a soil, or of pushing one block of rock along the surface of a neighbouring block, rearranges the particles of the soil, or grinds off the rough surfaces of the block, so that the strength of the material declines. Many materials, therefore, exhibit a higher initial *peak strength* but a lowered *residual strength* after the application of stress. The strength of rock and soil is, therefore, defined as being equal to the stress required to cause it to fail, but the magnitude of this stress is not a constant for that material. Strength changes as a result of the strain which has already occurred and as a result of the water content of the material.

Water and shear strength of soils

In a perfectly dry soil the mineral grains are

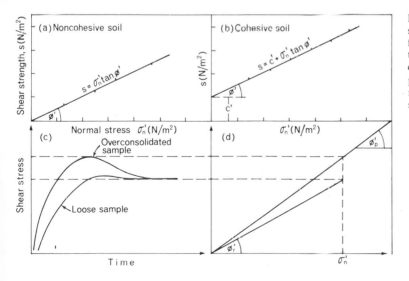

Fig. 6.4 (a, b) Plots of effective normal stress against shear strength at failure for two soils. (c, d) During strength testing an overconsolidated, or heavily compacted, soil may develop a high peak shear strength before its grains are reorganized into a lower residual strength arrangement.

supported entirely on the point contacts between them. The pores of the rock and soil fabric are filled with air at atmospheric pressure and pore water pressure is then zero. In a fully saturated soil, part of the weight of overburden is transferred from the soil fabric to the pore water and pressures in that water are then increased and said to be positive. The transfer of load from soil to water is equivalent to a buoyancy, or upthrust, effect. The result is that the normal force appears to be decreased and frictional strength is consequently reduced. For saturated soils the Coulomb equation has to be rewritten as:

$$s = c + (\sigma_n - u)\tan\phi \qquad (6.2)$$

where u is the pore water pressure derived from the weight of water filling the soil pores above that point. Alternatively this equation may be written as

$$s = c' + \sigma_n'\tan\phi' \qquad (6.3)$$

in which the superscript dash (′) denotes the *effective stresses* as modified by pore water pressure.

Because positive pore water pressures decrease soil strength, a mass of saturated soil on a slope is inherently weaker than a dry mass. Consequently landslides commonly occur during and after rainstorms which produce enough water to saturate the soil.

A partly moist soil is stronger than either a fully dry or a saturated soil. In this stage pore water pressures are said to be negative and a suction exists within the pores. This suction occurs because the available water is drawn over the surface of the mineral grains by capillary forces. These are at their strongest when water volume is low and they then have the effect of holding the grains together and creating an *apparent cohesion*. Apparent cohesion is lost both on drying and close to saturation. Through a deep soil profile, zones of positive, negative, and zero pore pressures may occur (Fig. 6.5). Decrease in soil strength is consequently associated with a rise of the water table so that the zone of saturated soil increases in depth.

The strength of intact rock, without fissures, may be described in terms of the Coulomb equation (6.1), but it will be weakened by positive pore pressures to a lesser extent than soil. A rock mass with all its fissures has a total strength which is more dependent upon the fissures than upon the cohesive and frictional properties of its constituent grains.

Strength of rock masses

A least eight features of a rock mass contribute to its total resistance to shearing forces. These are listed below, together with the approximate percentage significance of each feature:

(1) strength of intact rock (i.e. given by cohesion and friction)   20
(2) state of weathering of the rock   10
(3) the distance apart, or the spacing, of joints and fissures   30
(4) dip of the fissures with respect to a cliff face   20
(5) width of the fissures   7
(6) continuity of the fissures   5
(7) the amount of infill of fissures with weak soil   2
(8) movement of water in or out of the rock mass   6

Each one of these features may be assessed in the field and graded into five categories.

Intact strength may be assessed in the field either with a special instrument called a Schmidt hammer or, more readily, by its resistance to a blow with the pick end of a geological hammer (see Table 6.1).

Weathering state of a rock mass can be assessed by the degree of rock discoloration, ratio of rock to soil, and presence (or absence) of original rock texture. In a profile of soil developed upon bedrock it is often possible to recognize grades of alteration ranging from negligible at the bedrock surface (grade 1) to total alteration (grade 5) just below the organic soil layer containing many plant roots. Intermediate zones are recognizable by the percentages of rock and soil they contain (see Fig. 7.4 and Table 7.1).

Spacing of joints, or fissures of any kind, is important because all cohesive strength is lost along an open joint. The greater the volume of fissures in a rock mass, therefore, the weaker is that mass. The distance apart of joints along a line, or in a unit area of rock face, can be measured and graded (Table 6.2).

The orientation, or dip, of joints becomes increasingly unfavourable for stability as the dip out of a rock face becomes steeper. It is most favour-

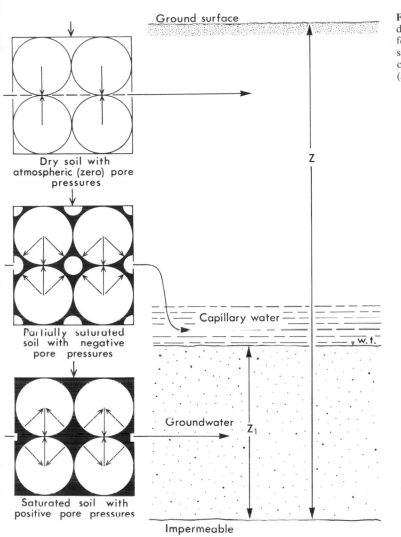

Ground surface

Dry soil with
atmospheric (zero) pore
pressures

Partially saturated
soil with negative
pore pressures

Saturated soil with
positive pore pressures

Capillary water

w.t.

Groundwater

Z

Z_1

Impermeable

Fig. 6.5 Pore water pressure within a deep soil profile. A water table (w.t.) forms the boundary between a saturated groundwater zone and the overlying capillary and dry soil zones (after Selby, 1982a).

able for stability against sliding where the dip is into the slope. Assuming that horizontal strata represent an intermediate condition the degree of stability can be graded (Table 6.2).

Width and continuity of joints influences frictional and cohesive strength as well as water movement in the rock mass. Wide joints have no cohesive strength and frictional strength can only develop at the point of contact of rock on either side of the joint. If a joint is limited in area it will have little effect on the strength of the rock mass, but if it is continuous it provides a ready-made plane along which shearing movements can occur. By measuring the width of the joint opening, and

assessing the continuity of the joint and the thickness of the infill of weak material in the joint, these features may be graded. Infill is important because if the whole joint is filled with weak clay, for example, the strength along the joint will be that of the clay, not that due to the rock.

Water flow in a joint is far more important than water in rock pores. Because it is the water in fissures or clefts which influences rock mass strength we speak of *cleft water pressures*. High rates of water flow and infilling of fissures with water can cause a buoyancy effect on the rock above a saturated joint. The only way this can be assessed, except by drilling a well into the rock

Table 6.1   A Strength Classification of Intact Rock

Description	Strength class	Examples of rock types
Very weak rock: crumbles under sharp blows with geological pick point, can be cut with a pocket-knife	5	chalk, rocksalt, lignite
Weak rock: shallow cuts or scratches may be made with a sharp knife, pick point indents deeply with firm blow	4	coal, siltstone, schist
Moderately strong rock: knife cannot scrape surface, shallow indentation under firm blow from pick point	3 ignimbrite	slate, shale, sandstone, mudstone,
Strong rock: hand-held sample breaks with one firm blow from hammer end of geological pick	2	marble, limestone, dolomite, andesite, granite, gneiss
Very strong rock: requires many blows from geological pick to break intact sample	1	quartzite, dolerite, gabbro

mass, is by measuring the amount of water flowing out of the rock.

If the features controlling rock mass strength are individually assessed and rated according to the scheme shown in Table 6.2, then the total rating can be obtained. The only items of equipment needed are a geological pick, a tape measure, an inclinometer (which can be made from a simple protractor and a plumb line), and a notebook. A rock face is divided into units which appear to be uniform and the eight features are measured for each unit over an area of about 10 m². If the rock slope profile is recorded then the relationship between hillslope profile and total rock mass resistance can be seen (see Fig. 6.6; Plate 5.23). Further examples are given in Figs. 8.16 and 8.17.

Flow of fluids

Laminar and turbulent flows

Moving liquid or gas within an extremely thin layer, adjacent to a plane solid surface, moves as layers of sheets, or laminae, shearing against one another. The layers of liquid slide over one another much as do the leaves of a book when it is placed flat on a table and a horizontal force is applied to the top cover. Movement of fluid in layers is called *laminar flow* and the thin layer in which it occurs is called the *laminar sublayer*. Laminar flow is readily seen in flows of viscous liquids such as treacle.

The most common form of flow in fluids of low viscosity is *turbulent flow*. It is characterized by local, ephemeral, rotational currents called *vortices* which cause a large increase in resistance to flow, compared with laminar flow.

In both kinds of flow the velocity of fluid movement is nil in the molecular layer at the solid boundary but increases linearly through the laminar sublayer away from the solid surface. Beyond the sublayer increases in the velocity are smaller (Fig. 6.7).

Experiment indicates that a combination of three factors determines whether the flow of a liquid within a pipe or channel is laminar or turbulent. This combination is known as the *Reynolds number*, R_e, and is defined as

$$R_e = \frac{VR}{v} \qquad (6.4)$$

where V is the mean forward velocity, v is the kinematic viscosity and R is the hydraulic radius of the channel (this approximates to the depth of flow). The Reynolds number is a dimensionless quantity and has the same numerical value in any consistent system of units. For a 1 m deep stream moving at 5 m/s, and with a water kinematic viscosity of 0·001 m²/s, $R_e = 50\,000$.

In streams, the Reynolds number for laminar flow has a maximum value of about 500 to 600 and above 2000 to 2500 (the actual value depends upon temperature of the fluid) the flow is turbulent. In the transition zone of 500–2500, the flow is unstable and may change from one type to another. In natural open channels laminar flow is rare, except near the channel boundary, although it may occur on hillslopes. The smooth glassy surface seen on some rivers does not indicate laminar flow.

In an open channel gravity forces are, under steady conditions, the only forces which cause flow. At small water depths in a rough channel gravity waves develop in the water. Such waves

Table 6.2   Geomorphic Rock Mass Strength Classification and Ratings (after Selby, 1980)

Parameter	1 Very strong	2 Strong	3 Moderate	4 Weak	5 Very weak
Intact rock strength	r:20	r:18	r:14	r:10	r:5
Weathering	unweathered r:10	slightly weathered r:9	moderately weathered r:7	highly weathered r:5	completely weathered r:3
Spacing of joints	>3m r:30	3–1m r:28	1–0.3m r:21	300–50mm r:15	<50mm r:8
Joint orientations	Very favourable. Steep dips into slope, cross joints interlock r:20	Favourable. Moderate dips into slope r:18	Fair. Horizontal dips, or nearly vertical (hard rocks only) r:14	Unfavourable. Moderate dips out of slope r:9	Very unfavourable. Steep dips out of slope r:5
Width of joints	<0.1mm r:7	0.1–1mm r:6	1–5mm r:5	5–20mm r:4	>20mm r:2
Continuity of joints	none continuous r:7	few continuous r:6	continuous, no infill r:5	continuous, thin infill r:4	continuous, thick infill r:1
Outflow of ground water	none r:6	trace r:5	slight $<25l/min/10m^2$ r:4	moderate $25-125l/min/10m^2$ r:3	great $>125l/min/10m^2$ r:1
Total rating (r)	100–91	90–71	70–51	50–26	<26

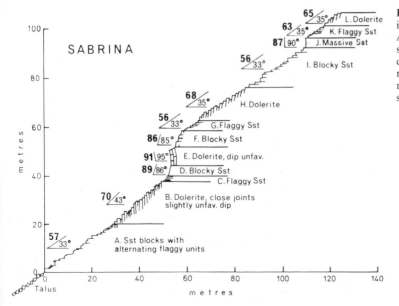

SABRINA

Fig. 6.6 Features of a rock slope profile in Sabrina Valley, Britannia Range, Antarctica. The slope is formed on sandstone (Sst) intruded by dark coloured dolerite. The bold numerals indicate the mass strength rating of each rock unit. The same slope is shown in Plate 5.23.

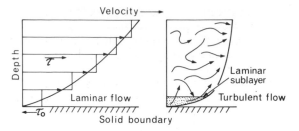

Fig. 6.7 Laminar and turbulent flow, with appropriate velocity profiles, and location of the laminar sublayer, in a stream channel. Tau (τ) symbolizes the shear stress.

are, essentially, local variations in the depth of flow. The ratio of the flow velocity to the rate of movement of the gravity waves is a characteristic of the state of flow. Since a gravity wave is propagated at a celerity $C = \sqrt{(gd)}$ where g is the acceleration due to gravity and d is the mean flow depth, then this ratio, represented by the *Froude number*, F_e, is:

$$F_e = \frac{V}{\sqrt{(gd)}} \qquad (6.5)$$

At $F_e = 1$ there is a critical transition between two types of flow (Fig. 6.8).

There are two kinds of turbulent flow because for a given discharge there are two possible combinations of depth and mean velocity at which flow can occur (i.e. the same discharge can be accommodated by slow deep flow or by rapid shallow flow). The faster shallower flow is termed *upper flow regime* or *shooting*, *rapid* or *supercritical*, whilst the deeper slower flow is described as *lower flow regime* or *tranquil*, or *subcritical*. The critical depth for the transition from tranquil to rapid flow is dependent on the velocity only.

Larger Froude numbers can be achieved in artificial channels than in natural ones because the banks of most alluvial channels cannot withstand high-velocity flow without eroding. This erosion causes an increase in the cross-sectional area and thus a reduction in the average velocity and the Froude number. Rarely does a Froude number exceed unity for any extended time-period in a natural stream with erodible banks, and alluvial channel beds are rarely stable when $F_e > 0.25$.

Rapid flow occurs in rapids and results in increased erosion. When velocity is decreased from the rapid to tranquil flow the water surface rises, causing a stationary wave or roller-type eddy called a hydraulic jump (Fig. 6.9). The reverse situation, where a drop in the water level accompanies increased velocity, is a hydraulic drop. Hydraulic jumps are noticeable in streams where the bed-slope decreases or where a reach widens or deepens. On wave-swept beaches they are common at the toes of backwash flows.

As the magnitude of the Reynolds number defines the turbulence at the channel boundary and the Froude number the nature of the flow, these two parameters are important values in the analysis of resistance to erosion of sediment lying on the bed of a channel or of soil resistance to erosion by wind (see the section on sedimentary bedforms in Chapter 9).

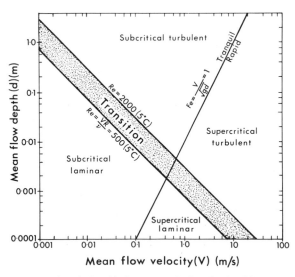

Fig. 6.8 The relationship between velocity, depth of flow, and flow regime, with the boundaries indicated by the Reynolds and Froude numbers (modified from Allen, 1970).

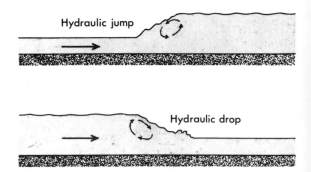

Fig. 6.9 Hydraulic jump and hydraulic drop.

Erosion of rock surfaces by fluids

The processes involved in bedrock erosion are: abrasion, fluid stressing, cavitation, corrosion, and plucking. Few of these mechanisms are well understood and none is simple.

Abrasion

Abrasion is the mechanical wearing, grinding, or scraping of rock surfaces by solid particles transported by wind, waves, running water, ice, or in free fall from a cliff. The term *corrasion* is a synonym of abrasion. Abrading particles may be of any diameter from boulders to silt, or even clay size. The rate at which rock and soil material is removed by abrasion is related to the power of the eroding fluid (i.e. water or air), or eroding solid (i.e. ice), and to the resistance of a rock surface. Resistance is given by frictional and cohesive strength and by the hardness of mineral grains. Erosive power is given by the velocity and turbulence of moving fluid and by the mass of abrading particles being carried.

When a hard particle collides with a rock surface stresses are set up, in the rock surface, which are concentrated in the centre of the contact area. As long as the magnitude of stresses does not exceed the strength of the rock only elastic deformation occurs. If, however, the elastic limit of the rock is exceeded the surface is destroyed by brittle fracture and fragments of it are removed in the fluid. The mass of rock eroded per impact is proportional to the mass of the impacting particle and to the square of the approach velocity of that particle, that is to the kinetic energy (e) of the impacting particle ($e = \frac{1}{2}mv^2$). For a stream of water or air the mass of the solid removed is approximated by:

$$m = \left[\frac{\pi}{12} D^3 (\sigma - \rho)(U_G \sin\beta - B)^2 \right] / E. \qquad (6.6)$$

where (Fig. 6.10)

- D is the diameter of the impacting particles (mm),
- σ is the mass density of the impacting particles (kg/m³),
- U_G is the approach velocity of the impacting particles (m/s),
- β is the angle of attack of the impacting particles to the rock surface (degrees),

- ρ is the density of the fluid stream bearing the particles (kg/m³),
- B is a coefficient whose value depends upon the velocity and properties of the impacting grains,
- E represents the energy needed to remove a unit volume of material from the rock surface and describes the behaviour and strength of the surface,
- m is the mass of the solid removed (kg).

The equation states that the mass eroded per impact is proportional to the cube of the particle diameter and the square of the approach velocity of the impacting particles: resistance to erosion is a characteristic of the bedrock. If B is very small compared with U_G the mass lost per impact varies approximately as the sine of the angle of attack and is at a maximum when $\beta = 90°$; abrasion is at a minimum below a certain angle of attack: in a natural stream the angles of attack will usually be very varied. The quantity ($\sigma - \rho$) indicates that abrasion increases with the number of impacting particles but declines as the density of the fluid increases. The density of a fluid is influenced by temperature and by the solutes, fines, and colloids in suspension. In a very dense fluid, such as a mudflow, large particles are buoyed up in the flow and erosion from abrasion is slight.

Fluid stressing

Resistant rocks cannot be eroded by strong currents of water that lack abrasive particles, but weak cohesive rocks such as muds can be eroded. The factors involved are the shear strength of the bed, the plasticity of the muds, the percentage of clay in the bed, and the fluid shear stress required for erosion. The fabric of the clay beds—edge to edge, edge to face, or face to face arrangements of

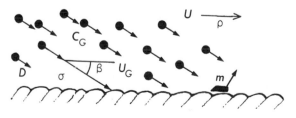

Fig. 6.10 Definition diagram for erosion by corrasion. U is fluid velocity, C_G is the number of particles in unit volume of the fluid, other terms are defined in the text (after Allen, 1971).

the particles—is presumed to have a considerable importance in determining the strength of the bed. The fluid shear stress is dependent upon both the velocity of the water and its turbulence.

Cavitation

Cavitation is a process well known to engineers because it causes severe erosion in hydraulic structures and machinery where a liquid flows very swiftly past a solid. There is little information available on the effect and occurrence of cavitation in natural channels, but it seems probable that it can occur in rapids and waterfalls.

In a very fast flow the pressure in a turbulent liquid may locally fall to that of the vapour pressure of that liquid and air bubbles will form. Cavitation erosion occurs when individual bubbles collapse close to a solid surface. In such a collapse an inward-moving microjet of liquid is projected across the imploding bubble towards the solid at velocities as high as 130 m/s to produce impact stresses as high as 60 MN/m². Such stresses may crack solids and prepare fractured particles for removal in the flow (Figs. 6.11, 6.12).

In stream channels cavitation will usually occur downstream of local obstructions on the bed, or at sharp changes of angle in the bedrock surface. At such sites local fluid velocity may exceed 20 m/s even where the mean velocity in a channel is 10 m/s. Large vapour-filled cavities develop in the flow downstream of the obstruction and clouds of bubbles are released. As bubbles at the rock surface collapse the rock is eroded by released shock waves.

Corrosion

Corrosion is a chemical process which results from the reaction of water and rocks. The chemistry of the process is discussed in more detail in Chapter 7. In natural conditions most water contains dissolved carbon dioxide, and sometimes other solutes derived from the air, vegetation, and soil, so corrosion is not a simple process of solution although this process is most readily explained.

When a soluble crystal is placed in a quantity of still solvent of uniform temperature the crystal undergoes a slow dissolution into its constituent molecules, which gradually diffuse into the solvent—this is solution. The solvent in immediate contact with the dissolving crystal becomes fully saturated with the solute, and at increasing distances from it the solvent contains less of the solute; thus there exists a concentration gradient of the solute in the solvent. Experience shows that solution continues either until the crystal is entirely dissolved or the solvent is fully saturated. A fully saturated solution has a uniform distribution of the solute within it (Fig. 6.13).

In a stream channel it is improbable that the water in contact with the bed will become saturated with solutes. As the rate of molecular diffusion from the rock into the water is proportional to the concentration gradient and the mean flow velocity, fast-flowing water with a low concentration of solutes is more aggressive than slower more concentrated flows. Solution processes by streams are most effective in limestones or rocks with a high carbonate content.

Plucking

Plucking is the direct removal of loose bedrock material by the force of water impact alone. Slabs of rock may be removed in this way where they are so fractured that there is no cohesive strength along the joints. The term 'plucking' is also used to refer to the freezing of bedrock to the sole of a glacier and its removal by ice. The process of entrainment of particles by a fluid is discussed below.

Entrainment and transport of grains by fluids

Uncohesive beds of soils and sediment are subject to erosion by fluids. The picking-up and initial

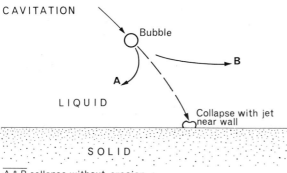

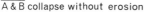

Fig. 6.11 Cavitation, collapse of an air bubble near a solid boundary.

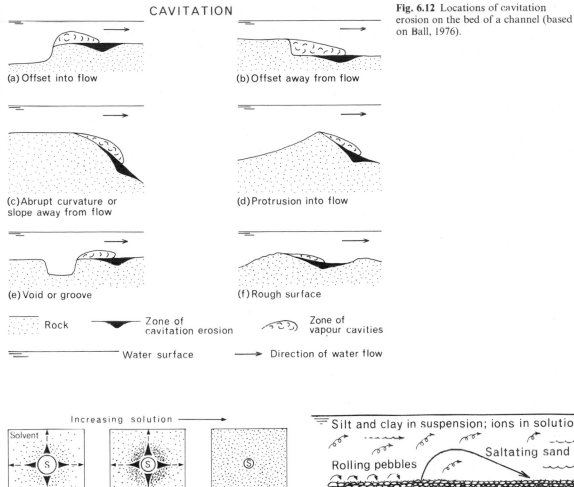

CAVITATION

(a) Offset into flow

(b) Offset away from flow

(c) Abrupt curvature or slope away from flow

(d) Protrusion into flow

(e) Void or groove

(f) Rough surface

Rock ⎯⎯ Zone of cavitation erosion

Water surface ⎯→ Direction of water flow

Zone of vapour cavities

Fig. 6.12 Locations of cavitation erosion on the bed of a channel (based on Ball, 1976).

KEY: S Solid ►--► Concentration gradient ⣿ Molecules

Increasing solution ⎯⎯⎯→

Solvent

(a)　(b)　(c)

Fig. 6.13 The process of solution. The dots and the width of the arrows represent the concentration of the solute.

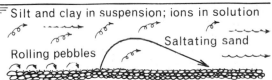

Silt and clay in suspension; ions in solution

Saltating sand

Rolling pebbles

Fig. 6.14 Modes of sediment transport in a fluid.

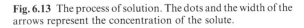

carrying of solid grains is called *entrainment*. It occurs when the stress applied to a particle at rest is greater than its resistance to motion. As the stress applied by a fluid is largely controlled by the fluid velocity many experiments upon entrainment have been carried out by engineers and geologists who are concerned with the stability of natural and man-made channels against erosion.

Five main processes, which act to entrain grains and keep them in motion, have been identified (Fig. 6.14):

(1) bed shear stress produced by a fluid flowing parallel to the bed surface and exerting a frictional drag on the particle (this drag, or tractive stress, is largely a result of the differences in fluid pressure upstream and downstream of the particle);
(2) a vertical lifting force;
(3) turbulent eddying in the fluid;
(4) collisions between particles on the bed and those which are *saltating* (i.e. hopping along the bed); and
(5) rolling of large particles.

Bed shear stress exerted by moving water, per unit area of channel boundary, is proportional to the product of hydraulic mean depth of the flow and the slope of the water surface. This relationship is expressed by:

$$\bar{\tau}_o = \gamma_w R \sin\beta \qquad (6.7)$$

where:

- $\bar{\tau}_o$ is the average boundary shear stress over the channel boundary (N/m^2),
- γ_w is the unit weight of water (9.8 kN/m^3),
- R is the mean depth of flow (m),
- β is the slope of the water surface (in degrees).

As the slope of the water surface and water depth increases so does the bed shear stress. It also increases as the density of the fluid increases so that fluids carrying large volumes of sediment (and hence having greater unit weights) are more erosive than clear water.

The resistance to entrainment of the particle is determined by its volume, weight, and friction with the bed. The average resistance to entrainment per unit area of channel bed is then approximated by:

$$n\gamma_{sub}d^3\left(\frac{\pi}{6}\right).\tan\phi \qquad (6\cdot8)$$

where:

- n is the number of particles per unit bed area,
- γ_{sub} is the submerged unit weight of particles (kN/m^3),
- d is the diameter of the particles (m),
- $\tan\phi$ is the coefficient of friction between particles and the bed.

Whether or not entrainment occurs depends upon the ratio of resisting forces to entraining forces or:

$$\frac{\text{resisting force}}{\text{entraining force}} = \frac{n\gamma_{sub}d^3\left(\frac{\pi}{6}\right).\tan\phi}{\gamma_w R \sin\beta}. \qquad (6.9)$$

If resisting forces exceed entraining forces then there is no transport. For entrainment by air the depth of flow is irrelevant and the dry unit weight is used. It should be noticed that the resisting forces in equation 6.8 are defined in terms of the Coulomb criterion for non-cohesive solids, with the unit weight and particle volume defining the normal stress exerted on the channel bed. It has been shown experimentally, however, that the

total stress required to entrain particles in a turbulent flow is only 10–20 per cent of that which would be required to cause sliding of the particles along the bed. This occurs because 80–90 per cent of the entrainment forces are provided by lift, turbulence and impact.

Laminar flow and low-velocity turbulent flow, exert only a very low shear stress, so particles have to be large enough to protrude through the laminar sublayer into the zone of high-velocity turbulent flow before entrainment can occur. A bed of clay-size grains is consequently resistant to entrainment because of both cohesion between particles and because they are not large enough to protrude through the laminar sublayer.

The tractive stress exerted by a fluid is directly proportional to the density of that fluid, and hence its unit weight. As fluid density is partly controlled by its temperature the threshold velocities for entrainment are temperature-dependent. Water in natural channels ranges in temperature from about $0°$ to $+30°C$, air ranges from about $-70°C$ in Antarctica to more than $+50°C$ in the Sahara. The difference in threshold velocities for entrainment of a 3 mm granule in air of various temperatures is shown in Fig. 6.15. At $-70°C$ the required wind velocity is 35 m/s but at $+50°C$ is 45 m/s. Note that the break in the curve at $0°C$ occurs because there is a slight difference in the

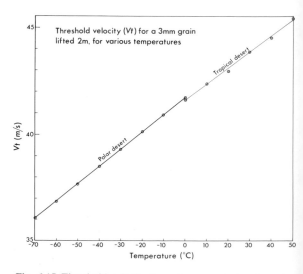

Fig. 6.15 Threshold velocity for a 3 mm grain in air, as influenced by temperature (after Selby *et al.*, 1974).

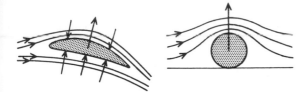

Fig. 6.16 Lift forces acting on a wing and on a spherical particle resting on the bed of a channel.

value of gravitational force between the poles and the equator.

A lift force is produced when fluid flows over a particle. In a flow of air over the wing of an aircraft the stream lines are compressed over the wing and hence velocity there is increased while velocity below the wing may be constant. The increased velocity over the wing is counterbalanced by an upthrust vertical pressure or lift force (Fig. 6.16). Flow of water over a particle resting on a channel bed, or of air over a particle on the ground, is faster than around the sides, and the fluid at the boundary between the particle and the bed is retarded by friction so that it is nearly stagnant. Consequently the particle is subjected to lift forces similar to those around an aircraft wing, and drag forces of fluctuating magnitude operating in the wake formed downstream of the particle.

A particle is set in motion because the laminar sublayer becomes thinner as the fluid velocity increases and more of the particle protrudes into the higher-velocity flow above. Lift forces at a stream bed or at the ground surface are at least equal to tractive forces, and may exceed them by a factor of two or three. As soon as the lift force exceeds the downward force provided by the weight of the particle, the grain will move up from the bed into the flow. The lift force declines rapidly above the bed as the streamlines become evenly spaced above and below the particle, which then sinks back to the bed. Lift forces are usually negligible a few centimetres above the bed.

Saltation (Fig. 6.14) is readily explained by lift forces. A particle rises to a height where lift is no longer effective but it is subjected to drag by the fluid. Consequently, instead of rising and falling vertically under lift forces alone, it has a parabolic trajectory caused by a combination of lift and drag, with a steep rise from the bed and a shallower angle of descent.

Turbulent eddying is an important process in lifting particles from a channel bed as it involves the transference of fast-moving fluid from higher to lower parts of a flow. The larger and faster the eddy the greater is the size of particles that can be transported.

Collisions between particles are very important in entrainment where the fluid is air. They are much less effective in water or mudflows because the submerged density of most mineral grains is only about twice the density of water, hence the impact of a grain, descending through water, upon another grain is slight. In air, however, a saltating grain can, by impact, move grains exceeding six times the size of those composing the saltation load. Wind can support in suspension grains of up to 3 mm diameter, but these may set in motion grains up to 18 mm diameter by impact. For wind, therefore, there is a critical threshold velocity for entrainment by drag forces and a much lower velocity for entrainment by impact (see Fig. 12.1).

Rolling of particles is the primary means by which large debris is moved along the bed of a stream. It is particularly effective in steeply sloping channels where the weight of the grain can operate downslope. Rolling is not an important transport process in air because the low mass density of air prevents it from exerting high tractive forces compared with the frictional resistance of a granular bed.

Velocity of flow and particle movement

As high fluid velocities permit entrainment of increasingly large particles above a critical size, a diagram may be drawn showing the relationship between mean fluid velocity needed for entrainment and particle size (Fig. 6.17). This diagram, known after its originator as the Hjulström diagram, is valid only for flowing water and quartz particles. It shows that, once a particle has been entrained by a flow exceeding the critical erosion velocity, that particle can be held in suspension and transported by slower moving water until a critically low velocity is reached at which deposition occurs. Very fine particles of clay size—such as quartz and feldspar crushed and abraded at the bed of a glacier—may not have the cohesion of clay mineral particles of similar size. Hence two curves are shown for entrainment of very fine grains. Because flow velocity varies with flow depth, a given grain size will vary in mobility with depth and discharge of water.

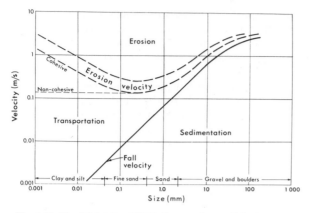

Fig. 6.17 The Hjulström (1935) diagram for particles in water.

The Hjulström diagram is a pioneer effort to relate fluid-transporting power to the size of particles which can be entrained and transported. Many diagrams of similar form, and only slightly different in numerical values, have been produced since (see, for example, Fig. 6.18).

Further reading

The nature of rock and soil behaviour is discussed by Johnson (1970) and the shear strength of rock and soil by any textbook on rock and soil mechanics. The mass strength of rock is less well understood and the only comprehensive review of

the topic is by Selby (1980, 1982a). Fluid flow is discussed in most textbooks on hydraulics and its application to geomorphology and sedimentology in such texts as that edited by Embleton and Thornes (1979) and in Blatt, Middleton, and Murray (1972).

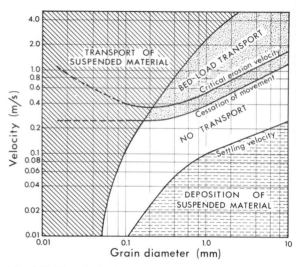

Fig. 6.18 The relation between flow velocity, grain size, and conditions of transport, entrainment, and deposition of quartz particles. The curve is not valid for high sediment concentrations which increase the fluid viscosity (modified from Sundborg, 1956).

7 Weathering and Landforms

Weathering is the process of alteration and breakdown of rock and soil materials at and near the Earth's surface by physical, chemical, and biotic processes. Igneous and metamorphic rocks, as well as deeply buried and lithified sedimentary rocks, are formed under regimes of high temperature and pressure. At the ground surface the environment is dominated by lower temperatures and pressures, and in humid climates by the presence of water for chemical reactions. Here rocks are altered by weathering to new materials which are in equilibrium with surface conditions

Weathering has four very important characteristics: (1) it is the process which renders resistant rock, and partly weathered rock, into a state of lower strength and greater permeability in which the processes of erosion can be effective; (2) it is the first step in the process of soil formation; (3) in some regions it accumulates, or releases in solution, lime, alumina, silica, and iron oxides, and when these are concentrated they form indurated shells on rocks, or layers in the soil —known as duricrusts—which become hard and resistant to erosion; and (4) it is also responsible for the formation of many distinctive minor landforms which develop on the surface of rock outcrops.

The alteration of rock by weathering occurs in place, that is *in situ*, and it does not directly involve removal processes. It may be characterized by the physical breakdown of rock material into progressively smaller fragments without marked changes in the nature of the mineral constituents. This disintegration process leads to the formation of a residual material comprising mineral and rock fragments virtually unchanged from the original rock. By contrast chemical alteration may induce thorough decomposition of most, or all, of the original minerals in a rock, resulting in the formation of a material composed entirely of new mineral species, particularly of clay minerals.

Biological weathering induced by biophysical and biochemical agencies is largely confined to the upper few metres of the Earth's crust in which plant roots are active.

It must be appreciated that physical, chemical, and biological processes usually operate together. Also erosion takes place from the surface of the ground, and within the soil by solution, almost continuously so that, although we speak of weathering as a process of decomposition *in situ*, transport of the residuum of weathering may be simultaneous and assists in the continuation of weathering.

Weathering processes

There is an extensive literature on weathering and soil formation, but there is still much debate about the mechanisms and rates of operation of many processes, and also much uncertainty about which weathering processes are responsible for many minor landforms. There are many reasons for these uncertainties, but the following can be readily identified. (1) Weathering rates are very slow and the time taken for landforms to develop is often of the order of many thousands of years: experiments cannot operate over, or simulate, such time-scales. (2) Most experiments have not simulated natural conditions of temperature, humidity, size of rock samples or columns of soil, or of size and abundance of joints and fissures. (3) Most experiments have not used standardized techniques, and reports have not specified fully all conditions of the experiment and properties of the rock and soil used. There is, consequently, much conflict in the results achieved and no opportunity to analyse the causes of the conflicts. (4) Landforms develop in environments in which physical, chemical, and biotic processes act in concert: experiments have to be carried out in simplified conditions in which one, or few, processes operate at one time, hence the results may be far removed from those of natural conditions. (5) During the thousands of years in which landforms develop climate is not stable, and the relicts which are the result of past processes may not be discernible from modern features. (6) Experiments on soil weathering have often been based on the assumption that the modern soil is formed *in situ*, but it is evident from many studies that windblown dust and additions by people, animals, and birds have added to, and disturbed, soil profiles.

Physical weathering

Mechanical breaking of rock can result from a number of physical processes. Its results are usually most obvious where rock is exposed at the ground surface, and are thus most characteristic of hot or cold deserts and of cliff faces. Physical weathering products range from large joint blocks many metres in dimensions to fragments of rock crystals.

Stress release is perhaps the most important of all weathering processes because it opens large fractures in rocks and permits other processes to operate. When a rock is produced from a magma, is metamorphosed under heat and pressure, or is buried by an overburden, that rock is compressed so that its mineral grains are forced together and the voids in the rock are reduced in volume. Once the original stresses are removed, as by erosion of the overburden, the rock can physically expand and, as it does, fractures form within the rock

mass. These may be large joints nearly parallel to the surface of the outcrop, or merely small and shallow fractures separating thin plates of rock from the surface of the outcrop (Plate 7.1a). Stress release can continue to a considerable depth into an outcrop and may be responsible for the separation or *exfoliation* of both large and small rock slabs. Because such slabs may be nearly concentric the shape of the original rock surface can be preserved by stress release exfoliation or *sheeting*.

Frost action and hydro-fracturing are two of the most important physical processes. They are dependent upon: (1) the presence of pores and cracks in rock; (2) the presence of water, with saturation and nearly pure water providing the most favourable conditions for the process; (3) a temperature regime in which crack and pore temperatures fall through the range of about -2 to $-7\,°C$ and hold that temperature for same hours in which freezing can be completed.

It was once thought that rock shattering in cold conditions is due solely to the formation of ice crystals in cracks, because when water freezes it increases in specific volume by 9 per cent, but it is now thought that such freezing cannot produce large enough pressures to overcome the tensile strength of strong intact rock. Such frost action, as it is called, is most effective in the heaving and pushing of pebbles and clods, as ice lenses form in soils. Frost action in this form is a major process in climates with very cold winters and can produce many kinds of patterned ground (see Chapter 14).

Water can enter fractures which may be little wider than the combined diameters of a few

7.1 (a) The surface of a granite outcrop in the Namib Desert. Plate exfoliation is occurring at the top left, granular disaggregation to the right and small tafoni are developing. This outcrop is near a salt deposit and salt weathering may be exploiting stress release cracks and causing disaggregation.

molecules of water. If ice seals off the end of such a fracture, liquid water in the crack may be forced towards the tip of the crack and so extend it. This process of hydrofracturing is probably responsible for much of the physical splitting of rocks, and the angular debris, which occurs in cold climates (Plate 7.1b).

It has become traditional to regard 'Icelandic' climatic regimes, with abundant moisture and frequent temperature changes of the order of $+8$ to $-8\,°C$ in 24 hours, as being more effective for frost action than 'Siberian' regimes with fewer crossings of the freezing-point, drier conditions, but lower temperatures. This view is changing now that it is recognized that the low thermal conductivity of rock may not permit water in a crack to be frozen, then thawed during a short freeze–thaw cycle, and also that water in narrow rock pores, or with salts in solution, may not freeze until rock temperatures fall well below $0\,°C$. (See also p. 396.)

In cold deserts, with suitable rocks such as slates, schists, or porous sandstone, the entire ground surface may be shattered to produce a surface of broken rock fragments called a *felsenmeer* (a German word meaning 'rock sea.) (Plate 7.1b).

Salt weathering results from the crystallization of salts in rock pores and fissures. Growing crystals can exert high pressure on the confining rock and cause both rock fragments to spall off and individual grains to fall out of the outcrop. The salts involved in this type of weathering may be derived from the sea and be carried inland in spray, or carried in snow and rain, or they may be derived from chemical weathering of rock. In desert areas particularly, drainage waters may evaporate to leave a salt-rich sediment which will be deflated by wind, and the salt redeposited on rock surfaces. The most obvious result of salt weathering is the *granular disaggregation* of an outcrop as grains fall out to leave an uneven, pitted or caverous surface (Plate 7.1c).

Insolation weathering is the result of alternate warming and cooling of rock surfaces under the direct influence of solar heating. Laboratory experiments on rock samples heated in furnaces have often failed to produce splitting and have caused some workers to doubt the existence of the process. The presence of cleanly split rocks (Plate 7.1d) in hot deserts, however, indicates that the process probably does occur, but it may be associated with the presence of water within the rock which, on heating, expands and splits the rock. It is also possible that splitting is produced when sudden rain showers fall on heated rock and produce a shock effect.

Alternate wetting and drying has been suggested as being responsible for the disintegration or *slaking* of fine-grained rocks. The nature of the process is not fully understood but it may involve the absorption of water molecules onto the negatively charged surface of a rock or clay crystal. Such adsorption within a crack may then force particles apart and detach them from a rock surface. Repeated wetting and drying may ultima-

7.1 (b) Physical disintegration of a schist, Antarctica. Note the elongate angular fragments derived from the schist. These fragments form a thin felsenmeer.

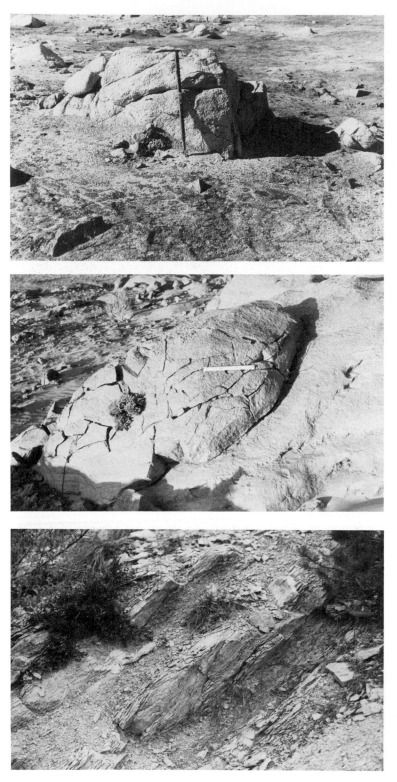

7.1 (c) Granular disaggregation of a marble outcrop by salt weathering, Antarctica.

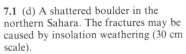

7.1 (d) A shattered boulder in the northern Sahara. The fractures may be caused by insolation weathering (30 cm scale).

7.1 (e) Slaking of mudstone, French Alps.

tely disintegrate rocks such as mudstone (Plate 7.1e). An alternative explanation is that during extreme desiccation air is drawn into the pores of a rock or soil. When water covers the outer surface of the rock the negative pore pressures within it create a suction, which draws capillary water into the pores and traps the air under pressure. This may then cause slaking by 'air breakage'. In some rocks held in water the slaking process can be seen to occur within seconds of immersion.

Some clays such as montmorillonite can increase in volume by 60 per cent or more, on wetting from a dry state and some clay-rick rocks are consequently severely affected by alternate swelling and shrinking associated with wetting and drying. The results of such processes are very obvious in the cracked beds of dried-up ponds.

Chemical weathering

The disintegration of rocks in the early stages of weathering usually involves both physical and chemical processes, for chemical weathering is commonly initiated along cracks between and across mineral grains, and not just at the rock surface. Major chemical weathering processes include solution, hydration, hydrolysis, oxidation, reduction, and carbonation. Their effect upon fresh rock is usually to etch pits into the more resistant minerals such as quartz, and more particularly the feldspars, so that under a scanning electron microscope the crystal appears to become increasingly honeycombed as weathering proceeds (Plate 7.2a). The progressive increase in porosity permits further chemical activity as ions are removed in solution from the feldspar crystal. One result of pitting and fracture of crystals is the formation of sand and silt-size grains. Clay minerals, as distinct from clay-size particles, may be released from sedimentary rocks or they may be created by intense weathering of rock minerals and crystallization of new clay minerals.

There are three main classes of clay minerals: (1) layer-silicates; (2) sesquioxides; (3) non-crystalline (amorphous) minerals.

(1) *Silicate minerals* have a structural form determined by the arrangement of their oxygen ions (O^{2-}). All silicates consist of groups of O^{2-} ions, sometimes in combination with hydroxyl ions (OH^-), that are organized into one or both of two fundamental units: one of these units contains four O^{2-} ions packed together in a four-sided *tetrahedron* (Fig. 7.1); and the other has six O^{2-} ions arranged in an eight-sided *octahedron*.

A Si^{4+} ion is centrally located in both the tetrahedron and the octahedron. The positive charge of this ion exerts an attraction for the negatively charged O^{2-} ions and thus provides the fundamental force which holds the units together: the strength of this bond largely determines the resistance of the mineral to weathering.

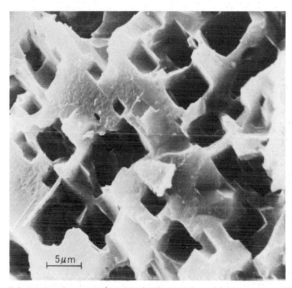

7.2 (a) An electron micrograph of a weathered feldspar crystal showing pitting (micrography by Dr M. J. Wilson, Macaulay Institute for Soil Research, Aberdeen).

7.2 (b) An electron micrograph of kaolinite in hexagonal platelets (micrography by Drs I. J. Smalley and J. G. Cabrera).

7.2 (c) Muscovite showing the multiple plates of its crystal structure.

7.2 (d) The two-dimensional structure of montmorillonite is shown by this electron micrograph of the edge view of the crystal units. The dark parallel lines, which are 100 nm ± 20 nm wide, are the units (micrograph by L. M. Barclay and D. W. Thompson, University of Bristol).

Because of its small size the silicon cation fits readily into the internal space of the tetrahedron and is the usual ion found inside this unit. Aluminium ions (Al^{3+}) may also occur in the tetrahedra but, because they are relatively large, they cause the unit to be less stable than a tetrahedron with a Si^{4+} ion. The substitution of

one ion for another in known as *isomorphous replacement*, and when it occurs it produces an imbalance of the electrical charges, within the structure, which is satisfied by the introduction of cations such as Na^+, K^+, Ca^{2+}, and Fe^{2+}.

Differences among silicate minerals depend upon whether tetrahedral or octahedral structural units are present, and on the way these are linked to the general structure of the mineral. Thus they may be linked as a chain, a sheet, or a ring, for example.

Most silicate clay minerals have their ions organized into plate-like units (platelets) which occur in stacks of varying thickness to form large crystals (Plate 7.2c). Such platelets are bound internally by shared oxygen ions (Fig. 7.2) of the tetrahedra and octahedra. The platelets making up the crystal occur in combinations such as of one tetrahedral and one octahedral sheet, when the mineral is said to have a 1:1 crystal lattice; or as two tetrahedral and one octahedral sheets when it is said to have a 2:1 crystal lattice.

Ionic substitution is characteristic of most 2:1 lattice minerals, although not of the 1:1 lattice minerals occurring in soils. In some 2:1 minerals a few Al^{3+} ions substitute for Si^{4+} ions in the tetrahedral sheets. In others Mg^{2+} or Fe^{2+} substitute for all, or part of, the Al^{3+} in the octahedral sheet. Whenever substitution occurs the platelets acquire a residual negative charge that is neutralized by the retention of cations on the surfaces of

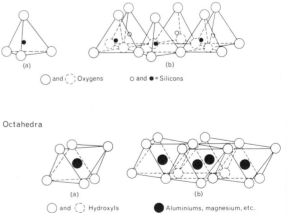

Fig. 7.1 The arrangement of ions in a tetrahedron. Four oxygen ions are symmetrically spaced around one silicon ion. In an octahedral unit magnesium or aluminium ions are co-ordinated octahedrally with oxygen or hydroxyl ions.

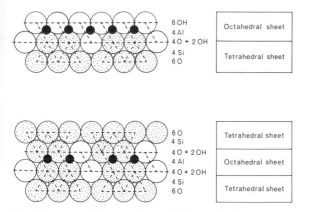

	6 OH	Octahedral sheet
	4 Al	
	4 O + 2 OH	
	4 Si	Tetrahedral sheet
	6 O	

	6 O	Tetrahedral sheet
	4 Si	
	4 O + 2 OH	
	4 Al	Octahedral sheet
	4 O + 2 OH	
	4 Si	Tetrahedral sheet
	6 O	

Fig. 7.2 A two-dimensional model illustrating the arrangement of the component ions in a 1:1 (top) and 2:1 (bottom) layered mineral structure.

platelets or between platelets. Cations between platelets bind the sheets together so that they make stacks of platelets of which the sheet silicate mineral is composed (Plate 7.2c).

The classes of layer silicate clay minerals are defined by the arrangement of the layers (1:1, 2:1, or 2:2) and by the thickness of one layer and one interlayer space. This thickness is a distance measured in nanometres (1 nm = 10^{-9}m) and may be determined when the crystal units are as close to one another as possible—this condition is said to be the collapsed state; or it can be a measure of the maximum possible separation when the lattice is swollen because ions, water or organic molecules have become included between the layers—this is the expanded or swollen state (Plate 7.2d).

The characteristics of clay minerals are determined by the strength of the bonds between their sheets, the spacing of the crystal units, and the extent to which they can swell (Fig. 7.3a,b,c).

There are three families of layer silicate minerals: (1) the 1:1 family which includes kaolinite and halloysite; (2) the 2:1 family which includes illite, vermiculite, and montmorillonite (since micas also have 2:1 lattices members of this family are also known as micaceous clay minerals or hydrous micas); (3) the 2:2 family which includes the chlorites.

(2) *The sesquioxides* may be either crytalline or amorphous. These are widely distributed oxide and hydroxide forms of iron and aluminium. In soils gibbsite ($Al_2O_3 \cdot 3H_2O$) is usually the dominant aluminium oxide; the common iron hydrous oxides include goethite ($Fe_2O_3 \cdot H_2O$) and limonite ($2Fe_2O_3 \cdot 3H_2O$). The sesquioxides occur in greatest abundance where weathering has been extensive and hence are particularly common in tropical soils, although they also occur in temperate regions where they may be intermixed with silicate clays.

(3) *Amorphous clay minerals* are those which do not appear to have a regular atomic structure when samples are examined by X-ray analysis. One of the most important amorphous minerals is allophane. Allophane is a hydrous aluminosilicate which does not have a fixed chemical composition (its SiO_2 varies from 25 to 34 per cent; Al_2O_3 from 30 to 37 per cent; and water from 25 to 40 per cent). Allophane is commonly formed in the soil by weathering of volcanic ash and rock and possibly as a result of frost weathering.

Clay mineral formation can occur by ionic substitution as when mica or illite suffer replacement of interlayer K^+ with other ions from soil solutions. The result is a new mineral, vermiculite, which has the original mica structure, but different physical characteristics in that it can expand when wet. The usual sequence is from complex to simple minerals such as:

feldspar→montmorillonite→kaolinite→gibbsite.

Such a sequence indicates stages of weathering from which something of the history of a site may be deduced, but there is also some control exerted by the primary minerals of the parent rock, with sesquioxides forming most readily from ultrabasic rocks, and montmorillonitic clays forming directly from basic and calcareous rocks. Mixed silicate clays are common in one soil, with halloysite and kaolinite occurring together and forming from most rock types.

Solution is commonly the first stage in chemical weathering and removes the most readily soluble minerals. It is, however, not a simple process because water in, or on, rock is never 'pure'. It contains ions which modify its pH, and as ions are released from the rock into the water these ions react with other ions or minerals to form new mineral combinations. The solubility of rock minerals has to be expressed in relation to the pH of the water; silica, for example, is slightly solube at all pH values, whereas alumina (Al_2O_3) is only readily soluble below pH4 and above pH9, i.e. under very acidic or very alkaline conditions which seldom occur in nature. Alumina, therefore,

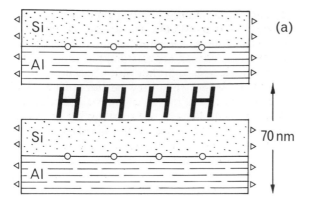

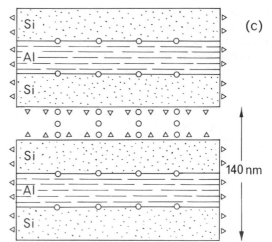

Fig. 7.3 (a) Schematic representation of two kaolinite crystal units. The silicon and aluminium sheets are bonded by shared oxygen ions (o) and the crystal units are joined by oxygen–hydroxyl linkages (H). The strong links between the two units prevent penetration by cations, or water molecules, between the units. Kaolinite is therefore non-swelling and its spacing is constantly 70 nm. Cation exchange is limited to the edges of the units (Δ).

Fig. 7.3 (c) Schematic representation of a montmorillonite crystal. Weak oxygen (o) linkages between the crystal units permit swelling to as much as 180 nm spacing and collapse to 100 nm spacing. The internal surfaces between units have a predominantly negative charge, consequently cations are adsorbed readily. These clay minerals are noted for their high cation exchange capacity.

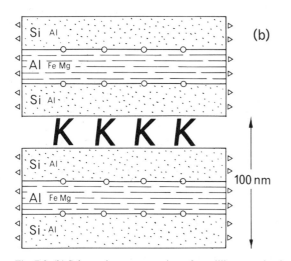

Fig. 7.3 (b) Schematic representation of two illite crystal units. Within the sheets some replacement of ions occurs with aluminium in the silicon layer and some iron and magnesium ions replacing aluminium ions in the aluminium layer. Potassium ions (K) link the crystal units and are strong enough to prevent illite being a swelling mineral. Illite has cation exchange properties which make it intermediate in character between kaolinite and montmorillonite.

tends to accumulate in the clayey residues resulting from chemical weathering in soils, but silica is slowly leached. Because of variability in solubility of minerals caused by pH, presence of organic compounds, variations in rock and soil permeability and in temperature, it is not possible to state a fixed order of rock or mineral solubility. The only invariable rule is that calcium, magnesium, sodium, and potassium are all more soluble than aluminium, silicon, and iron, and that rocks rich in the ferromagnesian and calcic plagioclase minerals are less resistant than those rich in orthoclase, quartz, and muscovite. Limestones are resistant to solution in pure water but, in the presence of dissolved carbon dioxide, calcium carbonate is replaced by calcium bicarbonate which is soluble in water. Limestones are consequently often regarded as the most soluble of the common rocks:

$$CaCO_3 + H_2O + CO_2 \rightarrow Ca(HCO_3)_2$$
calcium carbonate   calcium bicarbonate

Hydration is the addition of water to a mineral and its absorption into the crystal lattice. The absorption may make the mineral lattice more porous and subject to further weathering. Iron oxides for example may absorb water and turn into hydrated iron hydroxides:

$$2Fe_2O_3 + 3H_2O \rightleftarrows 2Fe_2O_3 \cdot 3H_2O$$
Hematite   Limonite

The hydration reaction is frequently reversible,

and because it involves a considerable volume change it is important in physical weathering.

Hydrolysis is a chemical reaction between a mineral and water, that is between the H^+ or OH^- ions of water and the ions of the mineral. In hydrolysis, therefore, water is a reactant and not merely a solvent.

The actual chain of reactions during hydrolysis is very complex but none-the-less effective. Coarse-grained and feldspar-rich rocks such as granite are very vulnerable and, as hydrolysis may involve expansion of silicate minerals, it can cause physical disruption of the rock and loosening of mineral grains.

It is often said that hydrolysis is a major process in the production of clay minerals but it seems more probable, from the rarity of pure water in soil solutions and the abundance of bicarbonate solutions, that carbonation is the cause of the decomposition of feldspars to clay minerals.

Carbonation can occur readily because bicarbonate is nearly always present in weathering solutions and is the major solution component of drainage waters. The bicarbonate ion is derived from the photosynthetic fixation of carbon dioxide and its subsequent respiration from plant roots and the bacterial degradation of plant debris. The carbon dioxide in the soil atmosphere dissolves and dissociates to produce bicarbonate:

$$H_2O + CO_2 \rightleftharpoons H_2CO_3 \rightleftharpoons H^+ + HCO_3^- \text{ bicarbonate}$$

The supply of acid soil waters is thus largely controlled by plant activity. The weathering of a feldspar may thus be represented by:

$$6KAlSi_3O_8 + 4H_2O + 4CO_2 \rightarrow$$
orthoclase
$$K_2Al_4(Si_6Al_2O_{20})(OH_4) +$$
illite
$$12SiO_2 + 4K^+ + 4HCO_3^-$$
solution

Oxidation is the process of combining with oxygen and results in an increase in positive valence or a decrease in negative valence, as where Cu^+ is oxidised to Cu^{2+} and S^- is oxidized to S. A particularly important oxidation process is the alteration of iron from the ferrous, reduced state, to the ferric, oxidized state:

$$4FeO + O_2 \rightarrow 2Fe_2O_3$$
ferrous oxide   ferric oxide

The mafic minerals such as amphiboles, biotite, olivines, and pyroxenes are subject to oxidation weathering and its effect can frequently be observed by the presence of brown iron-staining along cracks and between mineral grains.

The oxidation of pyrite and the formation of gypsum or jarosite may occur between fissile shale layers causing the rock to expand or heave. Where air is present, pyrite (FeS_2) is oxidized to sulphuric acid under aerobic conditions resulting in acidic ground water that reacts with available calcium carbonate (carbonate fossils, calcite, $CaCO_3$) to form gypsum ($CaSO_4 \cdot 2H_2O$):

$$4FeS_2 + 2H_2O + 15O_2 \rightarrow 2Fe_2(SO_4)_3 + 2H_2SO_4$$
$$H_2SO_4 + CaCO_3 + 2H_2O \rightarrow CaSO_4 \cdot 2H_2O + H_2CO_3$$

The calcite to gypsum transformation involves an, approximately, 99 per cent increase in volume. Where calcite is absent the sulphuric acid will react with other ions, such as potassium and sodium from micas, to form a basic sulphate, jarosite ($KFe_3(SO_4)_2(OH)_6 \cdot 2H_2O$). Gypsum and jarosite appear as white and yellow efflorescent salts, respectively, on rock surfaces.

Reduction is the opposite process to oxidation and usually occurs in waterlogged (anaerobic) conditions in the absence of free oxygen. The reduction of iron to the ferrous form renders it more soluble and mobile so water draining from bogs is frequently stained brown by oxides that have been reduced in the presence of anaerobic bacteria and then converted back to the ferric state as the drainage water is aerated.

Weathering profiles

Physical, and some chemical, weathering processes produce broken rock fragments, and chemical weathering produces new secondary minerals. Consequently a weathering product consists of a mixture of rock fragments and secondary minerals whose composition depends upon a number of factors including: mineralogy, porosity, texture and jointing of the original rock; the climate of the area, with warm humid climates favouring chemical weathering, and cold or hot desert conditions favouring physical weathering; the site itself, with steep slopes favouring erosion processes which can remove weathering products, but flat sites favouring their accumulation; and vegetation, which itself can be an active weathering process as roots widen fissures in rock, release CO_2 into the soil during respiration, and produce organic compounds which modifiy soil pH and produce complex chemical reactions. Vegetation may also protect soil from rainsplash and erosion by running water.

In favourable conditions of readily weathered rocks with closely spaced joints, a flat site, abundant vegetation, and time during which weathering processes can occur, weathering products can accumulate and form a layer or *regolith* of alteration products.

In most accumulations of weathering products it is possible to recognize varying degrees of alteration ranging from nearly fresh rock to completely altered material. The fresh rock is usually at the base of the profile where weathering processes are etching into the rock along planes of weakness, such as joints, and the most altered material is at the ground surface. It is often possible to recognize five grades of alteration products in the field and to use these grades as the basis for classifying rock weathering in the rock mass strength classification (see Table 6.1). The five grades may be distinguished according to the descriptions given in Table 7.1, and a characteristic profile derived from weathering of granite is shown in Fig. 7.4. It should be noted that the surface soil, rich in plant roots and humus, is not included in the classification as it owes its characteristics as much to plants and/or human interference as it does to weathering processes within the regolith. In a particular profile not all grades may be present: thus a profile may pass from fresh rock grade I, into moderately weathered grade III, and then into an agricultural soil.

Landforms produced by weathering

The development of many landforms involves the reduction of rock, by weathering, to a more easily eroded material and then the removal of this weaker product by erosive agents such as running water, ice, wind, or waves. Some minor landforms,

Table 7.1   Weathering Classification

Grade	Class	Description
—	Residual or agricultural soil	A pedological soil with characteristic horizons and no sign of the original rock fabric.
V	Completely weathered	All rock material is decomposed to soil but the original rock texture or fabric is largely preserved.
IV	Highly weathered	Rock is discoloured throughout; more than half of the rock material is decomposed; discoloured rock is present as blocks or rounded corestones.
III	Moderately weathered	Less than half of the rock is decomposed or disintegrated. Fresh rock fragments are present as blocks or corestones which fit together.
II	Slightly weathered	Rock may be slightly discoloured, especially along joints; rock is not much weaker than when fresh.
I	Fresh rock	Parent rock showing little or no sign of discolouration, loss of strength or other signs or weathering.

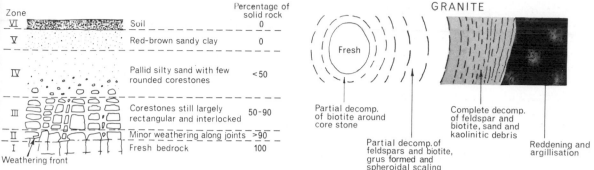

Fig. 7.4 Features of a full weathering profile developed on granitic rocks. The numerals refer to descriptions given in Table 7.1 (modified after Ruxton and Berry, 1957).

however, result more directly from weathering itself. These include the result of action within the weathering profile and also the small features on the surfaces of rock outcrops. Crusts of material such as iron oxides, silica, calcium carbonate, and alumina are produced by complex processes, but as they result largely from release of material in solution by weathering they are discussed here.

Corestones and tors

Within a weathering profile solution and other processes penetrate down joints and other fissures in the rock, consequently they tend to be most effective where joints are closely spaced, and least effective where joints are widely spaced. In many rocks joint spacing is variable and weathering penetration is irregular, with zones of shallow and deep penetration. Most joints have angular intersections, but weathering attack occurs on all faces of a joint block so that sharp corners are gradually rounded by weathering; even sharply irregular blocks are gradually rounded within the profile (Figs. 7.5, 7.6).

A feature of many boulders and joint blocks within the regolith is that they are surrounded by concentric shells of rock and weathering products (Plate 7.3a). There is still much controversy over the origin of such features with one group of hypotheses suggesting that the shells result from residual stresses contained in the rock, as it cooled or contracted, at the same time as major joints were formed. A second group of hypotheses suggests that the layering is a result of chemical reactions which cause solutions of minerals to separate and become concentrated according to

GRANITE

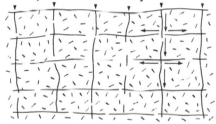

Fig. 7.5 The development of weathering zones around a granite corestone. Immediately around the corestone there is only limited weathering with biotite decomposing; beyond that both biotite and feldspars are decomposing leaving a grus of unaltered quartz crystals in a matrix of decomposed material. Stronger weathering produces kaolinite, but quartz grains still survive:farthest from the corestone, and along the joints, iron oxides stain all material a red-brown colour and clay formation (i.e. argillization) is advanced (after Selby, 1982a).

a. **Water penetrates down joints**

b. **Subsurface weathering guided by joint planes**

c. **Lowering of land surface; removal of debris; corestones exposed as tor boulders**

Fig. 7.6 The formation of boulder tors in a two-stage process (modified from Twidale, 1971).

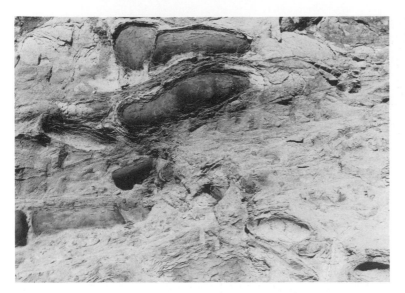

7.3 (a) Spheroidal weathering around corestones of basalt, Drakensberg, South Africa.

7.3 (b) A boulder field of corestones from a stripped deep weathering profile on granite, Cameroun.

species so that, for example, iron oxide-rich layers alternate with kaolinite-rich layers. Such layering is probably related to changes in pH and hence the solubility of certain minerals.

The formation of concentric weathering shells is called *spheroidal weathering*. It rounds off the edges of joint blocks and leads to the isolation of *corestones* of coherent rock within a matrix of weathering products. Granitic rocks are particularly susceptible to the production of corestones as the mixed mineralogy of the rock permits spheroidal weathering to develop, but many other igneous rocks such as basalt and andesite, and some arkosic sedimentary rocks also weather this way.

If erosion sets in, the fine-grained clays, silts, and sands of a weathering profile are washed away more readily than the corestones so that a *boulder field* may be left at the surface (Plate 7.3b). If the corestones in a profile are derived from a zone of very widely spaced regular joints the removal of the fine material may leave the corestones in place to form *boulder tors* (Fig. 7.7; Plate 7.3c). Such tors are particularly common in areas of granite bedrock where there is, or has been, a long period of stable ground surfaces with the development of very deep-weathering profiles. Consequently boulder tors are most common on the shield areas of the humid tropics or shields which, like central

7.3 (c) A boulder tor, Devil's Marbles, Australia.

7.3 (d) A tor in granite, Dartmoor. Note the angular joint intersections.

Australia, had a humid tropical climate during much of Tertiary time.

Tors of a different shape may develop other than by the two-stage process of deep weathering followed by stripping described above. They may occur almost anywhere there is a large contrast in the spacing of joints and erosion processes are powerful enough to remove the products of weathering from the zones of closely spaced joints, but not powerful enough to remove the rock where joints are widely spaced. Consequently tors may be produced by a variety of processes, on many kinds of rocks, and the term 'tor' is applied to any tower-like rock formation which is rooted in bedrock and stands conspicuously above its surroundings. Tors are commonly of the size of 5–20 m high.

The term 'tor' was originally applied to the features of Dartmoor, SW England. Most of these granite tors occur on the broad convex crests of the moor. It was proposed by D. L. Linton that they are two-stage tors, produced by deep weathering under a humid tropical climate in Tertiary times and then that the weathering products were stripped and carried away by soil flow (solifluction) and wash during the cold climatic periods of the last two million years or so. The Dartmoor tors are only subrounded granite masses, and are large joint blocks still in place and 'rooted' in the parent rock. Corestones are present, but the form of the tors is distinct from the boulder tors of tropical shields, partly because the joint pattern, rather than corestone shape, controls the general form, and partly because the tors may have been reshaped by hydrofracturing in cold periods (Plate 7.3d; Fig. 7.7).

Tors can also be produced in a one-stage process, especially in cold climates where hydrofracturing progresses along closely spaced joints but is less effective on the more massive rock with widely spaced joints. As long as hillslope processes can remove the broken rock fragments produced by hydrofracturing the massive rock will gradually become more prominent in the landscape, stand-

1. Deepweathering

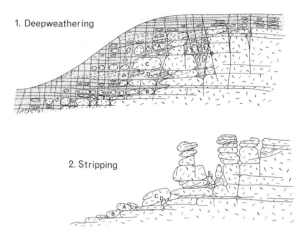

2. Stripping

Fig. 7.7 Linton's hypothesis of the formation of Dartmoor tors.

ing as a tor on a hillside which is retreating, or upon the crest of a hill which is being lowered. Tors of this type have been described from Antarctica, Tasmania, Alaska, Siberia, New Zealand, and Scandinavia. Rock types on which they form include quartzite, sandstone, granite, gneiss, schist, and dolerite.

Pits, pans, caverns, and rills

Many outcrops have a detailed relief formed by miniature depressions along either lines of weakness, such as joints, or on sites which appear to be uniform in resistance. The depressions are probably most common on very soluble rocks such as limestone, but can develop on any rock type, and in any climatic zone.

Weathering pits of flat surfaces range in size from a few centimetres to several metres in width and depth. The pits may be hemispherical, flask-shaped or flat pans; circular or irregular (Plate 7.4a). *Honeycomb* or *alveolar* weathering is characterized by numerous closely spaced pits and may occur on steep rock faces. *Tafoni* (singular, tafone) are hollows cut into steeply sloping rock faces: tafoni may have overhanging entrances called *visors*.

Most pits are assumed to be initiated at a depression or weakness and enlarged gradually, perhaps by consuming neighbouring pits. The rates at which they change are varied, but deepening at up to one centimetre a year can occur at coastal sites, with porous rocks, where alternate wetting and drying and salt weathering are active. Enlarging of pits usually involves granular disaggregation but also the exfoliation of rock. The removal of debris, which is necessary for the continuation of weathering, may be by wind, under gravity, or on flatter sites may be aided by overflow of water.

A variety of processes have been identified as causing cavernous forms. They may include chemical weathering of susceptible minerals such as feldspars, hydrofracturing, leaching of soluble minerals to weaken the rock fabric, salt weathering, and the action of algae, lichens, and fungi in releasing carbon dioxide and so causing carbonation. The pH of water in some weathering pits suggests that alkaline waters, in which solubility of

7.4 (a) A weathering pit, with soil and vegetation in it, on granite. Around the pit, plate exfoliation is occurring, Namib Desert.

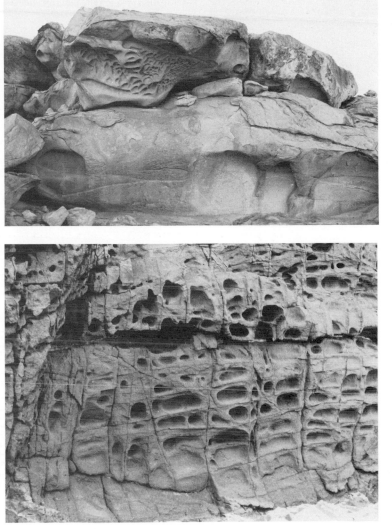

7.4 (b) Alveolar weathering of granite below a visor (upper boulder) and shallow tafoni in the face of the outcrop below, Namib Desert.

7.4 (c) Boxwork forms of tafoni in sandstone.

silica is greatly increased, are responsible for much of the enlargement. About the only generally valid conclusion, however, is that water is always involved in the weathering process. Consequently tafoni often form preferentially in damp, shady sites.

Case hardening or the production of a resistant weathering rind is a common feature of many outcrops. The minerals which cause the hardening are commonly iron and manganese hydroxides which may be associated with clay minerals such as goethite, and sometimes amorphous silica is the hardener. It is usually assumed that these minerals are dissolved from the rock interior, so weakening

it, and drawn by capillary attraction towards the rock surface where they are precipitated and dehydrated, sometimes to a less soluble mineral form. There is increasing evidence, however, that the interior of the rock has not been depleted of minerals, and many workers now recognize that the rind is composed of minerals carried there by wind. Case hardening may be involved in the development of visors over some tafoni entrances.

Desert varnish or *rock varnish* is not always thought of as being the same as the weathering rind associated with case hardening, but there is increasing evidence that the two phenomena have the same cause with the more lustrous appearance

7.4 (d) Rills on a granite outcrop and plate exfoliation over its surface, Namib Desert.

7.4 (e) Large rills, following the line of the bedding, in the arkose of Ayers Rock, Australia. The rills link a series of weathering pits.

of varnish being due to a smoother surface, or a polish caused by fine wind abrasion.

Boxwork weathering (plate 7.4c) is a special form of hardening caused by the precipitation of resistant minerals (usually iron oxides) along joints. The weaker parts of joint blocks may then be attacked by other weathering processes and hollows develop there.

Rills, grooves or gutters, 5–2000 mm wide and deep, form down the flanks of many rock outcrops (Plate 7.4d,e). They form the drainage network of the rock face but few contain enough debris to be regarded as miniature stream beds, and few are caused by abrasion. It is more probable that alkaline waters, especially on silicate rocks, cause solution. Some rills, like those on Ayers Rock, link

pits and form cascades from pool to pool during rain.

Duricrusts

An important group of landforms owes its origin to the presence and dissection of indurated crusts, of various mineral compositions, which have formed as a result of weathering processes or by the redistribution of weathering products. Such crusts frequently form resistant caps to hills and underlie extensive plains and plateaux, especially of the more stable continental platforms with tropical, subtropical, or temperate climates (Plate 7.5). The crusts, classed together as duricrusts, can also occur in other environments but seldom with

7.5 (a) Low hills capped by silcrete, near Coober Pedy, South Australia.

7.5 (b) A profile of a ferricrete. Nodular ferricrete forms the cap and the fallen boulders in the foreground, and the mottled zone lies below it, Darwin, Australia.

as great a thickness or forming major features of the landscape.

A duricrust is a product of processes acting within the zone of weathering to cause the accumulation of iron and aluminium oxides, silica, calcium carbonate, or, less commonly, gypsum and halite. The accumulation of these compounds may be either a result of removal of other materials to leave an enrichment of the crust-forming minerals, or it may be an enrichment caused by deposition from water, or of wind-borne minerals, which can then accumulate and harden to form a duricrust. This secondary or depositional origin can be of minerals derived locally, as material moving down a hillslope from a higher crust, or it may be a much more widespread redistribution by streams or wind.

Duricrusts are named according to their dominant crust-forming mineral.

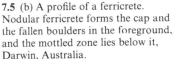

Silcrete	SiO_2, amorphous silica and quartz
Ferricrete	Fe_2O_3, hematite
Alcrete	$Al_2O_3 \cdot 3H_2O$, gibbsite
Calcrete	$CaCO_3$, calcite
Gypcrete	$CaSO_4 \cdot 2H_2O$, gypsum
Salcrete	$NaCl$, rock salt (halite)

It must be recognized, however, that these crusts are seldom, if ever, composed of a single mineral.

Many local names are also used for duricrusts and the literature can be confusing: a common use is 'laterite' for ferricrete. Laterite was originally used for the red iron-rich clay used for brick-making in India (Latin, *later*—a brick). This clay hardens on exposure to the sun and the term was

7.5 (c) Gibber plains of silcrete pebbles and boulders, Sturt Desert, Australia.

7.5 (d) A calcrete-cemented gravel deposit, Kuiseb Canyon, Namib Desert. The calcrete was derived from floodwaters.

later extended to the ironstone cappings found on the hills of central India. It is now often unclear whether it is being used for the clay or the crust.

Ferricrete and alcrete profiles which are developed in place usually have a number of horizons which vary in thickness, hardness, colour and composition.

0–2 m thick	*The soil zone* is often a red, sandy material containing hard, dark-red nodules. It may be eroded away.
1–10 m thick	*A crust* of reddish or brown hardened iron-rich nodules, or

blocks formed of nodules cemented together and with tube-like cavities filled with red or ochre clay. This crust has usually been leached of most minerals, except iron and aluminium oxides.

1–10 m thick	*A mottled zone* of white kaolinitic clay with patches (mottles) of yellowish oxides. The whole zone is strongly leached and weathered.
5–30 m thick	*A pallid zone* of bleached kaolinitic clay.

Up to 60 m thick *Weathered zone* of deeply weathered rock but still showing original rock structures.

At any site some horizons may be absent or thin and where the ferricrete, or alcrete, is not formed in place only a crust of recemented nodules (*pisoliths*) or blocks may occur on, or in, a soil or alluvial deposit. Such redeposited crusts occur where a capping is broken up by erosion, transported down a slope, and then recemented by mineral-rich soil water (Fig. 7.8).

Ferricretes are widespread in the humid tropics where deep weathering has caused leaching of the less stable minerals. The formation of a crust may occur upon exposure of an iron-rich nodular soil, and this has been aided by deforestation and soil erosion following excessive burning of natural vegetation and extension of agriculture. Formation of ferricretes (or '*laterization*' as it is sometimes called) has thus been aided by human activity, but many ferricretes are ancient. Those of northern Australia date back to at least mid-Tertiary times and were formed on an ancient, deeply weathered land surface which has subsequently undergone a drying of the climate, so that the ferricrete is now a relict of past processes. Alcretes occur in tropical rainforest zones and are economically very important as a source of bauxite.

Silcrete profiles consist of an indurated silicified layer, which may be up to 3 metres thick, with as much as 95 per cent silica. The rock has a white-to-grey, glassy, fine-grained appearance and, because many silcretes were formed by precipitation from ancient floodwaters which were rich in silica, the silcrete may be composed of cemented gravels or sand and be unassociated with a weathering profile. Silcrete is particularly widespread in east-central Australia where it dates from mid-Tertiary to early Pleistocene times. It forms extensive cappings on hills and silcrete debris stained red-brown by iron forms the stony *gibber* plains of the Sturt Desert (Plate 7.5c). Lenses of very pure amorphous silica are a source of opal in South Australia; isolated silicified boulders from parts of France and southern England where they are called *sarsen stones*—may be remnants of ancient silcretes.

Calcrete profiles, or *caliche* as it is called in much of North and Central America, are common in semi-arid climatic zones with annual rainfall of 500 mm or less. The lime accumulates within the soil as a powdery or nodular white material which can accumulate to form a thick hardpan. The lime is leached from rocks such as limestone and is redeposited in the soil profile when the soil water evaporates. Under higher rainfalls it is all leached away. Calcrete can also form from repeated evaporation of floodwaters and can then develop to thicknesses of tens of metres and consist largely of cemented soil and gravel (Plate 7.5d).

Gypcrete and salcrete profiles are exceedingly rare because of the high solubility of gypsum and halite. They are preserved only under arid climates and, because they are usually precipitates from the evaporation of sea water or saline desert streams, their profiles usually consist of a white or buff-coloured crust overlying lake or marine muds.

Climate and weathering profiles

It will be evident from the comments above that many weathering processes, and the landforms resulting from them, have a distribution which is strongly dependent upon climate (see Fig. 1.6). It must be recognized, however, that virtually every process can occur in nearly every climatic zone, but that there are zones in which particular processes are predominant. Two attempts to define the zones of predominant processes are shown in Figs. 7.9 and 1.6. It is clear from this also that depths of weathering profiles which could develop on normally drained, fairly flat sites are also dependent upon climate, with shallow profiles in hot and cold deserts, rather deeper profiles in sub-humid and temperate zones, and particularly deep profiles in the humid tropics where chemical weathering is at a maximum.

Such simple schemes, however, assume that climate is stable and land surfaces are stable.

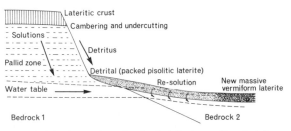

Fig. 7.8 The formation of a detrital ferricrete (laterite) from debris of a nodular ferricrete cap. The debris is carried downslope and undergoes cementation into a massive crust with large nodules and pores (vermiform laterite) (after Goudie, 1973).

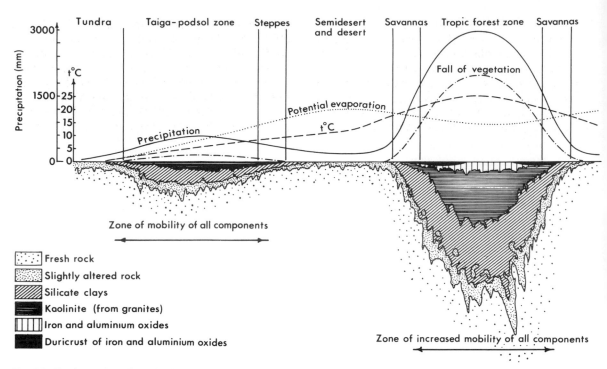

Fig. 7.9 The formation of weathering mantles in areas of tectonic stability and low relief. This scheme relates climatic factors, vegetation cover, depth of weathering, and dominant features of the regolith profiles. It does not consider relict effects (after Strakhov, 1967).

Neither is true, and there are many parts of the continents where the time available since the last climatic change is far too short for a weathering profile of much depth to form. This is clearly the case in much of northern Europe which had a cover of ice only 14 000 years ago. Even in the humid tropics the rainforests were severely decreased in area 20 000 years ago as arid and semi-arid conditions spread and permitted erosion to occur. Many weathering profiles, therefore, are weakly developed because of the lack of time for processes to act, and others are excessively deep for their present climate because they are relict from former climatic regimes. The deep-weathering profiles of south-eastern Australia may be 20–100 m deep and of mid-Tertiary age.

Conclusion

Weathering is of outstanding importance as the prelude to erosion and transport of weathered debris, and it produces some distinctive landforms of which duricrusts are probably the most important and widespread, especially in the tropics and subtropics. Weathering also has a negative economic significance for it attacks building stone as readily as rock in place, and in many industrial zones the release of sulphur compounds, which become sulphuric acid, has hastened deterioration of building materials, and soils have become acid. In a positive way ferricretes are a source of iron, alcretes of alumina, gypcretes of gypsum, and silcretes of opal. Calcretes are widely used for building materials in North Africa and ferricretes in south-east Asia. Most duricrusts, however, have soils of low fertility and hinder agriculture by their resistance to cultivation.

It has been stressed in this chapter that many landforms, such as pits and tors, can be produced by a variety of processes in a variety of environments. This convergence of landforms of various origins towards a similar shape demonstrates that it is seldom possible to study the shape alone, and so deduce the origin of the landform. Where detailed evidence is lacking it may be impossible to determine how a particular feature has evolved. The rates at which weathering proceeds are largely unknown and perhaps best gauged from the rates at which building stone has deteriorated, although this rate may owe more to industrial than natural environments.

Further reading

Ollier (1969a) has provided a fuller treatment of weathering than is common in most geomorpholo-gical works. Duricrusts are discussed by Goudie (1973), but this book is strongly oriented towards studies of calcrete. There are may introductory geology textbooks which give a comprehensive account of rocks and minerals; the book by Ernst (1969) is particularly useful. Clay minerals are discussed by Grim (1962), Mackenzie and Mitchell (1966), Gillott (1968), and by Millot (1970).

8   Hillslopes

The land surface may be thought of as being composed of a network of linking drainage basins. Within each basin precipitation, usually falling as rain, hits the soil or vegetation, seeps into the soil, or runs off its surface, and then enters a drainage channel. In most climatic zones small channels are tributaries of larger ones so that the water follows the branches of the drainage network until it reaches the sea, lake, swamp, or evaporation basin.

In its course water is involved in the processes of weathering, erosion, transport, and deposition and is the main cause of change of landforms. On hills the water is involved in splash erosion, surface wash, solution, gullying, landsliding, and other processes; on valley floors most of the work producing change is the result of water moving as streams within channels. This difference in the major types of processes is the chief justification for separating discussions of hillslopes and valley floors, but it should be remembered that this is a matter of convenience only. All parts of the land surface are linked in the flow of water, solutes, and sediment through the drainage basin system, and downslope areas receive what is transported into them from upslope.

The energy for drainage basin processes is that of gravity and received solar radiation, and the agent for change is water. Resistance to change is derived from the shear strength of rock and soil, with the assistance of vegetation.

Hillslopes in the hydrological cycle

The transport of water, evaporated from the sea, over the land, its descent as precipitation, drainage from the land and back into the sea is called the *hydrological cycle*. The scientific study of this cycle and the rates of water movement, and quantities of water storage, is called *hydrology* (Fig. 8.1).

Water reaches the ground as rain, snow, hail, or by the condensation of dew and fog—these forms of water are collectively called *precipitation*. Some of the precipitation may hit bare rock or impervious soil surfaces and run into channels, but in humid areas most of the rain hits vegetation and is *intercepted* on the leaves and stems. Some intercepted water may be evaporated directly into the atmosphere; in short duration rainfalls in dry seasons all of the precipitation may be lost in this way, but some rain may pass the leaves as *throughfall* and some will run down the trunks as *stemflow*.

Water reaching the ground surface may be retained temporarily in hollows in the ground and behind fallen plants as *depression* storage, but much is likely to seep into the soil as *infiltration* water. Once in the soil, water may reach an impermeable layer and be forced to move laterally as *interflow*, or it may percolate downwards to become part of the *groundwater*.

Water which is part of the impervious area *runoff* is often called *Hortonian overland flow* in honour of the American hydrologist R. E. Horton who first analysed much of the hydrological cycle. This type of flow is distinguished from water which has seeped through the soil, as interflow or reached the groundwater, before emerging at the base of a hillslope where high water tables may cause it to rise and flow over the ground as *return flow*. Rain falling directly onto this saturated area runs off as *saturated zone overland flow*.

This plethora of technical terms for components of the hydrological cycle has arisen because hydrologists have found it necessary to isolate the components in order to develop an understanding of the contribution each makes to the sources and storages of water on the ground. Water is, of course, the fundamental resource for life and understanding its availability is essential for plan-

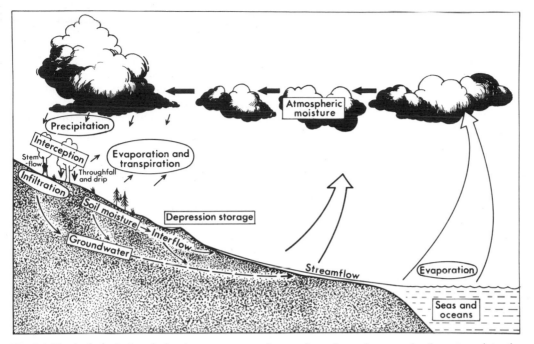

Fig. 8.1 The hydrological cycle involves movement of water from the main reservoir, the oceans, into the atmosphere and then onto the land and back again into the sea.

ning settlement, agriculture, industry, water transport, and, increasingly, recreation. The aim of the hydrologist is to understand in detail the amounts of precipitation (P), evaporation from ground surfaces (E) and transpiration (T) from plant surfaces, and storage (S) in the plant, soil, and groundwater. The *discharge* of water (Q) from a drainage basin may be represented by:

$$Q = P - ET \pm S.$$

Evapotranspiration is treated as one term, and storage may have a positive or negative value depending upon whether it is absorbing or yielding water in the short term. The overall relationship between the gains, losses, and storages in a basin is the *water balance* (Fig. 8.2):

$$\text{Water balance} = P - (Q + ET).$$

It is often very difficult to measure the quantities of water lost as evapotranspiration and of water held in the various stores within a basin; it is easier to measure rainfall in rain-gauges and stream discharge at gauging sites (see chapter 9). Consequently the use of these two equations makes it possible to assess water yields and water resources.

Vegetation and runoff

It is impossible to make universally applicable statements about the percentage of precipitation which is intercepted by vegetation, because interception depends upon many factors, including the duration, amount, intensity, and frequency of rainfall; windspeed; and the type of vegetation. It is usually found in detailed studies (Plate 8.1) that interception losses are greater beneath evergreen than beneath deciduous trees. Winter and summer losses beneath evergreens are usually about the same where rainfall is similar, but winter losses beneath evergreen trees are much larger than beneath deciduous trees. This arises because of the differences in density of foliage and spacing of the trees, hence there are big differences between tree species. During a single rainstorm upon previously dry plants a high initial loss is caused by evaporation but, as the plant surfaces become moist, losses decline to a lower level which is controlled by the water content of the air. Interception losses for a short duration rainstorm of a few hours can vary from 10–20 per cent on crops to 5–50 per cent on forests. Plants consequently have a major effect

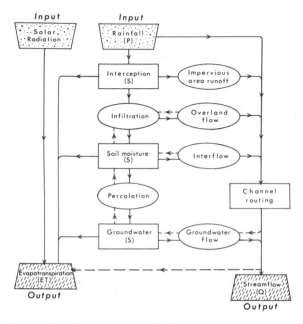

Fig. 8.2 The elements of a water balance analysis. Processes are shown as ellipses and stores of water as rectangles.

upon the supply of water to the soil and upon total runoff into streams during short duration storms, but once the vegetation surfaces are saturated interception is limited and the influence on runoff greatly reduced.

In addition to interception, plants influence slope processes by limiting the action of overland flow and soil moisture. They decrease the velocity of runoff, and hence the capacity to entrain sediment and cause erosion, as water is diverted around plant stems and roots; the energy of raindrops is absorbed and the direct splashing of soil particles into the air is reduced; roots give added shear strength to soil as they act rather like reinforcing rods in concrete; plants transpire water and so cause the soil to dry out; and plants insulate soil against high and low temperatures and so limit soil cracking and frost heave.

As a result of these controls a close vegetation cover provides the most effective protection against soil erosion and is the most common method used by soil conservators. Rates of erosion depend upon many factors such as climate, slope,

8.1 Instrumentation in a forest for measurement of throughfall and stemflow. The gutters around the trees and the troughs between them are connected by pipes to recorders.

and soil type, but it is common to find that erosion from bare soil is hundreds, or thousands, of times as fast as from forested land (see Table 8.1).

Soil and runoff

Infiltration is the process by which water enters the surface horizon of the soil. It may be controlled by a number of factors including the type of precipitation, especially its intensity; surface soil porosity; cracks; slope angle; freezing; cultivation; existing soil moisture; and plants. Of these factors porosity is usually most important because it is a measure of the void space in the soil through which water can pass. Porosity can be increased by the activity of soil organisms such as termites and earthworms, and by plant roots or cultivation, but it can be reduced by the compacting effect of machinery, treading humans or animals, ice in the soil, or soil grains splashed into the air by raindrops and then deposited so that they fill the pores at the soil surface. Existing soil moisture is important because water entering a soil passes through the film of water adhering to the surface of soil particles. Only when the films have thickened so that they almost fill the larger voids can water move at the maximum rate for that soil. This condition may be expressed by saying that the initial or *dry hydraulic conductivity* is low, and that *saturated hydraulic conductivity* is the maximum rate of water movement in the soil (Fig. 8.3a).

After the onset of rain the soil surface horizons form a transmission zone above a wetting front so that existing soil moisture is displaced downwards out of the large pores by newly infiltrating water. Surface horizons are commonly more cracked and

Table 8.1   The Relative Relationships between Erosion and Vegetation

Crop or practice	Relative erosion
forest ground layer and litter	0.001–1
pastures	0.001–1
poor grazing	5–10
hay fields	5
orchards and vineyards with ground cover crops	20
wheat stubble	60–70
orchards and vineyards clean tilled	90
row crops and fallow	100

porous than underlying soil horizons so initial infiltration rates may be high, but as soon as the surface pores are full of water infiltration can only occur at a rate controlled by the least permeable horizon. A steady rate is achieved when the entire soil profile is transmitting water at the maximum rate permitted by that horizon. A graph showing the rate of infiltration during a storm is called an infiltration curve (Fig. 8.3b): most such curves show initially high rates gradually decreasing towards a steady rate after a period of minutes or hours. The time taken to achieve a steady rate depends upon the factors listed above. It is usual for sandy soils to have higher rates than silty or clay soils and for undisturbed grassland, or forest soils, to have higher rates than compacted soils.

Water which cannot infiltrate becomes part of the depression storage, or Hortonian overland flow. Infiltrating water may percolate downwards to become groundwater but, unless the soil is very porous, some is usually impeded in its descent and forced to flow laterally through the soil at a level controlled by the least permeable horizons. Some soils, such as podzols, have dense horizons formed by accumulations of iron oxides and organic matter and these may be nearly impervious to water. Water moving laterally is called either *throughflow* if it occurs near the soil surface or *interflow* if it moves at a greater depth, but distinguishing these types of flow is very difficult and they are classed together as interflow in this text.

Interflow is the main carrier of material in solution and may cause grains to be displaced, and old root channels or soil cracks to be enlarged, so that a network of sub-surface water channels can form. These routes are called *percolines* where they are zones in the soil material with saturated subsurface flow during storms, but *pipes* when they become developed into conduits within the soil. Pipes can be an important route by which storm water can reach stream channels quickly.

Surface runoff

Water which cannot infiltrate may flow over the surface of the ground until it reaches more permeable soil, where it can infiltrate, or continue downslope as threads or films of water which may gradually merge until a defined channel is reached or formed. In theory we should expect the film of water to increase in thickness and velocity down-

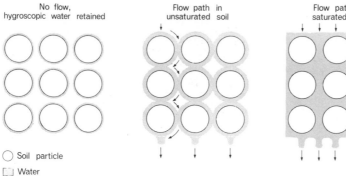

No flow,
hygroscopic water retained

Flow path in
unsaturated soil

Flow path in
saturated soil

○ Soil particle
☐ Water

Fig. 8.3 (a) Routes of water moving in a granular soil.

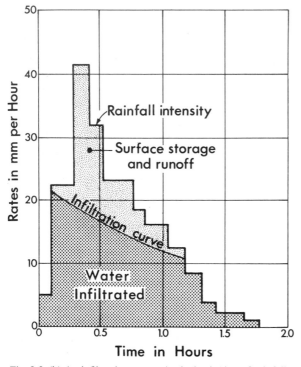

Fig. 8.3 (b) An infiltration curve. At the beginning of rainfall, on a moist soil, infiltration rates may be high, but they decrease to a steady state if the rainfall rate is maintained. If rainfall rate declines below the infiltration rate (as shown here) then actual infiltration will be controlled by the volume of available water.

slope because each segment of slope will shed water onto the segment below it, and raindrops falling into the surface flow will also be added to the discharge. Few observations support this theoretical deduction and water is usually channelled before it can develop a substantial depth of flow. It may, however, have sufficient energy to entrain soil particles and wash them downslope.

Hortonian overland flow is most common, and effective, in areas of intense rainfall and impervious surfaces free of vegetation. It is most common, therefore, in semi-arid climatic zones and on steep rocky slopes. In humid climates two other mechanisms are responsible for most of the water reaching streams: (1) *saturated zone overland flow* which develops at the foot of slopes, on floodplains, and in hillslope concavities, and produces the water which reaches streams most quickly; (2) *subsurface storm flow* which is derived from interflow but, because common soil permeabilities are 0.005–0.0005 m/s, it cannot contribute to streamflow early in a storm unless many macropores and pipes exist as rapid transit routes. Late in a storm interflow water can reach saturated zones and cause return flow.

Saturated zone overland flow is a characteristic of small valleys in humid climates where there are riparian (i.e. stream border) strips and hillside hollows with high water tables and, or, high soil moisture levels. These zones reach saturation and hence produce overland flow soon after the onset of rain. The saturated zone gradually expands up hollows, causing an increasing part of the landscape to become part of the storm-flow contributing area. At the end of a storm the contributing area again gradually contracts until only those areas receiving return flow are contributing storm flow. The *variable contributing area* concept has been an important recent contribution to our knowledge of the sources of stream flow.

The paths by which water may reach streams are indicated in Fig. 8.4. In Fig. 8.5 are indicated the influences of drainage basin characteristics on the storage of water in a basin, and in Fig. 8.6 the effects of the sources of runoff upon the stream discharge are indicated.

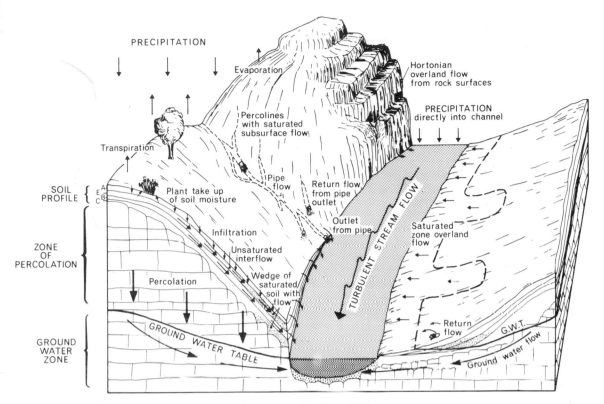

PRECIPITATION

Evaporation

Hortonian
overland flow
from rock surfaces

PRECIPITATION
directly into channel

Percolines
with saturated
subsurface flow

Transpiration

Pipe
flow

Return flow
from pipe
outlet

SOIL
PROFILE { E A B C

Plant take up
of soil moisture

Outlet
from pipe

Saturated
zone overland
flow

TURBULENT STREAM FLOW

Infiltration

Unsaturated
interflow

ZONE
OF
PERCOLATION

Percolation

Wedge of
saturated
soil with
flow

Return
flow

G.W.T.

GROUND
WATER
ZONE

GROUND WATER TABLE

Ground water flow

Fig. 8.4 Various forms of water movement in and on slopes (after Selby, 1982a).

Storage of water in a drainage basin can occur in the soil, as groundwater, or in surface ponding areas. Areas of gentle slope, deep soils, well-jointed rocks, and a complete vegetation cover tend to have a high storage capacity, but frozen or severely eroded areas with thin soil and vegetation covers have limited storage. As a result of infiltration and storage capacities the rate of discharge from catchments of the same size and shape can be very variable. Hortonian overland flow tends to develop rapidly during a storm and to decline rapidly at the end of the storm. The hydrograph consequently has a rapid rise and fall and a high peak. Streamflow from such basins is consequently likely to be characterized by severe floods, erosion of slopes, channel banks, and beds. Reservoirs built in such basins are often filled with sediment in a short period, and are alternately overtopped by floodwater and then depleted of water in dry periods.

Areas with subsurface storm flow have gradual rises and falls in the hydrograph with reduced flood peaks. Erosion in such basins is limited and is often predominantly by solution. Reservoirs have little sediment carried into them and water levels rise and fall relatively slowly. Such basins are therefore ideal for water storage and continuous supply.

Areas of saturated zone overland flow have storm hydrographs which may be intermediate between those of subsurface storm flow and Hortonian overland flow types, but, because of the storage capacity of the soils and rocks, both saturated zone and subsurface drainage tend to provide a dry period *baseflow* long after a storm has ceased. Such baseflow is the source of water for all areas drawing on channel flow. The regulation of flow provided by vegetation, soils, and permeable rock is thus of prime importance to plants, animals, and man.

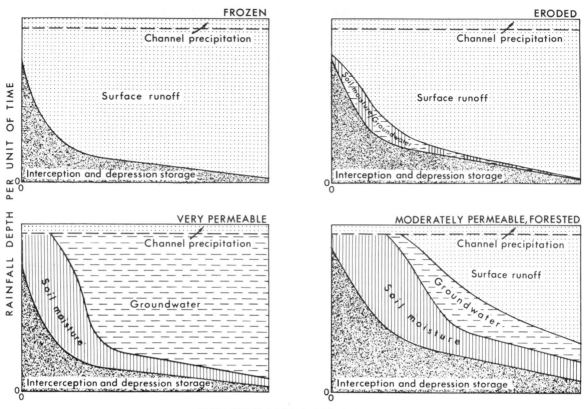

Fig. 8.5 The distribution of storm rainfall into its components on land from the start of rainfall (O). Interflow is not shown, but would be part of the groundwater (after Selby, 1970).

Water erosion on slopes

Erosion on slopes is a function of the eroding power of raindrops, running water, and sliding or flowing earth masses, and of the *erodibility* of the soil or rock.

The erosive power, or *erosivity*, of raindrops is controlled by the rainfall intensity, that is the amount of rain falling (mm) upon a unit area of land (m²) in a unit of time (s). Increasing intensity is a result of the rain containing an increasing proportion of large raindrops which, because of their size, have greater fall velocities and hence greater kinetic energies (e), as $e = \frac{1}{2}mv^2$. Large drops have diameters of 2 to 6 mm. In most temperate climates rainfall intensities seldom exceed 75 mm/h, and then only in thunderstorms, but in the tropics intensities of 150 mm/h are common and an intensity of 340 mm/h has been recorded in Africa.

Raindrops have the effect of breaking down soil aggregates, splashing grains into the air, causing turbulence in sheets of water moving as overland flow and carrying soil particles. On a perfectly level surface splashed soil particles would be evenly redistributed, but on a slope there is an increase in net downslope movement which is directly proportional to slope angle. The effect of splash erosion is limited by the thickness of the surface wash sheet, as once a sheet has a thickness greater than about three times the diameter of the impacting raindrops it effectively protects the soil from being displaced. Figure 8.7 illustrates the effect of increasing rainfall intensity upon surface processes.

Sheetwash transports uncohesive soil particles

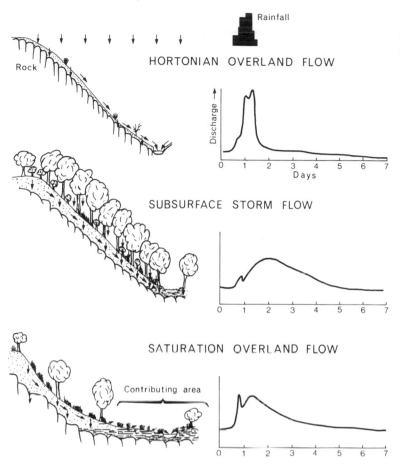

Rainfall

HORTONIAN OVERLAND FLOW

Rock

SUBSURFACE STORM FLOW

SATURATION OVERLAND FLOW

Contributing area

Fig. 8.6 Soil thickness and permeability, length of slope, slope angle and curvature, and vegetation cover, all affect the shape of the stream hydrograph for a storm of the same intensity, duration and amount (after, Selby, 1982a).

by rolling them along the ground. The total discharge of sediment is a function of the depth of flow, the roughness of the ground, the roughness of the particles, and the ground slope. Wash transport ceases either when the depth of flow falls below a critical value, or all of the available sediment has been removed and only a *lag* or *armour* of large grains is left.

Rills form where surface wash becomes concentrated as water is diverted around objects. The water in a rill has sufficient depth for turbulence to develop in it and rill flows can therefore entrain larger particles than sheet flows can. The head of a rill may not extend all the way to the top of a slope because flow there is insufficient to produce entrainment. Parallel rills on a fresh surface develop into a branching network as the divides between rills are broken down, and the water from one overflows into a neighbour, so that eventually rills become more widely spaced downslope. A master rill may capture all of the drainage of an area and become a permanent channel which holds water during every storm. A rill network has been known to form on a cultivated field during a single rainstorm, and is thus a major contributor to erosion.

Gullies are distinguished from rills largely by size: gullies are more than about 0.5 m in depth and width, and may be more than 10 m across. Gullies may be just enlarged rills but also they may form wherever the vegetation cover, on deep and mechanically weak soil, is broken. Many gullies consequently start at steep channel heads and extend downslope to permanent stream channels.

Gullies usually develop rapidly and as a result of changes in the environment such as burning of vegetation, overgrazing by animals, climatic changes causing a deterioration in plant cover,

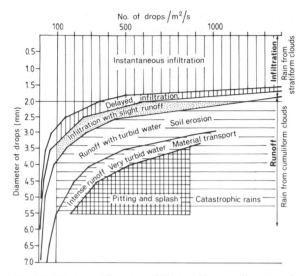

Fig. 8.7 Effects of different rainfall intensities on soil water and types of soil erosion in West Africa. The details should vary with soil type and vegetation cover, but the general relationship probably holds for most bare, and nearly bare, soils (modified from Barat, 1957).

deforestation, or the spread of cultivation. Erosion accelerates because either flood runoff increases, or the capacity of channels to carry the same runoff is reduced.

Gullies are nearly always steep-sided and flat floored. They carry water in wet seasons or during storms, and may be major sources of sediment which is deposited downslope or in a main stream channel. Gullies nearly always form in unconsoli-

dated regolith or river (alluvial) deposits, and their vertical extent is often limited where they cut down to bedrock. A second control on their development is their restricted catchment area and hence limited supply of water for further erosion.

Because gullies are often associated with land-use changes they are frequently studied by soil conservators and attempts are made to control them. By far the most effective way of controlling rill and gully erosion is in the use of vegetation to bind soil and to slow overland flow, but it may be necessary to build dams of brushwood, wire netting, or permeable rock to trap sediment, and to allow vegetation to establish itself (Plate 8.2).

Solution is a much more important process on slopes than is commonly realized. In many catchments more than half of the material removed in erosional processes is carried in solution. It is the transport of solutes from one part of the soil profile to another that permits continuation of weathering processes, the synthesis of secondary minerals, and the formation of pans of iron oxides or carbonates within the profile.

Solutes derived from the soil or rock of a drainage basin are the only true part of the erosional load of streams; other solutes may be carried in as wind-blown dust or salt, fertilizer, in rainfall, or in deep migrating groundwater. Where vegetation is stable the incorporation of nutrients into plant tissue should, over a year, be balanced by the return in biological decay, but ploughing up of pasture or burning of plants can produce a major increase in dissolved nitrogen. Sodium and

8.2 Rill and gully erosion has cut into unconsolidated sediments in this suburb of La Paz, the capital of Bolivia. Attempts to control the erosion have been ineffective because of the rapid runoff from the valley sides which have been denuded of natural grass vegetation, and cutting of trees for timber and fuel.

chlorides are commonly carried into basins from the ocean or desert areas, and nitrates and phosphates are often derived from fertilizer.

Total solute discharges are normally greatest in humid climates where solutions are not saturated and the flow of water through the soil is high. Total solutional erosion consequently tends to increase with rainfall and discharge, and to a lesser extent with temperature which controls the rate of chemical reactions. A rise of 10 °C may double or treble reaction rates. Published data suggest that there is a general increase in solute discharges from ancient igneous and metamorphic rocks to Tertiary sedimentary rocks, and that limestones are by far the most soluble rocks. If the data are expressed as the rate at which average ground-lowering could occur in a period of time (mm/1000 years) the yield of solute from different rock types can be compared (Table 8.2).

The composition of material leaving a catchment as solutes depends upon the rocks of the area, but silica, iron oxides, and the ions of calcium, sodium, potassium, magnesium, chloride, and sulphate are commonly present in areas of varied lithologies. Much of the solute load may, however, be held on the surface of fine sediment carried in suspension, so it is common for filtered water samples to under-represent the total load due to solution.

Erosion rates and people

Erosion rates have been greatly accelerated by the need to cultivate soils for food production. Careful conservation and terracing of soils on slopes can limit erosion, but its damaging effects can be

Table 8.2 Estimates of Ground Lowering by Solution alone

Rock type	Ground lowering (mm/ky)
Precambrian igneous and metamorphic	0.5–7.0
Precambrian micaceous schist	2.0–3.0
Ancient sandstones	1.5–22.0
Mesozoic and Tertiary sandstones	16–34
Glacial till	14–50
Chalk	22
Carboniferous limestone	22–100

Source: Waylen, 1979.

gauged from the differences between the natural soil production by weathering and loss by erosion each year. In Japan soil production on hill country is about 1 tonne/hectare each year. On a 10° slope erosion loss on bare land is 20–40 (t/ha)/year. On a 10° slope erosion loss on vegetated land is 10–20 (t/ha)/year. On a 24° slope erosion loss on vegetated land is 50 (t/ha)/year. In USA losses on cultivated land may reach 500 (t/ha)/year. Such losses due to cultivation cannot continue for long before the soil resource is destroyed.

Mass wasting and hillslopes

Mass wasting is the downslope movement of soil or rock material under the influence of gravity without the direct aid of other media such as water, air, or ice. Water and ice, however, are frequently involved in mass wasting by reducing the strength of rock and soil and by contributing to plastic and fluid behaviour of soils.

Mass wasting is a common phenomenon in all high and steep hill country and it can also occur on very low-angle slopes. Very large landslides are most common in the tectonically and seismically active belts of rising mountain chains where steep slopes, rapidly incising rivers and glaciers in valley floors, jointed and fractured rock masses on slopes, severe physical weathering, fluctuations in groundwater pressure, and many joints dipping steeply out of slopes, all contribute to instability. On lower and more gently inclined slopes covered with soil most landslides are caused, or at least triggered, by rises in water pressure within the soil during heavy rain. Exceptions to this last statement occur where vibrations caused by earthquakes, and soil saturation as snow and ice melt in cold climatic zones, are the trigger factors. These causes can be classified into two groups: (1) external causes that produce an increase in shear stress acting on the slope materials, but no change in the shear strength of those materials; (2) internal changes of shear strength without any change in shear stresses.

Types of mass wasting

There are many criteria available for distinguishing types of slope failure: they include velocity and mechanism of movement, material, type of mass wasting, shape of the moving mass, and water content of the material. There are thus many classifications in use: the one chosen here was published in 1975 and uses the type of movement and the type of material as the two criteria (Fig. 8.8, 8.9). In applying any classification it must be remembered that many landslides are complex and may change their nature as they move down-slope. Thus a large rockfall may descend into a valley and then turn into a flow as the rock

disintegrates and picks up water and soil from the valley floor. The classification distinguishes between rock moving as a mass, rock debris which behaves like a soil with discrete particles, and earth which is a fine-grained soil.

Falls and topples (Plate 8.4) occur as fractures spread through a rock and soil mass and separate blocks from a cliff or parent steep slope. Rockfalls

Type of Movement			Type of Material		
			Bedrock	Soils coarse	fine
FALLS			ROCKFALL	DEBRIS FALL	EARTH FALL
TOPPLES			ROCK TOPPLE	" TOPPLE	" TOPPLE
SLIDES	rotational	few units	" SLUMP	" SLUMP	" SLUMP
	translational		" BLOCK GLIDE	" BLOCK GLIDE	" BLOCK GLIDE
		many units	" SLIDE	" SLIDE	" SLIDE
LATERAL SPREAD			" SPREAD	" SPREAD	" SPREAD
FLOWS			" FLOW (deep creep)	" FLOW (soil creep)	" FLOW
COMPLEX			Combination of 2 or more types		

Fig. 8.8 A classification of mass wasting (after Varnes, 1975).

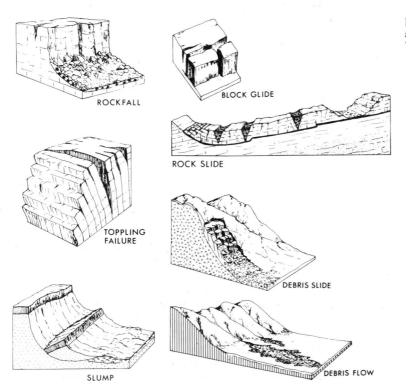

Fig. 8.9 Major classes of mass wasting according to the classification of Varnes.

ROCKFALL

BLOCK GLIDE

ROCK SLIDE

TOPPLING FAILURE

DEBRIS SLIDE

SLUMP

DEBRIS FLOW

8.3 Rill and gully erosion on one wall of a valley where few attempts have been made to control runoff but the terraces, built by Incas in the fourteenth and fifteenth centuries, are still in use and have prevented erosion. The effect of the terraces is to decrease the velocity and amount of runoff, by dividing the slope into short, low-angle segments, and to increase the time for infiltration. The stone terrace walls also trap much of the sediment. Upper Urubamba Valley, Peruvian Andes.

are features of all high and steep rock slopes and are one of the most important processes in rock slope erosion. They develop as physical weathering, especially stress release, joint opening, and hydrofracturing decrease the cohesion along joints. When fractures are sufficiently extended blocks can fall. In so doing they may dislodge other fragments and produce an accumulation of debris at the foot of a slope. Many falls, and some slides, involve large masses of rock which fall considerable heights and hence have large energies. As a consequence they may spread their

debris over a large area unless the depositional zone is confined by valley walls. There has been much debate on the internal mechanisms which permit large volumes of rock to travel considerable horizontal distances before internal frictional resistance causes it to halt, and on how flow-like behaviour can be associated with such materials. Because large rockfalls and slides occur on the Moon it seems that water and air are not necessary lubricants for such movement, and the most plausible hypothesis is that the rock particles are buoyed up in a flow of vibrating dust which keeps

8.4 (a) This column of sandstone is beginning to separate from the cliff. If diagonal cross-joints open, as in the lower central part of the cliff, a rockfall will occur, but if the column rotates about its base it will form a toppling failure. Darwin Mountains Antarctica.

the larger rock fragments from contact with one another until this vibration energy is dissipated, when the dry flow is brought to a halt as frictional contacts are re-established internally and along the bed.

Small debris and earth falls occur on cliffs and may result from the development of joints, from high water pressures or from the undercutting of the slope by waves and streams.

Topples involve the rotation of a block as it falls out of a cliff face: they are consequently most common where major joints dip out of a slope and where joint blocks are tall compared with their width. A fall or topple with a distinct triangular shape is a wedge failure (Plate 8.4b), which may occur where joint planes intersect so that they together form an angular failure surface. Wedge failures are often contributors to the saw-tooth form of some mountain ridges.

Slumps, glides and slides are related because they have a clearly defined plane along which movement takes place. Slumps always have shear planes

8.4 (b) A wedge failure in schist.

which are concave upwards, and the slumping mass has a rotational component in its movement. Glides and slides have predominantly straight, or planar shear planes, and the sliding mass usually breaks up into many blocks as it moves. In large rock slides the shear plane may be stepped at major joint intersections with the shear plane, and the rock mass may be broken up to become a debris slide (Plate 8.5).

Slumps are most common in thick regoliths and large mudstone rock units, and can also occur in hard rock which has been shattered by very closely spaced fractures. Glides are features in which a block of soil or rock remains largely undeformed as it moves over a planar slide plane and consequently the material on which the block moves is usually a uniform clay with a high water content.

Slides are by far the most common form of mass wasting. They nearly always have a length which is much greater than the depth of the moving mass and are, commonly, relatively narrow. Large rock slides may have volumes of tens of millions of

8.5 A rock slide has occurred on the bedding planes exposed on the flanks of a rising anticline. North-west Argentina.

cubic metres (10–100 Mm³), reach velocities exceeding 100 km/hour, and travel many kilometres from source. In the process the rock is broken up to small fragments.

Modern examples of devastating rock slides include that which moved down the slopes of the Vaiont Valley in the Italian Alps, and partly filled a lake impounded by a concrete dam in a steep gorge-like valley. The debris displaced 250 Mm³ of water which crossed the dam as a huge flood wave 100 m high and swept down the lower valley killing 2600 people. Much of the slide rock crossed the 200 m deep and 100 m wide gorge, and pushed its way 140 m up the opposite valley wall. This landslide was probably triggered by high water pressures as the lake behind the dam was filled, but the failure was made possible by open bedding joints dipping steeply, parallel with the upper valley walls.

Many debris and earth slides are, however, relatively small with lengths of a few tens of metres, depths of 1–3 m and widths of 10 m or so. Such features are essentially soil erosion phenomena and are usually triggered in storms, but deforestation, excessive grazing, and other forms of human interference can weaken soil and make it prone to failure. Because of the high water content of soil which fails during storms, many slides turn into flows downslope and valley floors may be covered with slide and flow debris (Plate 8.6).

Spreads and flows are characterized by very high water contents of their materials. Spreads result from the development of a flow at some depth and the rafting of more resistant blocks of material on the surface of a flow. They are not common in most parts of the world and are largely confined to lowlands in Siberia, Scandinavia, the St Lawrence Basin of Canada, and to coastal lowlands in Alaska. In these places glacial deposits may flow as a result of very high water pressures, causing soil to liquefy, or of earthquake motions. They usually occur near rivers and coasts where lateral spread-

8.6 Debris slides high on these slopes have turned into debris flows at the base of the slope and have infilled the valley floors. Wairarapa, New Zealand.

ing can move the debris towards the waterway. Lateral spreads may be devastating features and large enough to raft along houses, segments of roads, bridges, and railways (Fig. 8.10).

Flows may have a well-defined shear plane but, more frequently, saturated debris moves away from its original site over existing ground. Slow flows often have hummocky irregular ground surfaces and may flow at rates as slow as a few metres/day or year. Some flows develop as a result of very high water pressures in the soil, and others because their constituent clay minerals can absorb high proportions of water before becoming liquid. Montmorillonite has a particularly high capacity for water absorption and is often involved in flow behaviour.

Creep of rock or soil usually involves plastic behaviour, that is, slow deformation recurring once a critical stress is exceeded and continuing deformation thereafter. Rock creep may occur at depths as great as 300 m below the ground surface and involve rates of movement of < 1–10 mm/year. Large-scale ground ripples, or bulging of the toe of a slope, can result from depth creep.

Surface creep can occur in both soils and rock. In rock it is usually the result of fractures developing across small joint blocks near the ground surface, and then the tilting downslope of the separated block (Plate 8.7). In soil, creep is a slow downslope movement of material which occurs at rates of 0.1–5 mm/year in most climates but, in severe cold climates in which frozen ground thaws

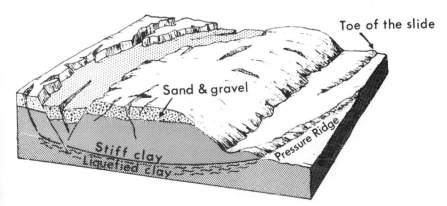

Fig. 8.10 A lateral spreading failure (after Hansen, 1965).

8.7 Curvature of steeply dipping shales, with fracturing across the beds, on the side of a steep hillslope.

in spring and becomes very mobile, slow flow (called *solifluction* or *gelifluction*) of several metres a year can occur (see Chapter 14, p. 405). It is, perhaps, debatable whether this process should be classed with soil creep. Soil creep is usually confined to the upper metre or so of most soils. Rates of soil creep are usually measured by placing pins or acrylic rods in the walls of trenches which are then refilled with soil; by inserting columns of beads, blocks, or tubes in the soil; and by using tilting bars pushed into the soil. Initial survey and then re-survey of the positions of the instruments indicates if movements have occurred.

Many types of evidence have been held to demonstrate the existence of soil creep including tree curvature, tilting of walls, fences and other structures, turf rolls, cracks in the soil, and flights of steps, called *terracettes* (Plate 8.8), on grassland slopes. Most of these phenomena can, however, also be produced by other processes such as wind, slope-wash, or tilting under the weight of the object. The only reliable method of determining the presence of creep is by direct measurement.

In environments with seasonal variations of soil moisture and soil temperature, soil creep is an episodic process caused by heaving and settling movements in the soil produced by solution, freeze–thaw, warming and cooling, and wetting

8.8 Terracettes in grassland soils.

and drying cycles. Creep may also be induced by the activity of soil animals and plant roots.

Measurements, using rods in the soil profile, over a 12 year period in northern England, have shown that there is a downslope component of movement of 0.25 mm/year in the top 20 cm of soil, but an inwards movement of 0.31 mm/year perpendicular to the ground surface, and towards the bedrock. This inward movement is interpreted as being caused by solution and settling of the remaining soil particles. If this study represents a general situation it may be that solution is the most important slope process in areas of complete vegetation cover.

Stability studies

Landslides tend to occur in groups, or repeatedly at the same site, where geological conditions, such as high water tables, weak rocks, steeply dipping joints outcropping in the faces of cliffs, or undercutting of a slope, produce suitable conditions for failure. In such areas the causes may be identified by mapping the slope forms, soil types, rock types, rock structures, and other features, and relating failures to these features.

It may, however, be necessary to determine the effect on a slope of raising a water table, depositing soil at the top of a slope or undercutting the toe of a slope. In such cases a stability analysis may be carried out to determine if the change will cause failure.

Factors of safety

The stability of a slope is usually expressed in terms of a factor of safety, *F*, where:

$$F = \frac{\text{sum of the resisting forces}}{\text{sum of the driving forces}}.$$

Where the forces promoting stability are exactly equal to the forces promoting instability $F=1$; where $F<1$ the slope is in a condition for failure; where $F>1$ the slope is likely to be stable. Most natural slopes upon which landslides can occur have *F* values between about 1 and 1.3, until earthquakes, undercutting, or high pore water pressures reduce this value and trigger a landslide.

Slides

Consider a joint block on a slope (Fig. 8.11a). The

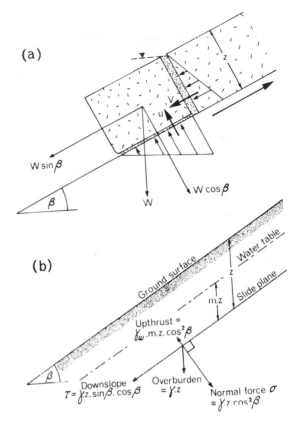

(a)

(b)

Fig. 8.11 (a) Stresses acting on a block on a rock slope.
(b) Stresses acting in a soil with a planar slide plane.

resisting forces are those described in the Coulomb equation, p. 178, with the weight of the block providing the stress normal to the plane of failure. In a two-dimensional analysis the resisting forces are composed of the weight (*W*) of the block multiplied by the cosine of the slope angle (*β*), the cohesion along the base of the block where the joint has not fully opened (c_j) and the frictional resistance between the block and the underlying rock. The resistance ($W.\cos\beta$) is offset by the uplift produced by water pressure in the joint (*u*). The force tending to drive the block down the hill is the weight of the block multiplied by the sine of the slope angle, plus the force applied by the weight of water in the joint (*V*) at the back of the block. At equilibrium then:

$$W.\sin\beta + V = c_j + (W.\cos\beta - u)\tan\phi$$

The situation for a slide in soil is exactly the same as that for the rock except that it is much easier to survey a landslide in soil by measuring the depth in a vertical plane. The rectangular form of the joint block of rock is then replaced by a soil block shaped like a parallelogram. To correct for this geometry, and the distribution of stress along the failure plane, the slope angle is multiplied by $\cos\beta$, then for effective stresses:

$$F = \frac{c' + (W.\cos^2\beta - u)\tan\phi'}{W.\sin\beta.\cos\beta}.$$

The value of W is given by the unit weight of soil (γ) multiplied by the vertical thickness (z) of the slide block. The length of the block can be ignored in this type of simple analysis and the slope length is considered to be infinite. The pore water pressure is given by the unit weight of water (γ_w) multiplied by the piezometric level (i.e. the height of water in an open standpipe at the moment of failure): this is given by $m.z$, where m is the fractional height. The shearing stresses (τ), the normal stresses (σ), and the uplift (u), are indicated in Fig. 8.11b. For an infinite slope then:

$$F = \frac{c' + (\gamma.z.\cos^2\beta - \gamma_w.m.z.\cos^2\beta)\tan\phi'}{\gamma.z.\sin\beta.\cos\beta}$$

which simplifies to:

$$F = \frac{c' + (\gamma - m.\gamma_w)z.\cos^2\beta.\tan\phi'}{\gamma.z.\sin\beta.\cos\beta}.$$

For example if:

$\phi' = 12°$, $c' = 11.9\text{kN/m}^2$, $\gamma = 17\text{kN/m}^3$, $\beta = 15°$, $z = 6$ metres, $m = 0.8$, $\gamma_w = 9.81\text{kN/m}^3$, then $F = 0.9$.

Slumps

Because the failure plane of a slump is curved the infinite slope assumption cannot be used. Instead, the cross-sectional area of the slump is divided vertically into slices and the forces acting on each slice are added together. The weight is assumed to act through the centre of gravity of each slice (Fig. 8.12), and angle of the failure plane for each slice is determined by constructing radii from the calculated centre of rotation. Because the lengths of the bases of the slices are different from each other, this length (l) is included in the analysis, and the overburden stress is distributed along it; the correction, $\cos\beta$, is therefore not needed and the stability equation becomes:

$$F = \frac{\Sigma[c'.l + (W.\cos\alpha - u.\,l)\tan\phi']}{\Sigma[W.\sin\alpha]}.$$

In this discussion it has been assumed that there was a piezometer in place before and during the landslide. Except in the case of engineering projects, such as canal banks and dams, this is

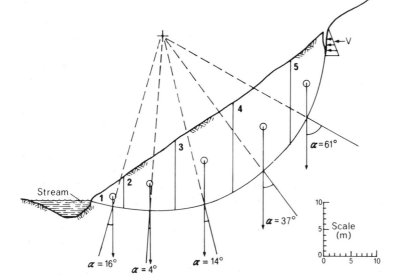

Fig. 8.12 Division of the cross-sectional area of a rotational failure into slices. The centre of gravity of each slice is shown, and a vertical through this centre and a radius from the centre of rotation define the average angle of slope of the failure plane for each slice. Note that because the basal length of the slice is used there is no need to correct the geometry by multiplying by $\cos\beta$.

Stream

$\alpha = 16°$
$\alpha = 4°$
$\alpha = 14°$
$\alpha = 37°$
$\alpha = 61°$

Scale (m)

unrealistic. Instead geomorphologists often assume that, at failure, F values were slightly below unity, and use the stability analysis to determine the piezometric height and hence to discover what processes caused the landslide.

Remedial and conservation measures

Mass wasting on hillslopes can be a cause of considerable economic costs. It has been estimated that the cost of slope failures in USA for 1978 exceeded $100 million. This included the direct costs of damage to buildings, roads, railways, fences, and other structures, and the indirect costs of loss of tax revenues on properties devalued as a result of landslides, loss of production on farmland or in forests, and loss of industrial production caused by landslides closing factories and blocking roads and railways. The indirect costs may exceed the direct costs. In addition to these costs are the threats to life, and loss of life, caused by landslides.

The mountains of Peru have experienced some of the most destructive landslides in this century. On the slopes of Mount Huascaran in 1962 a large slide of rock and ice killed 4000–5000 people living in villages in the valley below, and then in 1970 an earthquake, off the coast, triggered a larger slide into the same valley and buried two small towns, killing over 20 000 people. Many large landslides in the Andes and Himalayas are destructive because they block valley floors and impound lakes. When the temporary dam of debris is breached the lake may drain suddenly and a wave of water and rubble can sweep the valley below. The upper reaches of the Indus Valley experience frequent floods of this type.

There is probably nothing that can be done to prevent very large landslides, but threatened areas can be avoided if the hazard can be anticipated and if there are alternative sites available for settlement. The second tragedy below Mt Huascaran could have been avoided if the findings of a geological survey had been heeded.

Small landslides probably have a total effect upon property far exceeding the effect of the rare very large landslides, but cause a much smaller loss of life. Many small landslides can be avoided or prevented.

The most important preventive measure is the initial recognition of a hazard. This usually involves mapping old slides and geological features likely to cause them. Rockfalls threatening struc-tures can sometimes be prevented by tying loose blocks to mountain walls with steel cables or bolts, small falls can be reduced by placing wire mesh on slopes or by spraying the slope with concrete. Concrete or stone safety-walls may be built at the foot of slopes, or wire-mesh fences placed to catch rock debris. Roads and railways may be built in tunnels or under concrete shelters to avoid avalanches and rockfalls. It may, however, be cheaper to avoid the hazard by choosing a new route or site for the structure.

Improving drainage is one of the most important preventive measures. This may involve cutting ditches across a slope—with the danger of promoting gullying if the ditch overflows, or inducing a landslide if the slope above the ditch is undercut. Drainage waters, are, therefore, often piped away and, on rock slopes above large and expensive structures, drainage galleries or pipes may be cut inside the slope to reduce water pressure.

On farmland, or forested land, two measures may be used to reduce erosion hazards: (1) protection of the soil by a dense growth of vegetation; (2) the reduction of pore water pressures by drainage. In Plate 8–9a, b can be seen part of a valley in Hawkes Bay, New Zealand, after a severe storm and 20 years later, when soil conservation measures, involving control of grazing animals, establishment of a new ground cover and retirement from farming of the most affected slopes, had been carried out.

Man not only creates new slope faces by making cuttings and quarry walls, but builds slopes such as spoil heaps. Spoil heaps may be excess material from coal or other mines, or power station fly ash. Most such heaps were constructed by tipping debris and allowing it to lie at the angle at which it cames to rest. Such slopes are subject to gully erosion because it is difficult to establish plants on the waste. If pore water pressures build up inside the spoil heap dangerous flows or slides can occur. At Aberfan, in South Wales, in 1966 a flow from a heap descended upon the town and killed 166 school children and 28 adults. This failure occurred primarily because of high water pressures in the heap and the weathering, and loss of strength, of the tip material. As a result of this tragedy most new, and many old, tips are shaped with lower heights, flat tops, and lower-angled slopes.

In considering all forms of erosion it has to be

8.9 (a) The results of a severe storm in the Esk Valley, New Zealand, in March 1924.

recognized that man can often increase natural rates by a factor of 2–5 times. Land use and management are often the keys to slope stability.

Slope deposits

The general term for unconsolidated material deposited on slopes is *colluvium*. Colluvium is part of the regolith and may contain particles of any size derived from uphill. Most colluvium is unsorted and unbedded unless it is the deposit left by slope-wash processes. It may lie upon old ground surfaces in which case a buried soil with definite horizons may be present within it, and boulder or stone lines may be present if there have been distinct phases of deposition, with the stone line indicating an old ground surface (Plate 8.10).

Cliff faces formed of hard rock disintegrate by weathering and produce loose blocks which fall and accumulate at the base of the cliff as *talus* or *scree* slopes with angles in the range of 30–45° but commonly around 32–36°. Many talus slopes have straight upper segments and lower concave segments (Plate 8.11). It has been suggested that talus slopes in the early stages of development are usually concave, and that as the slope increases in height it becomes increasingly straight. Whilst this may be true of some slopes it is improbable that universally valid conclusions can be drawn at present, because of the lack of careful measurements made on taluses with varying length, slope angle, particle size, and particle shape, as well as varying heights of cliffs above them.

The theory of decreasing concavity is based on the observation that debris falling from a high cliff face will have greater kinetic energy than debris falling from a shorter cliff face, and the high cliff face should therefore produce an extensive thin-

8.9 (b) Thirty years later soil conservation works, control of grazing, and retirement from farming of some steep slopes have allowed plants to colonize the landslide scars and sediment loads of the river have been reduced (photos by courtesy of the National Water and Soil Conservation Organization).

8.10 Section through a soil profile showing a stony lag gravel at the surface, and a stone line at depth lying upon a darker and more structured horizon which may be a buried soil.

8.11 Long, nearly straight, talus slopes below peaks of shattered sandstone and argillite. Kaikoura Mountains, New Zealand (photo by courtesy of New Zealand Geological Survey, Crown copyright reserved).

ning 'tail' of debris at the lowest part of the talus. As the cliff face recedes by erosion, the height of rockfall is reduced and the extent of the tail should be reduced so that, on slopes from which the cliff face has been nearly eliminated, a straight slope with no basal concavity can form. Many talus slopes, however, have irregular profiles as a result of lobes of debris left by snow avalanches, debris flows, or shallow landslides in weathered talus.

Uniform talus slopes form against relatively straight cliffs but, if the cliff is notched by gullies or stream channels, the debris will accumulate as a debris cone or an alluvial fan, if deposition is from the stream (Fig. 8.13). Fans are discussed in Chapter 10.

Slope forms

Much of the geomorphology of the first half of the twentieth century was concerned with establishing the erosional history of landscapes by recognizing, in the field, what were thought to be remnants of old marine or river terraces and erosion surfaces, and also the extent to which slopes had been flattened by erosion. It was usually assumed, as a result of the urging of the American William Morris Davis, that slopes in humid climates progressively decline in angle. This study of denudation chronology, as it is called, has been gradually abandoned as it has been recognized that the shape of the land surface does not provide

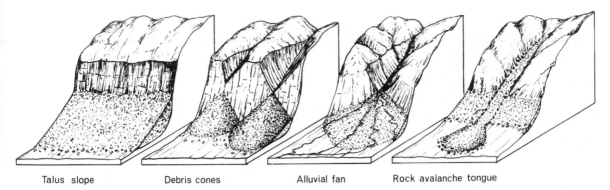

| Talus slope | Debris cones | Alluvial fan | Rock avalanche tongue |

Fig. 8.13 Forms of accumulation of rock materials beneath a mountain front (after Selby, 1982a)

adequate evidence on which to base a reconstruction of landform history.

It is now recognized that slopes, in most parts of the world, have been subjected to many variations in climate, and that nearly all long or high slopes result from the operation of a variety of processes over a long period of time. The study of slope history has now turned towards the effect of individual processes on slope form and of the influence of rock and soil strength on slope development. There is now no generally accepted single concept of slope evolution or change of form.

Slopes with soil cover

Most slopes with a soil cover consist of several segments which may be concave, convex, or straight when viewed down the steepest slope profile. It is commonly assumed that where slopes are forming as a result of a predominant process they develop a profile which is in *equilibrium* with that process. The shape of the slope is thus an expression of the process. Once equilibrium has been attained the slope will diminish as a result of erosion but will retain its overall shape. The main slope element consequently tends to form a *characteristic slope angle* if the segment is a straight (i.e. *rectilinear*) slope. The foot of the slope must at the same time be replaced by a slope of lesser angle. Possible changes in slope profile are indicated in Fig. 8.14.

Convex slope segments commonly occur on the upper parts of slopes near the drainage divide as a result of soil creep and raindrop splash. Splash is most effective in sparsely vegetated areas and

creep in well-vegetated areas. The principles applying to convex slopes were first stated by G. K. Gilbert who showed that soil eroded from the upper part of a slope has to pass each point below it and consequently that the volume of soil being moved increases with distance from the divide. If the transport rate for creep and splash is proportional to the slope angle then the slope angle must also increase with distance from the crest of the slope. hence convexity is the equilibrium form (Fig. 8.14a).

Since creep and splash act only slowly they can only control the form of slopes with angles which are below the characteristic angles for mass wasting.

Concave slope segments may be equilibrium forms, slopes of transportation, or depositional slopes, but they do not form where rapid downcutting is occurring at the base of the slope (compare Fig. 8.14e and f). A concavity is usually the result of either hydraulic processes in which the discharge increases downslope, or of talus deposition in which boulders with greater kinetic energy roll beyond the depositional zones of those with lesser energy.

As discharge increases downslope, because of the greater collection area for wash, water velocity can be maintained on lower slope angles. Particle sizes being transported tend to decrease downslope as a result of weathering and abrasion and any given flow can carry a larger load of fine than coarse material. There is thus an excess transporting capacity in the flow so that the slope angle can reduce while the power to transport is maintained. Consequently transportation slopes, and equilibrium slopes where wash is predominant, become

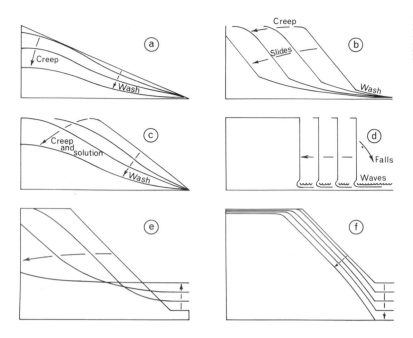

Fig. 8.14 Theoretical profiles of hillslopes in relation to the action of particular dominant processes (after Selby, 1982a).

concave as the transporting capacity and load carried are maintained in equilibrium, and depositional slopes are concave because large particles are dropped by the water flow before finer particles. Concave wash slopes are particularly well developed in areas of Hortonian overland flow.

Straight slope segments develop where the dominant erosional process is one of mass movement which removes soil to a uniform depth. Slopes dominated by shallow debris slides tend to have long straight profiles, as do steep slopes which are undercut by streams or waves. Talus slopes, which are controlled by the depositional form, have already been mentioned as commonly having straight segments.

Compound slope profiles are very common. On any slope several processes may operate and there may be zones of erosion, transport, and deposition. Fig. 8.15 shows a possible pattern of slope segments together with their dominant forms and processes. The nine-unit model does not imply that all of the segments will occur on one slope, or that the segments will always occur in the order shown. A slope with several outcrops of horizontal hard rock units, for example, may be undercut at the base so that starting at the top it may have units 1, 2, 3, 4, 5, 4, 5, 8, 9. A slope without rock outcrops may consist of units 1, 2, 3, 5, 6, 8, 9. The

use of the model assists in the recognition of the relationship between slope form and dominant process.

Slopes on rock

Slopes formed on bare rock have profiles controlled by the resistance of the rock, not by the erosional processes acting on it. This is because slope processes are of very limited power compared with the strength of rock. Even large landslides in rock have a form largely controlled by the rock structure. The important exceptions to this statement are those slopes which are undercut by rivers or waves, those which have inherited their form from some past tectonic events, and those which are controlled by structure. The great majority of rock slopes, however, have profiles which are in equilibrium with the strength of their rocks.

Strength equilibrium slopes owe their form to the operation of the eight features listed in Table 6.2. A slope profile like that shown in Fig. 8.16 has many units, each with a characteristic angle, controlled by the features of rock strength. On a plot of slope angle against rock mass strength, for each unit the data points all fall within an envelope. This strength equilibrium envelope both permits the recognition of strength equilibrium

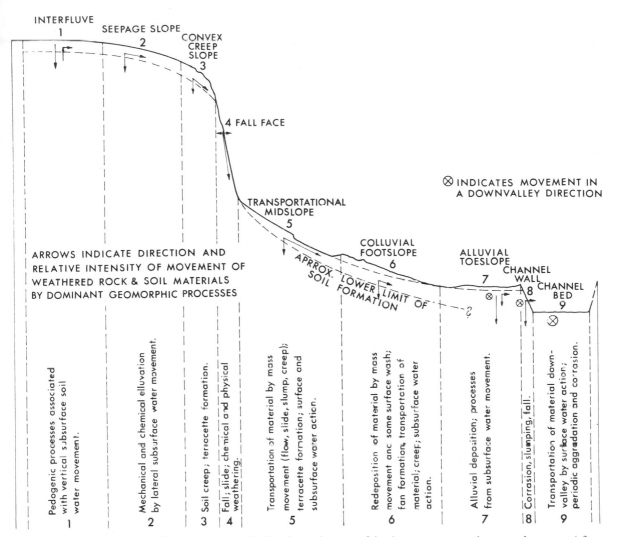

Fig. 8.15 A theoretical model of slope segments with direction and nature of dominant processes acting on each segment (after Dalrymple, Blong, and Conacher, 1968).

slopes and also the estimation of the safe angle of slope for rocks of a given mass strength. It can be seen from Fig. 8.17 that slopes on a certain rock type can have a variety of slope angles depending upon the influence of their features.

It is evident that, as strength equilibrium slopes are widespread, many rock slopes must retreat so that this relationship is preserved. Consequently we can assume that the rock faces retreat parallel to themselves as long as the strength of each rock unit remains constant. If only one of the features

of the rock changes with depth into the slope, then the characteristic angle will change.

In humid climates rock faces tend to be limited to high mountains or coastal cliffs, and even there many slopes have units of rock outcrop formed on resistant rocks, and soil-covered slope segments on intervening weaker rocks (Fig. 8.18). On the weaker rocks processes control the form of the slopes with soil and colluvium covers.

Structural controls on rock slopes are particularly evident where the dip of resistant beds or the

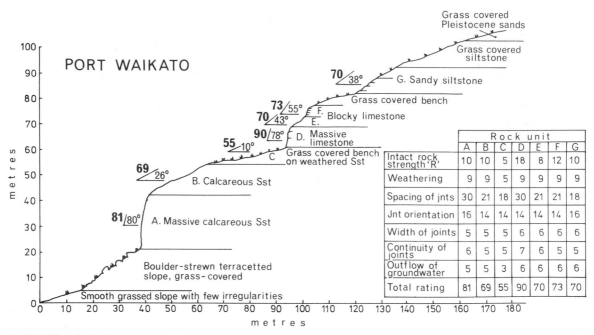

Fig. 8.16 The relationship between slope angle and the strength of each rock unit for a hillslope in New Zealand (after Selby, 1980). The method of determining strength values is discussed in Chapter 6.

The table from Fig. 8.16:

	Rock unit						
	A	B	C	D	E	F	G
Intact rock strength 'R'	10	10	5	18	8	12	10
Weathering	9	9	5	9	9	9	9
Spacing of jnts	30	21	18	30	21	21	18
Jnt orientation	16	14	14	14	14	14	16
Width of joints	5	5	5	6	6	6	6
Continuity of joints	6	5	5	7	6	5	5
Outflow of groundwater	5	5	3	6	6	6	6
Total rating	81	69	55	90	70	73	70

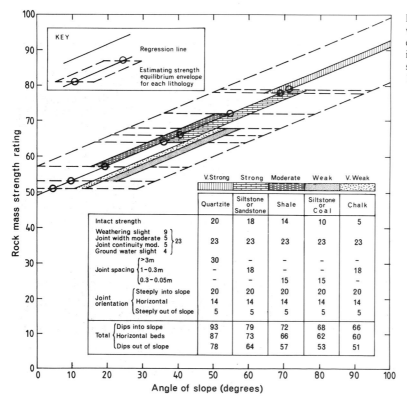

Fig. 8.17 Strength equilibrium envelope within which fall all data points for equilibrium rock slopes. The examples indicate a common range of values for strength of each rock type (after Selby, 1982b).

The table from Fig. 8.17:

		V.Strong	Strong	Moderate	Weak	V.Weak
		Quartzite	Siltstone or Sandstone	Shale	Siltstone or Coal	Chalk
Intact strength		20	18	14	10	5
Weathering slight 9 Joint width moderate 5 Joint continuity mod. 5 Ground water slight 4	23	23	23	23	23	23
Joint spacing	>3m	30	–	–	–	–
	1–0.3m	–	18	–	–	18
	0.3–0.05m	–	–	15	15	–
Joint orientation	Steeply into slope	20	20	20	20	20
	Horizontal	14	14	14	14	14
	Steeply out of slope	5	5	5	5	5
Total	Dips into slope	93	79	72	68	66
	Horizontal beds	87	73	66	62	60
	Dips out of slope	78	64	57	53	51

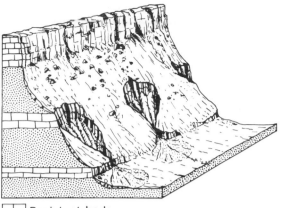

Resistant beds

Fig. 8.18 Resistant rock beds form outcrops because the rate of weathering and soil production is slower than the rate of soil erosion. Hard beds also limit processes acting in soil and colluvium.

inclination of folded rock units dominates the landscape (See Fig. 4.39). As these structural features are eroded their slopes gradually move towards strength equilibrium angles. This phenomenon is well illustrated by the domed outcrops of rock called *bornhardts*.

Bornhardts have many similarities with tors except that they are much larger. They are dome-shaped rock masses always more than 30 m high and frequently several hundred metres high (see Plate 5.24). They usually stand above erosional plains in the tropics and subtropics, but domed hills also occur within mountains and in high latitudes where severe glacial erosion has not occurred. Many bornhardts are formed in granite and gneisses, but some, like Ayers Rock and the Olgas in central Australia, are formed in coarse sandstones and conglomerate.

Discussions on the nature of bornhardts hinge around three main hypotheses of origin, and even more theories concerned with their joints.

(1) Their origin has been ascribed to a dome form originating with the shape of a granite intrusion. This hypothesis has some support from field evidence in the Namib desert where granite domes are being stripped of the schist which surrounds them, but is not commonly regarded as being a universally applicable hypothesis.

(2) The domes are the result of differential weathering at depth, with weathering proceeding in zones of closely spaced joints and only super-

ficially modifying the outer surface of massive rock which has few joints. Subsequent erosion, either by surface downwearing or by valley-side retreat then reveals the dome (Fig. 8.19). This hypothesis is related to stable continental shields with humid climates, in the past, in which deep weathering could occur. Granite domes have been found within deep cuttings through weathered soil in several parts of Africa.

(3) It has been proposed that bornhardts are up-faulted. This may be true of some features, such as the famous sugar-loaf mountains of Rio de Janeiro, but it is certainly not a valid general hypothesis.

What is usually clear are the sheets of rock some metres thick which exfoliate from the surfaces of domes. Loss of these sheets, which may be due to the relief of stresses induced by deep burial and tectonism, preserves the dome form, but causes the dome to become smaller. As the sheets are released by the opening of joints parallel to the dome surface, cross-joints open and subdivide each sheet until the dome, or its lower segments, becomes a strength equilibrium rock slope. In the final stages

Fig. 8.19 Stripping of a regolith, either by slope retreat as in the Australian example, or by downwearing and removal of soil, as in the example from Nigeria, leaves a rock dome outcropping as a bornhardt.

of dome decay, joints may so cut across the dome that it looks more like a large tor than a dome.

Decline, replacement, or parallel retreat of slopes

From the comments made already it will be evident that there is unlikely to be any universally applicable model of slope change. It has been suggested that combinations of creep, splash, and wash cause slope convexities and concavities to extend at the expense of straight slope segments (Fig. 8.14a, c), and to *decline* in angle with time. Slopes dominated by uniform retreat of straight segments, but with wash across lower slope segments, suffer *replacement* of the straight slope by an extending wash slope below (Fig. 8.14b). *Parallel retreat* of slopes implies that either processes are of constant intensity across the slope face and/or the resistance of the rock is constant. Parallel retreat (Fig. 8.14d) also implies that there is continuous removal of rock debris.

As will be seen from Chapters 16–20, the dominant processes and their intensity have varied as a result of the numerous changes of climate experienced in the last few million years. Consequently, although slopes are gradually removed from the landscape and replaced by extensive plains, as on the shields, it is impossible to envisage a constant process of slope decline, replacement, or parallel retreat. All three changes of form may have existed in the history of a particular area, and on ancient plains there stand now both steep and gentle slopes, reflecting both a complex history of erosion and the operation of varied structural and rock strength features. Simple uniformity of development is not a feature of most natural slopes.

Further reading

Although hillslopes are at the centre of geomorphic studies it is only recently that textbooks with detailed studies, have become available. The small book by Finlayson and Statham (1980) is a useful text for applied studies and the more advanced text by Selby (1982) has a strong emphasis on rock and soil strength. Of the older books Young (1972) is still valuable for its discussion of slope profiles, and the more advanced book by Carson and Kirkby (1972) provides a comprehensive treatment of mathematical modelling of slope processes.

9 Fluvial Processes

Rivers have a number of geomorphic functions: they carve the channels in which they flow and thereby form most of the relief available for slope process to act upon; they transport the debris of slope, and often of glacial, processes to the ocean; and they produce a suite of erosional and depositional landforms.

Because rivers are of such importance to human economies for transport, drinking water, hydroelectric power production, irrigation, and effluent disposal they have been studied in some detail by engineers over the last three centuries. As a result many empirical formulae have been developed for describing the relationships amongst stream velocity, depth, width, sediment-transporting capacity, and channel forms. Many studies were concerned with particular problems but from these general laws have been deduced.

Stream eroson, that is *fluvial* erosion, has three facets: the destruction of the bedrock of a channel, erosion of channel banks, and the *comminution* (that is the breaking down into smaller fragments) of the rock fragments carried into the stream by slope processes or derived from the channel bed and banks. The debris in a channel is moved in four main ways: *solution load* is composed of the solutes moved in the fluid; *suspension load*, which is sometimes called *wash load* when the particles are of colloidal size, consists of clay, silt, and colloidal material kept in suspension by turbulence in the fluid; sand and larger material moves by rolling or by saltation. The rolling and saltating materials together make up the *bedload*, while the deposits forming a channel bed are collectively known as *bed-material load*.

In the process of transportation rock debris is not only broken down by attrition and rounded, but also sorted according to particle size, particle shape, and particle density. The nature of flow in the fluid is such that these processes are accompanied by the creation of a series of *bedforms*, that is, ripples, dunes, and hollows—both in bedrock, where that is exposed, and in the bed sediments. The pattern of bedforms, in turn, has an effect upon the flow and also upon the gross morphology of the valley-floor landforms. An understanding of the basic principles of fluvial processes is therefore essential for the interpretation of fluvial landforms.

The mechanics of fluid flow and sediment transfer are far more complex than the static problems presented by mass movement of soils on slopes. Fluids are in constant irregular motion and it is impossible to determine exactly the forces acting along boundaries of a fluid and on objects protruding into the flow. Furthermore, much of the available stream energy is not always used for erosion and transport. Some energy is lost by friction, channel beds may have a cover, or *armour*, of particles too large to be moved by the available flow, or debris may not be available for movement. As little as 5 per cent of the available energy may be used in moving sediment and shaping the channel. Consequently many methods of determining the relations between fluids, their sediment load and bedforms are empirical, subject to revision, and very inexact.

Fluid flow in channels

Studies of flow in stream channels have to take into account the shape of the channel. Commonly used descriptors of channels are indicated in Fig. 9.1 in which:

- W is the width of water in the channel,
- P is the wetted perimeter (i.e. the boundary along which water is in contact with the channel floor and walls),
- A is the area of a transverse section of the stream,

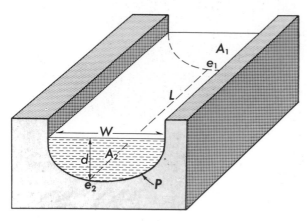

Fig. 9.1 Stream channel dimensions.

d is the water depth—for practical purposes, in wide channels, d is often taken as the same as the hydraulic radius R, but R is more fully given as A/P, or for a rectangular-section channel as $A/(2d + W)$,

S is the stream gradient or the drop in elevation between two points on the bottom of the channel divided by the projected horizontal distance between them or $(e_1 - e_2)/L$.

Discharge and velocity of flow in an open channel

If the mean velocity (V) for a given cross-section of flow (A) can be found, then the discharge (Q) through that section is easily obtained, where there is steady uniform flow, from

$$Q = AV.$$

Natural channels, however, are irregular with rough beds and walls. Engineers have developed a number of empirical formulae for determining the velocity of unsteady flow in rough channels. One of the first practical formulae was published in 1775 by Antoine Chézy who expressed velocity as a function of hydraulic radius (R) and the stream gradient (S). In the Chézy formula:

$$V = C\sqrt{(RS)}$$

where C is a coefficient whose value is dependent upon gravity and frictional forces.

The Irish engineer Robert Manning attempted to refine the Chézy equation by the formula:

$$V = \frac{1}{n} R^{\frac{2}{3}} S^{\frac{1}{2}} \text{ (m/s)}$$

where Manning's n is a roughness coefficient. Values of n can be determined by reference to Fig. 9.2 or Table 9.1.
Using the relationship $Q = AV$, then:

$$Q = \frac{A}{n} R^{\frac{2}{3}} S^{\frac{1}{2}} \text{ (m}^3\text{/s)}.$$

The mean velocities of rivers in flood usually vary from about 2 to 3 m/s. Rivers flowing over rapids reach much higher velocities than this, but none in excess of 9 m/s is known.

Variations of velocity and turbulence across a channel

Neither fluid velocity nor turbulence are uniformly distributed across a stream channel (Fig. 9.3). Maximum velocity in a symmetric channel occurs below the surface in the centre of a channel and decreases towards the boundaries. In an asymmetric channel the zone of maximum velocity if offset towards the steepest bank and deepest water.

As discharge is the product of cross-sectional area and velocity, the average velocity is derived by dividing the channel into a number of segments for each of which the area is measured, with a tape for the width and a rod or weighted cable for the depth. Velocity is determined for a number of depths: for shallow streams this may be at 0·6 depth or the average of $0·8d + 0·2d$ (Fig. 9.4), but for large rivers many more determinations may be necessary. Gauging sections should have straight, regular, and stable channels.

Under some conditions it is possible to build *weirs* and *flumes* across a channel so that precise measurements of discharge can be made at the controlled sections (Plates 9.1, 9.2). With such structures the cross-sectional area of the channel is known precisely and the flow is directly related to the head of water ponded behind the weir. It is thus possible to determine a relationship between the height of the water surface, known as the *stage*, behind the weir and the discharge. A simple measurement of the stage will then provide a value of the discharge; the stage is easily recorded continuously. A weir has a V-notch, a rectangular or some other geometrical shape and the water falls over the crest of the weir. In a flume the velocity and head are controlled by the passage of water through a constricted section. The constric-

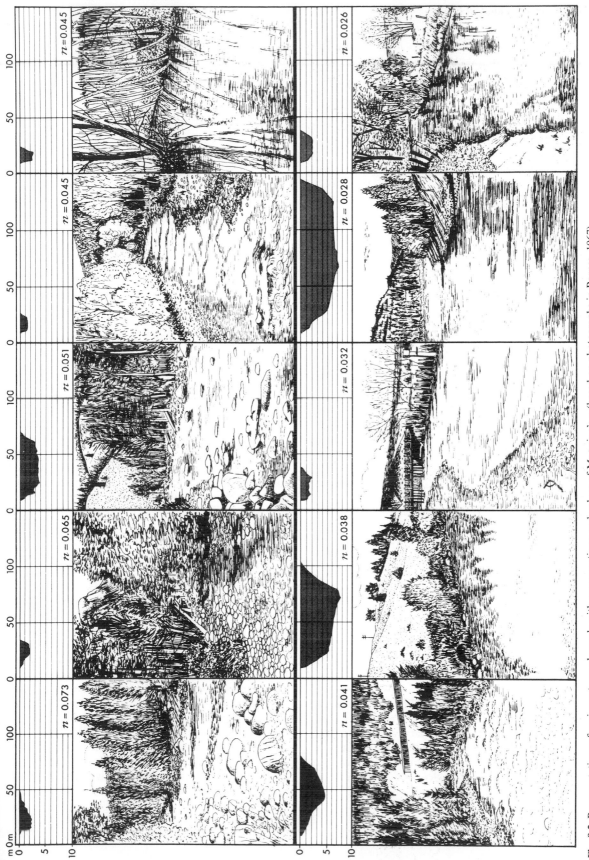

Fig. 9.2 Representations of various steam channels with cross-sections and values of Manning's n (based on photographs in Barnes, 1967).

Table 9.1   Values of Manning's *n* for Floodplains and Stream Channels

(a) Floodplains		
	short grass pasture	0.025–0.035
	mature crops	0.025–0.045
	brushwood	0.035–0.070
	forested	0.050–0.160
(b) Alluvial channels		
	smooth sand beds, no vegetation	0.014–0.035
	dunes on channel beds	0.018–0.035
	smooth beds with pools and water weeds	0.045–0.080
(c) Mountain streams		
	gravel and few boulders	0.030–0.050
	large boulders	0.040–0.070

Note: Values given are for straight channels. For winding
channels these values should be increased by up to 30 per
cent, depending on the degree of winding.
Source: Chow, 1959.

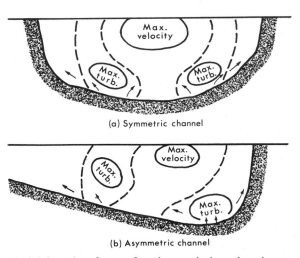

Fig. 9.3 Location of zones of maximum velocity and maximum
turbulence in a channel.

tion may be a narrowing of the channel, a hump in
its floor or a combination of the two. V-notch
weirs have the advantage that they can give more
precise data over a wide range of discharges but
flumes are less likely to trap sediment than weirs.

Zones of greatest turbulence tend to be away
from the zone of greatest velocity, and turbulence
is at a maximum near the bed of the channel where
obstructions cause the streamlines of water to
diverge round the obstacles and form *eddies*.
Eddies may be (1) intermittent *vortices*, (2) con-

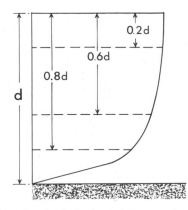

Fig. 9.4 Sampling depths for determining mean flow velocity at
a station. The figure also shows the velocity profile in a
downstream direction, with minimum velocity close to the bed.

tinuous spiral motions known as *helicoidal flow*
(Fig. 9.5), (3) *rollers* which have a cylindrical form,
or (4) *surge phenomena* which are pulsations in
water velocity. All fast-moving eddies tend to be
erosive and are important in the entrainment and
transport of debris. Slow eddies, by contrast, such
as slow rollers, contribute to the development of
depositional bedforms. Because of the preferential
development of maximum turbulence away from
the centre of channels, symmetric channels tend to
have erosion concentrated on the bed towards the
banks, and in asymmetric channels the zone of
greatest turbulence helps to undercut the bank on
the outside of a bend, while deposition occurs on
the inside of the bend.

Fluvial erosion of bedrock

The load that a river carries comes from a number
of sources: solutes and sediment supplied by slope
processes, the alluvial bed, the reworking of old
terrace and floodplain deposits, and sediment and
solutes produced by direct erosion of the channel
bedrock. The processes which are involved in
bedrock erosion are: corrasion, fluid stressing
(sometimes called *evorsion*), cavitation, corrosion,
and plucking. These processes are discussed in
Chapter 6.

In all cases the rate at which material of a solid is
removed in a fluid stream is related to the power of
the eroding fluid and the strength of the surface of
the solid, so that a slow-moving fluid, or a resistant
solid, cause rates of erosion to be low, while high

9.1 A V-notch weir showing the angles of the 3-notch forms, the pond behind the weir, and the recorder housing. The recorder is activated by a float, which is inside the tube below the recorder, and the float responds to the stage in the pond.

9.2 A rectangular flume. The recorder is in the box to the left above the end of the concrete wall. Note the coarse bed-material which would block a V-notch.

fluid velocities and weak solids are associated with high rates of erosion.

Erosion processes within stream channels leave behind irregular surfaces which may be the angular and broken edges of joint blocks, or scour marks smoothly abraded by numerous particle impacts; plane polished surfaces are rare.

Scour marks occur in a variety of forms, of which potholes (Plate 9.3) are the best known, and include fluted and scalloped, depressions (Plate 9.4). Collectively these features are known as *erosional marks*. Erosional marks are formed not only by abrasion in streams but also by solution in limestone caves, by ablation of ice and compacted

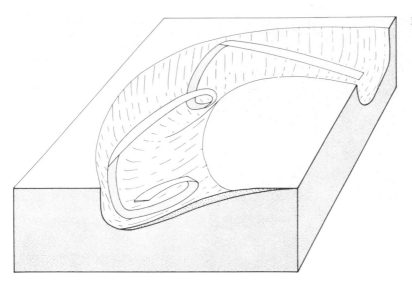

Fig. 9.5 Helicoidal flow in a channel.

9.3 Single potholes, up to 1 m diameter, cut in ignimbrite, and linked potholes forming a stream channel.

snow, by wind-blown sand and dust and also upon meteorites as they enter the atmosphere.

Erosional marks are depressions surrounded by rims which may be rounded or sharp; they are cusped in profile and have crests lying across the current path. Erosional marks are related to the formation of *separated flows*, that is, flow structures characterized by the detachment of the fluid shear layer from the channel bed and its reattachment some way downstream, with a recirculating flow occurring between the separation and attachment points (Fig. 9.6).

Potholes are widely recognized bedforms cut in the rock of stream channels. Some are very large—at Taylor's Falls, Minnesota, they are up to 18 m deep and 3.6 m in diameter. Three types of pothole may be recognized: (1) a plungepool type produced where waterfalls cut back along a narrow cleft to leave a channel with steep flared walls and a narrow slit-exit downstream; (2) shallow gouge holes in stream beds caused by a large separated flow; and (3) eddy holes which may be very deep, possess sharp edges, and contain spiral grooves on the walls as evidence of vorticular

9.4 Erosional marks cut in quartzite walls of a gorge. Marks are up to 1 m long. Kembé Falls on the Kotto River, one of the headwater channels feeding into the Zaïre River.

circulation. To this classification may be added linked potholes formed when neighbouring holes grow and breach the walls shared with one another (Plate 9.3).

It has been suggested that cavitation pits may be initial sites for the growth of potholes but the continuity in form between gouge holes and deep potholes suggests that they may grow at any site where flow separation can occur. Although most potholes occur in nearly horizontal channels they can also be formed in sloping or vertical rock faces if the flow conditions are suitable. Streams flowing within, or beside, glaciers may provide opportunities for large horizontal potholes to develop in valley walls (Plate 9.5) because very high velocities occur where confined streams under high pressure heads can jet water at rock faces. In such conditions cavitation may be a significant process.

A general condition relating to all erosional marks is that they may form both at sites where there are existing irregularities in a rock surface—as at a joint, a protrusion, a hollow, or merely the site of a more rapidly weathered rock crystal—or they may form in an apparently smooth regular rock surface. With all erosional marks in hard rock it is probably necessary for the flow to carry abrasive particles and to keep the mark clear of debris if the mark is to continue growing. There is no evidence that large potholes can be produced only by the swirling action of gravel and boulders, although these may be involved. Soft rocks such as mudstones can be pitted with erosional marks as a

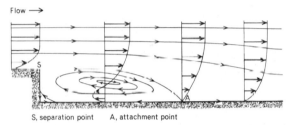

Flow →

S, separation point   A, attachment point

Fig. 9.6 A separated flow forming at a downward step in a channel bed, and a roller in the flow downstream of the step. The velocity profiles indicate the reverse flow in the roller (after Allen, 1971).

result of fluid stressing although this does not often occur alone in natural channels.

Sediment transport

Entrainment of sediment from the bed of a channel occurs only if particles small enough to be moved by the available shear stresses are present (see Chapter 6). Determining the concentration of suspended sediment in a flow is done with a sampler designed to limit disturbance of the flow (Plate 9.6).

With the most common depth-integrating sampler the collecting device is lowered to the stream bed and raised to the surface at a constant rate. Samples of suspensions are taken at a number of stations across the stream for which

9.5 A large erosional mark cut in schist alongside the Franz Josef Glacier, New Zealand. It is probable that the erosion was caused by a stream, within the glacier, jetting water at the rock.

velocities are also determined and the sediment discharge calculated from:

$$q_s = \frac{QC_s}{1000}$$

where q_s is the sediment discharge in kg/s,
$\quad\quad Q$ is the stream discharge in m³/s,
$\quad\quad C_s$ is the suspended sediment concentration in mg/l determined by filtering samples (mg/l = g/m³ = p.p.m.).

The sum of values for the stations across a stream provides a value for the total sediment discharge. Most samplers cannot be operated within 9–12 cm of the bed, and they therefore do not sample this zone of high concentration, and a correction factor for that type of stream has to be applied.

Assessment of bedload transport is an extremely difficult task. No single apparatus, technique, or theoretical solution has been universally accepted as being adequate for the determination of bedload discharge. Problems of direct measurement include:

(i)   interference with the flow by a necessarily large sampler;
(ii) difficulties with placing a device upon an uneven channel bed;
(iii) sampling efficiency;
(iv) irregular movement of bed material.

Methods used include grabs, steel-mesh baskets or trays in which to trap the particles, slot traps built into a channel bed, and acoustic devices which record the noise of bed material rolling. The volume of sound can then be related to other direct measurements.

Theoretical approaches usually rely upon the development of formulae from laboratory flume experiments in which flow velocities and particle sizes, densities and shapes can be closely controlled. Such approaches often work well when applied to canals and other engineering structures but natural channels are far more variable. Studies in the River Clyde for example demonstrated that measured rates of bedload transport were only about 0.1 per cent, or less, of those calculated by

9.6 Top left is a carriage, on a cableway, from which sampling instruments may be lowered into a large river. Below it is a D-49 depth-integrating sediment sampler; the intake nozzle is at the front and the stabilizing fins are at the back. Bottom left is a Helley-Smith bedload sampler used in sand-bed channels. Top right, a Helley-Smith sampler is being lowered on a rod; and bottom right a flow meter is being set to a required depth on the gauging rod.

theoretical formulae. The Clyde bedload is presumably not a capacity load.

Competence and capacity

Because of the uncertainties in describing flow in a precise quantitative manner it is extremely difficult to determine the competence and capacity of a stream. *Competence* is indicated by the largest size of particle that a stream can carry in traction as bedload. This size may be specified for a given flow at a certain channel cross-section, but discharge, slope, and other factors vary along the length of a channel and discharge varies at a point. Competence will hence vary almost continuously.

Capacity is the maximum amount of debris of a given size that a stream can carry in traction as bedload and, given a supply of debris, is related to channel gradient, to discharge, and to the calibre of the load (Plate 9.7). The maximum calibre (i.e. competence) is not simply determined because a suspension can increase fluid density and therefore capacity, whereas the introduction of large particles into a flow can decrease total capacity. Capacity therefore depends upon the size distribution of all particles in the load.

Deposition

The different modes of transport—suspension, saltation, rolling—result in different mobilities of the various size fractions which in turn result in sorting of the sediment according to particle size, shape, and density.

It can be readily understood that bedload will cease rolling when current velocities decrease below a critical value, and the saltation layer has a fairly uniform upper limit defined by the maximum height to which the lift forces can project a grain. Once lift forces fall below a certain value saltation of a given particle size will cease. As bedload is usually of sand-size or larger material, and saltation load is sand and coarse silt, it is found that the concentration of sand and silt decreases continuously and smoothly up from the bed.

Suspended loads of silt and clay supported by turbulence are usually distributed rather evenly through a flow. When the upward components of velocity (lift, drag, and turbulence) fall below the settling velocity for a given grain size the particle drops out of suspension and is deposited.

In a channel the water velocity curve is convex downslope and the sediment concentration curve concave downslope; as a result the sediment discharge curve is usually a gently convex curve in a downslope direction (Fig. 9.7).

Sorting

Hydraulic sorting is an inevitable result of the selective transport of a certain grain size by one process and by the varying efficiency of the transport processes. In most river systems, for example, grains larger than about 10 mm diameter cannot be moved in the seaward ends of channels, and hence they remain in the upper parts above the point at which channel slopes become too gentle

9.7 Boulders with dimensions of 10 m partly block the channel of the Haast River. They result from a rockfall. The largest debris which can be rolled along the bed has dimensions of about 3 m, but travel distances are short because of the channel roughness.

9.8 A, ripple bedding in sand; B, coarse bedload with no bedform structure; C, cross-bedding of the advancing faces of dunes; D, planar bedding. Scale is given by the hand lens (photo by C. S. Nelson).

9.9 Standing waves above antidunes of sandy gravels in a melt-water stream, Antarctica.

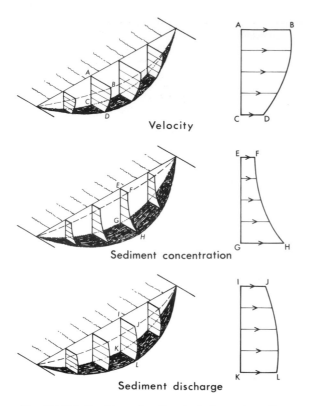

Velocity

Sediment concentration

Sediment discharge

Fig. 9.7 Relationships of sediment discharge to sediment concentration and flow velocity at stations across a channel (modified from Leopold, Wolman, and Miller, 1964).

for them to roll. Particles in the size range of 1–10 mm tend to move by rolling over sand grains and may be transported right through the system. Coarse to fine sand moves by traction and saltation and becomes temporarily stored in channel bed structures such as dunes and ripples. Silt and clay is carried in suspension either right through the system or is stored for a time in floodplains where it is left by overbank flooding.

As a result of sorting and attrition of particles, and of discharge and flow-depth variations down a river, the capacity to move bedload also varies. In most large rivers the coarse bedload of upper reaches can be transported only at flood stages, but in the lower reaches sand-size bedload is moved even at low discharges. It seems then that, even at low discharges, a large river is competent to move all, or most, of its bed material. Therefore it is the competence which controls sorting in small streams, but capacity controls it in large rivers. A change in capacity leads to a transfer of sediment from bedload to suspended load in a large river.

Sedimentary bedforms

Bedforms may be elongated parallel to the flow or aligned transversely across it. Those which are elongated parallel to the flow are *cuspate* and the cusp may face upstream—when it is called

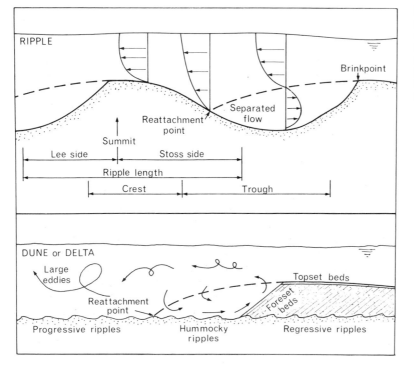

Fig. 9.8 The terminology related to ripples and dunes, and the flow conditions and bedding associated with them. Heights of ripples and dunes with respect to water depth are exaggerated. Note that ripples migrate upstream within the roller, but downstream where they are beyond its influence.

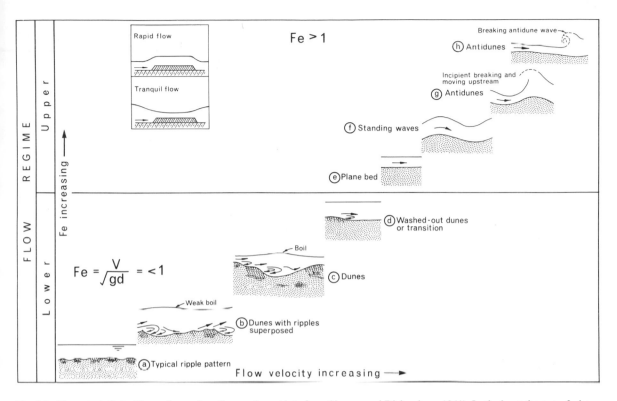

Fig. 9.9 Characteristic bedforms for various flow regimes (data from Simons and Richardson, 1961). In the inset the out-of-phase relationship of the bedform with the water surface is shown for tranquil lower flow regime, and the in-phase relationship for upper flow regime.

linguoid, or downstream—when it is called *lunate*. It is usually accepted that when depositional bedforms are less than 40 mm in height and 600 mm in wavelength they are called ripples; larger features are called dunes.

Migration of dunes and ripples occurs as material is carried up the stoss side (upstream side) of the feature and then avalanches over the brink down the lee side. The avalanching process consequently gives rise to characteristic cross-bedding in the sands or gravels (Plate 9.8, Fig. 9.8).

As soon as transport begins the flow starts to shape the bed. At low fluid velocities, just sufficient to move sand grains, ripples form; further increases in discharge lead to the formation of dunes. These two bedforms characterize the lower flow regime in which Froude numbers (i.e the ratio between the force required to stop a moving particle and the force of gravity) are substantially less than unity, and waves on the water surface are out of phase with undulations on the bed (Fig. 9.9). Further increases in discharge or slope result in the disappearance of the dunes and formation of a plane bed with linear depositional forms. On a plane bed resistance to flow is low and sediment transport is high.

The plane bed condition is transitional to an upper flow regime. As discharge continues to increase antidunes may develop when Froude numbers are about 0.8. Antidunes are in phase with the water waves (Plate 9.9) and resistance to flow is small. Antidunes migrate upstream as erosion from the downstream side of the antidune throws material into saltation and suspension more rapidly than it can be replenished from upstream. At very high flows there develop chutes with shooting flow and nearly plane bed; these are followed downstream by hydraulic jumps and a

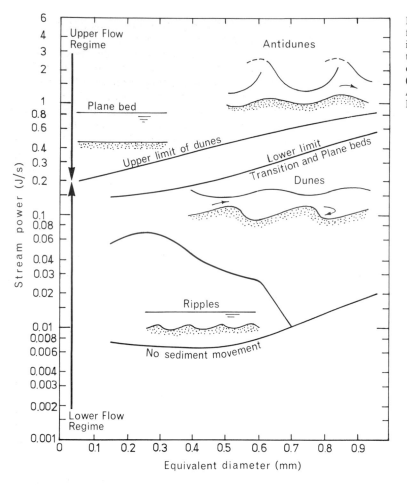

Fig. 9.10 Hydraulic criteria for the formation of bedforms. Stream power is derived from $P = \gamma_w DSV$, where γ_w is the unit weight of water, D is the depth of flow (m), S is the channel slope (m/m), V is velocity (m/s), (based on Allen, 1968, and Simons and Richardson, 1961).

deeper section known as a pool. There is thus a direct relationship between the stream power, the characteristics of the flow, the bedforms, the transport rate, and the particle sizes (Fig. 9.10).

Dunes are common bedforms in alluvial channels and in large rivers like the Mississippi or Niger they occur continuously along the bed for hundreds of kilometres. On average the height of river dunes is between 10 and 20 per cent of the mean flow depth, while the wavelength (Fig. 9.8) is several times the mean depth. In big rivers dunes may be 2–5 m in height and 50–200 m in wavelength.

Many bedforms become so large that they adjust only slowly to changes in the flow, so there is not always an equilibrium between flow regimes and bedforms. This is particularly so at low flows which may be accompanied by bedforms inherited from earlier periods of higher flows. The survival of bedforms in sediments makes it possible to reconstruct the conditions of flow and channel form under which the sediments were deposited (Fig. 9.11).

Discharge of water

Discharge is the water which runs off from a drainage basin. It is usually expressed as a volume (m^3), a volume in unit time (m^3/s), or as a depth of water spread over the whole catchment—thus a runoff of 40 mm is the average depth of water draining from the catchment. Water does, of course, not drain uniformly from the surfaces of a

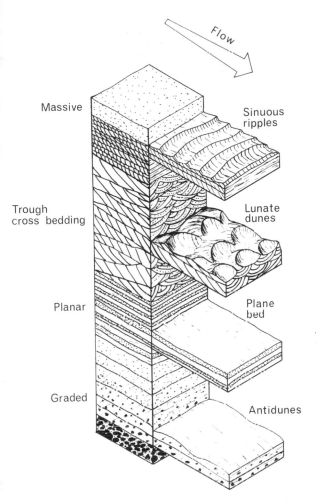

Massive

Trough
cross bedding

Planar

Graded

Sinuous
ripples

Lunate
dunes

Plane
bed

Antidunes

Flow

Fig. 9.11 Types of bedforms in alluvial deposits created by flows of various velocities. The coarse bed-materials become finer as depth of flow decreases (i.e. they are vertically graded). Planar beds are left by medium-velocity flows and cross-bedded deposits at lower velocities. A sequence of this kind is left as flow velocities decrease after a flood with graded, then planar, then cross-bedded sediments being deposited in sequence. The massive (i.e. unbedded) silts are overbank deposits of a later flood.

discharge hydrograph in which discharge is plotted against time (Figs. 9.12, 9.13).

To convert a stage to a discharge hydrograph a stage-discharge rating curve, or table, is used. This is produced by measuring discharge at a surveyed point on a stream channel of carefully measured cross-section, and then plotting the discharge for various heights of the water surface. The plot may be a straight line or a curve. It may also be a loop, for the relationship may differ between the rising and falling stages of the water surface as the slope of the water surface varies (Figs. 9.14, 9.15).

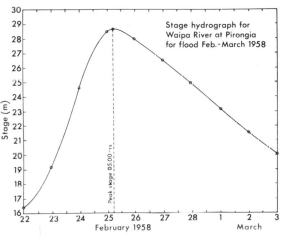

Fig. 9.12 A stage hydrograph. Data for Figs. 9.12, 9.13, and 9.14 were supplied by the Waikato Valley Authority.

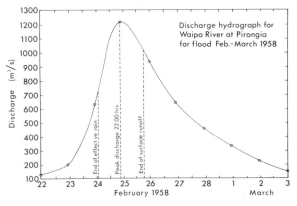

Fig. 9.13 A discharge hydrograph for the same storm as shown in fig. 9.12.

catchment but reaches the streams by various routes. The discharge from a catchment is extremely variable through time, and the graphical curve which represents this flow is called the *hydrograph*.

The hydrograph may be presented in two forms: a *stage hydrograph* in which the height of the water surface in the stream is plotted against time, and a

A hydrograph has a concentration curve or rising limb, a peak, and a recession curve or falling limb (Fig. 9.16). The shape of the hydrograph is controlled by the characteristics of the drainage basin and those of the rainfall.

The *concentration curve* is a function of the catchment area and the duration and uniformity of the rainfall. The normal shape is concave upwards. The time of concentration is the travel time required from the most distant part of the catchment to the gauging station, and is therefore the time from which runoff causes a rise in the water level in the stream to the peak flow at the station; the shape of the catchment therefore

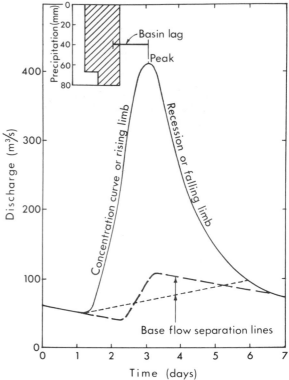

Fig. 9.16 Features of a hydrograph.

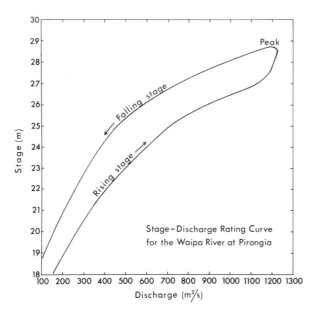

Fig. 9.14 A stage-discharge rating curve.

Fig. 9.15 Profiles of water surface during rising and falling stages of a flood. The varying slopes explain why stage-discharge rating curves may have a variety of shapes and slopes.

controls the shape of the concentration curve. The *peak*, as its name suggests, is the indicator of the maximum discharge during the storm. The *recession curve* is independent of time variations of rainfall and infiltration, and is therefore largely a function of the features of the drainage basin. The greater the storage within the basin the flatter and longer will be the recession curve.

Hydrograph shapes

During long periods of dry weather, runoff can come only from depletion of water stored in the soil and rocks of the catchment. This source is gradually reduced and the groundwater runoff hydrograph therefore shows a gradual decline in flow. The groundwater contribution to runoff is called *baseflow*.

If a storm occurs with sufficient intensity to increase the flow from the catchment, then storm runoff will cause an increase in flow which will show on the hydrograph (Fig. 9.17). The imme-

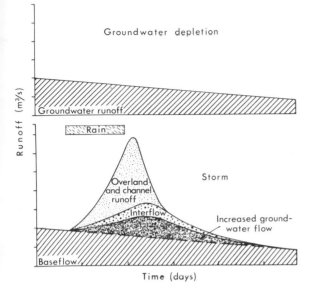

Fig. 9.17 Idealized stream hydrographs. Top: baseflow alone; bottom: the term overland flow covers both Hortonian and saturated zone runoff for a storm.

diate augmentation of runoff will be from rain which falls directly into the channel, but as time passes the other sources will begin to contribute. Storm flow may begin from the floodplain and the bases of slopes, then after a period groundwater rise and interflow will make their contribution. The shape of a particular hydrograph depends upon the features of the catchment and characteristics of the rainfall.

Rainfall with low intensity will cause only a low peak in the hydrograph compared with a high-intensity storm in which much rain is falling in a short period of time. The longer the duration of rainfall the greater will be the levelling off and extension of the peak flow. For a discussion on the influence of rock, soils, and vegetation on runoff see Chapter 8.

Flood frequency

Perhaps the greatest immediate practical use of hydrograph analysis is in the provision of information on the magnitude and probable frequency of floods. To predict flood recurrence intervals the maximum flood discharges for each year are ranked according to their magnitude with the

greatest flood being ranked highest. The recurrence interval is determined by:

$$\text{return period} = \frac{\text{number of years of record} + 1}{\text{rank of the individual flood}}$$

Each flood discharge is then plotted against its recurrence interval on special graph paper (Gumbel paper). From such a plot it is possible to predict the probability that a flood of a certain magnitude will recur in a particular number of years (Fig. 9.18). The method is only reliable where records span a number of years and it cannot be used reliably to predict for periods greater then the length of record.

A flood frequency analysis indicates a statistical probability that a flood of a certain magnitude will recur in a certain period of time. The reciprocal of the return period is the probability of obtaining a

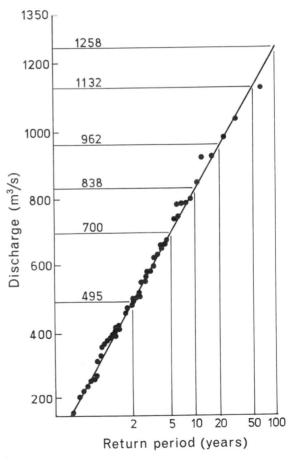

Fig. 9.18 Flood frequency relationships for the Trent River at Nottingham.

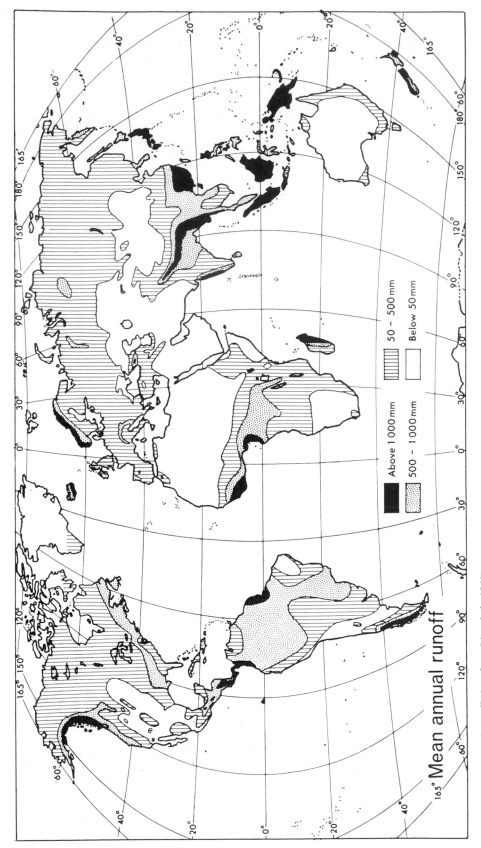

Fig. 9.19 Mean annual runoff (data from L'vovitch, 1961).

Mean annual runoff

Above 1 000 mm
500 - 1000 mm
50 - 500 mm
Below 50 mm

flood of that size. Thus the 100 year flood peak has a probability of 0.01 (or 1 per cent) of being exceeded in any one year.

It is a widely adopted convention to denote the 2.33 year flood as the mean annual flood; the 2 year flood is the median annual flood and the 1.58 year flood is the most probable annual flood.

Patterns of discharge

For any area the discharge is derived from the precipitation less the water lost in evaporation and transpiration together with that which is stored in lakes and groundwater. For the world as a whole, mean annual runoff is closely related to climatic patterns so that discharge is at a minimum in arid regions and at a maximum in humid areas (Fig. 9.19). Several factors, other than precipitation, affect river discharge patterns, notably, heat balance, vegetation, soils, rock structure; drainage basin shape, size, and relief, and hydraulic geometry. Of these only rock structure and drainage basin shape and size are independent of climate.

The effect of climate may be appreciated by a study of the annual discharge of selected rivers. The Tisza River rises in the Carpathians and flows south through Hungary to join the Danube north

of Belgrade. During the winter, baseflow is low and much of the precipitation is stored in the catchment as snow until the spring, when rainfall and snowmelt combine to produce peak floods. During the summer, baseflow declines from the spring maximum, but summer rainfalls produce moderate discharges. Late summer and autumn are seasons with low rainfall and low discharges. Strongly seasonal discharges are also evident for rivers in areas of seasonal rainfall and ice or snowmelt (Fig. 9.20).

Areas of high relief tend to be areas of high precipitation so that the mountains of the humid tropics have the highest runoff per unit area, but the subpolar mountains of Alaska, Norway, and Chile, which receive westerly air masses from large oceans, rival the Himalayas in water yield.

The mean annual discharge of some of the world's largest rivers is indicated in Fig. 9.21. It shows that the Amazon has by far the largest discharge and drainage basin area. It has been estimated that 82 per cent of the Amazon's sediment load comes from the Andes which make up only 12 per cent of its drainage area. The Nile has a very low discharge for its length because once it leaves the Ethiopian highlands it receives no major tributaries in its course across the Sudan

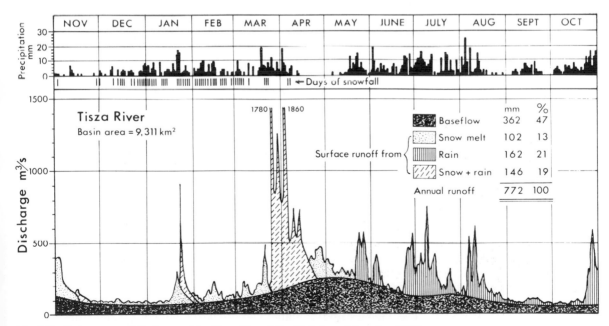

Fig. 9.20 Components of discharge of the Tisza River (after Kalinin and Szesztay, 1971).

MAJOR RIVERS

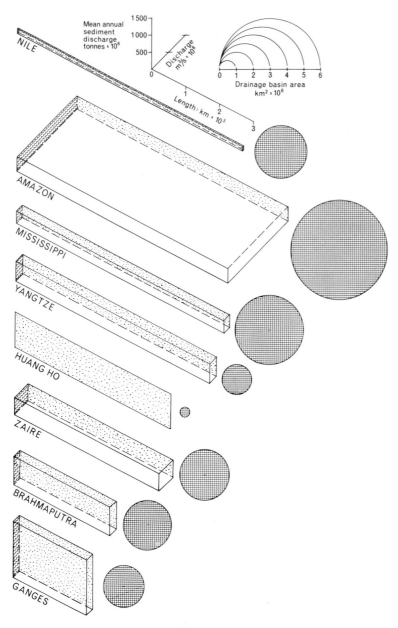

Fig. 9.21 Representations of the length, drainage basin area, and discharge of water and suspended load at the mouths of selected major rivers (data from Shen, 1971). (Note: (1) in the Pinyin system of transcription the official name of the Yangtze is Changjiang, and the Huang Ho (or Yellow river) is the Huanghe; (2) the river in Africa formerly known as the Congo is now known as the Zaïre along much of its length).

and Egyptian desert areas. The Huang Ho* has a very large suspended sediment discharge for its water discharge, because it flows through an area of intensely cultivated loess soils where silt-size particles are readily entrained. It is reported that, at times, discharges of the Huang Ho have been 40 per cent sediment by weight. The Ganges and

*Huang Ho is Huanghe in the Pinyin system of transliteration.

Brahmaputra carry very large suspended sediment loads derived from the Himalayas whose frontal ranges have extremely large monsoon rainfalls.

In considering figures for water and sediment discharges of major rivers, it has to be recognized that there are many errors and uncertainties in the data. The discharge figures for the Amazon, for example, could be in error by over 10 per cent, or over 17 000 m³/s—a water discharge greater than that of all but a few of the other large rivers. Estimates of sediment discharge are probably even less reliable, and for many large rivers there are few or no data.

Sediment discharge

Most of the available data on transport of load by rivers deals either with suspended load or with solution load because bedload is very difficult to measure or even estimate. It is clear that bedload movement tends to increase as discharge and flow velocity increase, but transport in solution and suspension is more variable in its relationship with changing discharge. Bedload is variously reported as comprising less than 1 per cent of the total sediment load for some large rivers, but more than 70 per cent in some alpine rivers. The proportions of total load that sediment load and solution load supply is even more variable, and in both cases ranges from less than 5 to more than 95 per cent, but is less extreme than this for large rivers.

Because load is far more difficult to record continuously than is discharge, attempts are frequently made to use discharges as a basis for estimating suspended and solution loads. Both suspended load and solution load increase as a power function of discharge, and suspended load usually increases far more rapidly than discharge (Fig. 9.22). For rivers like the Mississippi most of the suspended load is carried by floods with a frequency of at least once a year, so rare catastrophic events are not of great importance for total suspended sediment yield in this large river, and this is probably true for most river basins.

Solution load is largely carried by the water which makes up baseflow, for this is the water which has been in contact with bedrock and soils. Although dissolved load increases with discharge, it does so far more slowly than suspended load, so concentrations of solutes are frequently low during floods while those of suspended load are high. This is well shown by the Mississippi in which over 90 per cent of the dissolved load is carried by flows which recur monthly.

Although it is well established that during a flood the sediment discharge increases with water discharge, the two are not always directly related in time. The sediment discharge peak may occur before, at the same time as, or after the water discharge peak. The sediment discharge is in-

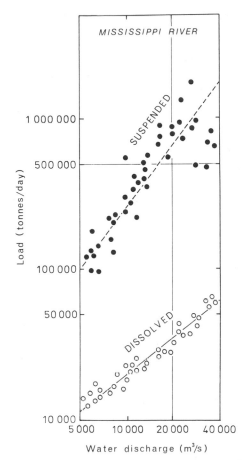

Fig. 9.22 Relationships between load and discharge for the Mississippi River (after Sedimentation Seminar, H. N. Fisk Laboratory of Sedimentology, 1977).

fluenced by the supply of sediment, the power of raindrops to erode soil, by river bank collapse, and by the capacity of streams to discharge it. It commonly occurs that all available erodible soil is removed by a flood before water discharge reaches its peak, and then the falling discharge of the receding flood is largely supplied by subsurface water that has not entrained soil before reaching the channels.

Total sediment yields are the best indicators of rates of denudation for different environments. These are discussed more fully in Chapter 20.

Further reading

The books by Gregory and Walling (1973), Leopold, Wolman, and Miller (1964), and by Richards (1982) amplify the topics discussed here.

10   River Valleys

Valleys display a remarkable association in long and cross profiles and in the processes of erosion, transport, and deposition occurring within them. The adjustment between landforms, the underlying structure and lithology and the processes occurring within valleys, however, is not always complete because the resistance to change varies through the valley, and the energy available to cause change varies not only in space but through time. As a result the degree of adjustment itself is variable in space and time. Any valley will have within it minor landforms which are relict from the last flood and most large valleys will also contain relict landforms inherited from past periods with different intensities of tectonic and climatic processes from those currently prevailing. A study of a river valley involves, therefore, not just an examination of current processes but an interpretation of relict features and deposits which may reveal something of the historical processes which have contributed to present-day landforms.

Valley profiles

As soon as water becomes confined in a channel it starts to modify the shape of that channel, both in cross-section and in gradient. This tendency is readily observed in sandpits where rills develop deep channels cut into steep slopes, but produce extensive fans at their bases, so that the long profile of the rill has a steeper gradient at the head and a gentler gradient on the fan than the original sand slope. If the flow of water is sustained the channel will eventually become adjusted throughout its length so that there is an approximate equilibrium between the energy of the water flowing on the sand, the amount of sand transported or deposited, and the size and shape of the channel. The idea that streams tend to equalize their work throughout their courses was first

enunciated by G. K. Gilbert (1877). When applied to river channels the concept helps to explain why the upper reaches of streams have characteristically steep, often irregular, channels with large and varied bedload materials, while lower reaches have lower gradients, bedload materials of smaller calibre, and more regular profiles, yet there is a continuity in the longitudinal profile.

Longitudinal profiles

The *longitudinal profile* of a stream indicates its gradient throughout its length. As channel slopes are commonly steep in the headwaters and gentle in lower reaches profiles are normally concave upwards with increasingly flat segments towards the mouth. There are exceptions to the general rule of concavity because streams in arid regions may lose capacity down-valley, as water evaporates or seeps away, and the profiles may then be convex.

Idealized longitudinal profiles are commonly depicted as being smooth curves throughout their length, and mathematical formulae are sometimes applied to these curves, but natural channels are seldom, if ever, smooth curves when considered either at a large or a small scale. Outcrops of resistant rock in channel beds or banks may produce waterfalls or rapids; tributaries bring into a main channel increased water and sediment so that channels adjust by steepening or decreasing their gradients; and changes in the effective lower limit of erosion—called *base level*—provided by the sea, or a lake, can cause increases or decreases in slope at the mouth (Fig. 10.1). Longitudinal profiles are thus essentially irregular, but have a general tendency to develop towards the idealized form provided that there are no changes in the controlling factors. In reality those factors are changing with varying speeds, so the adjustment of profiles to the controls is never complete and an

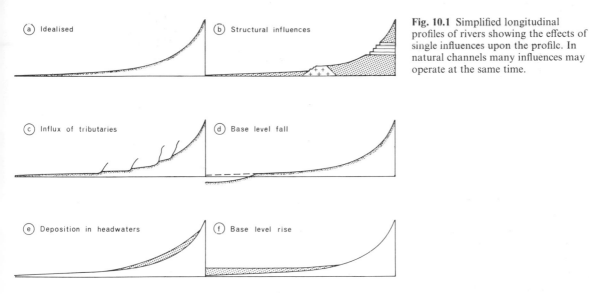

(a) Idealised (b) Structural influences

(c) Influx of tributaries (d) Base level fall

(e) Deposition in headwaters (f) Base level rise

Fig. 10.1 Simplified longitudinal profiles of rivers showing the effects of single influences upon the profile. In natural channels many influences may operate at the same time.

equilibrium between channel processes and channel slopes is never fully established. An approximate and partial adjustment of form and processes is, however, characteristic of streams and is denoted by use of the terms *quasi-equilibrium* or *dynamic equilibrium*.

Even at a very local level, long and cross profiles of channels are irregular. Along their lengths the line of deepest water, that is the *thalweg*, repeatedly crosses the channel between deeps and shallows alternating along the length and across the channel.

Base level is a term used to denote the effective lower limit to erosional processes. In his analysis of the term, W. M. Davis (1902) defined ultimate base level as being an imaginary level surface in extension of that of the ocean surface, and thus a level below which rivers cannot reduce the land surface. One might quibble that, in the estuaries and deltas of large rivers, flowing water erodes to below the ocean surface, and that wave action can reduce the land surface to the lower limit of effective wave attack. Another reservation is that in limestone terrains erosion can take place below sea-level, and in arid areas wind erosion may lower the floors of hollows down to the water table which may be below sea-level. In spite of such restrictions sea-level is the approximate effective lower limit of most erosional processes.

Temporary base levels are formed where resistant rock outcrops form a level below which land upstream of the outcrop cannot be reduced, and *local base levels* are provided by lakes and by trunk streams for their tributaries, for no tributary can cut below the channel of the stream into which it flows.

Neither ultimate nor local base levels have been constant in elevation over the last few million years. World sea-level has repeatedly fallen in each glacial period and risen in the following warm interval, with the maximum depression being commonly close to 100 m and the high sea-levels of warm periods being near to those of the present. Rates of change of sea-level may have been as high as several metres per century. As a result, rivers flowing to steep coasts with a falling sea-level sometimes developed waterfalls and rapids, which gradually migrated upstream. Such breaks in longitudinal profiles are known as *nickpoints*. On more gently sloping coasts stream adjustments may have involved channel steepening with the formation of sharp breaks in profile.

The interpretation of the history of base level change was once a major component of geomorphology, but it is now realized that repeated glacial to interglacial climatic fluctuations of approximately similar magnitudes, the repeated loading and unloading of the crust by the growth and wasting of ice sheets, and the uplift and depression of the crust by tectonic activity, together with varying discharges of water and sediment by rivers, all make any simple interpretation of base level change unrealistic.

Cross-valley profiles

Cross profiles are as variable as longitudinal profiles. In upland areas longitudinal profiles tend to be steep and irregular and cross-valley profiles are similarly steep with rock and talus slopes providing coarse material to the channel bed. Valleys tend to have narrow floors with little or no flat land beside the channel. Farther downstream as the channel slope is reduced and depositional processes become increasingly important the valley widens and floodplains and terraces take up more of the valley floor. Such broad generalizations have to be qualified because the effects of tectonic uplift, climatic change–especially of glaciation—the steepness of slopes, and the vegetation cover all have a strong effect upon valley forms, the rates at which they are modified, and the extent of inherited forms from earlier environments. Stream channels, however, are usually modified by contemporary processes far more rapidly than the valley which they occupy and channel forms are therefore the best indicators of current controls.

Bedrock channels

Stream channels are most commonly cut in bedrock in the upper segments of valleys, but many large rivers crossing uplands have channels deeply incised into rock even in their lower reaches. Far less attention has been directed at the features of such channels than towards those of alluvial channels. The influences of fold structures upon drainage patterns of bedrock channels have been discussed in Chapter 4: attention here will be focused on features of the channels themselves.

Irregularity in long profile seems to be a characteristic of incised valley floors. This is most noticeable where waterfalls and rapids interrupt the profile, and where variations in rock resistance form reaches with alternately confining and less restricted valley sides, but it is also a common feature of channels in uniform rock.

It has been shown from flume experiments in homogeneous material, and by field observations, that large hollows and risers may occur in longitudinal bed profiles so that locally the channel bed

gradient is reversed. In the Channelled Scablands of Western Washington (USA) closed rock basins over 50 m deep were formed by the Columbia River when it was dammed and diverted southwards by a Late Pleistocene ice sheet. Similarly large scour holes are found in the Snake River Plain, Idaho, where the sudden draining of Lake Bonneville produced a catastrophic flood which transported boulders over 1 m in diameter. Few rivers have scour holes or other irregularities as large as those of the Columbia and Snake spillways but irregular bottom profiles are normal (Plate 10.1). As long as the energy slope (the slope of the water surface) and the water depth is great enough it appears that both large bed-material can be moved, and deep scour holes can be formed, by stream erosion. There are so few studies of the long and cross profiles of bedrock channels that it is not known just how variable, and of what scale, most bedrock channel forms can be.

Waterfalls and *rapids* are amongst the most impressive fluvial landforms and are of value as scenic attractions or as sites for hydroelectric power generation. A waterfall is a site at which water falls vertically; a *cataract* is a step-like succession of waterfalls; *rapids* are part of the long profile in which the flow of water is broken by short steep slopes. Rapids may be completely drowned by the flow at high discharges when turbulence at the water surface is the only evidence of the underlying profile irregularity, but waterfalls and cataracts are seldom inundated.

Three general classes of waterfalls and rapids are commonly recognized: (1) those resulting from differential erosion of rocks with varying resistance; (2) those caused by vertical displacements in the longitudinal profile; and (3) those resulting from deposition in the stream channel (Fig. 10.2).

Horizontal or gently dipping strata may contain a particularly resistant bed—such as an indurated sandstone, massive limestone, or dolerite dyke —in a sequence of otherwise rather weak rocks. The resistant bed then forms a caprock which forms the lip of a waterfall while weaker underlying beds are more rapidly eroded, thus undermining the caprock. The Niagara Falls are an example of this type in which the resistant Lockport Dolomite forms a cap while shale and sandstone beds beneath it are more readily eroded. Recession of the falls has left a gorge about 10 km long. The rates of retreat in recent years have varied from 0.1 to 2.0 m/year (Fig. 10.3).

10.1 A section through a river terrace showing the scour holes cut in bedrock and later filled with bedload gravels. The channel deposits become finer in texture towards the surface and probably represent overbank silt and clay deposits. The modern soil is formed in windblown dust derived from exposed bars in braided channels.

Step falls develop in strata with a more uniform erodibility where the undermining occurs along individual bedding planes. Single falls also occur where vertically dipping beds juxtapose rocks of differing resistance—as at a dyke.

Vertical displacements in the longitudinal profile include those resulting from tectonic uplift, faulting, falls of base level—as at a steep coast—or the rapid incision of a trunk valley so that a tributary is left *hanging*, as in many overdeepened glaciated main valleys into which tributaries now cascade.

Deposition in a valley may be rapid and temporary as when a landslide blocks a channel, but more permanent as when the blocking is done by a lava flow. Deposition of calcite from saturated solu-

tions is a relatively common occurrence in limestone areas and it can occur in stream channels and form ledges, so that streams fall or cascade over low rims of calcite which can be several metres high.

Some of the largest falls in the world occur at the scarp edges of uplifted plains. The Victoria Falls on the Zambezi originated at such a break as did the 1000 m high Angel Falls at the edge of the Roraima Plateau on the Venezuelan Brazilian shield.

Many large tropical rivers are interrupted by waterfalls and cataracts. Some of these breaks are undoubtedly at the edges of uplifted plains, but many occur where resistant rocks outcrop. Breaks in the profiles of such rivers as the Nile, Zaïre, and

WATERFALLS

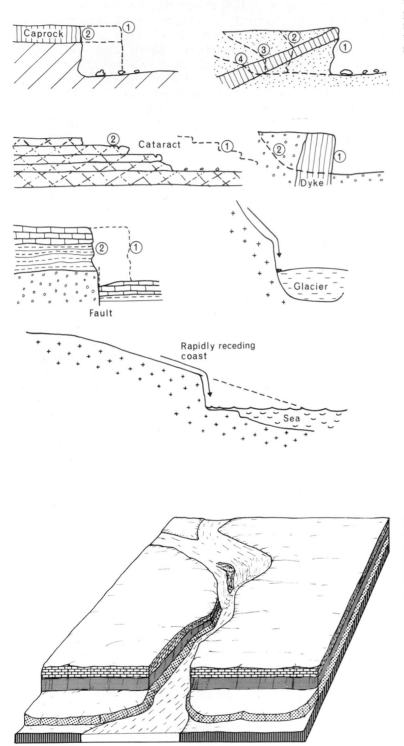

Fig. 10.2 Causes of, and controls on, the location of waterfalls. Successive positions of waterfalls are indicated by numerals. Note the plunge pools and the tendency to undercut the cap rock in the upper two figures.

Fig. 10.3 A representation of the form of the Niagara Falls. The escarpment edge indicates the location of the Falls about 30 000 years ago. The uppermost rock unit is limestone and it is underlain by shale and (at the lower river level) by sandstone.

Zambezi have greatly hindered their use for transport. One reason why so many tropical rivers have broken long profiles is that the rivers have relatively fine-grained bed-material loads, as a result of the predominantly chemical weathering processes in their catchments, hence abrasion is not rapid. Many temperate zone and cold climate rivers, by contrast, have large-size bed-material derived from physical weathering or glacial deposits. Some bedrocks periodically exposed in tropical river channels also become encrusted with iron and manganese oxides and silicates, but whether such coatings reduce abrasion is not always clear.

Rates of waterfall recession are very variable. Some tropical falls may have retreated little since the late Tertiary, but gully heads in unconsolidated materials may retreat many metres per day, and the heads of large waterfalls are commonly retreating at measurable rates of a few centimetres to several metres per year.

Structural controls on the shape of waterfalls are very common. Retreat is frequently along such lines of weakness as faults or joints. The Zambezi is a particularly good example as the gorges below the Victoria Falls have a zigzag pattern controlled by the intersection of two joint sets in the basalt over which the water cascades (Fig. 10.4; Plate 10.2). The gorges below the falls extend about 110 km downstream of the Falls which have receded at a rate of between 0.1 and 1 m/year during the last two million years. The duration of a waterfall or rapid is largely dependent upon structural controls, for resistant beds occurring in narrow outcrops will be cut through more rapidly than wide resistant zones. In sedimentary rocks the dip may also control the lifespan of a waterfall, as falls in beds dipping upstream will be gradually reduced in height, while strata which are horizontal or dipping downstream will maintain the falls.

All breaks in a long profile will, theoretically, be eliminated eventually, although the existence of breaks of profile in some of the world's largest rivers suggest that this may take much time, even on a geological time-scale. Elimination is aided by high velocity and erosive shooting flow, cavitation, undermining of caprocks, vigorous swirling or pothole erosion in the large *plunge pools* at the bases of falls, and by the constant wetting of rocks around the waterfall. At Niagara water and entrained rock debris strike the base of the 50 m high fall with a velocity of about 22 m/s, and much of this energy is available for erosion even though the water at the base of the falls is about 50 m deep.

Incised meanders

Rivers which are incising themselves into bedrock, either because of a lowered base level or tectonic uplift, may retain their channel directions and geometry including meander patterns. Where a meander is cut below symmetric valley sides it is said to be *entrenched* or *intrenched* (Plate 10.3). Where the valley is asymmetric with a steep slope on the outside of the bend and a gentler slip-off slope on the inside it is said to be *ingrown*. The ingrown type is the result of lateral erosion accompanying downward erosion. It is usually assumed that regular meanders are inherited from alluvial river meanders formed before downcuting commenced, but the evidence for such an origin is not always available. Lateral migration of meanders may cause undercutting of cliffs to form caves and even *natural arches* if two meander loops cut through opposite sides of a spur.

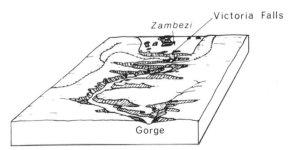

Fig. 10.4 The general alignment of the gorge below the Victoria Falls.

10.2 The Victoria Falls, and gorges below them, cut by the Zambezi River (photo and copyright: Aerofilms).

10.3 The entrenched meanders of the Fish River, Namibia. The river was photographed in the dry season so channel bars and point bars are exposed, but flow is also reduced by water being taken off upstream for irrigation.

Alluvial river channels

Channel-forming discharges

In a stable natural channel, with erodible banks and bed, transmission of the flow and bank stability must be maintained simultaneously. There is thus an approximate or dynamic equilibrium between erosion and deposition, such that the channel will be scoured to greater depths during floods and partly filled with sediment on falling discharges, or will migrate laterally by erosion of concave banks and deposition on convex banks, while still maintaining a similar form. Such channel changes do not occur at all flow conditions and many attempts have been made to define the flows which control the shape, slope, and size of a channel.

It is commonly asserted that the dominant channel-forming discharge is the *bankfull dis-charge*, that is, the discharge which is just contained within the banks (Plate 10.4). This idea has been extended into the statement that dominant discharge = bankfull discharge = discharge with a recurrence interval of 1.58 on the annual flood series (i.e. the discharge at the most probable annual flood).

This approach is an attempt to overcome the problem that defining bankfull discharge is seldom a simple matter. Not all rivers have clearly defined banks with sharp crests and, as a small increase in stage may produce a very large increase in discharge for a big river, there is some merit in attempting to define bankfull stage on a statistical rather than a field measurement basis. There remains, however, the problem of determining the discharge which predominantly controls the shape and size of a channel. There is evidence that some rivers have more than one channel (i.e. they have a

10.4 A river at slightly above bankfull discharge is just flowing onto its floodplain. Waipaoa River, New Zealand (photo by J. H. Johns, New Zealand Forest Service).

low-discharge channel inside a larger high-discharge channel), and many rivers in areas of strongly seasonal snowmelt or savanna wet-season floods have channels which are adjusted to floods more frequent than 1.58 years.

The diversity of views on the nature and size of dominant discharges is partly a reflection of our ignorance, but probably also a result of the variability of stream channels occurring in a variety of materials, in various climatic and vegetation zones, and with varying slopes.

Forms of alluvial channels

River channels are commonly classed as being either straight, meandering, braided, or anastomosing (Fig. 10.5). This scheme ignores much of the complexity of channels both because along its length one river may have several different types of channel and also because one type of channel tends to grade into another type. The classification shown in Fig. 10.6 is concerned only with alluvial channels, but it emphasizes the variability of channel forms and their relationship with streambed gradient and sediment supply and calibre. The relationships between these factors indicate why channel forms can vary down-valley and through time. As the gradient, supply, and calibre of load change, so does the form of the

channel (Table 10.1). A relatively wide and shallow channel is associated with the movement of a high proportion of bed-material load, whereas a narrow deep channel is associated with the transport of a predominantly suspended load.

Rivers with *straight channels* are rare, although many rivers have short reaches which are straight. Straight reaches may have a range of sediment sizes forming the load or may have a predominantly suspension load. Although the channel is not sinuous the thalweg meanders back and forth within the channel, and low mud or silt bars are often deposited along the channel edges.

Meandering rivers have sinuous channels (sinuosity (P) is the ratio of channel length (l_c) to the length of the meander belt axis (λ) (Fig. 10.7). Truly straight channels would have a sinuosity of 1 and strongly meandering channels sinuosities of 3 or more. A sinuosity of 1.5 is usually regarded as being the dividing point between meandering and straight channels, but one type clearly grades into the other.

Braided streams have several channels which, except at high flood, divide around coarse-grained sediment bars. Each channel may be sinuous but the whole channel system is relatively straight and the width of the channel system is large compared with channel depth (Plate 10.5).

The term *anastomosing* is usually confined to

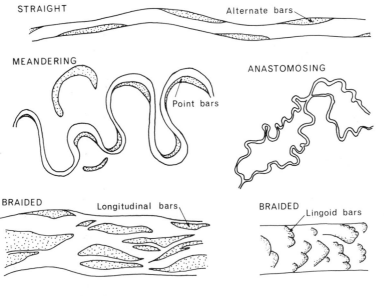

Fig. 10.5 Four major types of alluvial channel with characteristic channel bars.

STRAIGHT   Alternate bars

MEANDERING

ANASTOMOSING

Point bars

BRAIDED   Longitudinal bars

BRAIDED   Lingoid bars

Bars (covered in flood stages)

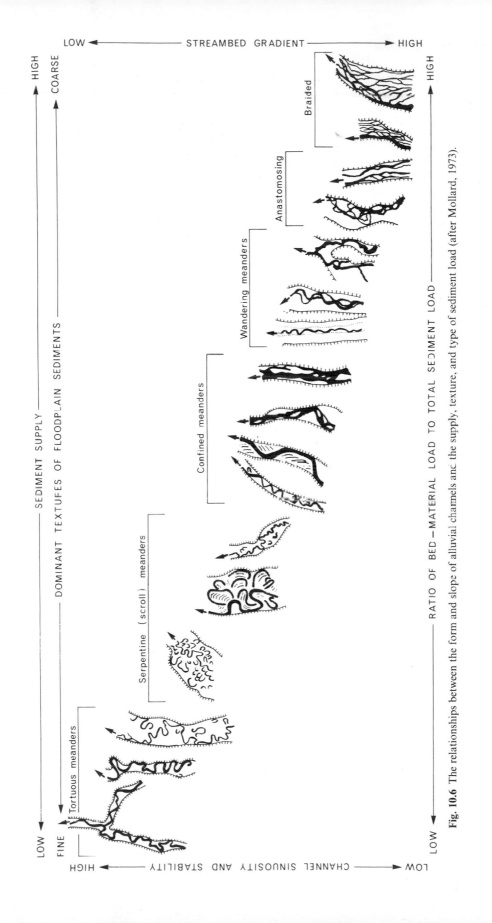

Fig. 10.6 The relationships between the form and slope of alluvial channels and the supply, texture, and type of sediment load (after Mollard, 1973).

Table 10.1   Classification of River Channels

Type	Morphology	Sinuosity	Load type	Bedload (% of total load)	Width/ depth ratio	Erosive behaviour	Depositional behaviour
Meandering	single channels	> 1.5	suspension or mixed load	< 11	< 40	channel incision, meander widening	point-bar formation
Braided	two or more channels with bars and small islands	< 1.3	bedload	> 11	> 40	channel widening	channel aggradation, mid-channel bar formation
Straight	single channel with pools and riffles, meandering thalweg	< 1.5	suspension, mixed or bedload	< 11	< 40	minor channel widening and incision	side-channel bar formation
Anastomosing	two or more channels with large, stable islands	> 2.0	suspension load	< 3	< 10	slow meander widening	slow bank accretion

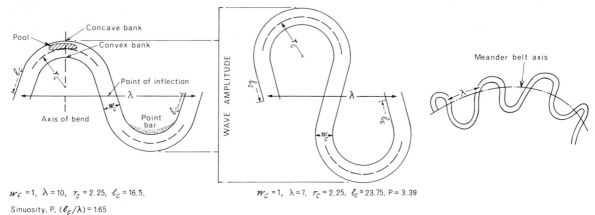

$w_C = 1$, $\lambda = 10$, $r_C = 2.25$, $\ell_C = 16.5$,

Sinuosity, P, $(\ell_C / \lambda) = 1.65$

$w_C = 1$, $\lambda = 7$, $r_C = 2.25$, $\ell_C = 23.75$, $P = 3.39$

Fig. 10.7 Descriptors of the form of meandering channels.

rivers with relatively permanent and stable systems of very sinuous channels with cohesive banks. The channels diverge and converge around large, stable, vegetated islands (Plate 10.6).

The features which are responsible for many of the characteristics of river channels are the pools, riffles, and bars which alternate across and along the channel.

Pools and riffles

The thalweg has a zigzag or meandering course between bars which develop at each bank alternately down the channel, or through low-gradient deep pools which alternate with shallower riffles. *Pools* are topographically low areas along a bed and they usually have a surface of relatively fine

10.5 Braided channel of the Waipawa River, New Zealand. In the background is the Ruahine Range which supplies large quantities of gravel to the river. Note the slip-off terraces on the right.

bed-material. A *riffle* is a topographically high area formed by a lobate accumulation of coarse material. It may be symmetric or asymmetric so that the thalweg may diverge around it or pass along one side. In the latter case the line of the thalweg meanders down valley (Fig. 10.8). The tendency for thalwegs to meander is demonstrated in flume experiments in which a perfectly rectangular channel is occupied by a stream which gradually forms alternating bars and pools with the thalweg meandering through them. The same meandering tendency can be seen if water is poured down a sheet of dust-coated glass. A meandering form thus appears to be a characteristic of streams.

In alluvial channels pools and riffles are made up of sediments, but even in channels cut in bedrock deeper sections alternate with rapids. In the Grand Canyon of the Colorado River rapids are spaced at intervals of 2.6 km in a pattern which appears to be independent of the bedrock. About 50 per cent of the decrease in elevation along the Colorado takes place in the rapids which occupy only about 10 per cent of the channel length.

Pools in bedrock channels probably develop below rapids were water velocities are high and turbulence is at a maximum. In alluvial channels sorting of coarse and fine material is necessary for the riffles to be maintained. It has been suggested that riffles exist as a result of the interaction of the riffles with the flow velocity. Once coarse material is in motion it will stay mobile until either an obstruction prevents bedload transport or water velocity decreases. As soon as deposition starts bed roughness increases, velocity falls, and coarse material accumulates. The result will be a series of riffles downstream which will become organized into a spacing which is related to the water velocity

10.6 Anastomosing channels in the lower Okavango during a flood, northern Kalahari (photo by K. Thompson).

and thus to the channel width. Movement of the riffle materials can only occur above certain flow velocities, and at very high velocities all bed-material may be moved, so riffles may be formed by flows which are exceeded only 10–20 per cent of the time. At these intermediate discharges the riffle materials may move as a kinematic wave from one riffle to the next, like vehicles between sets of traffic lights.

It has been commonly observed that the spacing between pools and between riffles in alluvial channels is five to seven channel widths (Fig. 10.8b). As a stream increases its sinuosity by development of meanders, and hence its channel length, new pools and riffles form and keep the spacing constant. It would apear that a spacing of greater than about seven channel widths is inherently unstable and breaks up into two pool and riffle sequences initially spaced at three to five

channel widths. This new spacing subsequently increases to five to seven widths.

Alluvial bars and stream flow

Three types of alluvial channel bar are commonly recognized: alternating bars, point bars, and braid bars (Figure 10.5). Bars may be considered as large bedforms which develop at certain sites along a channel. *Alternating bars* form in straight channel segments within the curves of the meandering thalweg, and as such they are analogous to *point bars* which form in the area of relatively low stream power on the inside of a channel meander. Because meandering rivers usually have rather fine-grained bed-material, alternating and point bars are commonly of silt and sand, but this is not always the case; both types of bar also occur in braided channels where they may be composed of

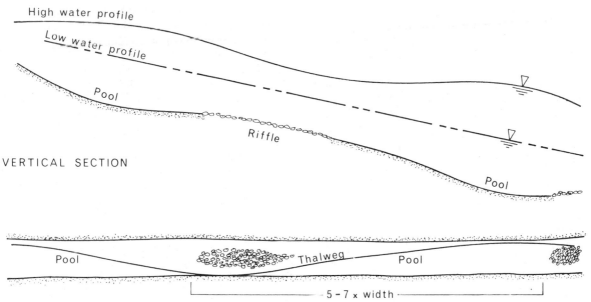

High water profile

Low water profile

Pool

Riffle

VERTICAL SECTION

Pool

Pool

Thalweg

Pool

5 – 7 x width

PLAN

Fig. 10.8 (a) A vertical section through an alluvial channel shows the location of pools and riffles and their relationships with the profile of the water surface. At high stages water velocity increases over the riffle, and can have a depressed surface compared with deeper, but slower, flow over the pools. This profile preserves the relationship between bed-material and flow. The plan shows the line of the thalweg.

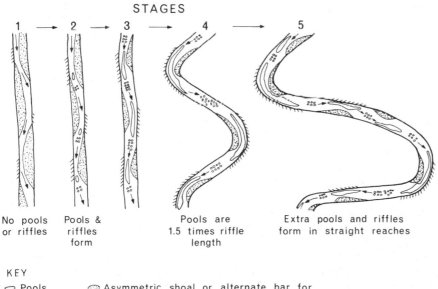

STAGES

1 → 2 → 3 → 4 → 5

No pools
or riffles

Pools &
riffles
form

Pools are
1.5 times riffle
length

Extra pools and riffles
form in straight reaches

KEY

⬭ Pools

⫶⫶⫶ Riffle

⌇⌇⌇ Erosion

⬭ Asymmetric shoal or alternate bar for
stage 1 & 2, point bar for stages 3,4 and 5

Fig. 10.8 (b) The arrangement of pools and riffles in straight channels and those of increasing meander curvature (after Keller, 1972).

gravel and cobbles and they may also be of coarse-grained materials in some meandering channels. *Braid bars* are most commonly of coarse material, but they also may be formed of sand and silts. They form within the channel and grow there as the flow diverges around them. They have several characteristic shapes but most commonly are diamond-shaped aligned parallel to the flow, when they are called *longitudinal bars* (Plate 10.7). Such bars are the dominant type in coarse materials. In sands *linguoid bars* form transverse to the flow; these have upper surfaces which dip gently upstream towards the preceding bar. Braid bars are commonly large features many metres long and wide. The linguoid type are often covered by dunes and ripples.

Bars develop in response to the pattern of stream flow. It has been recognized for several centuries that this tends to be helicoidal and such a pattern is clearly seen inside meander bends. The spiral in the bend is towards the outer bank at the surface, vertically downwards at the outer bank, and towards the inner bank along the bottom (Fig. 10.9). As a result the outer bank is eroded because flow velocity, and hence bed shear stress, is high there. Along the bottom towards the inner bank velocity decreases in the shallower water and bed shear stress is low, hence deposition can take place. Sediment from the erosion and collapse of the outer banks is moved as bedload by the flow on to the inner bank of the next bend downstream, or is thrown into suspension until it can settle in the shallow water close to the inner bank. The result is the formation of a crescent-shaped point bar at the inner bend which slopes into the channel at an angle of a few degrees. The surface of the bar is

10.7 Braid bars during a high stage.

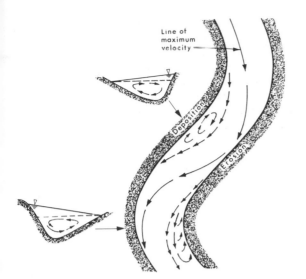

migration of dunes on the lower part of the bar, with small-scale cross-bedding from ripples on the higher part of the bar. Plane bed lamination in upper parts of the sequence may be evidence that a high flood has swept across the bar on some occasion (Figs. 10.10, 10.11).

Braid bars may have similar internal structures to point bars and alternating bars, although the formation of dunes and ripples in very coarse materials is not so well developed as in sands. In general the materials become finer upwards in the sequence, as the coarsest debris is moved along the thalweg and not over the bar. This fining-upwards sequence also suggests that deposition occurs predominantly on a falling stage.

Fig. 10.9 Streamlines showing directions of water movement in meanders, and the resulting sites of erosion and deposition. The vertical cross-sections show the relative height of the water surface in the bends but exaggerate the differences. In many channels the cross-over point of the flow is delayed towards the next loop, thus producing asymmetric meanders.

Straight and meandering alluvial channels

Straight alluvial channels may have steep or low gradients but always have low bedload discharges. Unless they are incised in bedrock straight channels usually develop features which lead to meandering, and even the straightest channels have a sinuous thalweg.

Meandering channels have geometric forms which are very varied, but certain ratios amongst their properties are characteristic of them regardless of the size of the river. Meander wavelength, amplitude, and channel width are related to the square root of water discharge in power functions with the form:

$$\lambda = k_1 Q^{0.5}; \quad A_m = k_2 Q^{0.5}; \quad w_c = k_3 Q^{0.5}$$

covered by dunes and ripples (if it is a sand or silt bed) which are active during floods. Alternating bars have similar origins and forms.

Migration of a point bar produces a sequence of sedimentary structures approximately equal in thickness to the maximum depth of the channel. At the base of the bar channel bed deposits may be overlain by materials with plane bed lamination and then by trough cross-bedding formed by

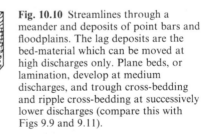

Fig. 10.10 Streamlines through a meander and deposits of point bars and floodplains. The lag deposits are the bed-material which can be moved at high discharges only. Plane beds, or lamination, develop at medium discharges, and trough cross-bedding and ripple cross-bedding at successively lower discharges (compare this with Figs 9.9 and 9.11).

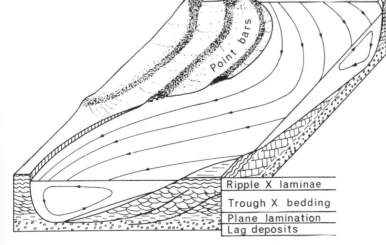

| Ripple X laminae |
| Trough X bedding |
| Plane lamination |
| Lag deposits |

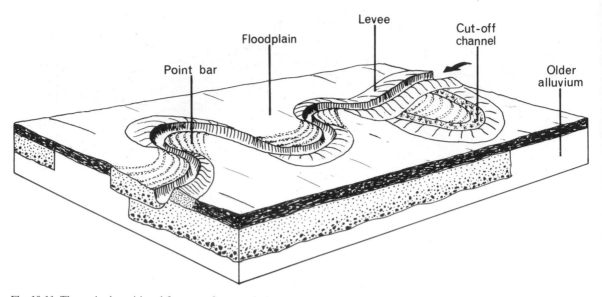

Fig. 10.11 The main depositional features of a meandering channel (modified after Allen, 1970).

where λ is the meander wavelength, A_m is the meander amplitude, w_c is the channel width, and k_1, k_2, k_3 are coefficients whose value varies from locality to locality. Such relationships indicate that meanders are shaped to discharge the flow occurring within them, but it is not yet clear whether the meander form is related to a channel-forming discharge, such as bankfull discharge, or to more frequent discharges; nor is the relationship between wavelength and bank and channel sediment fully understood.

Many hypotheses have been advanced to explain the development of meanders. Most start from the assumption that meandering channels begin with a straight channel which becomes sinuous. It is true that straight channels do develop into sinuous channels in flume experiments, but there is no necessity for this to be the unversal sequence. It is a feature of many newly initiated rills in the field that they are sinuous from their inception. Amongst commonly advanced ideas of cause are the Earth's rotation; initial deflection of the current by an obstacle; transverse oscillations in the flow giving rise to secondary flows, that is, flows transverse to the dominant downstream direction of flow; minimization of energy loss along the channel, and equalization of energy expenditure. Of these hypotheses that which attributes meandering to the Earth's rotation and Coriolis effect faces the objection that

meandering happens in channels of all directions and sizes, not just those most suitably aligned, and nearby channels are not all similarly affected. The idea of local obstructions may be relevant in a few cases but is not universally applicable. The most plausible hypothesis is that meanders result from secondary flow giving rise to helicoidal motion and that they are maintained along reaches because the meander form is the most efficient for the maintenance of stable discharge of water and sediment.

It is a feature of most meander loops that they are asymmetric with the cross-over zones of the thalweg, and high-velocity flow, being delayed in a down-valley direction beyond each meander loop so that it hugs the down-valley bank. This is to be expected because meanders migrate down-valley. Consequently, many meander patterns are successions of straight reaches followed by small meander loops. An explanation of delayed cross-over flow may prove to be basic to an understanding of meander mechanisms. As yet no theory has proved to be adequate.

Braided channels

Braided streams have channels which, at low water, diverge around islands or bars (Fig. 10.12). Braiding usually begins when a river is overloaded with bed-material so that deposition occurs and

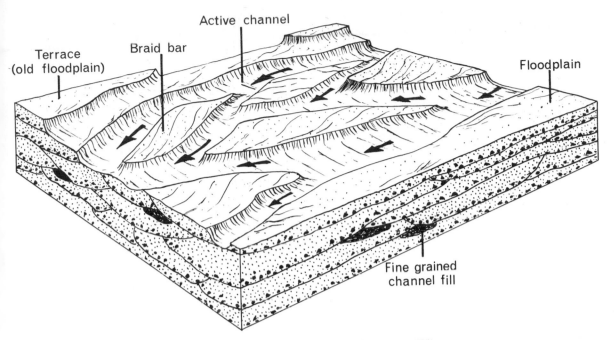

Fig. 10.12 The main depositional features of a braided channel (modified after Allen, 1970).

the resulting braid bar increases in size by extending from its downstream end. At high stages braid bars may be drowned and fine-grained material can then be deposited on the coarser material of the bar. Many large rivers are braided along extensive parts or along the whole length of their course. Such rivers can occur in any climatic regime.

The chief characteristics which are identified as being the causes of braiding are (1) overloading which causes the stream to deposit part of its load; (2) steep slopes which produce a wide shallow channel in which bars and islands form readily; (3) easily eroded banks which permit the channel to be widened at high flow; and (4) a large bed-material load, in comparison with its suspended load, so that the large material is immobile except at high flood stages. The presence of vegetation may have a considerable influence upon the stability of channels, for root systems can considerably increase channel bank and braid bar stability by adding apparent cohesion to the sediment. Thus a rapidly changing and unstable channel can be made more stable by plant colonization of its banks and in some situations the tendency to braiding may be reduced. This has happened as a

result of climatic change and of human interference.

Braided channels are notoriously unstable and difficult for engineers to control, especially where the braiding is caused by rapid deposition (i.e. by aggradation). The Kosi River, India, for example, receives most of its load from the Himalayas and its abundant supply has permitted the river to migrate laterally over its own deposits a distance of 112 km in 228 years. Diversion of the flow in high discharges causes the catastrophic erosion of new channels and the abandonment of old ones. This condition is aided by the sparseness of vegetation on the piedmont plains of northern India. The Brahmaputra also shows severe channel migration in short periods with rates of lateral movement of 900 m/year being quite common. The most significant bankline modifications take place during falling flood stages when excess load is deposited, as bars form within the channel, causing a change in local flow direction and migration of the thalweg.

Many braided streams have very low flow depths compared with the channel width (width/ depth ratios may exceed 300), consequently only the very largest braided rivers like the Brahmaputra, which is 3–10 km wide, have flow depths in excess of about 7 m; braided reaches in this river have a maximum scour depth during floods of about 25 m, but in meandering reaches scour depths are proportionally considerably deeper.

Some dunes in sand-bed braided channels are so large that at low discharges they become braid bars of considerable size. Some in the Brahmaputra are 8–15 m high and 180–900 m in wavelength. Such dunes migrate as much as 600 m in 24 hours in flood discharges.

Long-term channel changes

It has been shown by Leopold and Wolman (1957) that braided channels have a steeper slope than meandering channels for the same discharge when the flow is just filling the channel (i.e. discharge is at bankfull) (Fig. 10.13). On their plot a line can be drawn separating data points of braided channels above it, from points representing meandering channels below the line. Straight channels may be either of high or low steepness. The graph indicates that for any discharge meanders occur at lower slopes then braiding, and for a given slope meanders occur at a smaller discharge than does braiding.

Experiments and field observations also indicate that the proportions of a stream load which move as suspended load and bedload, and the size of the particles in the load also influence channel

sinuosity and meander stability (Fig. 10.6). As a result of tectonic and climatic changes, discharge of water, sediment supply and particle size, and channel slope can change. Consequently it is to be expected that channel forms will change in response to tectonic and climatic effects.

Mississippi River

During a glacial period sea-level around the world falls as water is retained in the ice sheets which grow on the Northern Hemisphere continents. At the same time climate becomes generally colder and drier; vegetation cover then changes or decreases, and in areas on the fringes of glaciers the sediment supply to rivers is greatly increased in volume and calibre, as material is washed from moraines into the stream channels. These environmental changes have very great effects upon the rivers themselves.

In the Last Glacial the North American ice sheet supplied great quantities of meltwater and sediment to the headwaters of the Mississippi, and a fall of sea-level, of over 100 m at its seaward end in the Gulf of Mexico, lowered the base level and caused erosion of earlier alluvial deposits and scour of the valley floor.

Fisk (1944) showed that during the maximum of the Last Glacial, about 20 000 to 18 000 years ago, when sea-level was at its lowest, the Mississippi near its mouth had a steeper gradient than now

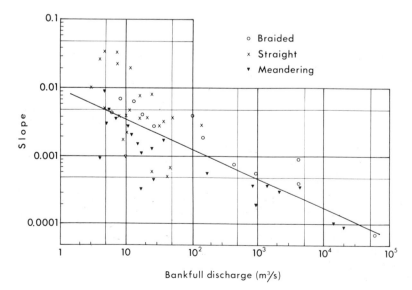

Fig. 10.13 Braided channels have steeper slopes, for similar discharges, than meandering alluvial channels (redrawn after Leopold, Wolman, and Miller, 1964, © W. H. Freeman & Co. All rights reserved.).

and that the slope was steepest close to the sea (Fig. 10.14). As the ice sheet wasted and sea-level rose during the period 14 000 to about 5000 years ago deposition in the valley increased, first with coarse detritus and then with progressively finer sands. The upward decrease in particle size resulted from the progressive decrease in channel slope and the northward retreat of the ice front which was supplying much of the load until about 11 000 years ago. The wave of alluvial deposition was probably greatest in the lower reaches of the valley where the effect of sea-level rise was most immediately felt. Throughout the valley the tributaries were carrying relatively coarse debris and many of them built fans out into the main valley. Silts and clays are almost absent from these late-glacial period deposits, suggesting that they were carried into the Gulf of Mexico as suspended load while the Mississippi had a braided channel.

Towards the end of the period of rising sea-level, possibly in the period of about 8000–5000 y BP, the supply of coarse sands decreased and the load was predominantly of fine sand, silts, and clays. Its volume was still great enough to cause deposition on the reduced valley slope and the silts and clays were often spread across the aggrading floor during floods. Eventually the fine-grained deposits formed such cohesive banks, and slopes were so low, that the Mississippi became a meandering single-channel river.

The Murrumbidgee River

The Murrumbidgee River drains from the highlands of New South Wales, Australia, towards the west. It crosses the alluvial Riverine Plain and joins the Murray River. On the floodplain of the modern river are traces of old abandoned channels (paleochannels). The youngest set of paleochannels has been called the ancestral channels and the older set prior stream channels. The characteristics of the three sets of channels are given in Table 10.2.

Schumm (1968) has interpreted the prior channels as being adjusted to a drier climate than now with semi-arid characteristics and hence reduced vegetation cover, so that, although discharge of water was not high, erosion was severe and the sediment supply to the channel was large. The channel was nearly straight and hence had a steep gradient, great width, and little depth. As the climate became wetter the vegetation cover in-

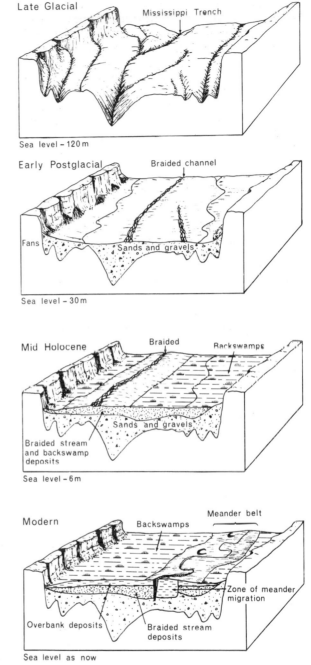

Fig. 10.14 Changing channel patterns and deposits, resulting from changing sea-levels, supply of debris and discharge, in the last 20 000 years for the lower Mississippi (redrawn after Fisk, 1944).

Table 10.2　Features of Murrumbidgee River Channels in Modern and Former Times

	modern	ancestral	prior
size	small	large	large
sinuosity	sinuous	sinuous	straight
width	narrow	moderate	wide
depth	deep	deep	shallow
sediment transport	low	high	very high
climate compared with present	—	wetter	drier
sediment supply compared with present	—	larger	very large

creased, water discharges also increased, but the sediment supply was reduced. The ancestral river then underwent a change to a narrower but deeper channel with greater sinuosity so that both gradient and meander wavelength decreased. Towards the present the climate became somewhat drier, but vegetation cover did not decline, so the modern channel remained rather like that of the ancestral channel, except that it is smaller and consequently has a much lower meander wavelength, and transports little sediment (Fig. 10.15).

Underfit streams

Many meandering valleys contain floodplains over which modern streams flow with meander belts much narrower, and meander wavelengths far smaller than those of the valley. Such streams are commonly called *manifestly underfit*, the implication being that the windings of the valleys were produced by much larger meandering rivers than exist now. The basis to this argument is the finding that for many streams channel geometry is well related to modern discharges. The implication, of course, is that climatic changes have caused a decrease in discharge over late Quaternary times.

An *Osage type* of underfit has pools and riffles spaced along the stream in accordance with its modern width, but this spacing is not related to the meander wavelength, for far more pools and riffles occur in the bed than would be expected within one meander bend. In other words the channel is still adjusting to a reduced discharge and has not yet reduced the amplitude of its meanders to fit the spacing of its pools and riffles. Several modes of transformation of channels may be imagined (Fig. 10.16). Further discussion will be found on p. 541.

Short-term channel changes

Hydraulic geometry of stream channels

A graphical plot of stream discharge (Q) against each of mean water velocity (v), mean depth (d) and width (w) of the flowing water, slope (S), channel roughness (n), and suspended sediment

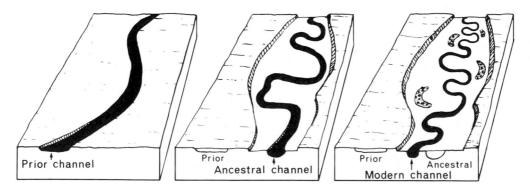

Fig. 10.15 Changing channel forms and dimensions of the Murrumbidgee River.

(L_s) shows how the channel at a site accommodates the rise and fall of flow during a storm. This relationship is called the hydraulic geometry at-a-station. Studies by members of the US Geological Survey indicate that the resulting curves show a power-function relation of Q to w, d, v, S, n, and L_s in the form:

$$w = aQ^b; \qquad S = gQ^z;$$
$$d = cQ^f; \qquad n = pQ^y;$$
$$v = kQ^m; \qquad L_s = rQ^j;$$

where $a, c, k, b, f, m, g, p, r, j, y, z$ are numerical coefficients. As the cross-section area $A = w.d$ and $w.d.v = Q$

then $aQ^b.cQ^f.kQ^m = Q$, or $Q = ackQ^{b+f+m}$
hence $a.c.k = 1.0$ and $b+f+m = 1.0$

Determination of the values of b, f, and m is useful because it indicates how streams with different bed-materials and in different environments adjust to varying discharges. Streams with uncohesive bank materials, for example, are seen to increase their width, compared with the deepening of channels in cohesive bank material. These general relationships at-a-station and in a downstream direction are indicated in Fig. 10.17. Similar patterns of change for channel roughness and bed-material size have also been established.

Many recent studies have broadly validated the concept of hydraulic geometry, but also shown that there is a large variation amongst rivers in the manner of their adjustments to changing discharge.

Graded reaches

Recognition that an alluvial river tends to maintain itself in a stable condition with an efficient channel configuration is not new; it has been known under the term *grade* since the beginning of this century, but its most concise definition is probably that of Mackin (1948) who said:

A graded river is one in which, over a period of years, slope and channel characteristics are delicately adjusted to provide, with available discharge, just the velocity required for the transportation of the load supplied from the drainage basin. The graded stream is a system in equilibrium; its diagnostic characteristic is that any change in any of the controlling factors will cause a displacement of the equilibrium in a direction that will tend to absorb the effect of the change.

For a river to be graded it must flow on adjustable materials, hence channels in bedrock cannot be graded. Grade is thus first achieved at the downstream ends of channels in alluvial reaches and it gradually extends headwards. A trunk stream may thus be at grade while its

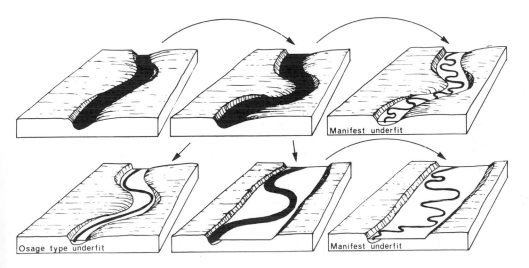

Fig. 10.16 Types of underfit streams (redrawn after Dury, 1970).

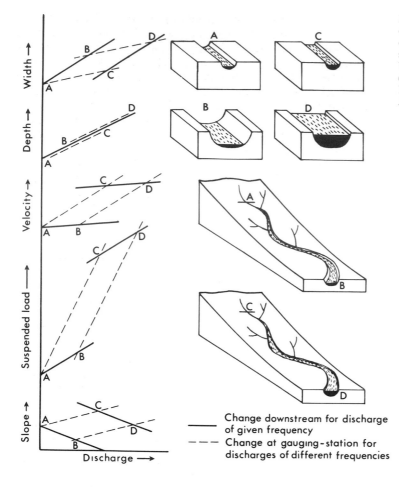

tributaries are still influenced by bedrock or local base levels. A *graded reach* may exist between two controls such as rock bars.

An alternative view of a graded condition is that it tends towards a condition in which energy expenditure along a channel is at a minimum and is evenly distributed along the channel. These concepts have already been recognized as applying to meanders and to the mutual adjustments between water velocity, depth, and width in a channel. It follows then that rivers tend to develop graded long profiles. Such a profile would, throughout its length, be on alluvial materials and could only occur where base level is constant, and discharge of water and sediment is relatively uniform over a period of time. In practice, tectonic and sea-level adjustments modify base level; and climate, and hence vegetation cover, is not constant. An ultimate graded condition may thus never be attained so that a quasi-equilibrium is a closer approximation to reality.

Floodplains and terraces

Alongside the channel of most rivers is a relatively flat surface extending to the base of the valley walls. It is appropriately called the *floodplain* because it is periodically inundated (Plate 10.4).

Parts of the floodplain close to a channel may be occupied by river water every one or two years, but rarer and more extreme floods will occupy larger areas and with deeper waters. Floods with a return period of a few years may have a depth of less than a metre to 1 or 2 m, but some rivers have much deeper floods—the River Lena, Siberia, for example, 420 km above its mouth can have a flood depth of 32 m—and many rivers have depths of several metres for the 10 year or greater floods.

Floodplain deposits

A floodplain accumulates its sediments in two ways: by lateral migration of channels across the floodplain; and by over-bank deposition from floodwaters (Figs. 10.10, 10.11; Table 10.3). Each process gives rise to a distinctive suite of deposits and related landforms. The character of these features changes down-valley.

In upland areas the alluvial deposits alongside rivers are often narrow and relatively thin. They are composed of poorly sorted mixtures of silts, sands, and gravels with angular or sub-rounded particles and a high proportion of feldspathic or other minerals with a moderate-to-low stability against weathering and abrasion. Even overbank deposits may be relatively coarse. As rivers emerge from uplands their channels decrease in slope and alluvial fans may be formed with a high proportion of the bedload being deposited. In mountainous reaches a river may carry 5–10 times as much suspended load as bedload material, but as a result of abrasion and deposition on fans the suspended load may be 10–20 times that of bedload in the lowland reaches and in floods it may be 100–1000 times that of bedload.

Table 10.3   Valley Sediments

Place of deposition	Name	Characteristics
Channel	Transitory channel	Composed largely of bedload materials in transit and similar sediment forming bars which move less frequently
	Lag deposits	Large particles 'armouring' the channel bed; they may be moved only at very high discharges or if derived from rockfalls etc. may be immobile
	Channel fills	Accumulations filling abandoned channels; particles may be from bed load or of finer-grained overbank flood deposits, or organic
Channel margin	Lateral accretion deposits	Point- and alternating bars which have been preserved as a result of channel shifting
Overbank deposits of floodplain	Vertical accretion deposits	Fine-grained sediment deposited from suspended load of overbank floodwater; includes levee and backswamp deposits
	Splays	Bedload materials spread on to floodplains when levees are breached by flood waters
Valley margin	Fan deposits	Predominantly bedload with coarsest material at the fan apex; lenses of mudflow deposits may be included
	Colluvium	Deposits at the base of hills derived largely from slope wash, and soil creep
	Mass movement deposits	Landslide debris at the base of hills, often interbedded with colluvium

The alluvial deposits of lowlands are largely composed of well-rounded quartz sand with silts and clays. The rate of transport of material within the channel depends upon how it is carried. The silts and clays held in suspension move at the velocity of the water, or at about 1–2 m/s, but sands move with the bedform dunes and ripples at about 1 m/day or 5000 times slower than the suspended materials.

Lateral migration of channels occurs with both braided and meandering channels. Braided rivers usually change course when overbank flow spreads on to the floodplain and becomes channelled so that a new braided system of channels is formed, while the abandoned system becomes fossilized and may eventually be filled in as organic matter and flood silts fill the depressions. Such channel changes tend to be catastrophic rather than gradual processes.

Meandering channels migrate as the outside of a bend is eroded by undercutting and collapses, and on the inside of a bend point-bar deposition occurs to maintain the distribution of sediment along the channel and to preserve a nearly uniform channel width. Point-bar accretion results in the formation of a cross-stratified deposit with a subdued relief of ridges separated by swales that are the evidence of the channel migration. The crests of point bars are usually close to the level of the floodplain, unless the stream is notably entrenching or building up its bed, so the floodplain level may not change its altitude during lateral migration of the channel.

The ages of point-bar deposits may be deter-mined from the ages of trees on the floodplain, from ^{14}C dates of organic matter infilling the swales, or from aerial photographs. Such studies show that many channels migrate at rates of metres per year, yet the overall size of the floodplain does not change greatly. The floodplain is, rather, a deposit of alluvium which moves into an area, is stored for a while, is weathered, and then moves on down the valley.

The high rates at which natural channels may migrate laterally is indicated by the Kosi River of northern India whose channels have migrated laterally an average of 760 metres per year over several centuries, and the Mississippi near Rosendale with migrations of 49–190 m/year. Similarly high rates are common for the rivers of Bangladesh.

In the process of migrating, a meander loop becomes increasingly sinuous until it is cut off either at the neck or by the formation of a *chute* along a swale between two point bars (Fig. 10.18). The cut-off loop is then abandoned and becomes an *oxbow lake* when the ends of the loop are plugged with bank deposits alongside the newly cut shorter and straighter channel. Oxbow lakes are temporary features which are eventually filled with organic matter and sediment deposited from overbank flows.

Vertical accretion of a floodplain occurs during floods. These may last for only a few days or for several months and be either very shallow or deep inundations. As a consequence, the deposits of a single flood can range from a millimetre to a metre or more in thickness. It is common for the greatest

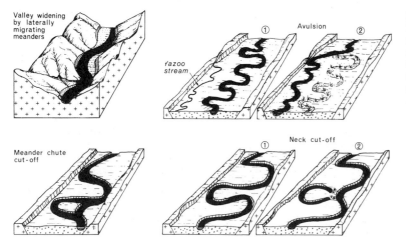

Fig. 10.18 Meanders and floodplains. Note that valley-widening tends to trim the ends of spurs and creates cliffs, thus decreasing the interlocking of spurs from either side of the valley, and it also develops a floodplain. Neck cut-offs leave oxbow lakes.

thickness of sediment to be laid down on the *levees*. As a result levees may grow to an elevation of several metres above the floodplain. Areas between the levees and valley walls then become poorly drained so that back swamps are commonly formed and tributaries may flow parallel to the main stream for considerable distances before joining it. Such tributaries have been called *Yazoo* tributaries from the Yazoo River, a tributary of the Mississippi. When levees extend well above floodplains they may be breached during floods. The breach cut through the levee is called a *crevasse* and the delta-like deposit formed on the floodplain is called a *crevasse splay*. Large breaches may cause the main channel to be diverted on to the floodplain were a new levee system is built up and the lower reach of the old channel is then abandoned. Such channel diversion is called *avulsion* (see Figs. 10.11, 10.14, 10.18).

At least three processes have been recognized as being involved in the breaching of the levees of the lower Mississippi: (1) seasonal deepening of the scour pool at the outside of meander bends during high river stages; (2) over-steepening at the base of an outer bank causing subaqueous landslides of the bank material; and (3) subaqueous landslides in part of the bank lead to failures in contiguous areas. The largest failures occur in sandy and point-bar deposits, while the clay plugs of former oxbow lakes are more resistant and may impede free development of migrating meanders.

Vertical accretion has been widespread in the lower reaches of many rivers during the last 12 000 years or so because world sea-level has been rising. It is improbable that such excessive accretion would occur under conditions of a stable base level.

Alluvial plains

Extensive plains formed of alluvial deposits occur round the bases of many mountain ranges. Such plains are partly low-angled alluvial fans, but also they have some of the characteristics of floodplains formed by braided rivers. The High Plains of the central United States are veneered by sands and gravels of the Ogallala Formation which covers an area 1 300 km long and 500 km wide. This formation is up to 150 m thick and has been formed as a result of severe erosion, in the Rocky Mountains, which accompanied tectonic uplift

and the frequent climatic changes of the late Cenozoic. Similar piedmont alluvial plains form the Argentine Pampas on the eastern side of the Andes, the Canterbury Plains of New Zealand to the east of the Southern Alps (Plate 10.8), the Indo-Gangetic Plain south of the Himalayas, and the Po Valley Plain south of the European Alps.

Many of these plains have been built up spasmodically during periods of severe erosion and glaciation in the uplands when sediment supply was at a maximum. The thickness of sediments below the Indo-Gangetic Plain and the Po Valley has also been greatly increased because these are areas of crustal subsidence. The nature of the sediments depends upon the type of river which has built up the deposit and much of the history of changing flow and sediment discharges may be interpreted from the sedimentary record (Fig. 10.10).

Terraces

Fluvial terraces are benches, approximately parallel to the channel or valley walls, which usually represent former levels of floodplains or valley floors. Terraces may be discontinuous or continuous down the valley; it is common for the lowest terraces in a sequence to be more continuous and younger than those at higher elevations, which are older, more eroded, and dismembered. Terraces are separated from one another by steep risers and have nearly flat surfaces, which sometimes have old floodplain features on them, and which usually dip down-valley at an angle which may be similar to or different from that of the modern floodplain.

River terraces, like floodplains, may be cut across bedrock so that they are essentially rock cut-surfaces bearing a veneer of fluvial sediments, or they may be formed on thick alluvium (Fig. 10.19; Plates 10.1, 10.5, 10.9).

Rock cut-terraces result from stream incision into underlying resistant rock. The incision may result from either uplift of the land or from a fall in sea-level; in either case it is the relative fall of base level which causes incision. The term *strath terrace* is sometimes applied to rock cut-terraces after a Scottish strath, or wide and long valley, with a flat rock cut-floor. A special type of cut terrace occurs in alternating beds of resistant with less resistant rock so structurally controlled terraces are formed by differential erosion.

10.8 The alluvial plain of Kaikoura, New Zealand with the Kaikoura Ranges in the distance and the bedrock peninsula in the foreground. Note the terraces, on the peninsula, formed by wave action and now uplifted (photo NZ Geological Survey).

Terraces formed in fluvial sediments are very common. They may result from infilling of a valley with sediment, as a result of an increase in load, a decrease in discharge, or a rise in base level, and therefore a decrease in channel slope. Because of the frequent changes of climate which have characterized the late Cenozoic, and the accompanying changes of sea-level, many river valleys have experienced alternating phases of infilling with sediment and latter incision of those sediments to leave flights of terraces.

Infilling of valleys, that is aggradation, results from an oversupply of debris compared with the capacity of the river to transport it. Aggradation may result from a phase of accelerated erosion on slopes resulting from a severe storm, a fire, devegetation, a longer-term climatic change, in-

creasing rates of denudation, or changes in the mode of erosion as when severe frost action or glacial erosion occur in a cold climate. It may also occur after severe tectonic events, or a volcanic event which produces much debris. Decreases in discharge are most commonly the result of a climatic trend towards aridity. Incision of the aggradation deposit occurs when the stream increases its capacity to transport debris relative to the supply. Incision therefore usually follows a tendency towards a wetter climate, a reduction in the rate of denudation, an increase in the vegetation cover, or an increase in channel slope. An increase in slope may be caused by tectonic tilting but it may also result from more local adjustments to removal of channel obstructions, readjustments of stream gradients, as when meandering channels

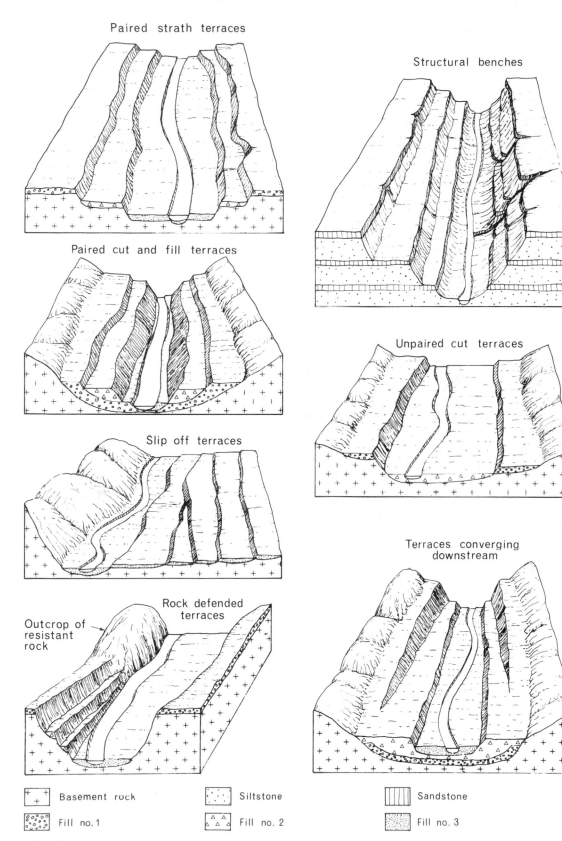

Fig. 10.19 Types of valley terraces and structural benches.

10.9 Terraces of the Rangitikei River, New Zealand, have been cut in mudstone bedrock during the last 250 000 years as the land was uplifted. Each terrace has a veneer of alluvial deposits. Note the point bar and floodplain in the foreground and the channel bars farther downstream.

develop a braided habit, or the capture of a drainage system by a river at a lower elevation. Shortening of channel lengths, as when a main channel migrates laterally and reduces the length of its tributaries, may cause incision and terrace formation along the shortened reach.

The significance of tectonic warping is evident where uplift rates have been measured with an extreme of 5–7 m/1000 years or up to 7 mm/year. Thus in 10 years a change of slope of 70 mm per kilometre of channel may occur and this is sufficient to change the direction of the energy slope of some rivers.

Repeated episodes of aggradation and erosion are so common that valley deposits show clear evidence of alternating phases of *cut and fill*. With each phase of alluviation the older cut terraces have been buried, and with each period of incision some of the fill is eroded away, so a valley may have experienced far more episodes of cutting and filling than are represented in either the terraces or the infill deposits.

Both *paired* and *unpaired* terraces may result from channel incision. Paired terraces occur either where there is a structural control on terrace elevations or, more commonly, where downcutting clearly dominates lateral cutting. In such a situation the terraces on either side of the valley match each other in elevation and slope, i.e. they are paired.

Unpaired terraces form where a channel is predominantly migrating laterally across a valley—particularly across a wide valley—so that incision is of less significance. Conditions of slowly continuing sea-level fall, or tectonic uplift, may also favour the development of unpaired terraces, whereas episodic falls of base level may favour the formation of paired terraces.

Terraces often occur in flights, and, seen along a short segment of a valley, they may appear to have similar slopes with respect to the horizontal, but when they are precisely surveyed it is commonly found that they differ in slope and may even cross one another; some examples of the controls on terrace slopes are given in Fig. 10.20. In a mountainous area the rate of uplift is commonly greatest in the area of the highest peaks (that is why they are highest) consequently the headwaters of streams have experienced greater uplift per unit of time than the middle reaches. As a consequence the oldest (i.e. highest) terraces have a steeper slope than the youngest, and the terraces are separated by the greatest differences of height in the headwaters and they tend to converge downstream.

Terrace flights may also be disturbed by uplift in the middle reaches of a valley—as in the Rhine Gorge where the older terraces have been up-arched—or by downwarping, as in the Little Hungarian Plain where the Danube is infilling this area of its basin—so that the older terraces are buried beneath younger sediments.

Sea-level changes have been particularly influential in the seaward reaches of valleys. During glacial periods ultimate base level has repeatedly fallen by 100 metres or more and in interglacials has risen to approximately present levels or a little above them. In glacials, therefore, terraces in the seaward ends of valleys are aligned to a low base level while, in many areas, the upstream ends of valleys have suffered aggradation as a result of glacial deposition, an excess of frost-weathered colluvium, or decreases in stream discharge. In interglacials, by contrast, incision takes place in upstream reaches as vegetation cover increases, loads decrease, and discharge increases, while in

TERRACE LONG PROFILES

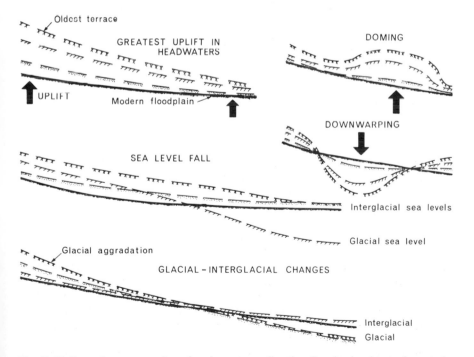

Fig. 10.20 Controls on terrace slopes in a downstream direction. Base level and tectonic controls may maintain vertical separation or cause terraces to converge down-valley. More rarely, local divergence can occur down-valley where channels are maintained across a rising structure. These diagrams indicate a continuity in the profiles which is not always the case.

seaward ends of valleys deposition occurs because of lowered channel slopes. Glacial floodplains and terraces are therefore steep and interglacial ones less steep, and they cross somewhere close to the upstream limit of ultimate base level influence.

Terrace sequences are seldom complete and simple to interpret. In its lateral migrations a river may undercut one valley wall and erode away the terraces above it, while leaving the sequence complete in other reaches. Resistant outcrops may protect some terrace remnants downstream of the outcrop to form *rock-defended terraces* with a limited downstream extent, and a channel migration down the dip of rocks may leave *slip-off slope terraces* on one side of the valley but a river cliff on the other side. Such unpaired and incomplete sequences can easily create problems for the geomorphologist, who seeks to reconstruct past erosional and depositional events by interpreting the record of change from terrace forms and deposits. Some workers have created difficulties of correlation for themselves by referring to terraces by altitude or by number—as in 'the 50 m or 3rd terrace'—when it would be safer to give it a geographical name, e.g. the 'Frasertown Terrace', so that correlations can be attempted only when the full sequence has been recognized.

Terraces are not always clearly defined features, for their form may be obscured by erosion, by partial burial by fans, by wind-blown sand, loess, or volcanic ash, and in active tectonic areas flights of terraces may be offset or displaced by faulting. Terrace chronologies are, therefore, not always easy to establish and, because of the multiplicity of causes of terrace formation, are often very difficult to interpret without a very detailed study of the sediments and the floras and faunas they contain.

Alluvial fans

An alluvial fan is a fluvial deposit whose surface forms a segment of a cone that radiates downslope from an apex where the depositing stream leaves its upland source area. The source area is usually in a mountain and the depositional area is in a valley, a basin, or at the margin of a plain (Plate 10.10). The long profiles of the stream channel are concave and are continuous from the mountain reaches into the alluvial fan. Alluvial fans grade into steeper debris cones and into erosional *pediments* which are cut across the edge of an upland. Pediments nearly always have lower angles of

slope than fans, and their alluvium forms a thin veneer across a cut surface whilst fans are thick depositional features (Figs. 10.21, 8.13).

The lateral coalescence of many alluvial fans along a mountain front is commonly called a *bajada* in North America. Along bajadas many small fans lose their fan shape where they are restricted by adjacent larger fans. An alluvial lowland that lacks the form of coalescing alluvial fans is best called an *alluvial slope* or alluvial plain.

Fans vary greatly in size from a few metres in length to more than 20 km and the thickness of their deposits is likewise variable. Where deposition is occurring in a sinking basin, such as a graben, fan deposits may be hundreds or even thousands of metres thick; as a general guide fans may be distinguished from thinner deposits, such as those of pediments, by the thickness of their deposits exceeding 1/100th of the length of the landform from apex to toe.

Alluvial fans have a great diversity of sizes, slopes, types of deposits, and source-area characteristics. They are most common in arid and semi-arid parts of the world although they do also occur in humid areas, but their dominant areas of occurrence are tectonically active mountains where the upland source area is being raised by contrast with the depositional zone. Such tectonic environments permit the deposition of very thick fans. Fans are thus common in mountains of humid regions which have been glaciated and, because a ready supply of debris is necessary for fan formation, they develop in terrains of readily eroded rock such as mudstones or highly fractured hard rocks. Massive resistant rocks seldom provide suitable environments for fan formation because trunk streams in valleys can remove the small quantities of debris carried by tributaries.

Fan processes and deposits

The dominant influence upon fan morphology and processes is the stream which transports debris from the eroding source area to the depositional zone. Deposition is caused by decreases in stream depth and water velocity. This may occur at the apex of the fan or, where the stream is entrenched in its own apex deposits, some distance down the fan. Here the channel opens from its confined single trench and spreads out on the fan to form numerous distributaries as water infiltrates into permeable fan sediments. The loss of

10.10 An alluvial fan formed during a severe storm accompanied by much landsliding and sediment transport. Since the storm, rivers have incised and left low terraces in the deposits, Ruahine Ranges.

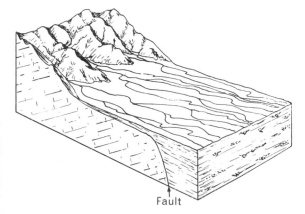

Fig. 10.21 Sub-surface profiles of a pediment which has been cut across a mountain front. The front is faulted.

discharge causes streams to drop their sediment load. Inevitably the largest-calibre debris is dropped first and the size of material progressively decreases down the fan. This simple relationship, however, is often disturbed because rare large floods carry coarse material beyond the apex and deposit it amongst the finer sediments characteristic of the middle and toe sections of the fan.

The deposition of coarse debris near the head of the fan causes the stream to braid, and continuing infiltration may prevent it from maintaining even a braided system, so a channel network may be replaced by sheetflow in the middle part of the fan.

The proportion of a fan that is flooded by a rare storm flow varies with the size of the source and fan area. Fans larger than 100 km^2 may have only

5 per cent of their area affected by the flood, but smaller fans of 1 km^2 may have more than 20 per cent of their area covered by a flood of the same occurrence interval.

Climate is an important variable affecting fans. Periods of accumulation of debris often coincide with accelerated erosion in the source area, perhaps as the result of a decrease in vegetation cover, a lowering of the tree line, an increase in storminess, or more severe physical weathering. Alternatively, accumulation may occur because of decreased competence of transportational processes at the mountain foot. This may result from decreased flow or a reduction in flood flows.

Lakes

Lakes are bodies of still water occupying depressions in the ground and having no direct opening to the sea. All lakes are temporary features of the landscape on a geological time-scale, but their duration may vary from a few hours to millions of years. At present lakes are most common in northern latitudes and in high mountains which have been recently glaciated. Such lakes have fresh water, but the rarer lakes of sub-humid regions, or those close to the sea, may be saline.

Lake basins may be formed by tectonic, volcanic, landslide, glacial, fluvial, eolian, or marine processes, or be the result of solution, accumulation of organic matter, or meteorite impact.

The tectonic movements which form lake basins may be gentle warping, folding, or faulting. The East African region provides excellent examples of such lakes with Lake Victoria occupying a shallow basin with upwarped margins and Lake Kyoga a river valley system which has been drowned by backward tilting of the plateau surface. Large elongated lakes like Tanganyika and Malawi occupy rift valleys (Fig. 4.31). In volcanic regions lakes may form in calderas, craters, or behind lava barriers.

Shallow, and usually short-lived lakes may be impounded when landslides block valleys and river deposits may themselves impound water when levees of main rivers rise above floodplains, block tributaries, or create depressions between levees. Such fluvial features are common in the lower Mississippi and Yangtze valleys.

Glacial activity creates lakes in several ways. Erosion carves hollows in bedrock, glacial debris forms dams across valleys, and melting ice masses in moraine or deeply frozen ground produce hollows which become small lakes. Most of the large valley lakes of glaciated mountains occupy hollows carved in bedrock and many such hollows are increased in depth by an impounding dam of moraine. The Italian alpine lakes like Como and Garda are clearly impounded by glacial moraines and the great depth of the lakes has been attributed to deep fluvial incision when the Mediterranean virtually dried up in Miocene times. In this case the significance of glacial erosion is debatable.

Lake basins of many sizes have been formed by solution in limestone areas, by wind deflation of hollows in arid regions, between sand dunes in coastal and arid areas, and behind spits and bars in coastal areas. Less common are those created by meteorite impact such as Ungava Lake, Quebec, by beavers building log dams, or men mining minerals underground and allowing subsidence to occur into the excavation. Quarries are also common in industrial regions.

Because they have so many origins lakes are not readily classified, but they are of considerable geological importance. The presence of lake beds and shorelines in areas now well drained, or arid, is often a valuable indicator of past geological or climatic events, and they are particularly valuable in arid regions where carbonate or organic matter in the lake deposits can be dated and used as an indication of formerly more humid climates.

All lakes are temporary features of the landscape which are gradually infilled with sediment, especially by the growth of deltas and the gradual deposition of muds and organic matter (Fig. 10.22). The infill material may itself be a valuable geological indicator, for peat can be dated and the annual sediment layers, or varves, of lakes close to an ice front may be counted and used for establishing a chronology of glacial retreat. Old lake shores around the periphery of a lake basin may indicate the greater extent and size of former lakes and hence of wetter climates in the past.

Drainage basins as units

The drainage basin is the fundamental unit in geomorphology within which may be studied the relationships between landforms and the processes which modify them. The basin is the collecting ground and storage container for precipitation, the system of routes by which water and sediment are transported to the ocean, and the expression of

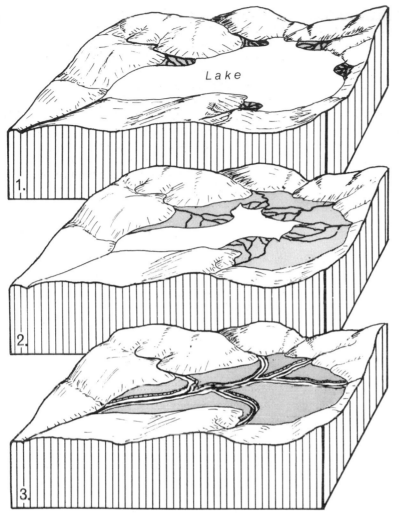

Fig. 10.22 Progressive infilling of a lake by river deposits.

the underlying geological structure, while its surface is the zone of interaction between atmosphere, lithosphere, pedosphere, and biosphere. Its geometry changes in a complex way to the forces acting on it both by long- and short-term responses.

The study of the geometry of the basin and the way in which it changes in response to processes, has become a major part of modern geomorphology. Research has been particularly concentrated upon (1) the form of the basin and its pattern of channels; (2) changes within the basin both in time and space; and (3) the budget of water, sediment, and energy inputs and outputs.

Drainage basin forms

The forms within drainage basins can be considered from the point of view of the basin as a whole, the network of channels within it, the channel reaches, and sections across the channel: for each of these units of study the area, length, shape, and relief may be determined (Table 10.4).

The size of a drainage basin influences the amount of water yield; the length, shape, and relief affect the rate at which water is discharged from the basin and the total yield of sediment; the length and character of the stream channels affect the availability of sediment for stream transport and

Table 10.4   Units of Study of Drainage Basins

	Basin	Drainage network	Channel reach	Channel cross-section
Area	Basin area	Area tributary to stream channels	Area of channel	Cross-sectional area of channel
Length	Basin length and perimeter	Drainage density, stream length	Channel length ↓ sinuosity ↑	Width
Shape	Basin shape	Drainage pattern, network shape	Channel shape	Shape
Relief	Basin relief, slope	Network relief, slope	Channel relief, slope	Depth

Source: after Gregory and Walling, 1973.

the rate at which water and sediment are discharged. Such considerations have, therefore, given rise to many attempts to relate basin and network geometries to the transmission of water and sediment through the basin. Because basins are complex entities the search for useful descriptors of their forms has given rise to many proposals.

Area and *shape* of basins may appear at first sight to be readily defined, but in practice this is not always so. On areas of low relief the perimeters of basins are often not clear and may occupy a broad featureless zone, and where the drainage is largely underground, as in many limestone areas, the surface channel network and the underground drainage network may be very different from each other. Definition of areas always depends upon the accuracy of the maps, air photos, or ground surveys used as a basis and, especially in forested regions, these may be surprisingly inaccurate. Once a basin perimeter has been defined it is a relatively easy matter to determine its area with a planimeter or by counting squares on a transparent overlay, but shape properties are so variable that several methods are in common use which attempt to compare the basin outline with a circle or with basin length (Table 10.5).

The significance of the shape of a basin may be most readily appreciated from a study of storm hydrographs of basins with very different shapes (Fig. 10.23). In the elongated catchment (A) water from the head of the basin takes a long time to reach the outlet from the basin, while the area near the outlet contributes very quickly. The hydrograph is thus flattened. In a nearly circular catch-

ment (B) the travel times from the basin perimeter are more nearly equal and the hydrograph is more peaked.

Network shapes have been most commonly described in relation to structural controls (see Fig. 10.29). The simplest way of characterizing network shape is to divide each component of the network into arbitrary measured lengths selected according to map scale, measuring the azimuth of each and then plotting the result as a circular frequency distribution. The orientation of the channels is thus obtained. An alternative is to measure the angle of the junctions between main channels and their tributaries.

Density of the stream network is commonly recognized as being a valuable indicator of the relationship between climate, vegetation, and the resistance of rock and soil to erosion. Density of the drainage network was defined by Horton (1932) as the length of streams (L) per unit of drainage area (A) (drainage density $= \Sigma L/A$). This again seems to be a very simple concept, but can be difficult to apply because of the uncertainties in the accuracy of maps or surveys, and difficulties in air photo interpretation. A particular problem lies in the distinction between the network of channels which actually carry water and the linear depressions towards the heads of basins which may become part of the network during a storm. The significance of the headward growth of a network during a storm has been pointed out already (Fig. 8.4). Attempts to overcome these difficulties have involved the definition of channels by V-shaped notches in map contours, but such indices vary in value from one survey to

Table 10.5 Some Proposed Measures of Basin Shape

Shape factor	Definition	Source
Form Factor (F)	$\dfrac{\text{Basin area}}{(\text{Basin length})^2}$ or $F = \dfrac{A}{L^2}$	Horton (1932)
Shape (S)	$\dfrac{(\text{Basin length})^2}{\text{Basin area}}$ or $S = \dfrac{L^2}{A}$	US Corps of Engineers
Shape (S)	$\dfrac{\text{Basin length}}{\text{Basin width}}$ or $S = \dfrac{L}{W}$	
Circularity ratio (C)	$\dfrac{\text{Basin area}}{\text{Area of circle with same perimeter}}$ or $C = \dfrac{4\pi A}{p^2}$	Miller (1953)
Basin elongation (E)	$\dfrac{\text{Diameter of circle with same area as basin}}{\text{Basin length}}$ or $E = \dfrac{2\sqrt{(A/\pi)}}{L}$	Schumm (1956)
Lemniscate ratio (K)	$\dfrac{(\text{Basin length})^2}{4\,(\text{Basin area})}$ or $K = \dfrac{L^2}{4A}$	Chorley *et al.* (1957)

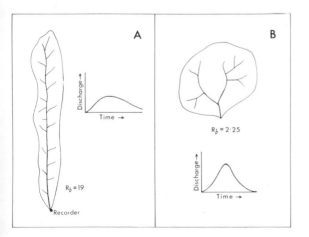

Fig. 10.23 Theoretical relationships between the shape of drainage basins and the flood hydrograph (redrawn after Strahler, 1964).

another and with the scale of the maps used. Forested terrains clearly present problems for map-makers and ground surveyors. It is common experience to find that drainage density determined from maps with a scale of 1:50 000 is less than half that determined from maps with a scale of 1:25 000. In making statements about drainage density, it is therefore essential to indicate the methods of analysis, and only to compare data derived from similar methods. Particular care must be taken to distinguish between the basic channel network which contributes flow during much of the year, and the expanded channel network which carries runoff from storms occurring only a few times a year.

It is probably premature to attempt a comparison of drainage densities for different climatic and lithologic areas as too few reliable data derived by comparable methods are available. It would be expected that under the same climate impervious rocks would support a high drainage density compared with permeable rock. Given the same lithology semi-arid areas should have a very high drainage density because of the rapid runoff and rill erosion occurring on slopes with little vegetation (Plates 10.11, 10.12). Such a situation seems to be evident in the literature as far as a comparison between humid and semi-arid areas is concerned, but it is still not clear whether less extreme climatic contrasts can be recognized in the drainage densities. Many parts of the world have drainage densities ranging from 1 to about 30 km/km^2 but more extreme values exceed 100 km/km^2.

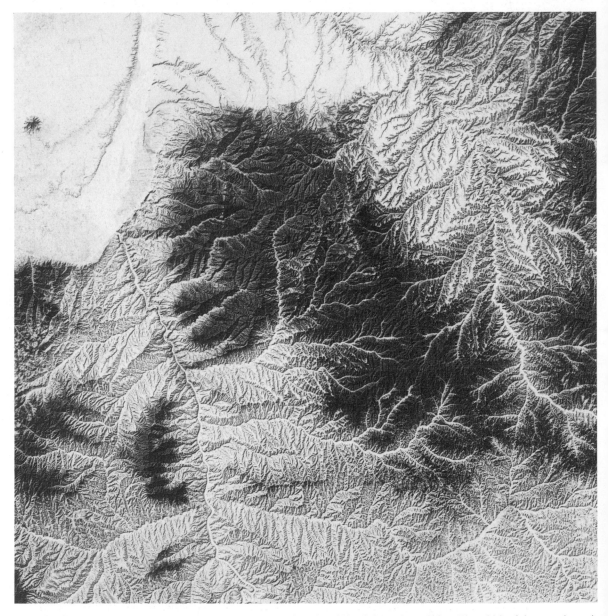

10.11 Drainage pattern cut in deep loess (wind-blown dust) deposits of Shenhsi Province, China. The width of the area shown is about 180 km. Note the area of much lower drainage density (satellite photo by courtesy of NASA).

10.12 Fine-textured drainage pattern cut in uniform mudstone, Wanganui, New Zealand. The order-one channels are mostly landslide scars and carry runoff only in severe storms. Note the nearly uniform hillslope angles. The ranges in the background are composed of greywacke (photo NZ Geological Survey, Crown copyright reserved).

The effect of drainage density upon drainage basin relief may be appreciated from Fig. 10.24. If relief is constant a high drainage density is accompanied by steep slopes, but if drainage density is constant the areas of greatest relief have the steeper slopes. There is thus a close relationship between valley slope, channel slope, drainage density, and discharge from a basin.

The area of land needed to maintain a certain length of stream channel depends upon climate, vegetation, and upon soil and rock permeability. The greater the permeability and the denser the vegetation cover the larger is the area required. Changes in land use consequently have a major effect upon drainage density with deforestation, destruction of soil structure, and the sealing of surfaces all tending to increase drainage density. As a result of such a change, flood water is conducted more rapidly to lowland areas and flood hazards there are increased.

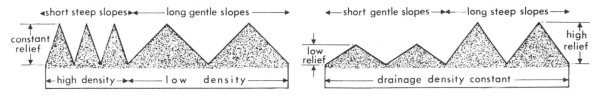

Fig. 10.24 Relationships between drainage density, hillslope angles, and local relief.

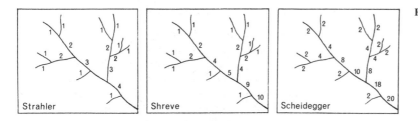

Fig. 10.25 Methods of stream ordering.

Channel orders are defined in several ways: the most commonly used system is that of Strahler (1952) who designated all finger-tip tributaries as order one, two first order channels produce a second order segment, two second orders provide a third order and so on (Fig. 10.25). In the Strahler method of ordering, the order of a trunk stream is not changed by the addition of tributary streams of lower order. This is a limitation because one purpose of stream ordering is to provide an index of scale and also to indicate the discharge which can be produced by a particular network. For the latter a method of ordering is required in which the order grows as each tributary enters the network and provides an increase in discharge. As a result, the methods of Shreve (1967) and Scheidegger (1965) were introduced. Shreve used segment ordering in which each successive link has a magnitude equal to the sum of all first-order segments which ultimately feed it. In Scheidegger's system the finger-tip tributaries are order two and the orders of the segments are progressively summed downstream. A measure of the branching pattern is the bifurcation ratio (R_b) which, in the Strahler ordering system, is the ratio of the number of streams of one order to the number of streams of the next higher order (see Fig. 10.23).

One weakness of any ordering system is that the sum of all lower order basin areas does not equal the area of the next higher-order basin (Fig. 10.26). In effect much of the initial contributing area for storm flow is excluded from the lower-order basins. The effect of this upon studies relating basin areas and forms to discharge has yet to be fully analysed.

The 'Laws' of drainage basin geometry, that is, the statistical correlations between basin properties, have been established by a number of workers. Using the Strahler method of ordering (see Fig. 10.27):

(1) The law of stream number states that the number of streams of different orders in a drainage

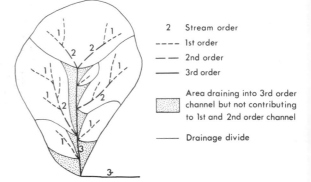

Fig. 10.26 An hypothetical drainage basin showing drainage contributing directly to the main channel alone (redrawn after Leopold, Wolman, and Miller, 1964, © W. H. Freeman & Co. All rights reserved).

basin tends closely to approximate an inverse geometric series in which the first term is the number of streams of order one.

(2) The law of stream length states that the average lengths of the streams of each of the different orders in a basin closely approximate a direct geometric series in which the first term is the average length of streams of the first order.

(3) The law of stream slopes states that slopes of streams of different orders are related to stream order by an inverse geometric series.

(4) The law of drainage basin area states that the mean drainage basin areas of streams of different orders tend closely to approximate a direct geometric series in which the first term is the mean area of the first order basins.

(5) The law of contributing areas states that the drainage basin areas of streams of each order and the total stream lengths contained within and supported by these areas is a direct logarithmic function.

(6) The discharge of a stream is a direct logarithmic function of the area of the drainage basin of the stream above the point at which discharge was measured.

The major characteristics of drainage basins thus show a close adjustment to one another which has the general relationship of:

$$O \propto \log N$$
$$O \propto \log l$$
$$O \propto \log s$$
$$O \propto \log A$$
$$Q \propto A^k$$

where O denotes stream order, N is the number of streams of a given order, l is the stream length, s is the stream slope, A is the drainage basin area, Q is the discharge, and k is an exponent (see Fig. 10.27).

There have, as yet, been few attempts to relate the processes acting in a drainage basin to the form of the basin. Moreover few of the 'laws' have been verified by studies carried out in a variety of tectonic and climatic zones. A study of the relationship between drainage area and mean annual runoff (Fig. 10.28) shows that climatic influences are worthy of much more study. An approximate relationship between bankfull discharge (Q_b), basin area (A), and drainage density (D_d) is:

$$Q_b \propto A.D_d^2$$

Growth of channel networks

The regularity of most drainage basins, and especially the consistent relationship between the area of a basin and the number of channel segments it contains of order one, has given rise to two theories of channel network growth. (1) The first theory suggests that such regularity is only possible if river networks develop in phases of growth in which new order-one channels are being added at a rate which is proportional to the size of the drainage basin as a whole. This is known as allometric growth. (2) An alternative view is that the drainage system develops at random, but that this very randomness produces a maintainable uniformity. The second hypothesis appears to be supported by experiments using computers to develop network patterns.

The processes of channel extension may involve the formation of landslide scars and their subsequent inclusion in the contributing area of drainage network during storms; the extension and collapse of subsurface pipes; the headward erosion of springs; and, perhaps more commonly in semi-arid areas, the formation of rills during extreme erosional events.

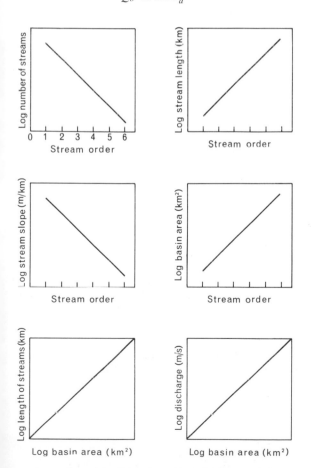

Fig. 10.27 (a) The form of graphs showing the relationships from which the laws of drainage basin geometry were derived.

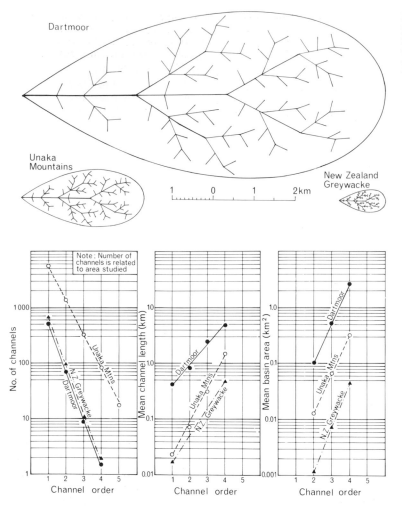

Fig. 10.27 (b) Diagrammatic drainage patterns and basin sizes for three sites in England, USA, and New Zealand, together with semilogarithmic plots of their characteristics (Dartmoor and Unaka data from Chorley and Morgan, 1962).

Drainage patterns

The design of most drainage networks, and especially of the lower order channels, is controlled by regional geological structure (see Fig. 10.29).

Dendritic patterns are most commonly formed on horizontally bedded and uniform sediments or on uniformly resistant crystalline rocks. The pattern is reminiscent of the spreading branches of an oak tree (Plates 10.11, 10.12).

Parallel drainage usually develops on moderate-to-steep regional slopes, but also where regional structure, such as outcropping resistant rock bands, are elongated and parallel. All forms of transitions can occur between this type and dendritic and trellis patterns.

Trellis patterns occur most commonly on dipping or folded sedimentary or weakly metamor-

phosed sedimentary rocks; also in areas of joints or faults which intersect at right angles, and old sand dunes with a parallel alignment.

Rectangular forms are less orderly than the trellis pattern but produced by similar influences.

Radial drainage occurs on and around domes or cones, and is particularly common in volcanic areas.

Annular patterns also develop around domes, and especially where there exist alternating resistant and weak beds so that major channels can be incised along the strike, and the low order tributaries follow the dip of the rocks.

Multi-basinal drainage can occur in a variety of conditions where local hummocks and depressions inhibit a continuous channel network. Irregular glacial deposits or erosion hollows, subsi-

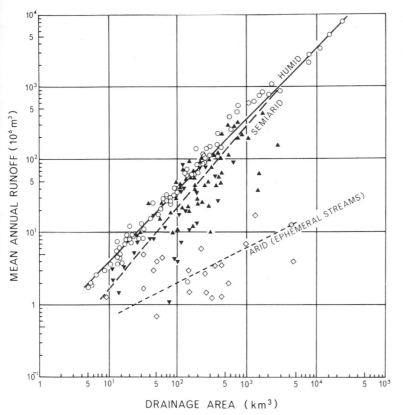

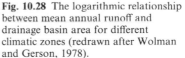

Fig. 10.28 The logarithmic relationship between mean annual runoff and drainage basin area for different climatic zones (redrawn after Wolman and Gerson, 1978).

dence hollows due to solution of underlying salt or limestone rocks, irregular thawing of frozen soil, coastal dunes, delta plains, and wind eroded hollows, can all produce such relief.

Contorted drainage is usually incised into rocks with very complicated structural patterns. It is consequently associated with crystalline and metamorphic rocks with a history of intense folding, faulting, intrusion, jointing, and alteration.

To these well-defined patterns could be added *centripetal* drainage which converges on a crater of depression; *anastomotic* which develops on nearly horizontal sediments (Plate 10.6); and *distributary* drainage which is the result of channels diverging from the apex down an alluvial fan.

Some of the largest rivers in the world have courses which have been influenced by large-scale structures in cratons. The Amazon, for example, follows an arm radiating from a triple junction which developed during the opening of the Atlantic (Fig. 4.34), and its lower reaches still occupy a complex rift (Fig. 10.30). The river itself flows over its own sediments which have accumulated throughout much of Mesozoic and all of Cenozoic time. The structural control is consequently a very ancient one which no longer influences the channel of the Amazon or its tributaries, but only the general alignment of the river valley. The same is true of the Mississippi (Fig. 10.31) which, below Cairo, follows an ancient rifted and subsided block now infilled with Mississippi sediments which have been accumulating since the Paleozoic. The present channel owes its characteristics to far more recent influences (Fig. 10.14).

Some river systems are very ancient and persistent. Others date only from the last glaciation, the last marine transgression, the last significant tectonic upwarping or major uplift, the last volcanic outpouring, or the last arid period.

Note

The relationship between drainage and structure, and superimposition and antecedence of drainage is discussed more fully in Chapter 4, p. 123. Estuaries and deltas are discussed in Chapter 13, p. 369.

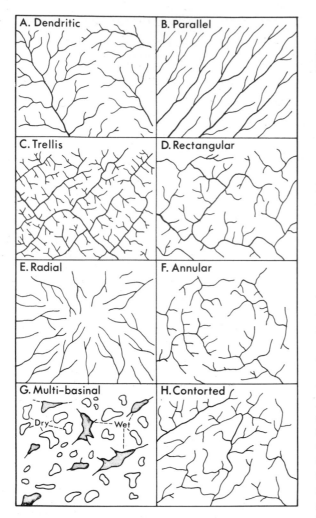

Fig. 10.29 Types of drainage pattern (redrawn after Howard, 1967).

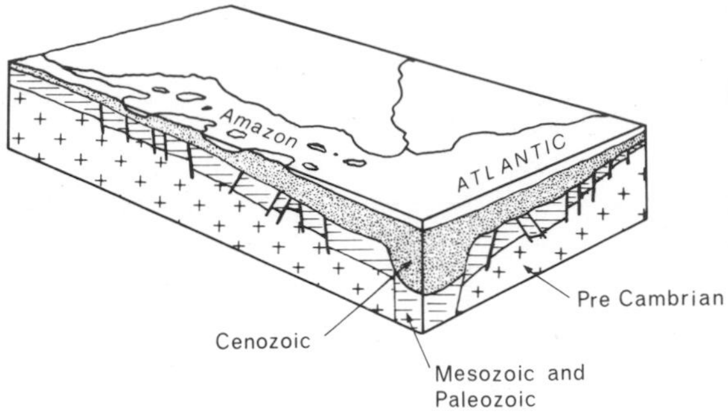

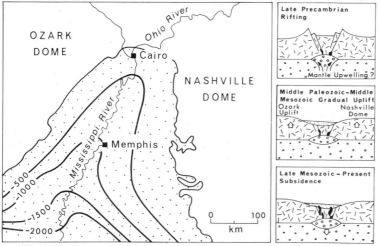

Fig. 10.30 Schematic relationship of the present channel of the Amazon, at its mouth, with the underlying structure.

Further reading

General textbooks on fluvial processes and landforms are few: that of Leopold, Wolman, and Miller (1964) is still one of the best and has been only partly replaced by Schumm (1977), and Richards (1982); that of Douglas (1977) gives a useful review of rivers in tropical environments. Amongst the more readable texts dealing with sedimentary environments are those of Allen (1970) and Blatt, Middleton, and Murray (1972). Of the books directed towards engineers Shen (1971) is particularly useful. Gregory and Walling (1973) give valuable discussions of methods of measurement and data analysis. Chow (1964) is still a valuable source of information; Gregory (1977) provides discussions of the controls on river channel changes.

Fig. 10.31 The present course of the Mississippi, in a structural embayment, and the history of that embayment (redrawn after Potter, 1976).

11   Karst

Karst landforms are characterized primarily by underground drainage in areas of massive limestone, and the formation at the ground surface of hollows and pits where water enters the rock and enlarges joints and fissures by solution. The word *karst* is the German form of the Slovene word *kras*, meaning crag or stony ground and especially the bare rock surfaces of what is now northern Yugoslavia. This area is now reafforested but is still regarded as the Classical Karst in contrast to the Dinaric Karst which stretches south-eastwards to Albania.

From the word karst are derived: *karstic*, an adjective used for karst-type landforms; *karstification*, a word used to describe the processes by which limestone is dissolved and underground drainage established; and *karstland*, an area with karst landforms. Other words derived from northern Yugoslavia include *doline* and *polje*, which will be defined later. Many other terms used in studies of karst are derived from local names for features, and the vocabulary is consequently developed from many languages and contains many synonyms.

In a true karst landscape no permanent water occurs on the surface so there is little surface washing, hollows are enlarged and not filled in, soil mantles are thin or non-existent, joint patterns have a strong influence on the alignment of landforms, and at zones of contact with other rocks solution of the limestone frequently produces very flat plains or depressions. Karst development is most effective where a very thick, hard, and well-jointed limestone occurs in an area of high relief with a humid climate. Thin limestones do not permit the development of underground drainage; soft limestones such as chalk do not permit the survival of deep surface depressions or the formation of caves and passages underground; and in arid regions there is too little water

for much solution to occur. The limestone must, of course, outcrop for landforms to occur on it, but underground drainage can occur where the limestone is covered by beds of more insoluble rocks. In areas like the northern flank of the South Wales Coalfield large cave systems have developed in the Carboniferous Limestone which underlies Millstone Grit. A karst landscape has developed in the Millstone Grit as a result of cavern collapse and the diversion of drainage underground, along joints in the Grit.

Limestones are widely distributed, and karst features are associated with many areas of outcrop, but the extensive limestone surfaces of the central Sahara and Arabia occur in areas too dry for karst development and the chalk of south-eastern England and north-western France is not a true karst because the rock is soft.

Limestone

The term 'limestone' is applied to a great variety of rocks. The usual definition is that at least half of the rock consists of carbonate minerals, of which calcite ($CaCO_3$) is the most common and the other forms are aragonite, dolomite, and magnesium carbonate. Limestone may also contain impurities of silica (especially in the form of flints and quartz sand), clay minerals, and minor impurities such as iron oxides.

Calcite deposition occurs mainly in deep-sea conditions in middle and low latitudes where limestone forms from the accumulation of the calcareous remains of animal and plant plankton or other marine life-forms. The voids between skeletons become filled in by muds or chemical precipitates. In shallow seas, lime mudstones may develop where there is an input of fine-grained sediment from the land, or pure limestones may form from algal, coral, or bryozoal accumulations. Shelly limestones form where shellfish remains have collected; *oolites* form as spheroidal precipitates around a foreign nucleus; and reef limestones form from coral.

Aragonite is precipitated in sea water by organisms where the sea is warm and highly supersaturated with $CaCO_3$. Aragonite is unstable and normally turns into calcite over a span of geological time. Most corals consist of aragonite, but old coral limestones have been largely converted to a mosaic of calcite.

Dolomites are those varieties of limestone containing more than 50 per cent carbonate, of which more than half is dolomite, $MgCO_3 \cdot CaCO_3$, $[CaMg(CO_3)_2]$. The origin of dolomites is not well understood but most appear to be secondary in origin through the replacement of calcite, consequently they are predominantly crystalline. In calcareous marine shells the magnesium content varies according to the temperature of the sea water during the life of the organism: the higher the temperature the greater is the proportion of $MgCO_3$ in the shell skeleton. The carbonate rocks consist, therefore, of the two stable minerals calcite and dolomite. There is, however, a wide range in the degree of recrystallization, porosity, development of joints, and lithification of limestones and these characteristics have a major effect upon the details of karst landform development.

Solution of limestone

Because they contain two major minerals only, calcite and dolomite, which are both soluble in natural water containing dilute carbonic acid, limestones have a solution chemistry which can be summarized in the following form:

$$CaCO_3 + H_2CO_3 \rightleftharpoons Ca^{2+} + 2HCO_3^-$$
$$CaMg(CO_3)_2 + 2H_2CO_3 \rightleftharpoons Ca^{2+} + Mg^{2+} + 4HCO_3^-$$

The carbonic acid, H_2CO_3, is derived by solution of CO_2 from the air and its reaction with water:

$$CO_2 + H_2O \rightleftharpoons H_2CO_3$$

The solubility of carbon dioxide depends upon (1) its concentration in the air, measured in terms of its partial pressure; and (2) the temperature of the solution.

(1) The solubility of $CaCO_3$ in water at various partial pressures is indicated in Fig. 11.1. The

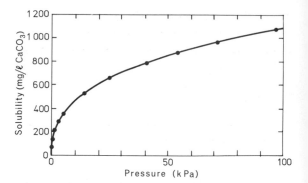

Fig. 11.1 Solubility of $CaCO_3$ in water at various partial pressures of CO_2 when temperature is 16 °C (after Adams and Swinnerton, 1937).

partial pressure of CO_2 in an aerated aqueous solution is only 30 Pa and the related solubility of $CaCO_3$ is very low at about 63 mg/l. Under anaerobic conditions in a soil, however, the partial pressure of CO_2 may increase a thousandfold to 30 kPa and result in a tenfold increase of $CaCO_3$ in solution to 700 mg/l. Groundwater in karst areas contains dissolved $CaCO_3$, or $CaCO_3 + MgCO_3$, in concentrations of about 40 to 400 mg/l, with values in excess of 63 mg/l being most common. Limestone solution, and therefore karstification, is strongly dependent upon CO_2 produced in the soil by plant root respiration and by the decay of organic matter. In the decay process (humification) organic carbon is slowly oxidized to carbon dioxide by soil micro-organisms. Soil air commonly has a carbon dioxide content of 0.25–4.5 per cent, compared with a normal open atmosphere of 0.03 per cent (or 10–200 times the concentration).

(2) Water at 10 °C dissolves about twice as much atmospheric CO_2 as water at 30 °C and water at 0 °C dissolves almost three times as much as at 30 °C. Cold water should, therefore, be more *aggressive* than warm water in dissolving limestone. The importance of temperature in controlling solution rates is far less than that of organically produced CO_2, as is indicated by the available data on concentration of $CaCO_3$ or *hardness*, in karst waters (Table 11.1). The well-vegetated temperate and tropical zones have much higher concentrations than arctic and alpine areas which have limited vegetation cover.

Water moving underground is subjected to a hydrostatic pressure induced by the depth of the

Table 11.1 Hardness of Karst Waters

Climatic zone	Observations	Mean hardness mg/l (CaCO₃)
Tropical	34	174
Temperate	154	211
Arctic/alpine	43	83

Source: Smith and Atkinson, 1976.

water column above it. In theory, then, ground-water should absorb more CO_2 in proportion to the partial pressure graph (Fig. 11.1) and consequently should dissolve more $CaCO_3$. In fact, however, once water has seeped through the soil and into rock very little additional CO_2 is available for absorption and water pressure has a negligible effect on solution. When water that has been under pressure in permeable rock enters a cave the pressure will be that of the free atmosphere; this decrease in water pressure will cause dissolved carbonate to be precipitated as *dripstone* or *flow-stone*. A second cause of dripstone production is the loss of dissolved CO_2 in turbulent streamflow in a cave.

The non-linear solubility of $CaCO_3$ in carbonated water has an important effect where two unlike water masses mix. Water which has percolated through a soil, and into limestone, becomes saturated with $CaCO_3$ in a few hours of contact with the rock if it is moving through pores or joints

in the limestone (it may take up to 10 days if it is moving through cave passages). The two unlike water masses may have been saturated with $CaCO_3$ at different CO_2 partial pressures. Because of the curved relationship between calcite solubility and CO_2 partial pressure, any mixing of two saturated but unlike water masses results in an undersaturated solution (Fig. 11.2). This mixing effect is probably the reason why most caves develop just below the water table, and without it the opening of joint passages by solution could not occur unless undersaturated water can penetrate deeply into the rock mass. As undersaturated water can be produced only by rapid runoff in storm conditions, or from non-limestone areas, solution along joints is thought to occur primarily in storms, but again it could not explain the origin of open joint passages without the mixing effect. Undersaturated rapid runoff may be important in enlarging caves when a stream, which normally drains out of limestone, is back-flooded during high stages.

Erosion rates

Water hardness is not a direct indicator of the amount of rock which is taken into solution or of the rate at which landforms change. A more useful indicator is the erosion rate which is the product of hardness and rate of runoff for a unit of time. Most methods of calculating the erosion rate have the form of the relationship:

$$E = \frac{Q}{A} \cdot \frac{T}{10^6 \rho} \cdot \frac{1}{a_l}$$

where E is the erosion rate (m³/km²)/(year),
Q is the mean annual runoff (m³/year),
T is the mean hardness (mg/l of $CaCO_3$, or $CaCO_3 + MgCO_3$),
ρ is the bulk density (i.e. mass/volume, g/cm³) of limestone,
A is the catchment area (km²),
a_l is the fraction of the catchment area occupied by limestone.

One of the main difficulties in applying this relationship is in determining the catchment area where most of the drainage is underground. It may involve many tracer studies to determine the source of water in caves or emerging as springs. The most common methods involve the use of fluorescein dyes or spores of the club moss *Lyco-*

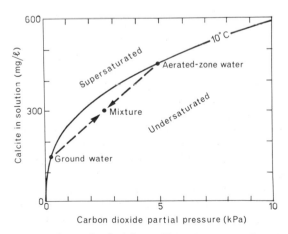

Fig. 11.2 The result of mixing unlike water masses, that are each saturated with respect to calcite, is a water mass that is undersaturated (diagram of G. W. Moore in Bloom, 1978).

11.1 (a) Rainpits, small pans, and overflow rillenkarren on limestone in the southern Namib desert (scale given by notebook). (b) Small rillen and corrosion hollows produced by algae on a marble boulder. (c) Rinnenkarren developing into rundkarren, Nelson, New Zealand (photo by P. W. Williams).

(a)

(b)

(c)

(d) Rundkarren and a grike, Craven, Yorkshire. (e) Limestone pavement of clints with enlarged joints (grikes), Craven. (f) Section through a small doline in chalk, filled with residual clay and flints, NW France.

podium. These spores have the same density as water and are carried through limestone at approximately the same speed as water. Water moving through pores of the rock has a velocity of 10–200 m/day, but in joints and caves one of up to 6 km/day.

It has been calculated that, in the Peak District of England, enough $CaCO_3$ is removed to erode the entire cave system in a single year and it is assumed, therefore, that nearly all erosion is from rock at the base of the soil. The rate of erosion is influenced by a considerable number of factors including (1) species composition of the vegetation; (2) soil type and texture; (3) organic matter and root content of the soil; (4) annual soil moisture regime; (5) the nature of the drainage, including the porosity of the rock, and fissuring of the limestone; and (6) the presence of solutional cavities, caves and conduits. Factors 1–4 influence the CO_2 supply to percolating water and the other factors influence the runoff rate. Together the factors produce a wide variety of erosion rates in most catchments and militate against any simple relationship between erosion rate and one factor—such as climate. Most published erosion rates fall in the range of 2–140 (m^3/km^2)/year. The exceptions to this are in high mountains where runoff is so great that it dominates all other factors.

Erosion of 1 (m^3/km^2)/year is equivalent to an average ground surface lowering of 0.001 mm/year or 1 mm/1000 years. Average ground lowering may be a rather meaningless figure in areas of unequal erosional activity but it does provide an indication of the rate at which the land surface can change.

Surface landforms

As in nearly all branches of landform study classification is difficult because one form or process tends to grade into another. We may, however, broadly class karst features into minor and major types. Minor features are small landforms created by solution of the surface of the rock and major features include enclosed depressions, landforms produced by fluvial erosion, and underground landforms of caves.

Minor solution sculpture

The German word *karren* and the French word *lapies* are commonly used to refer to small-scale solutional sculpture. There is no English synonym for karren. Karren range in size from a few millimetres to several metres.

The most important control on karren is the presence or absence of soil and vegetation cover on the rock. Where it is absent sculptural forms are usually sharply etched, and where it is present forms are rounded. A second control is that of the rock. Thus a porous, soft, chalk, such as that of southern England and north-western France, is unfavourable for the development of sculptured forms, and harder limestone, which disaggregates into a gravel, like that of the eastern Cooleman Plain, New South Wales, produces rounded forms in outcrop. A third control is climate. In cold climates freeze–thaw processes may be too rapid to allow solutional forms to develop, and in arid areas solution may be inactive. Areas with snow patch cover, however, may be subject to formation of solution hollows beneath the snow, which acts rather like a patch of soil (even though it is biologically sterile).

Rainpits (Plate 11.1a) may occur where raindrops fall on rock and produce individual pits 1–2 cm in diameter. They normally occur only on horizontal surfaces and are rare phenomena. If rain is ponded on the surface small solution pans can develop and *solution flutes* (*rillenkarren*) may radiate from the pits and pans.

Most flutes develop on steeply sloping rock (Plate 11.1b,c), where they form in sets with sharp crests between the flutes. Several flutes may converge and produce wider (5–50 cm) and more rounded features (*rinnenkarren*). Limestone which has been covered with soil may also be fluted, but the flutes (also called *runnels* and *rundkarren*) lack sharp edges and may be smoothly rounded or have a rough surface texture (Plate 11.1d).

The greatest amount of solution in compact rock takes place along lines of weakness, so joints are progressively widened to produce clefts (*kluftkarren* or *grikes*). Very large clefts, 2–4 m wide and 1–5 m deep, extending some tens of metres are known as *bogaz*. Between the grikes joint blocks are left as pyramidal or flat-topped blocks, depending upon the dip of the joints and the amount of fluting of their faces. Individual blocks are called clints and the nearly level surfaces formed on tabular limestones are *limestone pavements* (or *karrenfeld*) (Plate 11.1e).

While it is probable that most minor solutional

sculptured forms have been produced in the last 10 000 years, or so, there may well be exceptions to this statement, especially in arid areas; furthermore many features may have had a complex history. Glaciers covered the limestone areas of Yorkshire, for example, and left striae, or grooves, cut in the rock. Where the pavements have since been exposed by soil erosion the striae have been destroyed by solution and the edges of the clints have sharp-edged flutes. In many places, however, soil and vegetation covered the pavements and rounded forms developed on the rock. Vegetation clearance and soil erosion since human occupation of the area has allowed many rounded flutes to be modified and now soil is preserved only in the floors of the grikes.

Enclosed depressions

Dolines are closed hollows which are well-, cone-, or bowl-shaped, with rocky or vegetated sides, circular or elliptical planforms and dimensions of 2–100 m depth and 10–1000 m diameter. They can occur in isolation or in groups. The pitted relief which is so characteristic of karst is caused by the presence of innumerable dolines; they may be regarded as the fundamental features of karst, replacing river valleys of a fluvial terrain (Plate 11.2).

There are many different kinds of doline but most fall into one of four classes: *solution dolines*; *collapse dolines*; *subsidence dolines*; and *alluvial streamsink dolines* (Fig. 11.3). The loosely defined English terms like sinkhole, swallet, and swallow hole are applied to dolines but are less useful than the terms describing the dominant process contributing to doline formation.

(1) *Solution dolines* develop where pronounced solution occurs at a particularly favourable site, such as the intersection of major joints. The solutes and some residues move down solution-widened fissures and create a surface depression. This permits the collection of more of the surface runoff and hence the progressive enlargement of the doline. The presence of insoluble residues permits a soil to form and many solution dolines

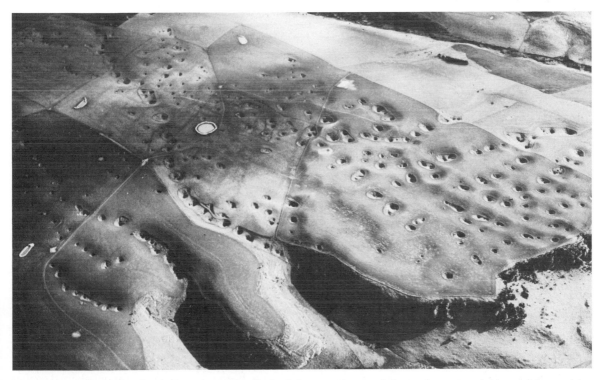

11.2 Dolines at Craigmore, South Canterbury, New Zealand. Some are due to solution and some to subsidence of the overlying loess (photo by E. Thornley, NZ Geological Survey).

DOLINES

Fig. 11.3 Types of doline, an uvala and cockpit karst.

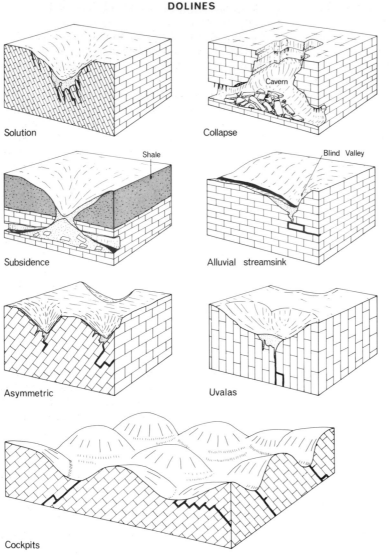

Solution

Collapse Cavern

Subsidence Shale

Alluvial streamsink Blind Valley

Asymmetric

Uvalas

Cockpits

have sloping vegetated sides and may have swampy floors if impervious residues collect. Other dolines may have cave entrances or open shafts at the base and hold temporary ponds during heavy rain or in wet seasons. In the Classical Karst the red soils, *terra-rossa*, which accumulate in the floors of dolines are important for agriculture in areas which are otherwise largely composed of bare rock. Solution dolines can form in soft limestones and in north-western France many, formed in the chalk, are filled, or partly

filled, with residual deposits of the last cold climatic period (Plate 11.1f).

(2) *Collapse dolines* usually form where the roof of a cave falls to leave nearly vertically walled, and often angular, shafts. The Velika doline in Yugoslavia is 164 m deep and 500 m wide. The steep walls may persist for a great length of time in dry climates, but in humid areas erosion of the walls (by solution, freeze-thaw, rock-fall etc.) causes the shaft to become increasingly funnel-shaped so that the collapse origin may no longer be discernible

from surface inspection. Where collapse has been into a water-filled cave, or where there has been a subsequent rise of the water table, the collapse doline may contain a lake. The cenotes of the Yucatan Peninsula have such an origin.

(3) *Subsidence dolines* form where superficial deposits lie upon the limestone. A sudden collapse may involve the cover beds in the formation of a collapse doline, but if spasmodic subsidence, or more continuous piping and solution, occurs then the cover beds will be gradually lowered or washed into the cavity to leave a conical or bowl-shaped hollow. The 'shakeholes' of Craven, in Yorkshire, are subsidence dolines formed in glacial moraines upon limestone. Subsidence dolines are also common in calcrete, which has a thin wind-blown cover of sand and soil in the inland Namib desert. In the northern part of the South Wales coalfield the Pennant Sandstone, of Millstone Grit age, lies upon Carboniferous Limestone. Numerous crater-like dolines have formed at the surface, in the 50 m thick sandstone, where large caverns have collapsed in the limestone.

(4) *Alluvial streamsink dolines* form where a stream flows into a doline and then down into the rock to form a shaft or cave. Most such dolines become trench-like and often have floors covered with limestone boulders from which the soil has been washed.

As dolines grow they may encroach on one another and link up to form *uvalas*. Size is not a criterion for recognition of uvalas, just the merging of two or more dolines. Where dolines are aligned along a joint or fault uvalas may be elongated in form, but others are irregular or lobate.

In the humid tropics roughly circular dolines are replaced by irregular star-shaped hollows with rounded hills surrounding them. The irregular depressions are now often called *cockpits*, after the name given to them in Jamaica, and the intervening regular hills are known as *cones* or *kegel*. The whole assemblage is known as a *cockpit karst* or *kegelkarst*.

Jamaican cockpits are commonly about 100 m deep and have side slopes of 30–40°. Their bases contain a brown residual soil and their sides are chemically weathered with tafoni and have a talus of fallen limestone blocks. The floors of most cockpits contain a streamsink, and during the wet season channels flowing into it mould the cockpit into a star shape. Cockpits, according to this view,

are consequently dolines which have been modified by fluvial processes.

All dolines and cockpits have a shape which is influenced by the dip of the limestone bedding and most dips exceeding about 10° tend to produce asymmetric hollows. Such hollows may also form at erosional steps, or at sites where snow can accumulate preferentially on one side of a doline. The snow cover may then either increase, or decrease, freeze–thaw processes or increase solution of the rock. The snow-covered face may develop either a steeper or gentler angle than bare faces.

Polje is a term used by peasants in the Yugoslavian karst to refer to areas of flat land in the floors of large depressions. The term has been extended by geomorphologists to indicate large enclosed basins, with flat floors, of karst regions (Plate 11.3). The largest polje in the Dinaric karst has an area of 400 km^2, but poljes range in area down to 2 km^2. At least one side, and commonly all sides, of the depression have steep slopes of 30° or more rising sharply above the polje floor. Water drains into the polje in a stream channel, especially where the stream passes over impermeable rocks forming one edge of the polje, or as flow from springs emerging at the polje edge. Water drains from the polje either into streamsinks, called *ponors*, or from open-sided poljes drains through a river valley or gorge penetrating one of the walls. The floor of a polje is usually covered with a layer of impermeable alluvium and its relief may be broken by steep-sided limestone residual hills, called *hums*.

Discussions of the classification and origin of poljes are centred round two topics: the geology and hydrology of the features and their surrounds. The earliest theories suggested that they are the end-product of a progressive sequence of depression enlargement–doline→uvala→polje. This idea is now rejected.

A major feature of most poljes, and especially those of Yugoslavia, is that they are aligned along structural trends, especially along axes of folds, along faults or fault troughs, or along margins of contact with impermeable rocks. Some poljes may be tectonic features such as graben, but the majority are caused by laterally directed corrasion, or planation, and the geological influences merely guide erosion, not control it (Fig. 11.4). As can be seen from the map of poljes in the Dinaric karst (Fig. 11.5) many are border poljes, that is

11.3 Lake Disappear, near Aotea Harbour, New Zealand. The basin is a polje which is periodically flooded (photo by P. W. Williams).

POLJES

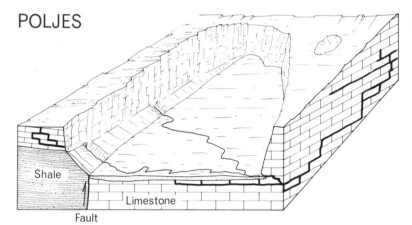

Shale

Limestone

Fault

Fig. 11.4 Structural influences on the location of poljes.

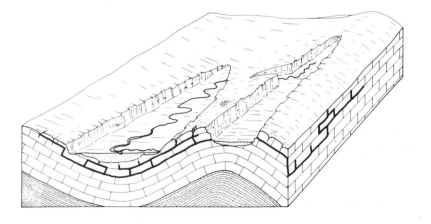

they form at a geological contact between impervious rocks and limestones. Runoff from the impervious rocks is aggressive and so causes rapid corrosion of the limestones. Lateral planation by surface streams may cut a bench across the impervious rocks as well as the limestone and seepage into the limestone may be reduced or prevented by the alluvium on the polje floor.

Lateral erosion may also be enhanced by the fact that many poljes are periodically inundated by floodwaters. This is most common in the winter wet season in Yugoslavia and Italy, but can occur at any time of the year when rainfall and runoff exceed the capacity of the ponors to drain the polje. Flooding may also be increased by groundwater rise and the ponors may then feed water into the temporary lakes of the polje floor. Close to the sea, where water levels are always high, poljes may have permanent lakes. In the Dinaric karst it is possible to recognize several types of hydrological regime. Dry poljes have no lakes because ponors can drain them adequately at all times. Overflow poljes have temporary lakes as a result of impermeable alluvial covers. Many poljes are large enough to have zones which are dry, zones which periodically flood, and also permanent lakes.

Most poljes have a long and complicated history. They may have originated in middle to late Tertiary times and been subjected to severe alternations of climate in the last 2 million years or so. Consequently some, in high mountains, contain

glacial moraines, talus materials from freeze–thaw processes, and alluvia of many periods. Poljes occur in the humid climatic zones of both temperate and tropical latitudes and may have been strongly or weakly influenced by geological structure.

Labyrinth and tower karst is a feature of many limestone areas, especially in the humid tropics; it is characterized by the presence of rock towers, 30–200 m high, standing above wide, flat valley floors, or plains, with alluvial covers. Many of the towers are indeed spectacular landforms with nearly vertical walls and only gently domed or serrated summits. They were formerly ascribed to the effect of intense solution in tropical climates with steep walls due to rapid evaporation, and hence deposition of a protective calcite on the rock faces, and to undercutting of the cliffs by aggressive waters from alluvial plains, swamps, and peat bogs, and the consequent development of notches and caves at the slope base. Towers in this view were regarded as the residual hills of intense humid tropical erosion in thick limestones, and a distinct tower or pinnacle karst (or *turmkarst*) was recognized in subtropical southern China, Malaysia, Sarawak, Cuba, Puerto Rico, and at other sites in the low latitudes.

The recognition of karst towers in the Mackenzie Mountains of Canada, at 61–62°N, has altered the focus of attention from climate to geological influences. In the Mackenzie Mountains water

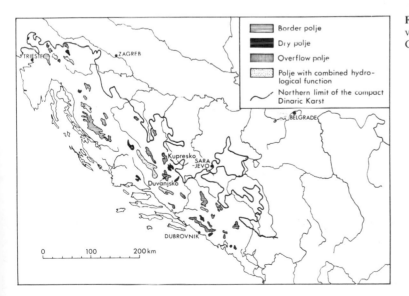

Fig. 11.5 The distribution of poljes of various kinds in the Dinaric karst (after Gams, 1969).

drains underground through deep dolines and along open joints and bogaz. By enlargement of these features, and their coalescence along joints and faults, deep box canyons or *karst streets* are created, and the intersections of streets and blind box valleys forms a *labyrinth*. Individual streets are up to 185 m deep and 9 km long. As the streets widen and extend, the intervening rock ridges are dissected and ultimately destroyed to leave closed depressions with vertical walls, called *platea*. Where insoluble sediment is carried to these depressions by streams from nearby impermeable rocks (i.e. *allogenic streams*) underground drainage becomes impeded and there is flooding after heavy rains. Floors of platea become covered with alluvium; lateral corrosion, and sometimes corrasion, can occur and the platea eventually become poljes. As platea and poljes expand rock towers are left rising above karst margin plains (Fig. 11.6). Features similar to those of Canada have been described in the Celebes, New Guinea, Sarawak, Arnhem Land in northern Australia, and in Tanzania.

Deep labyrinths are restricted to massive carbonate rocks of low porosity that are cut by widely spaced, but vertically and horizontally extensive, fractures. Local relief of many tens of metres must be available. The rarity of labyrinths is caused by the lack of suitable rock masses, and karst pavements are more common as these can develop on thinner limestones with less available relief. Labyrinth and tower karst is possibly most common in the humid tropics because such areas have not suffered interference of labyrinth development by glaciation, and especially by the deposition of impermeable glacial deposits.

Karst landforms of fluvial erosion

Rivers crossing karst areas have at least three main features which distinguish them from rivers on impermeable rocks: (1) they commonly develop gorges; (2) they may sink underground and leave dry valleys, which may be occupied by streams in flood periods only, or may be permanently dry; and (3) sections of their channels may be alternately across alluvial lowlands, and in caves through uplands where cave roof collapse may produce natural bridges (Fig. 11.7).

(1) *Limestone gorges*, like those of the Lot, Tarn, Jonte, and Dourbie of the Grandes Causses of France, or the smaller Cheddar Gorge of

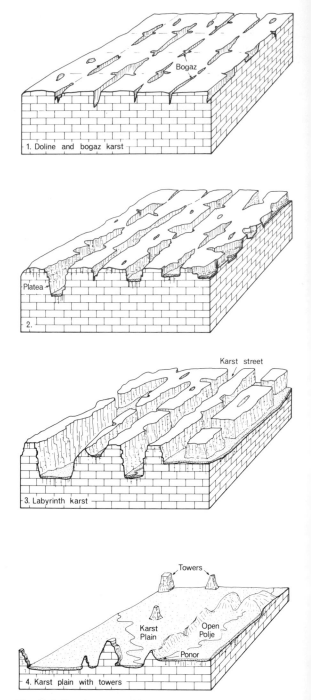

Fig. 11.6 Development of karst streets and tower karst of the Mackenzie Mountains type.

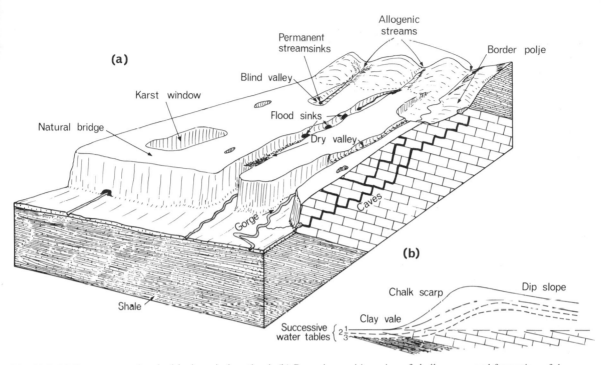

Fig. 11.7 (a) Features associated with rivers in karstland. (b) Recession and lowering of chalk scarps and formation of dry valleys.

England, have steep walls because valley slope processes are ineffective compared with the power of corroding water to incise into the bedrock channel and to undercut the gorge walls. Thin soils, resistant rock, and lack of surface drainage keep slopes stable. Lateral corrosion at stream level permits caves and undercuts, particularly at the outside of meanders.

(2) *Dry valleys*, and blind valleys, result from loss of water into streamsinks either at one point or at successive points along a channel. Gradual enlargement of the sinkholes can lead to complete diversion of flow underground. Blind valleys form where the upstream section of a valley has been lowered to leave an older continuous, and now dry course, hanging above the level of the blind valley which has an enclosed terminus at the streamsink.

Dry valleys are not restricted to karst areas and may form networks on permeable sandstones and deep pumice, but they are most common on limestones. In the Craven, Yorkshire, the Watlowes (Fig. 11.8; Plate 11.4a) is a craggy valley once occupied by a stream from Malham Tarn which cascaded over the 75 m cliff of Malham

Cove (Plate 11.4b). The stream at the base of the Cove derives its water from both Malham Tarn and Smelt Mill.

The dry valleys of the chalk of south-eastern England and northern France have smoother more gently sloping sides (Plate 11.4c). In the weak, porous, rock closed depressions and surface drainage are rare. Allogenic rivers pass through the cuestas (Fig. 4.40) in water gaps, but only the lower ends of major valleys have permanent streams. Springhead, steep-sided valleys are particularly common along the faces of cuestas and scarp slopes.

Some chalk dry valleys may be drainage lines inherited from impervious Tertiary cover rocks and erosion during periods of high water tables, but this is an improbable explanation for dry valleys closer to the uppermost areas of cuestas. It has been suggested that during glacial periods the ground would have been frozen and spring snow meltwater, and summer surface runoff, could then flow over impermeable rock and soil, with ice-filled pores, so cutting the valleys. The presence of rock rubble (Plate 11.4d), thought to result from

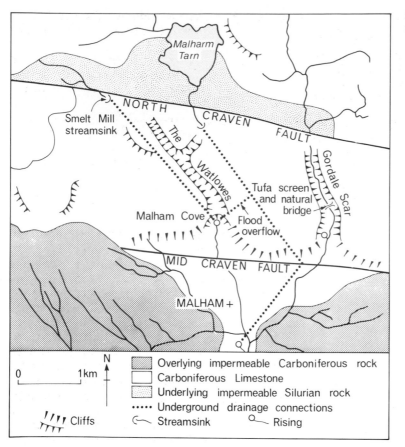

Fig. 11.8 The Craven area of Yorkshire (based on Jennings, 1971).

freeze–thaw processes, in many valleys supports this hypothesis. Against it, however, is its assumption that the chalk had no valley network before the glacial period.

Another theory is that water tables were once higher in the chalk and streamflow on the surface was therefore normal. As incision occurred in the lowlands cut across interbedded weaker rocks, then scarp-valleys with springheads would cut back into the chalk and lower the water table so that drainage from the upper reaches of the dip slope valleys would be diverted underground (Fig. 11.7b).

(3) *Surface rivers* are usually few in karst; in the whole Dinaric karst only four rivers cross it to reach the sea, and all are allogenic. In areas like the Nullarbor Plain of Australia all drainage to the sea is underground; this is, of course, a climatically semi-arid area so water is in limited supply. To cross karst on the surface both allogenic rivers, and those which emerge from springs within the karst (*autogenic* rivers), either have to flow in areas of very high groundwater levels or upon impermeable alluvium.

While many rivers flow in gorges, or across plains, others pass through limestone hills by entering caves. Collapse of cave roofs in sections leaves either *karst windows* where the roof is mostly intact, or *natural bridges* where collapse is predominant.

A feature of some stream channels is the formation of calcite dams across channels. The calcite is called *tufa* and is usually precipitated because CO_2 is lost from the stream water, or because calcite is secreted by certain species of moss and algae. Tufa barriers may cause rapids or waterfalls and can impound lakes of considerable size.

Underground water, caves, and springs

In a profile from the ground surface through soil into bedrock it is possible to distinguish zones with distinct water contents. Immediately below the surface is the *vadose* zone in which the soil or rock pores contain either air or water. In this zone flow is normally unsaturated, capillary rise can occur, and water contents fluctuate with percolation from the surface. Below the *water table* is the *phreatic* zone in which all pores and cavities are permanently filled with water. The vertical extent of groundwater varies with the type of rock, its geological structure, and mass porosity. The ultimate depth is about 30 km where rock ductility will virtually eliminate porosity; more commonly pores in sedimentary rocks are closed at depths greater than 10–15 km, and at 2–3 km in igneous and metamorphic rocks.

Rocks such as porous chalk, sandstones, fractured limestones, and igneous rocks, which contain economically significant quantities of water are called *aquifers*. *Aquicludes* are rocks with low rates of water transmission. Within a vadose zone local or *perched aquifers* can exist where impermeable strata prevent vertical seepage and localized zones of saturated flow can also occur below *perched water tables* above impermeable beds (Fig. 11.9). Most economically significant aquifers lie within 500 m of the ground surface but, in synclinal structures with interlayered permeable and impermeable strata, recharge areas and aquifer water tables may be higher than well-heads so that confined water is under sufficient pressure to give rise to *artesian* conditions. Such a condition existed in the London Basin until excessive abstraction of water from the chalk depleted the artesian head.

Water flow beneath the water table occurs because the water table has a relief which follows the ground surface, although in a subdued fashion, and there is movement of ground water in accordance with the slope of the water table, ie. down the pressure gradient. In soil or rock with interconnecting pores the permeability is a measure of capacity of the medium to transmit water through a saturated sample of defined length in a unit of time. *Darcy's Law* states that the flow through a permeable medium is proportional to the pressure gradient, or head loss:

$$Q = \frac{k.A\,(h_1 - h_2)}{l}$$

where

Q		is the volume of water flowing through the sample in a unit of time (m^3/s),
k		is the coefficient of permeability peculiar to the medium,
A		is the cross-sectional area of the sample,
$(h_1 - h_2)$		is the difference in water table elevation or between the ends of the sample (m),
l		is the length of the sample (m).

Water flow in limestones

It is often assumed that water in limestones obeys Darcy's Law just as it does in other rocks or soils. This assumption is probably valid in soft rocks like chalk, but is often inappropriate for hard limestones in which solution along fissures has created numerous cavities and caves. Wells and tunnels driven into limestone frequently reveal dry and water-filled cavities close together, and tracing experiments have shown that underground water connections can cross one another and pass under surface streams without interference. Some flooded poljes exist above rock containing air-filled caves. In such conditions concepts of regional water tables and uniform saturated flow are meaningless. It is then more valid to think of independent underground conduit systems in a three-dimensional space. Hydrostatic pressure can build up in conduits and water may consequently be driven uphill. The two opposing views of karst hydrology can, in some areas, be resolved by considering a developmental sequence: where limestone is first exposed to surface drainage by erosion of cover beds, or uplift, stream flow occurs at the surface and flow in rock pores at depth. As solution opens fissures, drainage disappears underground, and conduit flow becomes increasingly important so that the concept of a water table has little application. When conduits are widely open, and frequent, the rock is honeycombed with caves and water levels in them equalize so that a general water table becomes established again.

Phreatic flow below a water table can occur to depths of at least 100 m and so corrosion can occur below a water table, but such flow is probably most common in areas of high relief, and hence high water pressures. In rocks with low dips, and in areas of low relief, water pressures are also low

11.4 (a) Dry Valley of The Watlowes, Craven. (b) Malham Cove, with a stream emerging from a cave at the base, and The Watlowes above the Cove.

(a)

(b)

(c) A dry valley of a chalk dip slope, Paris Basin. (d) A soil of shattered chalk; this soil may be a residual from the last cold period, NW France.

and phreatic flow and corrosion are probably limited to a few metres below the water table.

Rates of water flow in rock are dependent upon the size and connection of pores and cavities. In the Berkshire Downs, near Oxford, England, water issuing from springs is at least 15 years old even though it has percolated through only 80 m of chalk. Water passing through cavernous limestone, by contast, may emerge in springs only hours after falling as rain.

Chalk and limestone are important aquifers yet it is not often realized that the water being pumped from them is old. Some of the water from the chalk

beneath London is at least 25 000 years old, and overpumping can seriously deplete this resource. More significantly, much of the groundwater feeding oases in the Sahara and Arabia fell as rain during the last wet period 35 000 years ago, and this water is not being replaced under the present arid climate.

Limestone Caves

Limestones vary in capacity to contain caves because of their chemical composition and mechanical strength. Impure, weak, and highly

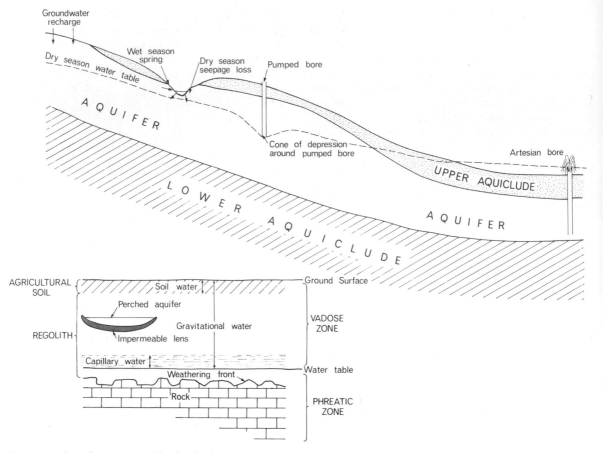

Fig. 11.9 Sub-surface water and its distribution.

fractured rocks have few or small caves. Water is also necessary for cave formation and limestones of the Sahara and Nullarbor Plain have few caves in spite of the fact that the Nullarbor may be the largest area of continuous limestone in the world.

Karst caves vary in length from a few metres to over 100 km in the cases of Flint Ridge Cave, Kentucky, and Hölloch, Switzerland; in depth they reach to 1311 m in Gouffre St Pierre Martin in the French Pyrenees. Gouffre Berger in the French Alps has a volume of 3 km³. Cave systems vary from single rooms, shafts, or passages to intricate connecting cavities of all shapes, with sizes of single chambers as great as the $400 \times 230 \times 100$ m of Big Room in Carlsbad Cavern, New Mexico.

Some caves are dry and inactive, others are filled with water; some are periodically flooded; some contain streams or lakes. In a mountain or upland

it is common for upper-level caves to be dry and connected by shafts to lower, periodically inundated, systems (Fig. 11.10).

Caves are predominantly due to solution but mechanical abrasion can also be important, especially where sediment and pebbles are carried through the limestone by streams. Tectonic influences, such as faults or stress release, may be of local significance. Roof collapse, and rockfalls into caves, are often important in cave extension and modification. Structural lineaments, such as joints, often strongly influence the alignment of vadose caves which often contain features characteristic of surface streams, such as fluted and scalloped erosional bedforms, pot-holes, waterfalls, meanders, pools and riffles, and incised or trough-shaped channels.

Water moving under pressure in the upper part

of the phreatic zone is in contact with the rock over all surfaces of the cave and tends to produce a set of tube-like caves with roughly circular or elliptical cross-section. Phreatic tubes most commonly develop along bedding planes and form continuous networks at a level controlled by the bedding. As water levels in a cave system are lowered due to falls of base level, or cave opening, a phreatic tube may come to contain a vadose stream so that the cave can have the form of an upper arch with an incised active channel cut into its floor. Very large caves are probably of multiple-process origin, but the evidence is usually destroyed during their creation.

Cave deposits are of importance to geomorphologists and archaeologists because they preserve evidence of cave history, climatic change, and human occupation. The most obvious deposits are the decorative *speleothems* produced as CO_2 diffuses from water into the cave air with a resulting deposition of calcite. *Stalactites* cling to ceilings and grow downwards as water seeps from roof fractures and drips from the ends of the tube, or cone, of the speleothem. *Stalagmites* grow upwards from where drips splash. Many stalactites start as tubes around a droplet, and grow downwards to form straws about 5 mm in diameter. Straws are usually blocked after a time and the speleothem turns into a cone, which thickens as water flows over its surface, and also it lengthens from the tip (Fig. 11.10). *Helictites* grow out from walls or form from the ends of stalactites and curve upwards as calcite crystals grow on the upper surfaces of tubes, as solutions are fed along capillaries to the tip of the helictite. Flowstone and dripstone formations form curtains, sheets, and other beautiful features as water runs down rock and calcite surfaces.

Sediments in caves include stream deposits, lake silts, soil, volcanic ash, and other inorganic material washed into a cave or which has fallen through cavities. Together with animal skeletons, cave ice, and crystals of minerals, sediments preserve a record of events at the surface of a karst. One of the most remarkable results of cave deposit studies is the record of past climate which they may yield. Stalactites have been analysed for their oxygen-isotope composition and dated by carbon and uranium-thorium isotopes. Because shallow, closed caves have air temperatures which are the mean annual temperature of the surface air above the cave, and the oxygen-isotope composition reflects this temperature, stalactites can be used as paleothermometers giving mean annual temperatures: several records now go back to nearly 400 000 years.

Cave deposits of all kinds are best preserved in dry caves from which streams are excluded, and speleothems can only develop in dry or vadose caves which are not inundated.

Springs

Water draining through limestone commonly emerges in streams from caves or as upwelling springs. Because water can circulate beneath the water table, where one exists, under pressure, the upwelling or *resurgence* can have a large volume and high velocity. Many cave exits above the dry season exits are occupied by water only after floods or during wet seasons.

Karst evolution

Limestone frequently forms high relief because it is structurally resistant, it produces little sediment which can be used for stream abrasion, and it has little surface water. It is evident, however, that limestone masses are gradually eroded and that lateral extension of sediment-covered karst plains reduces areas of limestone uplands. In the early part of this century it was thought that direct sequences of landform evolution can be recognized and that any present-day landform can be interpreted as being in a stage of evolution. There is no doubt that landforms do evolve, but the complexity of climatic change and its influences, and the effects of tectonism, may be so important in landform evolution that few geomorphologists now accept that there is a simple and direct sequence of landforms which can be recognized as developing according to a particular invariable pattern.

Two evolutionary schemes of development of karst landforms have been put forward. That of Grund, published in 1914, assumed that a limestone surface existed above the water table and that in the initial stages dolines develop at points favouring solution. Most European karst is at this early stage of widespread doline development. He thought that dolines would progressively increase in number, merge as uvalas, and destroy intervening ridges until a cockpit karst was created. Grund envisaged that once depression floors reach the

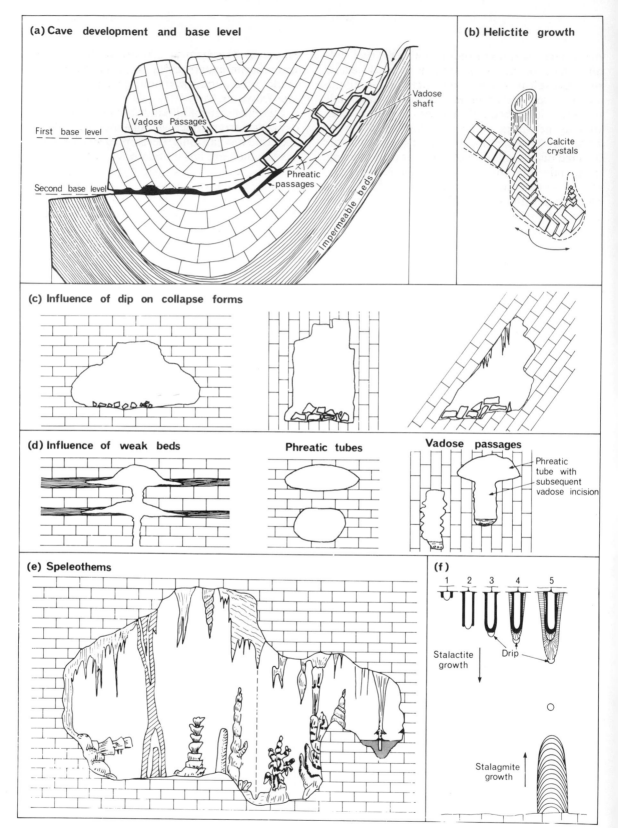

Fig. 11.10 Cave locations, forms, and features: (b) is after Prinz (1908) and (f) is partly after Jennings (1971).

water table a corrosional plain would form with residual hums upon it. He thought that poljes are not part of a general evolutionary sequence because they result largely from tectonic and structural influence; in that he was correct, but the idea that cockpit karst necessarily develops from a doline and uvala karst is questionable.

Cvijić (1918) attempted to explain the Dinaric karst and confined himself to a thick limestone unit between to impermeable rock units, all of which lie above sea-level. He thought that once an upper impermeable rock unit is stripped then a doline karst, with poljes in suitable tectonically controlled sites, would form,. Elaborate cave systems would supply water to karst margins and limestone residuals would finally diminish as a normal river system crossed impermeable underlying rock, with residual limestone hills or interfluves.

Simple schemes of development have been questioned by authors who believe that tower karst and cockpit karst are not exclusively features of the humid tropics, and by those whose interest is increasingly focused on the influence of lithology, base level change, and structure upon karst landforms. The evidence that many karst features may be old—the karst towers of southern China may date from the early Tertiary—and have suffered many climatic changes and influences, makes simple schemes hard to validate or accept. Working in New Guinea, however, Williams (1972) has been able to show, by using morphometric methods, that there is a recognizable sequence of sink and doline formation to a form which is very similar to that of Grund's cockpit karst stage. William's work is outstanding in that he arrived at his conclusions by field measurement and inductive reasoning rather than by *a priori* deduction, and showed that many styles of karst landform can exist in related climatic environments.

Further reading

The books by Jennings (1971) and Sweeting (1972) provide a review of the subject, but the papers of Williams (1972) and Brook and Ford (1978) should also be consulted.

12   Eolian Processes and Landforms

The work of wind in eroding, transporting, and depositing material is called *eolian* activity (also spelt *aeolian*). Wind is an effective geomorphic agent in a variety of natural environments that lack a protective cover of vegetation; it is most important in deserts and semi-deserts, coastal dune areas, exposed mountain regions, and in agricultural areas where bare soil is exposed to high winds during dry periods. Wind is not confined in its effects by channels and so can operate over extensive bare surfaces. The results of its activity are most evident in the great sand seas, called *ergs*, and vast gravel-covered plains of the world's major deserts.

Eolian erosion

The kinetic energy of wind acts on the ground surface where some of this energy is transformed into heat, but some detaches particles and moves them by rolling or saltation (refer to Chapter 6, p. 183 for a discussion of the processes). The action of removing material from a surface and lowering it is called *deflation* (Plate 12.1).

The pressure exerted by wind against a flat surface ranges from 2 Pa at 5 km/h, to 100 Pa at 50 km/h, and 600 Pa at the hurricane velocity of 120 km/h, but velocity alone does not control the capacity of a fluid to entrain particles and to cause abrasion. This is influenced by the density of the fluid, and nature of the fluid motion, the character of the bed materials, and the impact of saltating grains (see Figs. 12.1 and 6.15).

Clays and silts may not be entrained, by wind shear, from a smooth surface, but impacts may throw them into saltation or release them from aggregates, and roughened surfaces may allow them to protrude into the zone of turbulent air above the laminar sublayer which may be up to 0.05 mm thick; furthermore violent eddies may break through the laminar sublayer and entrain fine particles.

Sands lack cohesion and hence offer only frictional resistance to entrainment, but clays and silts have cohesion, and when wet have added apparent cohesion which increases their stability. Aggregation, ground roughness, plants, the presence of humus and soluble salts all increase the stability of silts, but a dry smooth surface of low density and low cohesion materials, such as peaty soil, may be readily eroded by turbulent air.

Eolian erosional landforms

Deflation is a most effective process where extensive non-cohesive materials are exposed; consequently river deposits, dry lake beds, dry river beds, and beaches are the main sources of material for wind transport. The repeated removal of sands and silts from alluvium leaves a coarse lag deposit protecting mixed coarse and fine material. Surfaces with lags are called *desert pavements* or *desert armour*. In some places surface wash may contribute to their formation, but in the extremely arid areas of the Namib, Atacama, and ice-free areas of Antarctica there can be no doubt that deflation alone is effective. Once a lag has formed, further deflation can occur only where weathering has produced new fine-grained material, or a new fluvial deposit has been formed.

Stony desert pavements are often major surfaces of large deserts and, in dune areas, they may form all surfaces between the dunes. In Libya they are called *serir*, in Algeria *reg*, and in Australia *gibbers*. Many gibbers are composed of silcrete fragments which are stained dark red by iron oxides (Plate 12.2a).

The importance of deflation for creating distinc-

12.1 A deflation hollow on a mountain in the New Zealand Alps. The material being removed is mostly loess which was originally derived from the river channels.

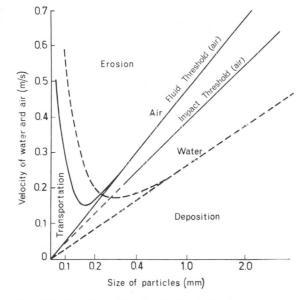

Fig. 12.1 Variation of the threshold velocity for movement, with grain size, for air and water. Data for water are derived from Hjulström (1935) and those for air from Bagnold (1941).

tive landforms has been contested by many scientists, but study of satellite photographs has shown that very large enclosed hollows are major features of many deserts. Many of these hollows cannot be explained by tectonism, subsidence, solution, or compaction. Big Hollow, near Laramie, Wyoming, is about 5 km wide, 15 km long, and about 100 m deep, and the Qattara Depression, west of the Nile delta in the Egyptian Sahara, is over 250 km long, 100 km wide, and has a floor which reaches

134 m below sea-level. Many other large hollows exist in the Sahara. It is probable that scarp retreat has produced much of the fine-grained debris which floors the Qattara Depression. The effective base level for deflation is the regional water table.

Many shallow hollows in the Kalahari, on the Pampas of Argentina, in Australia west of the Flinders Ranges, and in central Kansas are now occupied by shallow lakes, at least in wet seasons. Many of them were undoubtedly formed by deflation in dry climatic conditions, and the Australian examples frequently have dunes, called *lunettes*, downwind of the hollows, which add support to this theory.

Deflation is most effective where it is cutting into unconsolidated or weak soils and sediment. Abrasion of rock is, by contrast, a much slower and more limited process.

Wind abrasion is a form of natural sandblasting. Its effectiveness is related to wind velocity, the hardness of the dust and sand carried by the wind, and to the hardness of the rock being eroded. It is probably most effective in limited areas of polar regions where cold, dense air can carry large particles at very high velocities and where, at winter temperatures, even ice has the hardness of some rock minerals (Fig. 12.2).

Boulders and pebbles may be fluted, scalloped, or faceted by wind abrasion (Plate 12.2 b,c,d,e). Flute and scallop forms vary in length from a few millimetres to several metres, but large flutes are uncommon in hard rocks. It is thought that they are produced by turbulent and helical flows carrying dust and sand, and that the scallops grow downwind. Where they are cut on large boulders

12.2 (a) A gibber on the eastern edge of the Simpson Desert, Australia. (b) A wind-fluted boulder of granodiorite, Antarctica. The wind has come from the side of the boulder where there is an axe. (c) Detailed sculpting of a block of fine-grained dolerite in Antarctica. The sides have been polished smooth, and flutes have been cut from the keel and edges by vortices carrying dust and fine sand. On the upper surface weathering pits have been elongated and turned into scallop forms.

(d) Ventifacts from a deflated lag in Antarctica. (e) Large-faceted boulders of dolerite in the central Namib. The polished faces are to the left; scale is given by the open knife. (f) Yardangs in the central Sahara; the rock is sandstone.

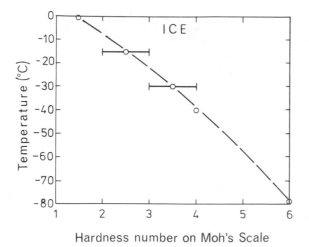

Fig. 12.2 The hardness of ice as a function of temperature. Values reach the equivalent of those for orthoclase feldspar (hardness 6) in the Antarctic winter (data from Selby, 1977).

they are clear indicators of the strongest, or dominant, wind directions. For this reason they are most commonly reported from places such as formerly glaciated valleys in Antarctica, where katabatic winds from the Polar Plateau have had a constant direction for thousands of years, and from deserts like the Namib where dominant winds have also had a constant direction for much of the Quaternary, at least.

Some of the best-known wind-cut features are *ventifacts*. This is a general term for wind-faceted pebbles and boulders which may occur within an area of sand cover, or as the components of a desert lag. A rock face abraded by wind may be pitted if there is a range of hardness in the minerals of the face, or it may be smooth where the rock is fine-grained or composed of one mineral only. Thus rocks like granite and coarse-grained dolerites have pitted surfaces, but quartzite and fine-grained dolerites have smoothly polished surfaces. Ventifacts have a great range of surface shapes with plane or curved faces and two or multiple facets. The German term *dreikanter* is frequently used for three-faceted pebbles, and a ventifact with one sharp edge, or *keel*, forming an intersection between two facets is an *einkanter*.

The occurrence of einkanter, dreikanter, and weakly faceted pebbles in one lag, and orientations of the face which vary from one ventifact to

another have caused some writers to deny the possibility of a wind abrasion origin for these features. Instead it has been proposed that the faceting is caused by splitting of the rock by weathering processes, or faceting is caused by exfoliation or chemical weathering of a pebble face in contact with the ground, and its later exposure as the pebble is rolled over. The origin of some features is uncertain, but flutes, large boulders, and smaller pebbles of a lag all having facets facing a dominant wind are certainly produced by wind abrasion. Abrasion increases with height up to about 1 m, so most facets are inclined away from the wind and ultimately the facet is reduced in angle towards the horizontal. The keel is the line of separation between a face being abraded from one direction and a face which is either not abraded, is abraded from a different direction, is caused by a process other than wind abrasion, or is a relict from some earlier phase of erosion. An actively developing facet is transverse to the wind direction. The presence of einkanter and dreikanter forms in one area is probably the result of the dreikanter forms being rotated or rolled over as one side of the pebble is undercut by deflation.

In weak rocks, wind abrasion may be so effective that it cuts long grooves or channels to leave upstanding elongated residuals called *yardangs* (Fig. 12.3). The presence of joints probably has an important influence upon the orientation of some yardangs, but the extensive yardang fields around the Tibesti Massif in the Sahara are aligned parallel with the dominant Harmattan winds which are deflected by the Massif. Yardangs are also reported from the southern Lut of Iran, where some crests stand 200 m above the troughs, from Turkestan, Peru, and the Mojave. Most yardangs have pitted and etched surfaces where the softest materials have been selectively removed, and where resistant beds are interlayered with weaker beds the resistant ones protrude or may form caps on pedestals of weaker rock; such elongated tabular and mushroom forms are called *zeugen*.

Little information is available on the rates of eolian abrasion, but it is probable that small ventifacts can be faceted in a few tens of years, deflation hollows have formed in soils bared of vegetation in a few days, but yardangs and large ventifacts probably form over tens of thousands of years.

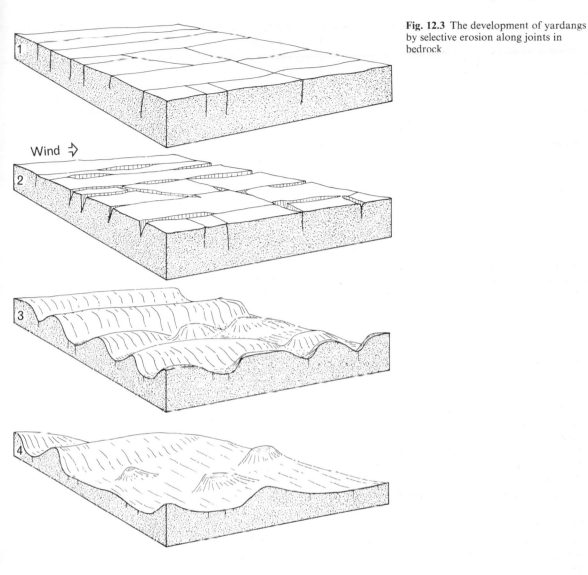

Eolian transport and deposition

Most of the material transported by wind is composed of quartz because this is both chemically resistant and so is a residual material of rock weathering, and also it is resistant to abrasion. Other minerals are of course also transported, but most are progressively eliminated from eolian deposits so the great ergs are almost entirely of quartz sand grains. The sorting of grains which is initiated while grains are set in motion by wind is continued as they are transported, and again when they are deposited and make up depositional landforms.

During saltation most sand grains rise only 1–2 cm above the ground but a few bounce 1–3 m. Their trajectories are long compared with the height and the grains have an angle of 10–16° with the horizontal as they land and collide with other grains. Collisions splash grains of a range of sizes into the air. The smallest grains are of clay- and silt-size, but especially the latter. The processes of repeated impact, fracture the sharp edges of sand grains, and these broken fragments provide further silt particles which are of irregular shape. Sand grains thus become progressively rounded and smooth (Plate 12.3 a,b). Silt and clay grains

12.3 (a) Sand grains: the angular grains have been washed from a glacier and then incorporated into dunes in the Victoria Valley, Antarctica (see plate 12.7a). Many of the grains have been subsequently rounded during eolian transport. (b) An electronmicrograph of the edge of a silt grain from a loess deposit. Note the angular broken edge and the irregular shape (photo by courtesy of I. Smalley).

(a)

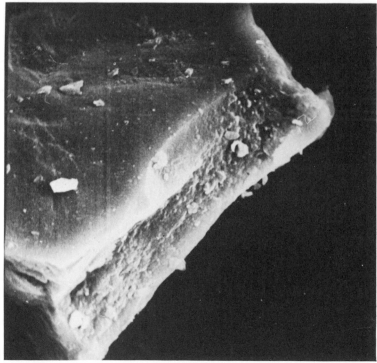

(b)

have very low settling velocities so, once they have been set in motion, air turbulence can maintain them in suspension. In turbulent air grains smaller than about 200 μm (Fig. 12.4) can be maintained in suspension almost indefinitely. This material consequently becomes separated from saltating and creeping sand grains. The sand becomes increasingly free of fines as transport causes sorting downwind from the source area. The fines are carried to heights of 1–2 km and after periods of high winds may form a continuous haze for hundreds, or thousands, of kilometres downwind

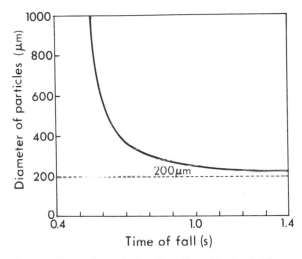

Fig. 12.4 The settling velocity of small particles in air. Those with very small diameter may remain in suspension for considerable periods (data from Bagnold, 1941).

from the source (Plate 12.4). The dust settles during calms and is also carried to the ground in rain; the latter may be the more important of the two processes.

At present the Sahara is the chief source for silt- and clay-size particles which are carried north-wards into Europe on several occasions in most years. During the northern hemisphere winters, the steady Harmattan winds from the north-east blow across the south-west Sahara and the Chad basin and carry an estimated 200–400 Mt/year of dust into the Atlantic. Approximately two-thirds of this is deposited within 2000 km and forms a major part of the deep-sea muds, but 50 Mt/year reaches the Caribbean Sea. Dust from Africa has formed a 10 cm layer of silt in the coastal areas of southern France and northern Italy in the past 300 years, and similar dusts from central Asia have formed deposits over 100 m thick in the North China plains and hills during the Quaternary. Such extensive deposits are called loess and are composed predominantly (60 per cent or so) of silts with a size of 10–50 μm (see the last section of this chapter). The best estimates suggest that up to 1000 Mt/year of dust, from all sources, is transported in the atmosphere.

Ripples and dunes

With fine particles there is a clear distinction between transport and deposition; but with sands transport is often intermittent with sand moving through a system of bedforms in which it may be stored temporarily. A feature of most large sand

12.4 A dust haze near Kano, Nigeria, resulting from the transport of material from the southern Sahara, towards the Atlantic, by the Harmattan. The photograph was taken at 10.00 a.m. looking directly at the sun.

deposits is the existence of three distinct bedform groupings in separate size classes: ripples, dunes, and megadunes (also called *draa*).

(1) Ripples are between 5 cm and 2 m apart and 0.1 to 5 cm high.
(2) Dunes are 3 m to 600 m apart and 0.1 m to 15 m high.
(3) Megadunes are 300 m to 3 km apart and 20 to 400 m high.

The differences between these three classes are created by differences in the mechanisms of transport.

Ripples are created by saltation. Where there is an irregularity in the sand surface more sand will be thrown up from the surface facing the wind than from the lee face which is protected from bombardment. If all sand grains are of similar size then, in a wind of constant velocity, they will bounce an equal distance downwind from the initial irregularity and so create a new mound (Fig. 12.5a). The process is thus perpetuated downwind and becomes spread out laterally until numerous ripples are created transverse to the wind direction. Most dune and sand sheet surfaces are covered by ripples with varying degrees of regularity (Plate 12.5). Variations in spacing and height are largely due to the size of the sand particles: large particles bounce farthest and move only at higher wind velocities, so increasing ripple spacing. Ripple forms travel downwind as grains are transported from the back of each ripple to that of the next, while some grains creep down the lee face and across the inter-ripple zone.

Dunes and *megadunes* have a variety of shapes but can be broadly classed into those which are longitudinal, that is aligned along the dominant wind direction, and those which are transverse, that is aligned across, or at right angles to, the wind direction. Individual dunes travel across rock or pebble surfaces with the smallest dunes travelling fastest; there is consequently a tendency for small dunes to merge with large ones to form *megadunes* (Plate 12.6a). Small crescent-shaped dunes may travel at rates of 50 m/year, but large dunes and megadunes seldom advance at rates greater than 1–2 m/year.

Longitudinal dunes are perhaps the most common dune forms of desert plains, and in some areas, like most of Australia, the Namib, and parts of the Kalahari, they are the only major dune form. They have long narrow ridges which may be

remarkably straight or slightly sinuous with local hook forms. In north Africa the sharp-crested longitudinal dunes of moderate dimensions (up to about 2 km long) are called *seif* dunes, but large, active longitudinal megadunes of similar form may be up to 50 km long and 100–200 m high and have numerous smaller dunes migrating along their flanks (Plate 12.6b). Most longitudinal dunes are separated by rock and pebble lag surfaces of nearly uniform width for considerable distances, although transverse dunes can form on such areas and longitudinal dunes do sometimes converge to form Y-junctions, or 'tuning-fork' junctions. Studies carried out using smoke bombs and balloons have led to the hypothesis that longitudinal dunes are created by long vortex airflows which pass down the corridors between dunes and control the spacing between them: vortices may also control the migrations of sand along and up the dune flanks (Fig. 12.5b). Y-junctions occur where a vortex pair lifts off the ground and the vortices of flanking dunes then push their ridges closer together. There is still much doubt about the origin and size of vortices but it seems, most probably, that they result from regularly spaced thermal convective cells which are pushed downwind, so that they initially control dune ridge spacing, but later conform to the pattern of ridges they have already made. An alternative view is that vortices are usually too large to account for dune spacing and hence play little or no part in their formation.

There is clear evidence that some seif dunes are formed by the expansion of one arm of a crescent-shaped dune (Fig. 12.5d), but large longitudinal draa may originate as narrow sand humps and grow as they capture small migrating dunes.

Cross-winds may regularly shape the crests of a longitudinal dune (Plate 12.6c) and produce local dunes on top of the main longitudinal ridge. In the Namib the prevailing wind is from the south, but a 'berg' wind blows several times a year from the east and creates a local dune crest with a slip face transverse to that wind. This local dune form is then modified or eliminated by the prevailing wind. Most of the longitudinal dunes of Australia and the Kalahari are now partly or wholly vegetated so that only the crests are active. These dunes are mostly small (< 20 m high) compared with the active longitudinal draa of the Sahara and Namib.

All active longitudinal dunes and draas are aligned roughly parallel with the prevailing wind,

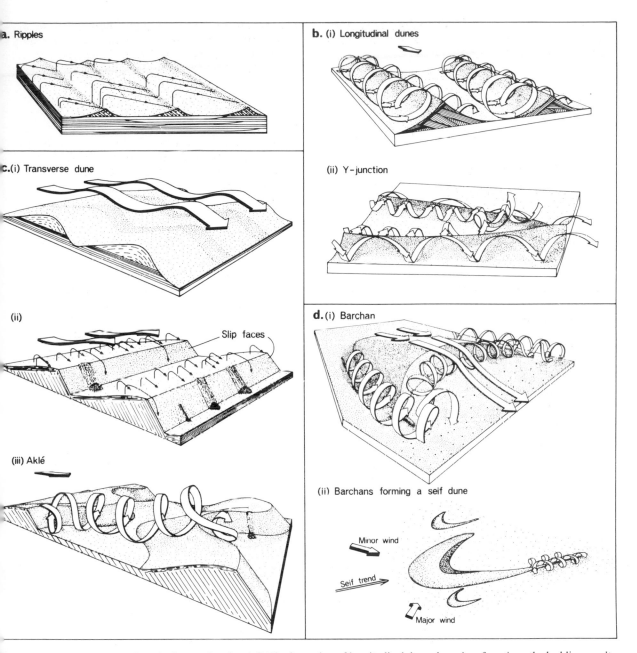

a. Ripples

b. (i) Longitudinal dunes

c.(i) Transverse dune

(ii) Y-junction

(ii)

Slip faces

d.(i) Barchan

(iii) Aklé

(ii) Barchans forming a seif dune

Minor wind

Seif trend

Major wind

g. 12.5 a. Ripple formation by saltating sand grains. b(i) The formation of longitudinal dunes by pairs of vortices: the bedding results om migration of transverse dunes along the flanks. b(ii) A Y-junction develops where one pair of vortices lifts off the ground. c(i, ii) ypothetical stages in the development of transverse dunes. c(iii) A simple transverse pattern like that of c(ii) is transformed into a ɛtwork of sinuous ridges by vortices which distort successive dunes. d(i) Airflow around an isolated barchan. Note that the wave eates a roller in front of the slip face and that this carries sand to the arms of the dune. d(ii) The theory of Bagnold (1941) that seif ɪnes are formed by elongation of one arm of a barchan which is extended in the direction of a stronger wind (these figures are based ɪ the work of P. Knott and A. Warren, 1979).

(a)   (b)

12.5(a),(b) Ripple patterns on the backs of transverse dunes. The irregular ripples are formed in less well-sorted sands than those of the regular ripples.

but the influence on alignment exerted by the rare strong wind is not clear. Some dunes appear to be most closely aligned to the average or prevailing wind; some parallel the resultant of all winds exceeding the threshold for sand movement (about 16 km/h); some show the influence of strong cross-winds. Much of the debate is, however, difficult to assess, as few records of wind direction and speed are available from the largest ergs.

Transverse dunes develop wherever the air above a sand deposit has a wave motion. Such motions develop downwind from a ridge, an obstacle, a dark rock surface which has been heated by solar radiation so that air rises above it, or some other disturbing influence. Where the air in the trough of a wave touches a surface with a sand cover it will set that sand in motion, but beneath the crest of the air wave sand transport

will be less. The sand from the high transport zone moves into the low transport zone until a mound is formed of such size that it can create its own wave pattern in the air (Fig. 12.5c). The upwind side of the mound becomes covered with active ripples in which sand is transported to the mound crest which increases in height until the sand is unstable and slides as a dry sandflow. The slide face of most dunes has an angle of about 33° but this varies a little, depending upon the size and shape of the sand grains.

A pattern of straight, parallel dunes transverse to the wind is very rare (Plate 12.7a). Commonly local vortices develop and sweep sand from some parts of ridges and cause local zones of converging and diverging air and sand flow. The original sand sheet consequently becomes a sea of large and small intersecting dunes. Under a steady wind direction the vortices develop a regular and repeat-

(a)

12.6 (a) Draa in the Namib Desert. The small barchans to the left have formed in a fluvial deposit and are travelling towards the larger dunes and megadunes to the right. (b) A view along a large longitudinal megadune in the Namib Desert. Transport is predominantly along the ridge and flanks, but a cross-wind from the right has pushed small dunes up the megadune. (c) 'Smoking' megadunes in the Namib during a 'berg' storm. The main ridges trend towards the camera, but the berg wind is blowing from left to right.

(b)

(c)

12.7 (a) Transverse dunes formed of sand washed out of the Victoria Glacier, Antarctica, which is in the background. The ridges are developing sinuosity but are not yet network forms. (b) Aklé in the Great Western Erg, Algerian Sahara.

(a)

(b)

ing pattern and unite in pairs, one on either side of a sand ridge. The sand ridges consequently become *aklé* or networks of bulbous advancing ridges and curved re-entrants, with each ridge having a ripple-covered slope upwind and a slide face in its lee (Plate 12.7b,c).

In some parts of Arabia coarse sand deposits develop large transverse regular ridges without slip faces: these are called *zibar*.

Where sand is so sparse that diverging vortices sweep it entirely onto areas of convergence, crescent-shaped *barchans* are formed. Barchans result from both wave-like motions of air passing over their crests with rollers in the lee, and also from vortices spiralling along the arms of the crescent

(Fig. 12.5d). Barchans advance in the direction of the dominant wind or the resultant of the annual wind pattern. In areas of increasing sand supply they may merge and eventually form aklé, or merge *en échelon* to create seifs.

Megadunes are nearly all of the longitudinal type with predominantly transverse dune forms migrating along their flanks. A few barchan forms are of such size that they are megadunes with small barchan dunes passing over their backs.

A feature of some longitudinal dunes and aklé in both Arabia and the Sahara are star-shaped *rhourds* (Plate 12.7d). Most rhourds appear to be part of a crossing pattern of sand flow where several sand ridges converge. It seems most pro-

(c) Aklé in Saudi Arabia, ridges are more than 1 km apart (ERTS-1 satellite photo by NASA). (d) Large rhourds on longitudinal draa in the Libyan Desert (photo by US Air Force).

(c)

(d)

bable that they form where sand is repeatedly blown back and forth so that it accumulates rather than travels on. They may also be so large that they are self-generating features which produce their own sets of vortices.

There is still much controversy about the origin, development, alignment, and rate of travel of dunes. Recent research has questioned many long held ideas—including those of vortices—but it also suggests that the following hypotheses may be generally valid: (1) barchans occur where there is very little sand and nearly unidirectional winds; transverse dunes occur where sand is abundant and the wind is moderately variable; longitudinal dunes occur where the winds are more variable but there is little sand; rhourds occur where sand is abundant and wind variability is at a maximum (if all of these contentions are true then the juxtaposition of different dune types implies great local variation in wind direction and sand supply); (2) the spacing of dunes is directly correlated with their height and width, with spacing increasing with size; and (3) dune spacing becomes greater as sand grain size increases.

Most dune and megadune complexes are part of ergs which are separated from other ergs by rocky uplands or by regs. If the sand in an erg were spread to an even depth over the area of that erg, its depth would be 2–30 metres. It is a feature of all active ergs and their dunes that they are aligned with the dominant wind even though major cross-winds produce local patterns and divergences of a few degrees from perfect alignment. It has been calculated that a large erg would take some few thousand years (1000–4000) to become adjusted to a new wind direction, thus suggesting that the modern active ergs have been under relatively constant wind conditions for several thousand years, or that our knowledge of modern wind patterns is so poor that we are unable to detect minor divergencies in alignments.

Many inactive dunes and ergs are aligned with the assumed dominant wind of the last dry period which is thought to coincide approximately with the peak of the Last Glacial, about 18 000 to 20 000 years ago. They are thus valuable indicators of paleowind directions. Since they became inactive they have been eroded by rain and sheetwash and many are probably smaller than when they were active (see Chapter 19).

Sand shadows and drifts are accumulations of sand in the lee of an obstacle (sand shadow), or downwind from a valley or depression which funnels wind into a high-velocity stream; downwind of the funnel the air can diverge and deposition takes place. Other local sand accumulations are curtains of sand which is trapped at the foot of a slope facing the wind, and local parabolic dunes around deflated hollows. These (unlike barchans) have trailing arms around the sides of the hollow and a thicker, and higher, rounded nose. Parabolic dunes are particularly common in coastal areas where a local deflation basin or 'blowout' forms the source of sand (see Plate 13.7b).

A feature of all deserts is the abrupt boundary between different kinds of surface—bare rock, reg, isolated dune, or erg. These abrupt boundaries are nearly all caused by wind action which, because of its sorting effect, related to critical threshold velocities for setting in motion silt and sand, and stripping them from surfaces of various degrees of roughness, is able to create a mosaic of landforms of widely varying size from ripples to ergs. The largest erg is the Rub al Khali of Arabia, with an area of 560 000 km^2.

Loess

Much of the loess of the world (Fig. 12.6) is now fixed by vegetation and is a relict from drier, and mostly colder, climates than exist now. The greatest thicknesses (> 250 m) and most extensive loess sheets occur in China. The basal beds were formed at least 2.4 My ago, according to paleomagnetic measurements. There is controversy concerning the sources of this material. The most probable hypothesis is that it was derived from fans, lake beds, and stream channels around the Gobi and northern and eastern ends of the Tibetan Plateau. The Plateau may have been uplifted as much as 3500 m in the last 2 My, and it has certainly caused rain-shadow aridity in central Asia. Some small component of the Chinese loess may have come from glacial outwash deposits but, because of the dry climates, this is probably not a major source.

In Europe, North America, and New Zealand, dust has been carried from the seasonally dry beds of rivers which drained from Pleistocene ice sheets. This relationship is particularly clear in Fig. 16.12 where it can be seen that most of the loess of Europe is a cover upon the extensive plains and hills south of the former Scandinavian ice sheet and is closely associated with the major rivers

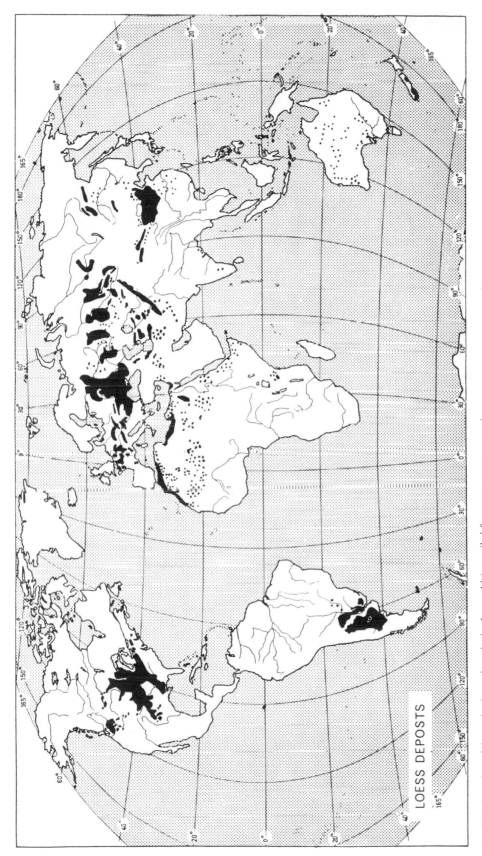

LOESS DEPOSTS

Fig. 12.6 Location of the major loess deposits in the world (compiled from numerous sources)

which drained it. Similarly in USA most of the loess is in the Mississippi and other valleys draining south from the former Laurentide ice sheet. Much loess, however, has had a complicated history. It has been washed as 'glacial flour' from ice sheets and later exposed on the river beds at times of low discharge. From river beds it has been carried onto local river terraces, or farther away, by the wind. Much of this loess was trapped by local vegetation and stabilized in place, where it became the parent material for soil, but a considerable proportion was eroded again and carried back into rivers and, either deposited in the sea, or again removed from river beds and deposited on land. Thus much loess is of secondary eolian origin (Fig. 12.7).

Loess is readily recognizable in many parts of the world as a buff-coloured, predominantly silt-size deposit with considerable cohesion, so that when dry it can form steep cliff faces. When wet, however, it loses this cohesion and gully erosion and landslides are very common in it. Loess forms a recognizable deposit on at least 10 per cent of the land surface, but is present in the soils of a far larger area. Much of the prosperity of the great wheat-growing areas of the northern hemisphere is attributable to its presence, and it has also been widely used for making bricks; it is called *brick-earth* in many parts of south-eastern England.

Further reading

The standard text is still that of Bagnold (1941), latter work is reviewed by Cooke and Warren (1973), Warren (1979), and by Mabbutt (1977). Many recent studies of dune and erg forms are summarized in the volume edited by McKee (1979). A useful volume on dune studies is *Zeitschrift für Geomorphologie Supplementband*, 45 (1983).

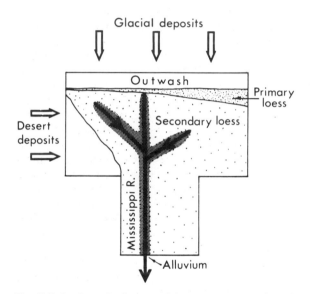

Fig. 12.7 A schematic diagram of the sources, transport, and deposition of the loess of the Midwest of USA (after Selby, 1976, from an idea by Smalley, 1972).

13   Coastal Processes and Landforms

About two-thirds of the world's population lives within a narrow belt of land at the ocean edge. Such a concentration of people, with their associated industrial, recreational, and transport activities, creates pressure on coastal resources so that there has been a demand for scientists, engineers, and planners to develop an understanding of coastal processes and the obviously dynamic nature of beaches, dunes, cliffs, estuaries, and deltas. In spite of the scientific effort expended over the last 100 years there is still much to be learnt and many attempted solutions to problems of erosion, deposition, and pollution are now recognized to have been inadequate.

The energy available for causing change along coasts is derived both from the Sun, which creates weather systems and hence wind and waves, and the gravitational pull of the Moon and the Sun creating tides. Both tides and waves produce the mass drifts of water which are coastal currents, but deep ocean currents are caused by the consistent winds of atmospheric pressure systems and by variations in ocean water temperature and salinity.

Tides, waves, and currents

Tides

Tides are the periodic rises and falls of sea-level which occur, in most parts of the sea, twice daily. They are important because they increase the vertical height over which wave action can affect the coast and they extend the horizontal surface of activity even more. Currents generated by tidal ebbs and flows are particularly important around the entrances to bays and lagoons, where they are responsible for sediment transport into and out of the entrances, and for keeping those entrances open. Only in exceptional locations are tidal

currents capable of eroding coastal rocks. Currents moving coarse bedload materials off Mont Saint Michel, Normandy, and in the Bay of Fundy, Canada, can bring about such erosion, but these sites are areas with the largest tidal ranges in Europe and the world, respectively.

Sir Isaac Newton first demonstrated that tides are the result of the gravitational pull of the Moon and Sun on the Earth. This pull affects the land masses as well as the oceans, but the lower viscosity of water allows it to respond far more quickly, and to a greater magnitude, than does rock. Each element of Earth material is attracted by the Moon with a force of slightly different magnitude from that which attracts other elements, because of their different distances from the Moon. An element of Earth on the surface facing the Moon is attracted more strongly than an element of the same mass on the opposite side of the planet, away from the Moon. It is these differences from the average force of attraction which causes tides.

The attraction between the Earth and the Moon causes them to orbit one another, each moving in an approximate circle about a common centre of mass which lies within the Earth at about 4500 km from its centre.

If the Earth were completely covered with a layer of water of uniform thickness, that water would be drawn out into an elliptical bulge, with its long axis pointing towards the Moon, because the difference between the attraction of each water element by the Moon and the overall attraction between the Earth and Moon is directed away from the Moon and from the Earth's centre. The result is a tidal bulge on either side of the Earth. (Note: the bulge is not due to a centrifugal force resulting from rotation of the Earth.)

The two tidal bulges form huge waves, each with a length extending half-way round Earth's circumference, which move at a celerity governed by the depth of the ocean. This celerity (C) is given by

$$C = \sqrt{(gd)}$$

where g is the acceleration due to gravity (9.81 m/s^2) and d is the depth of water (metres). The Atlantic Ocean has an average depth of about 4000 m so the wave celerity is about 200 m/s, or 380 knots. The speed of the tidal bulge can be compared with that of the Earth's rotation at the equator of 450 m/s. As the bulges migrate during the Earth's rotation on its axis, each point on the surface of a uniform ocean-covered Earth would experience two high and two low tides each day.

When the Sun, Moon, and Earth are aligned, at the time of a new and full Moon, the gravitational forces produced by the Moon and Sun reinforce each other and produce *spring tides*. These occur every two weeks throughout the year (more precisely every $14\frac{3}{4}$ days). *Neap tides* are formed when the Sun and Moon appear to be at right angles and the gravitational forces are opposed to each other thus producing counter-bulges at the first and third quarter of the Moon (Fig. 13.1). The Moon's elliptical orbit about the Earth also causes a variation in gravitational forces so that tides are about 20 per cent higher and lower every 27.55 days when the Moon is closest to Earth.

The simple pattern described above is broken by irregular water depths and, more importantly, by the blocking presence of continents which form nearly closed ocean basins. If we consider a rectangular ocean basin in the northern hemisphere the tidal bulge would advance from east to west until it met the ocean boundary. There it would be reflected and move in the opposite direction, to the east. If the reflection is perfect it will combine with the other tidal waves moving to the west so that a stationary or standing wave is produced. The motion of such a wave is shown in Fig. 13.1c. The water oscillates about a *nodal line* where no amplitude changes are observed, but where strong periodic oscillating currents are felt. The largest sea-level changes occur at the walls of the basin. The natural period of oscillation of any basin is controlled by its east–west length and its depth in the relationship

$$T_o = \frac{2l}{\sqrt{(gd)}}$$

where T_o is the period of oscillation (s), l is the

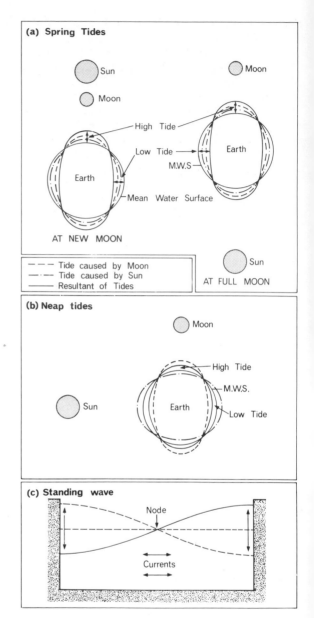

Fig. 13.1 (a,b) Schematic cross-section through the equator showing effects of Sun and Moon in causing tides. (a) Spring tides caused by alignment. (b) Neap tides caused by Sun and Moon in opposition. (c) A standing wave in an enclosed basin with a single node. The horizontal currents are strongest beneath the node but the vertical motions are greatest at the end walls.

east–west length of the basin (m), and *d* is the depth of the basin (m).

The tidal pattern, however, is not simple. Due to the rotation of the Earth, the Coriolis deflection acts upon the standing wave oscillation causing the water body in the northern hemisphere to veer to the right, and in the southern hemisphere to the left, rather than simply to travel between east and west. As a result the water rotates around a point, called the *amphidromic point*, rather than across a nodal line (Fig. 13.2). The high water of the tidal bulge now progresses around the tidal basin in a counter-clockwise direction in the northern hemisphere, and clockwise in the southern hemisphere.

At the amphidromic point the tidal range is zero, and greatest around the edge of the basin. The boundaries of tidal basins are not just coasts. Many enclosed water bodies behave as several separate units (Fig. 13.2.f). In the North Sea, friction of the water with the sea floor has caused the amphidromic points to move away from the source of the tidal energy. In the northern area the point is now nearly on the coast of Norway, so the coasts of Norway, Denmark, and northern Germany have small tides but the east coast of Britain relatively large tides. For similar reasons the Isle of Wight is near the point and has a low tidal range, but the northern French coast has the highest tides in Europe. In the Irish Sea the amphidromic point is just south of Dublin.

The tidal range in mid-ocean is small, about 50 cm, but its height is increased as it reaches the shallow waters of the continental shelf. Low tidal ranges occur most commonly on the open coasts of the world's oceans and in landlocked seas like the Mediterranean, Black, and Caspian Seas (Fig. 13.3). The highest tides occur in relatively limited areas of gulfs and deep embayments. In the Bay of Fundy the range increases progressively from the mouth, where it is about 3 m, to the head where it is 14–15 m at spring tides. The high tides of embayments are partly due to convergence of shores wedging water masses so that they deepen, but also to formation of local oscillations within the Bay caused by reflection and resonance of the tidal water body.

As a tidal bulge proceeds up an estuary and river it may oversteepen and break, producing a *bore*. The largest bore in the world is on the Amazon, where it moves upstream, at 10 m/s, as a 2 km wide waterfall some 5 m high.

Because of their lengths and depths the basins within the Atlantic Ocean and south-western Pacific have periods of oscillation which correspond roughly with the 12 hour tide-generating gravitational forces and produce *semi-diurnal tides* (two high and two low waters of about the same height). The resonance period of the Gulf of Mexico is closer to 24 hours so that it has *diurnal tides*, with only a single high tide per day. The natural period of much of the central Pacific Ocean does not correspond very closely to either 12 or 24 hour tide-generating forces and produces *mixed tides* with elements of diurnal and semi-diurnal patterns.

Waves

Wind-generated surface waves are the main source of energy along a coast. They are responsible for erosion of the coast and for the formation of beach features. Waves, in turn, may generate coastal currents which are responsible for the drift of sediment along beaches. An understanding of waves is therefore essential to an understanding of beach processes.

The process by which wind, blowing over the surface of water, transfers energy to the water is not fully understood, but is related to variations of pressure across the water. Near a storm centre where waves are being generated the waves have complex forms and confused patterns. Individual waves have steep irregular forms with smaller waves and ripples superimposed on them. Such waves are called *sea waves*. Sea waves are propagated in the direction of the wind and as they leave the storm centre, or move away from the winds that generate them, longer period waves absorb those with shorter periods thus producing regular sinusoidal waves, called *swell*. Swell waves have long, rounded crests, have nearly uniform height and spacing, and travel for thousands of kilometres across the oceans without losing much energy (Plate 13.1).

With the exception of tropical cyclones, which are of rather limited extent, strong winds are confined to relatively high latitudes north of 30°N and south of about 35°S. Within the tropical and subtropical belts the trade winds blow at 3–10 m/s, and even where there is a *fetch* (distance of wind blowing over which a swell is generated) of 1500 km waves are seldom more than 3.5 m high. By contrast, an easterly gale blowing across the North Sea, from Holland to East Anglia, may produce

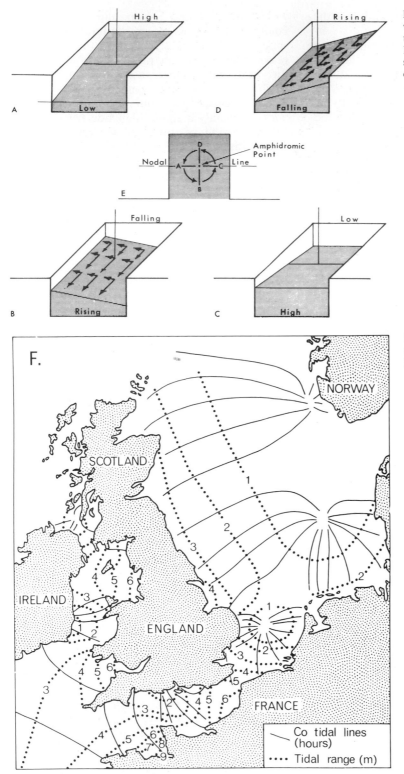

Fig. 13.2 A–E. Oscillation of a standing wave around a nodal line as a result of the Coriolis deflection induced by the Earth's rotation: northern hemisphere situation; the rotation is in a clockwise direction in the southern hemisphere.

F. Amphidromic points and lines of equal tide heights in the North Sea, Irish Sea and English Channel.

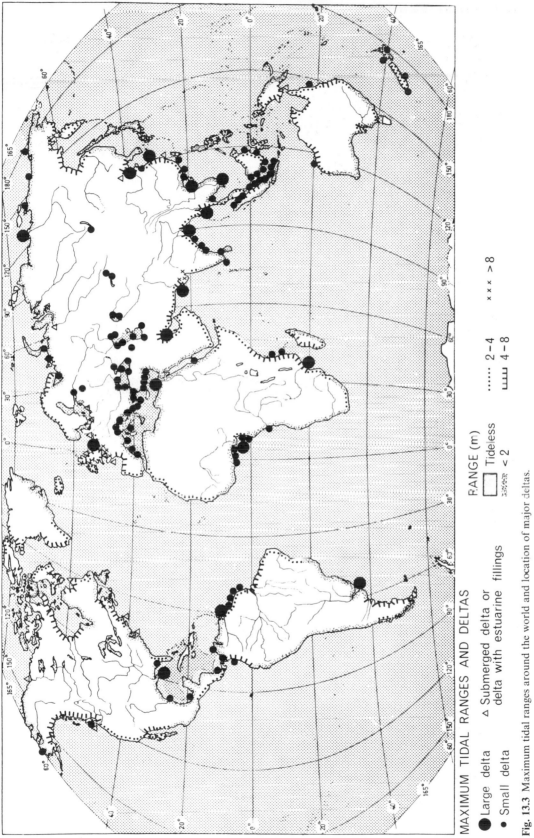

MAXIMUM TIDAL RANGES AND DELTAS

● Large delta   △ Submerged delta or

● Small delta   delta with estuarine fillings

RANGE (m)

☐ Tideless

░░ < 2

⋯⋯ 2 – 4   × × × > 8

⊥⊥⊥ 4 – 8

Fig. 13.3 Maximum tidal ranges around the world and location of major deltas.

13.1 Regular swells arriving at the coast are transformed from sinusoidal forms into breaking waves, separated by broad shallow troughs, with progressively closer spacing as the leading waves slow down in the shallowing water.

waves 7 m high with eight times the energy of trade wind swells. Swells generated in the circum-Antarctic Southern Ocean develop great height and energy and travel up the Atlantic to reach the shores of western Europe, and up the Pacific to reach Alaska.

Gentle puffs of wind brushing across water can be observed to ruffle its surface, but regular ripples are not produced until waves have a velocity greater than about 1.1 m/s. As wind velocity increases so waves can increase in height. The *height* (vertical difference between the crest and the trough) and the *period* (time taken for successive crests to pass a fixed point) of the waves are governed by the wind velocity, the duration (or time that it blows), and the fetch. The fetch controls the time during which individual waves can move under the controlling wind and hence the time available for the transfer of energy.

In a pond waves may have a period of 2–3 seconds, but ocean waves of 10–15 seconds and more rarely up to 20 seconds. Wave heights off the

coast may reach 15 m during storms. The highest wave reliably measured on the open ocean was seen from the USS *Ramapo* when the watch officer observed a sequence of waves 24, 27, 30, and 33 m high: he lined up the wave crests with the crow's nest and the horizon and used trigonometry to calculate the wave heights.

The relationships between wave height (H in metres), fetch (F in km), wind velocity (U in km/h), wavelength (L in metres), and period (T in seconds) are empirically derived from observations. High winds will generate waves of a height related to fetch by

$$H = 0.36\sqrt{F}$$

and H is generally proportional to the square of wind velocity

$$H = 0.0024U^2.$$

These relationships can be used to plot a forecasting set of curves (Fig. 13.4).

An observer standing on a beach can see that, at

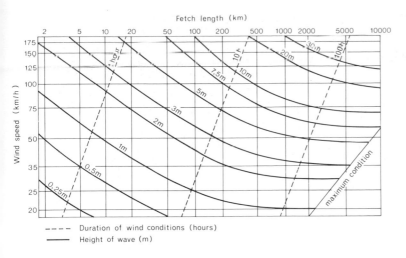

Fetch length (km)

Fig. 13.4 Deep-water wave forecasting curves relating storm conditions to the generated significant wave height and period. To use the graph start at the left side and follow a wind speed line to its intersection with the fetch length line or the duration line, whichever comes first. This is governed by whether the wave generation is limited by duration or fetch (greatly simplified and transformed from an original by Bretschneider, 1959).

- - - - Duration of wind conditions (hours)
——— Height of wave (m)

a particular time and place, most of the waves have about the same height, but that there is a range of heights over a brief interval. This is because the wave record is formed by combinations of waves from several different sources, with local waves and long-distance swells. Where two waves combine they form a higher wave in which

$$H = \sqrt{(h_1^2 + h_2^2)}$$

where h_1 and h_2 are the heights of the original waves.

Patterns of waves can be examined statistically and average and maximum wave heights determined. The typical value of wave height is given by the square root of the average of the squared values

$$H_{typ} = \sqrt{[(h_1^2 + h_2^2 + h_3^2 \ldots h_n^2)/n]}.$$

Much of the work done by waves along a beach is accomplished by those waves that, through their size (or magnitude) and frequency of occurrence, generate the most energy over a period of time. These waves are normally taken to be the mean height of the highest one-third of the waves in a particular record and approximately

$$H_{1/3} = 1.5 H_{typ}.$$

As waves move into shallow water and break the wave period (T) does not change, but the wavelength progressively decreases as leading waves decrease in celerity:

$$L = T\sqrt{(gd)}$$

where d is the water depth. The water particles do not advance with the wave, but move roughly in a circular path with a diameter equal to the wave height. Some movement of the water extends to a depth about as great as the wavelength but the wave form is first distorted from its sine curve form by the bottom when the depth is equal to half the wavelength ($d = \frac{1}{2}L$) (see Fig. 13.5). As the water becomes shallower the movement of water near the bottom increases and the wave form is increasingly altered.

Where the wave first causes water near the bed to move, it will be able to move only mud which is already in suspension, but as water shallows first sand and then shingle can be moved. Shingle is only transported where the depth is no more than the wave height. Consequently coarse sediment can only be moved close to the shoreline. Few waves stir the bottom at depths greater than about 10–15 m.

As a single wave advances into water of decreasing depth, the wave height increases until the wave front is too steep for stability and it breaks into *surf* in which the crest is thrown forward and disintegrated into foam. There is no simple rule defining the depth at which a wave becomes so steep that it will break. The depth is approximately equal to the wave height, but it also varies with wavelength and beach slope. On a very gently sloping beach, the ratio of breaking wave height to depth of water (H_b/d_b) is 0.78; on steeper beaches it is close to 1.0 or even above 1.2.

Three types of *breaker* are commonly recognized: *spilling*, *plunging*, and *surging* (Fig. 13.6). In general, spilling breakers tend to occur on beaches

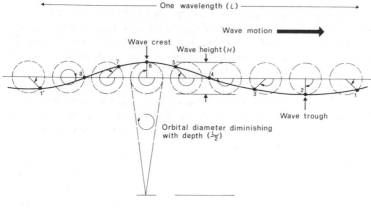

One wavelength (*L*)

Wave motion

Wave crest

Wave height (*H*)

Wave trough

Orbital diameter diminishing with depth ($\frac{L}{2}$)

Fig. 13.5 (a) The motion of surface water particles during the passage of an idealized sinusoidal wave. The orbital diameter decreases with depth beneath the wave and is eliminated at a depth which is about one-half of the wavelength.

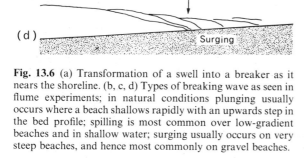

Deep water swell   Zone of transition   Breaking

(a)

(b)   Spilling

(c)   Plunging

(d)   Surging

Fig. 13.6 (a) Transformation of a swell into a breaker as it nears the shoreline. (b, c, d) Types of breaking wave as seen in flume experiments; in natural conditions plunging usually occurs where a beach shallows rapidly with an upwards step in the bed profile; spilling is most common over low-gradient beaches and in shallow water; surging usually occurs on very steep beaches, and hence most commonly on gravel beaches.

of very low slope with steep waves; plunging waves develop on steeper beaches above a steeply shoaling bottom; surging occurs on very steep beaches usually of gravel, with waves of low steepness.

Once a wave breaks the energy is rapidly

dissipated as the turbulent *swash* runs up the beach carrying with it some of the beach sediment. The water runs back as *backwash* until it is overtaken by the swash of the next wave. As some of the swash water will percolate into porous beach materials the backwash volume is less than that of the swash. For waves of a particular height and steepness, the work done by the swash and backwash adjusts the beach slope until it is in equilibrium with the forces acting on it. Consequently alternating periods of storms and calms cause the beach slope to change repeatedly.

The energy (*E*) brought by waves to a beach is a function of wave height (*H*) and water density (ρ for sea water is 1025 kg/m³):

$$E = {}^{1}/_{8}\rho g H^{2}.$$

E is expressed in units of J/m² per unit wave crest length. As energy varies primarily with the square of wave height, high waves bring very large levels of energy to the beach compared with low waves.

Swell can be reflected from a seawall or cliff with almost no transfer of energy because the mass of water is not moving forward, but a breaking wave does throw its water forward. The greatest pressures are exerted by plunging waves that trap air between the wave and a cliff so that the air is compressed. The compressed air can exert pressures of over 600 kPa against a seawall for 1/100 second and move large blocks of rock. The wave energy against the most exposed face of Bikini Atoll has been calculated as reaching 20 kJ/s along every metre of reef front.

Wave refraction occurs as waves enter shallow water because their celerity declines with decreasing water depth, $C = \sqrt{(gd)}$. All waves bend towards the zone in which they are travelling slowest. That part of an ocean wave, approaching a coast, that reaches shallower water first will go slower than that part still in deep water. The section of a wave in deeper water will catch up with the part in shallower water until the whole wave is nearly parallel to the bottom contours. For straight coasts with parallel bottom contours the wave crests therefore become nearly parallel to the shoreline. Wave refraction may cause either a spreading out or convergence of the wave energy. This can be seen most readily by reference to Fig. 13.7. In the zone between the beach and the headland the orthogonals (lines drawn at right angles to the wave crests) A and B diverge near to

the shoreline and so the energy available along a metre length of the crest is least at the beach. By contrast orthogonals B and C, and C and D converge towards the coast so that the energy available along the crest at 3 has become concentrated along 3′. The waves refract towards the headland because of the shallow offshore zone. The wave energy is, therefore, concentrated on the headland and wave heights there may be several times as large as in the middle of the beach where energy is lower. The same effect may occur on a straight beach where a deep water channel runs into the land: over this waves will diverge.

Even after refraction few waves are exactly parallel to a beach, but approach it at an angle which causes the swash to move sediment diagonally up the beach in the direction of the wave approach (Fig. 13.8). The backwash will run more directly down the beach so that the sediment undergoes net transport along the beach. The longshore movement of waves and the movement of sediment along the beach together create a *longshore* or *littoral drift*. On most coasts there are dominant waves from one direction and sediment is transported predominantly in that direction.

Currents

Ocean currents are drifts of water responding to prevailing wind patterns and to variations in water temperature and salinity. True ocean currents have little significance for coastal processes and landforms, except where they influence the distribution of coastal ice, coral reefs, or mangrove swamps.

Tidal currents, produced by the ebb and flow of tides, repeatedly and regularly reverse in direction. In the open ocean their velocities rarely exceed 1 m/s, but where the flow is channelled through narrow straits, or entrances to estuaries and lagoons, they may exceed 4 m/s locally. The velocity of currents in these confined channels is related not only to the capacity of the channel but also to the tidal range, being greatest in areas of large range (sometimes called *macrotidal* areas, as opposed to *mesotidal* and *microtidal* areas with medium and low tidal ranges). In the Raz Blanchart, between Alderney and Cap de la Hague in NW France, and around Melville Island in northern Australia, very fast tidal currents occur.

Along coasts of certain areas like East Anglia local tidal currents may reverse along the shore

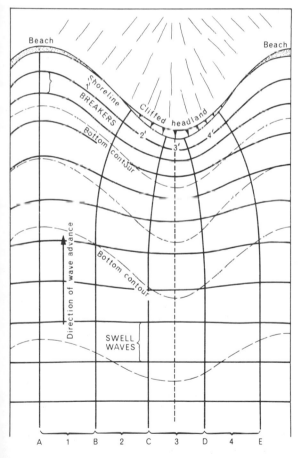

Fig. 13.7 Wave refraction over an irregular shoaling bottom. Orthogonals drawn at right angles through the swell crests indicate divergence and convergence (modified from Bloom, 1978).

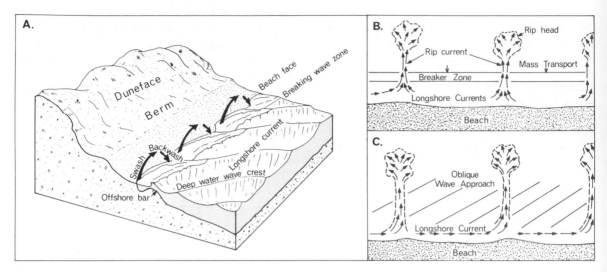

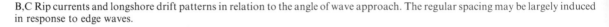

Fig. 13.8 A. Refracted waves breaking on a beach set up a longshore current and induce littoral drift of beach sediment (after Healy, in Soons and Selby, 1982).

B,C Rip currents and longshore drift patterns in relation to the angle of wave approach. The regular spacing may be largely induced in response to edge waves.

during the few hours preceding and after high tide, but such local effects are usually minor compared with the littoral drift induced by wind-driven waves. Wind-driven masses of surface water can increase or decrease the tidal range, locally and temporarily, as they are piled up by onshore winds or driven away from land by offshore winds.

In the nearshore zone water carried onshore by waves is often carried away again as *rip currents* (Fig. 13.8B, C.). Rip currents may be regularly or irregularly spaced. On some beaches rips develop in zones where breaker height is greatest and are fed by water from the highest breakers, but their origin and mechanism is still not well understood.

The power of currents to erode the coast is very limited and is usually confined to the transport of sediment which has already been placed in suspension by wave action or by a river. The sea floor, however, can be eroded and silts and sands set in motion. Many hollows, shallow troughs, and large dunes in shallow seas have been attributed to current action, but few convincing data are available.

Storm surges and tsunamis

Storms have the effect not only of generating waves but also of increasing the height of the sea surface. A drop in atmospheric pressure of 100 Pa (1 mb) causes the sea surface to rise by 1 cm: as hurricanes and typhoons frequently have atmospheric pressure in their centres 100 mb below normal, the sea surface may rise 1 m from this effect alone. Of greater importance is the pile up of water along the coast by onshore winds which can raise local sea-level several metres. The combined result is a *storm surge*. In surges, storm waves can travel inland far beyond the normal shoreline because of the deepened water offshore. During the 1953 storms on the Lincolnshire coast, England, floods of sea water swept more than 6 km inland. Along the Gulf of Mexico coasts of USA, hurricane-induced surges have raised local sea-level by 5–8 m. In 1970 a storm in the Bay of Bengal produced a surge that drowned 700 000 people in Bangladesh.

A *tsunami* is a large sea wave which may be generated by any of several mechanisms. Most are formed as a result of a large earthquake (magnitude > 5.6) in which either there is an upward movement of part of the sea floor, or a submarine landslide, displacing a large volume of water in a few seconds. Most tsunamis are created in shallow earthquakes around the rim of the Pacific Ocean. Other mechanisms include: submarine volcanic explosions; air blast from a large lateral volcanic eruption, as at Krakatau in 1883; explosion of an ignimbrite flow which has been rapidly rafted across the sea upon steam (such a flow must have a volume of several cubic kilometres); very large landslides off coastal cliffs in enclosed bays, such as fjords; and calving of very large icebergs from glaciers in fjords.

The displacement of water creates a single wave—or rarely two or three waves—which travels at about 800 km/h in the open ocean. A tsunami 'feels' the bottom all the way across the Pacific and loses energy by friction with sea mounts, ridges, and islands, and by refraction around them. It travels fastest along trenches. Its arrival time at a point around the Pacific can be predicted from a knowledge of the source and the ocean-floor topography.

The coasts which are most severely affected are those nearest to the source, and those with steep continental slopes and narrow continental shelves which cause the water to pile up into high waves. The Krakatau tsunami had a maximum height of 30 m. Even a small tsunami may cause much damage on coasts with funnel-shaped inlets and estuaries which it enters as a bore. A large tsunami may have an energy of 1 GJ.

Tsunamis are recorded about once in 5 to 10 years in Hawaii. An earthquake off Chile, in 1960, generated a tsunami which was damaging along the coast of Japan, especially in areas which are parallel to the line of the Chilean coast, and which was observed around much of the rim of the Pacific (Fig. 13.9).

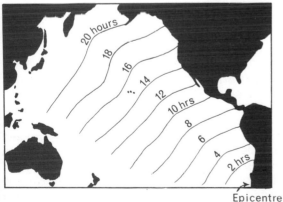

Fig. 13.9 Progression of a tsunami across the Pacific Ocean after being generated by an earthquake off Chile.

Long-term sea-level change

World-wide changes in average sea-level are described as *eustatic* sea-levels to distinguish them from local influences, such as apparent sea-levels which result from tectonic uplift (Plate 13.2) or subsidence (Plate 13.3). Eustatic sea-level changes result from the two main causes: (1) changes in the volume of the ocean basins; and (2) changes in the volume of sea water.

Ocean-basin volumes

A study of plate tectonic theory (Chapter 3) shows that the ocean basins are not constant in shape or volume over geological time. The basin volumes can be affected by: (1) sediment deposition; (2) trench formation; (3) subsidence of the oceanic crust; (4) uplift of mid-ocean ridges; and (5) an increase or decrease in the length of the ridge systems. Factors (1), (2), and (3) are virtually balanced by subduction and can be ignored. It is variations in ridge volume which are important.

It has been calculated that, in the late Cretaceous, rapid ridge spreading would have been accompanied by a sea-level rise of over 300 m. Such a rise almost certainly accounts for the widespread flooding of the continents to form shallow seas in which, for example, the chalk of much of Western Europe and the sedimentary rocks of much of the Sahara were formed. A gradual fall in sea-level is evident through much of the Tertiary as spreading rates from ocean ridges decreased and the height of the ridges declined. The falling Tertiary sea-levels have certainly left their mark on some landscapes where extensive old shorelines and erosion surfaces were left at high elevations (see Chapter 17, and Fig. 4.2).

Volume of sea water

The geological evidence suggests that the mass of water at the surface of Earth has not changed significantly since Precambrian times. The volume of sea water does vary both in the short term of a few years, or even during a year, and to a much greater extent over the tens of thousands of years associated with advances and retreats of ice sheets.

The short-term changes which are effective during a year include volume changes associated with variations in the density of sea water resulting from inflows of fresh river water and in changes in

13.2 Raised beaches of a tectonically emerging coastline. The beach ridges in the foreground were raised during earthquakes which occurred in AD 1855 (the lowest raised beach) and AD 1460. The next higher beaches were probably raised about 3100 BP (this beach was raised 8.2 m in the event) and about 4900 BP (raised 2.7 m). The more steeply tilted terrace on the headland is about 150 m above sea-level and is thought to have formed in the Last Interglacial (125 000 BP). The higher terrace behind it is at 200 m and may have been cut in an earlier interglacial about 250 000 BP. The level on the higher spur may be a remnant of an even older terrace. Cape Turakirae, near Wellington, New Zealand (photo by NZ Geological Survey).

water temperature. It is common for seasonal variations of 5–6 cm to be recorded on tide gauge records from these effects. Local build-ups of water by onshore winds, local low atmospheric pressures, and high tides also cause local sea-level to rise, but the opposite conditions cause it to fall. None of these effects is cumulative but all are part of variation about a mean sea-level.

Another detectable short-term change is that attributable to variation in glacier volumes over periods of a few tens of years. During the 'Little Ice Age' of the seventeenth and eighteenth centuries glaciers advanced, and there is some evidence that

sea-levels fell a few centimetres. At present many glaciers are retreating and sea-level around many coasts appears to be progressively rising by 1 to 2 mm/year.

The effects of such slow changes may be severe. It has been postulated that, if a beach and nearshore sea bed profile can adjust to storm events and periods of low wave energy, as sea-level rises sediments are eroded from the beach and deposited on the nearshore sea floor; in turn the nearshore floor is elevated in direct proportion to the elevation increase in sea-level, and the volume of sediment eroded from the beach is equal to the

13.3 The Marlborough Sounds, New Zealand. The Sounds are old river valleys drowned by tectonic downwarping and postglacial sea-level rise. Note the absence of cliffs (photo and copyright: Whites Aviation).

volume of sediment deposited in the nearshore. The horizontal distance of shoreline retreat (X) is inversely proportional to the angle $(\beta)°$ of the nearshore slope seaward from the breaking waves, and directly proportional to the rise in water-level (a), thus:

$$X = a/\tan \beta.$$

Around the coast of northern New Zealand, for example, tide gauges indicate that mean sea-level is rising by between 0.17 m/100 years and 0.35 m/100 years. The nearshore slope angle seaward of the breaker zone of some beaches is about 1°; shoreline recession from this cause alone is thus likely to be between 10 and 20 m in the next century, if sea-level trends continue.

By far the most important control on sea water volume during the Late Cenozoic is the amount of water which is locked up in polar ice sheets. Ice shelves do not affect this situation because the ice is floating and displacing its own mass of water. If all of the ice in Antarctica were to melt, world sea-levels would rise between about 60 and 75 m, and the Greenland ice cap would add about 5 m. The added load of water on the oceanic crust, however, would cause some mantle rock to flow towards the continents as isostatic compensation, causing the ocean floor to sink and the land to rise, and so would reduce the total rise of sea-level from ice melting to about 40–50 m. The figures are rather uncertain because neither the volume of ice nor the extent of isostatic adjustment is known precisely.

The Antarctic Ice Sheet developed during the

middle and late Tertiary (see Chapter 16) and added its effect to that of ocean-basin volume changes. About 3 to 4 million years ago extensive ice sheets developed for the first time on the northern hemisphere continents and since then ice volume and ice extent have fluctuated widely. At its most extreme, sea-level may have fallen 200 m below that of the present, and risen 5–8 m above present levels during the particularly warm peak of the Last Interglacial at about125 000 BP.

If ice volume alone controlled sea-levels uniformly around the coasts, then the sea-level evidence would uniformly follow a pattern imposed on it by climate. The sea-level curve for the last 15 000 years would show a progressive rise from about 100–130 m below present levels to about the present level 6000 to 4000 years ago. Some published sea-level curves have been calculated from probable ice volumes and mean temperature changes and have the form of the melting curve shown in Fig. 13.10d (compare with Fig. 16.30). This curve shows an initial steep rise of about 8 mm/year until about 7000 BP and then a slowing down to 1.4 mm/year until 6000 BP. A few sharp rises and falls in this progression can be related to minor readvances of the ice. Considerable controversy continues about whether sea-level has risen above that of the present in postglacial time.

The climatically derived ice-melting curve is verified in general form by the discovery of datable shallow-water molluscs, salt marsh peat, coral, and lagoon organic deposits in boreholes through estuarine and lagoon sediments. These materials reveal the altitude of sea-level for the time at which they formed.

The concept of a eustatic sea-level change is complicated by isostatic and tectonic effects. As ice builds up on land an ice sheet develops a dome-like profile and very large ice sheets, like the Laurentide Ice Sheet which covered much of Canada, achieve thickness of 3000–4000 m. Ice has a density of about one-third that of rock and can depress the continent beneath the ice dome by up to 1000 m, and at its periphery by 100–300 m. This depression is recoverable when the ice melts. The ice load effect extends from the edge of the ice for a few hundred kilometres, and beyond that the flow of mantle rock from beneath the ice-loaded crust causes an upward bulge, or forebulge, of a few metres to a few tens of metres (Fig. 13.10a).

When the ice sheet melts, water will be returned to the ocean and the formerly depressed continent will isostatically recover and rise. The initial rate of isostatic uplift is rapid with changes of between 3 and 10 m per 100 years. The best evidence for uplift comes from old shorelines of coastal plains which are being uplifted (Plate 13.4). These leave beach ridges which often contain mollusc shells, driftwood, fish, whalebone, and peat remains which can be dated by radiocarbon. As the ice load is removed the rate of uplift declines, but will continue for several thousand years after all of the ice has melted because viscous mantle rock, and crustal rock of low elasticity, cannot respond as quickly as the melting occurs. In northern Canada the maximum isostatic uplift in the period of melting from 14 000 years ago to 4000 years ago was about 400 m in areas which were below the thickest ice.

From detailed studies carried out in Scandinavia some data on magnitudes and rates of isostatic adjustment have emerged. The centre of uplift started to rise about 13 000 BP and the total absolute uplift is 830 m. During this uplift, upper mantle material flowed from below the forebulge (at depths of 50–200 km) and caused it to sink 170 m. The highest rates of uplift reached 50 mm/year at about 10 000 BP and the rate has since declined. It is believed by some workers that the glacial isostatic factor died out some 2000–3000 years ago, and that the uplift which still continues is of tectonic origin. Present uplift, however, still follows what would be expected from glacio-isostatic controls, with rates of emergence of up to 10 mm/year in the north of the Gulf of Bothnia, declining away from this area and producing a submergence of 1–2 mm/year in Denmark (Fig. 13.10c). Note that the terms emergence and submergence are used to distinguish rates with respect to modern sea-level.

Rising sea-level, from melting, imposed an added load on the ocean floors and continental shelves (*hydro-isostasy*) which caused variable effects depending upon the strength of the crust. It is probable that mid-ocean islands, far from the ice sheets, suffered least depression by water loading and most clearly record the sea-level rise due to meltwater (*glacio-isostasy*).

The pattern of sea-level and isostatic adjustments can be described most readily by reference to Fig. 13.10d. Sites near ice sheets have old beach levels at high elevations and younger beaches at lower elevations near the present coastline (e.g.

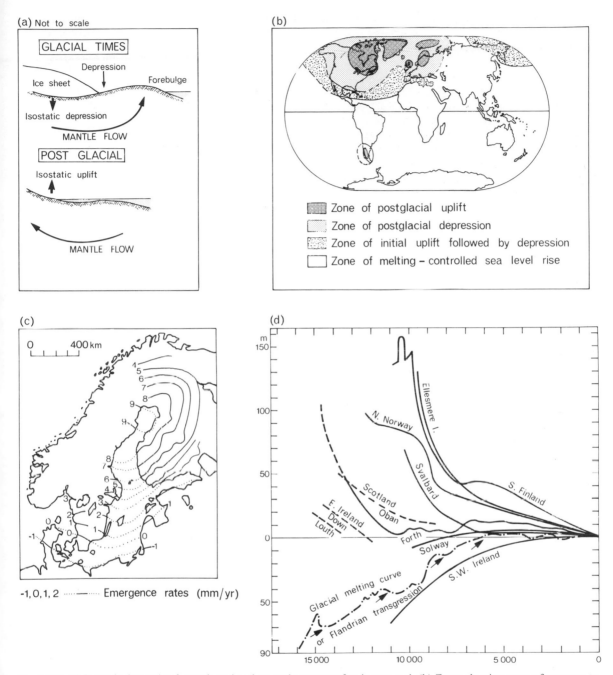

Fig. 13.10 (a) Isostatic depression beneath an ice sheet and recovery after its removal. (b) Zones showing types of response to isostatic and melting effects to produce various degrees of emergence and submergence of the land with respect to sea-level (after Walcott, 1972). (c) Modern emergence rates, in the Baltic region, with respect to sea-level (after Flint, 1971). (d) Altitude of late glacial and postglacial shorelines at various localities with respect to modern sea-level. It can be seen that the notion of a world-wide Holocene, or Flandrian, transgression from glacial melting applies only to areas not greatly affected by isostatic adjustments (modified from Synge, 1977).

13.4 A sequence of raised beaches on the frontal slope of a glacial outwash delta at the north end of Haukdal on Varangerfjord, north Norway. At O is the highest marine limit, marked by a notch at the base of a cliff. The ridges marked by dashed lines are upper and lower (e) shingle bars formed on an outwash terrace surface; they formed sometime between 12 000 and 10 000 BP (photo by F. M. Synge).

Ellesmere Island and Svalbard). In areas immediately around the thickest parts of ice domes the land was relatively depressed, and the contrary effects of isostatic uplift and sea-level rise from melting are sometimes almost in balance, so local sea-level appears to have been nearly constant over the last 10 000 years (e.g. Solway Firth). In areas outside this zone the land was first uplifted quickly, but melting-controlled sea-level rise caught up with it, until about 7000 years ago when the melting effect was again exceeded by isostasy (e.g. Firth of Forth). In areas well beyond the isostatic effect (like the Pacific Islands, New Zealand, and Australia) the sea-level curve from melting alone describes the record fairly closely unless local tectonism is predominant.

In many parts of the world clearly defined old shorelines are found at depth on the continental shelf, and old coastal deposits occur in the sedimentary accumulations of basins like the North Sea and deltas, like those of the Rhine and Mississippi rivers, which are sinking at rates of 1–10 mm/year under the weight of their sediment.

Where tectonic uplift is predominant spectacular flights of coastal terraces occur. Those which are composed of coral have great importance because they can be dated by uranium–thorium methods and each terrace is thought to represent a period at which sea-level stood still. Hence the uplifted terrace sequences in New Guinea, Bermuda, and Barbados have been used to determine periods of both high and low still-stands, even though all terraces are now uplifted. The pattern of still-stands over the last 400 000 years has been related to the pattern of warm and cold periods during the same time (see Fig. 16.21).

The origin of many flights of coastal terraces is obvious, for the lower, younger, benches have slopes characteristic of beaches, and covers of beach sands and gravels, and they are backed by old sea cliffs which may contain caves. The older, higher benches may be less clear after erosion has dissected them, colluvium from cliffs behind them has obscured their form and beach deposits, and tectonic tilting has altered their slope.

During glacial times rivers extended their courses about half-way across the continental shelves which are, in consequence, partly drowned coastal plains. Many of these rivers were entrenched in gorges. As sea-level rose coastal sites occupied by early man were inundated and the shorelines were cut back at rates of up to 10 m/year in areas where the plains were 100–150 km wide. The land links between Britain and the continent of Europe, Britain and Ireland, Australia and New Guinea, Japan and China, Alaska and Siberia, and the two main islands of New Zealand, were all inundated. In the Persian Gulf region there was a shoreline displacement of around 500 km in only 4000–5000 years, a rate of 100–120 m/year. Such rates are easily recognizable in the life of an individual and may be the source of such legends as Noah's Flood. If coastal erosion were still occurring at such a rate there would be a cry for governments and engineers to build huge protective works!

The postglacial rise of sea-level is known as the Flandrian Transgression. Much of the debate about the effects of the rise has been centred on events of the last 6000 years. Deposits like the Carse Clay of the raised mudflats of central Scotland, and the layered marine, brackish, and freshwater sediments of the Dutch coast, the Somerset Levels and Fens of East Anglia, indicate a fluctuating sea-level rising to a maximum of a few metres above that of the present. Coasts elsewhere show various patterns of emergence and submergence depending upon local isostatic effects.

In some localities coastlines have changed so rapidly during the Flandrian Transgression that sand supply to beaches has been insufficient to maintain them. The lower reaches of many rivers have suffered sedimentation with the formation of estuaries, deltas, and lagoons built of muds, sands, and peats (Plate 13.5).

It will be evident from this discussion that coastal terraces, raised beaches, and other coastal features of a particular age are likely to occur at various altitudes above or below modern sea-level. The 12 000 BP strandline, for example, is now around +200 m near Montreal, +50 m in Maine, at −30 m near New York, and at −25 m around the Delaware and Chesapeake Bay regions. Most coastlines have been tilted and many isostatic rebound adjustments have been achieved along fault lines, rather than by simple flexure of the lithosphere. It is, consequently, unwise to name a feature by its altitude: the '8 m terrace' at one site may have a different age and origin from an apparently similar terrace elsewhere, and the altitude of the one feature may vary in short distances along the coast (Fig. 13.11). A general classification of influences on coastline altitudes is shown in Fig. 13.10b.

Depositional landforms of the coast

The general word *coast* applies to the geographical region close to the margin between land and sea;

13.5 Lower reaches of a valley drowned by the Flandrian Transgression and now being silted up and colonized by mangroves. Puhoi River, New Zealand.

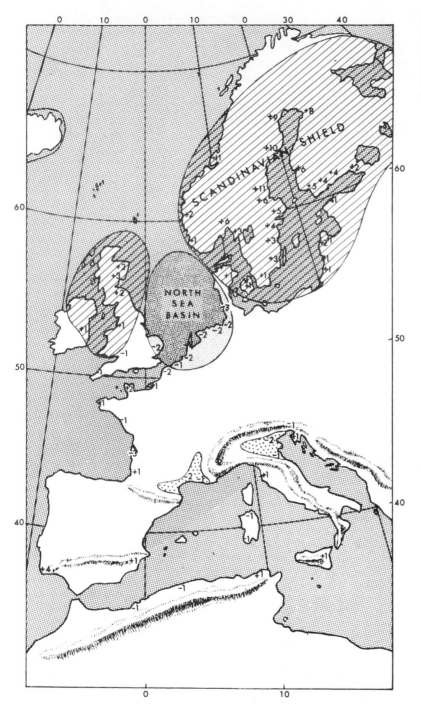

Fig. 13.11 Modern emergence and submergence around the coasts of Western Europe (mm/year) as a result of ice melting, glacio-isostasy, tectonism, and basin sedimentation (after Fairbridge, 1960 © Scientific American Inc.

the *shoreline* marks the boundary between them. The shoreline fluctuates in position with the tides and waves so that the word *shore* is commonly used for the zone affected by wave action. The shore may be subdivided into: the *offshore* seaward of the low tide breaker line; the *nearshore* between the low-tide breaker line and the high-tide shoreline; the *foreshore* between high- and low-tide shorelines, and the *backshore* between the highest high-tide line with storm effects to the base of the permanent land (cliff or vegetated dune) (Fig. 13.12).

Beaches

Beaches are accumulations of sediment deposited by waves and longshore currents in the shore zone. They are typically composed of sand or shingle. Most beach sediment is well sorted with the size range at one site being very limited so that sand beaches, for example, contain little silt or gravel and the sand mean grain size is coarser than the median size. This is in contrast to fluvial, eolian, and lagoonal deposits which usually have a median size which is coarser than the mean size. The well-sorted character is a result of repeated winnowing as material is transported by waves.

There are four main sources for beach sediment: (1) local cliffs and promontories which are weathered and eroded to release grains of a size which is characteristic of the source rock and the extent of comminution occurring during transport; (2) the offshore zone from where sand may be derived from depths of 10–20 m by modern wave action, or may have been pushed landwards as sea-level rose during the Flandrian Transgression; (3) calcareous sand and shell fragments which may provide beach material where coral or mollusc debris is

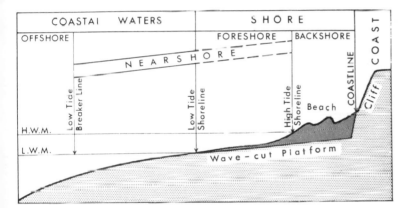

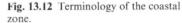

Fig. 13.12 Terminology of the coastal zone.

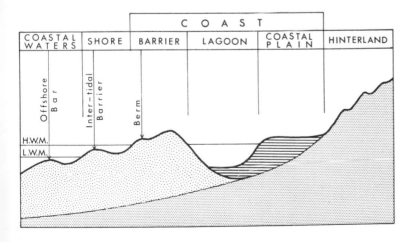

13.6 The Northland peninsula, New Zealand. Note the long straight beaches of the west coast (left) where there are few rocky headlands but extensive Quaternary barriers with dune systems forming the hinterland. By comparison the east coast has a deeply crenulated coast with rocky headlands and curved bayhead barrier beaches formed by refracted swell waves from the Pacific (satellite photo by NASA).

abundant; and (4) local river bedload material (on coasts with littoral drift this is usually the major source). An offshore source is sometimes the only possible one where a coast lacks local rivers and bedrock exposures (Plate 13.6). The colour and mineralogy of beach material is obviously dependent upon the source, with beaches like those of Hawaii and parts of North Island, New Zealand, being composed of volcanically derived sand and

therefore being black or grey as a result of the ferromagnesian mineral grains, or yellow to white where quartz, coral, or shell debris forms the beach. Pebble and boulder beaches may be formed either because local rivers carry large volumes of course bedload to the coast, or because, locally, large fragments are yielded from coarse sedimentary or highly fractured rocks. Areas where glaciers have reached the coast in the recent past commonly have pebble and cobble beaches of reworked glacial debris.

The size of beach sediment largely controls the slope of an equilibrium beach. Large particles tend to give rise to steep beach gradients because of their great permeability and the reduced backwash is incapable of moving the beach particles. Fine sand beaches have gentle slopes because of the more effective backwash. Mud beaches are rare and are usually only found in enclosed lagoons and embayments where wave heights are low: on the open coast wave action would take muds into suspension, bury them with sand carried in the littoral drift, or carry them offshore.

A *beach budget* (Fig. 13.13) can be calculated from a knowledge of the sources of sediment and from a knowledge of losses. To do this beach and offshore profiles to depths of about 20 m have to be surveyed regularly and the movement of sediment traced. Tracing is made possible by using grains labelled with fluorescent dye or radioactive materials. A budget is very important because not only are beaches recreational resources but also they protect the hinterland from wave attack. There have been many examples of human interference with one or more sources of coastal sediment with disastrous consequences. On the south coast of England, for example, a breakwater was built to protect Newhaven Harbour and this intercepted the eastward drift of shingle so that the beach eastwards of Newhaven was eroded with severe depletion at Seaford. At Bournemouth the progressive extension, since 1900, of a promenade on the sea front has halted cliff erosion and cut off the beach from its sand supply; without replenishment of sand lost by littoral drift the beach is diminishing. In California and parts of Australia attempts are now being made to pipe sand around man-made obstructions so that littoral drift can continue and maintain the coast, and sediment is being pumped from the offshore zone to replenish beaches at Bournemouth and many other localities.

Beach profiles are related not only to the size of their particles but also to the wave conditions. In calm weather spilling breakers tend to push sand or shingle up a beach and form a berm parallel to the shoreline. In rough weather higher and steeper plunging breakers with a more effective backwash comb down a beach. In addition heavy rainfall, and higher local sea-levels produced by low atmospheric pressures and storm surges with onshore winds, raise the water tables in beaches. The sands are consequently saturated and lose strength and they are rapidly moved by waves into an offshore bar. Many sandy beaches consequently have a storm, or storm season, profile contrasting with a calm, or calm season, profile as a result of alternating cut and fill processes. Such beaches respond rapidly to predominant wave energies and maintain an approximate equilibrium form (Fig. 13.14a).

The offshore bars act as constantly changing reservoirs from which, and into which, littoral drift and beach material is transported. The equilibrium beach concept is also thought to apply over long periods so that, along coasts experiencing long-term submergence, the foredunes behind the beach are eroded and the sediment is incorporated into the beach and nearshore zone, so water depth remains nearly constant as long as the sediment supply is maintained. Dredging an offshore bar for building sand, or to maintain a harbour navigation channel, may lead to severe beach erosion.

Failure to recognize that frontal dunes are part of the active beach sediment system has often caused property losses where buildings have been placed on what appear to be stable, vegetated dunes which are destroyed in later severe storms. The current rise of sea-level of about 1 mm/year suggests that erosion of frontal dunes is a probable occurrence on many coasts in the near future.

Where berms survive storms, and there is an excess of sediment being supplied to the beach, a new berm may be built in front of the existing one in a period of calm weather. In this way a coast may *prograde*, or build out towards the sea. The hinterland is then formed of successive berm and beach ridges. These may be relatively low features if the beach material is shingle or shell material, but if it is sand onshore winds may form dunes which bury the beach ridges as they are formed (Plate 13.7a,b). Shingle berms may be built up by storms which would be erosive on sand beaches.

Beach plan forms tend to conform closely with the shape of the refracted wave front. On coasts with alternating headlands and beaches, and a dominant swell which approaches obliquely, sediment is drifted preferentially in one direction to form a curve which is known variously as *half-heart*, *zetaform*, or *log-spiral form* (Fig. 13.14b,c). Where the beach faces the incoming swells directly, material is moved from the ends towards the middle so that the beach becomes straighter and more nearly parallel to the dominant swells. Each beach is a nearly closed system of sand supply and movement, unless sand can pass round a headland to the next beach.

Many beaches are cusped, especially at high tide (Plate 13.8). The cusp is a scoop-shaped hollow in the beach front and is maintained by variations in water depth in which waves diverge on either side of spurs, and the backwash is concentrated in the hollow. Cusps are usually found in association with the almost circular set of local currents in which there is a mass transport of water towards the shore between rip currents which carry much of the water seaward again (Fig. 13.15). Cusps

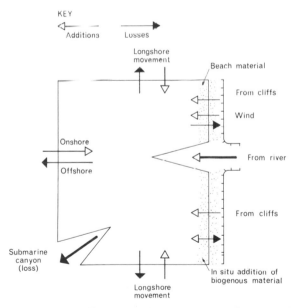

Fig. 13.13 Identification of the components of a coastal sediment budget. On some coasts there is also a loss of the sediment deposited in estuaries and lagoons; on barrier coasts this may be the major loss (based on an original figure by R. M. Kirk).

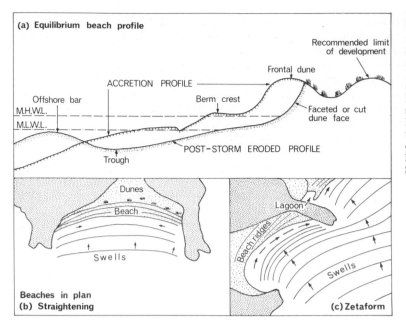

Fig. 13.14 (a) Sections through one beach for a period following accretion under a direct wave approach. (b) Beach straightening by accretion under a direct wave approach. (c) Development of a zetaform beach by obliquely approaching waves. The headlands act like groynes trapping sediment moved against them by the littoral drift, but depriving the beach in their lee; hence lagoons have their outlet channels in the lee of the headland (a, after Healy, in Soons and Selby, 1982).

cannot, however, be explained wholly by rip current circulations and there is some evidence that small standing fluctuations of higher water level, called *edge waves* may occur in association with the cusps. In spite of their relatively common occurrence and great range of sizes (with long-shore lengths varying from a kilometre or more to as little as 2 m) and their spectacular regularity, there are still no entirely satisfactory theories for the origin of beach cusps.

Barriers and spits

It has been estimated that sand, and more rarely shingle, ridges of various kinds extend along 13 per cent of the world's coasts. These ridges enclose tidal mudflats or lagoons and may be close to the mainland or many kilometres offshore. The type known as a barrier island is usually elongated and extends for a few hundred metres to as much as 100 km along the coast; such islands may be several kilometres wide, but more commonly are much less, they have beaches on their seaward side and may be capped by wind-formed dunes. The eastern seaboard of USA and the Gulf of Mexico coast of that country are largely fronted by barriers and they are also common along the southern coast of the Baltic Sea and the south-east

coast of Australia, although in both the Baltic and Australia they are most commonly tied at one or both ends to a rock headland (Plates 13.9, 13.10; Fig. 13.16).

Closely related features are *spits* which usually diverge from the coast and commonly have hooks or curved ends which turn back towards the land. *Tombolos* are barriers or spits which join offshore islands to the mainland and the term *foreland* is sometimes used to distinguish triangular lowlands of sand or shingle which are attached on one side to a mainland (Fig. 13.17).

There are many theories of the origin of barriers and related features; many were proposed before modern dating methods, and before borehole data from the nearshore zone, became available. It is now widely believed that many barriers developed during the Flandrian Transgression. As the sea advanced across the formerly exposed coastal plain the waves sorted an abundance of soil, weathered rock, fluvial, dune, and beach deposits. The silts and clays were entrained and carried offshore in suspension, but sands and shingle were moved landwards and incorporated into offshore bars and barrier ridges. The rising sea-level caused most barriers in shallow water to be overtopped and the sediment washed into the lagoons and embayments on the landward side. Barriers consequently have a complex history. During the last 4000–6000 years sea-level has been fluctuating within a much more restricted range than was occurring in the earlier postglacial period. Perhaps because of the nearly stable sea-level and perhaps because the supply of sediment is now being less rapidly augmented, or maintained, many barriers are undergoing erosion. In areas of the eastern and Gulf coasts of USA the erosion is commonly associated with hurricanes and other very severe storms in which storm surges, often at times of spring high tides, contribute to waves washing across ocean beaches, destroying foredunes, and even reaching the lagoon side of some barriers. Such waves carry sand into the lagoon and cause the barrier to retreat landwards; eventually barriers may become fixed to land along much of their length, and take on the character of normal beaches. In the process of overwashing lagoon muds may come to form the shore if sand is carried inland.

Sand is lost to the barrier system by being washed into tidal channels and then carried out through the barrier into subtidal deltas or lost into deeper water offshore (see Fig. 13.16). Alternatively it may be lost in the littoral drift and carried into a river mouth or it may be incorporated into coastal dunes or lagoon sediments. This natural process of barrier decline would not be a matter of concern if barriers were not so heavily populated. But the demand for coastal recreation has encouraged settlement so that erosion now threatens property and lives. Hurricane Camille destroyed $1.4 billion of property in 1969 in south-eastern USA when 250 km/hour winds generated a storm surge that drove water more than 8 m above mean sea-level. Many barriers are retreating at annual rates of 1.5 m on the Atlantic coast, and the building of seawalls, groynes, and other defences is reducing littoral drift and causing even greater potential property losses.

Spits grow in the predominant direction of longshore sediment transport under wave influence. The hooked ends of spits are formed either by the activity of sets of waves which arrive from different directions or by wave refraction around their seaward ends (Fig. 13.18a). Old hooks frequently survive and form slightly higher ridges than the marshes and lagoons which they partly enclose (see Plate 13.10 of Scolt Head Island). Where spits are driven inland, or littoral drift constantly removes sand along the seaward face,

lagoon deposits eventually outcrop and become subject to wave erosion. Some spits prograde and form *barrier spits*.

In Europe and North America the history of barriers, spits, and forelands is often traceable from comparisons of modern and old survey maps and it is evident that severe storms produce a marked periodicity in the changing outline of barriers, and that tidal channels and entrances change position in response to storms, floods, and pulses of sediment transport, as well as to progressive accumulations and losses. Dating of deposits has shown that many barriers have formed in earlier periods of high sea-level than the Flandrian, and then been stranded as sea-level fell, only to form the inland section of modern barrier systems as sea-level reached modern levels again. This type of record is particularly clear in New South Wales.

Lagoons, mudflats, marshes, and *mangrove swamps* are sites of mud and organic deposition. Clay- and silt-size particles are flocculated into loose, large aggregates in sea water and settle out of suspension in quiet waters as mud. Mud in the shore zone is carried into lagoons, estuaries, and embayments by tidal currents and rivers, and flocculated mud then settles at slack water of the flood tide. The tendency of mud to be carried away from the surf zone explains the contrasting environments of clean sandy beaches and muddy lagoons and marshes.

Accretion on mudflats is partly a biological process, as many burrowing animals pass the mud through their bodies and excrete it as fecal pellets which require relatively high water energies to move them. Salt-tolerant (*halophytic*) plants, ranging from initial colonizers such as algae and then grasses, sedges, and rushes, fix the mud and encourage further deposition so that mudflats tend to grow upwards to a height where they are eventually no longer submerged at every high tide. At this stage freshwater peats and shrub and woodland may form and produce a closely vegetated coastal lowland. Mudflats are usually separated by tidal channels with a dendritic pattern.

In the tropics and subtropics the most effective colonizers are mangroves of various species. Mangroves cannot encroach on mudflats where there is a large tide range and strong tidal scour, so they are most common in sheltered bays of meso- and micro-tidal areas. Along the Malacca Straits coast of Sumatra they may encourage progradation at

(a)

13.7 (a) Shingle ridges of a prograding shoreline, Marlborough, New Zealand. (b) Parabolic sand dunes, now partly stabilized by vegetation, are aligned along an axis determined by the resultant of the local winds. The wave pattern shows the interference of two wave trains: the long swell arriving from directly offshore, and locally wind-generated short-wave swell travelling parallel to the shore and refracting against it, so producing a littoral drift towards the bottom of the photo, Manawatu coast (photo by NZ Geological Survey).

rates of 50–100 m/year. As with marshes of grasses and sedges, mangrove swamps gradually emerge from the tidal zone, as a result of vertical accretion, and there is a succession of mangrove species and then coastal trees which are adapted to the various coastal environments.

Lagoons are embayments wholly or partly enclosed by depositional barriers of sand or shingle. Well-known examples include those of the Landes of France, the coast near Venice, and extensive coastal areas in south and eastern USA and the south-east coast of Australia. A much studied one is The Fleet, which is enclosed by Chesil Beach, in the south of England.

Lagoons often have three zones: a zone close to the mouths of rivers where fresh water spreads over a wedge of salt water during high tides; a salt-water tidal zone, close to the entrance through the enclosing barrier, where halophytic plants stabilize the mud; and an intermediate zone of brackish water where few plants exist and wave action and longshore drifting are active. In this intermediate zone small spits, hooks, and forelands may form and they may segment the lagoon physically into nearly independent lakes. The fresh-water zone commonly has deltas forming in it where streams deposit their load and fresh-water plants colonize the mudflats, and the salt-water zone has an entrance kept open by storm floodwaters, river runoff, and tidal scour. Entrances are often unstable as longshore drifting tends to carry sand into the entrance channel at the seaward end.

(b)

Where lagoons are used for shipping, or are surrounded by lowland which is flooded in rainstorms, it is often necessary to maintain a permanent entrance and many lagoons are kept open by dredging and the construction of groynes to control the littoral drift.

Lagoons are essentially temporary features which will be filled by sand washed over large barriers, or blown from the dunes of higher barriers, and by river deposits. They eventually become coastal plains which may contain freshwater lakes and peat swamps in surviving hollows.

Coastal dunes

Coastal dunes are formed where sand deposited on the shore dries out and is blown to the back of a beach. A large tidal range and broad intertidal sandy beaches assist in creating suitable sources for dune sand, but by far the most important control is an adequate supply of sand. On some coasts dunes are of very limited extent and height and do not penetrate inland but, in areas such as south-east Australia and the north-west coasts of North Island, New Zealand, coastal dunes are up to 100 m high and in some areas they are formed of many ridges, with the oldest ridges being farthest inland and several hundred thousand years old. Very old dunes have strongly weathered surface sands and well-developed soils and vegetation cover. Closer to the modern beaches weathering and soil formation is less. Old, weathered dune sands may also contain iron accumulations, iron pans, humus-iron pans, or zones of calcium carbonate cemented sand. As coasts are cut back by erosion such resistant beds may be exposed or stand out in the weak cliffs in former dune

13.8 Beach cusps showing the backwash.

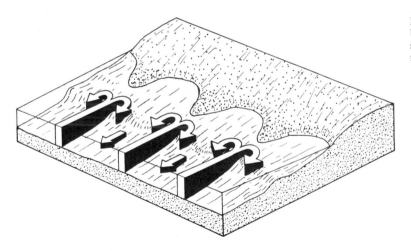

Fig. 13.15 Beach cusps with an offshore bar and the associated water motion for small cusps. The arrows represent the mass transport of the water.

deposits. *Beach rock* is a similar material found on some tropical and subtropical beaches.

Where dunes are not fixed by vegetation, or where fires, animal, human, and vehicle pressure destroy the vegetation, wind blowing may cause hollows, or *blowouts*, in the dunes and drive fresh sand inland as parabolic dunes with advancing noses and trailing arms. Traces of such dunes can be seen in Plate 13.7b.

Organic reefs

Corals and calcareous algae are the most important builders of organic reefs although molluscs,

13.9 Tauranga Harbour, New Zealand. Most of the harbour is enclosed by the barrier island of Matakana Island. In the foreground another barrier has joined the rhyolite dome of Mount Maunganui and so become a tombolo. Note the edge waves normal to the beach (photo and copyright: Whites Aviation).

sponges, and other organisms also build reefs in special situations. Coral and coralline algae reefs are widespread in tropical seas where water is less than 100 m deep, water is clear so photosynthesis can occur, a rocky foundation is present, and water temperatures are not lower than 18 °C during winter (Fig. 13.19). Fresh water and mud from rivers frequently prevents reef growth and causes breaks in coastal reefs. Coral reefs are major landforms of the tropical coasts.

A coral reef is a mass of limestone which has an upper surface controlled at the mean low-tide level. On the ocean side much coral debris is broken off by wave action and sinks to form a submarine talus. On the leeside storm debris, which has been washed over the reef, covers the sea floor and may nearly fill a lagoon so that the leeside consists of shallow tidal lagoons and intertidal reef flats. Debris piled up by storms can

be cemented by precipitated calcium carbonate and forms high-tide platforms (which are not necessarily evidence of higher Flandrian sea-levels).

Emergent reefs raised by tectonism, or those exposed by glacial low sea-levels, are subject to karst weathering and solution, so many raised reefs, and submerged ones, are riddled by caves and deep dolines which are sometimes called *grottoes*.

The most commonly recognized reef forms are the *barrier reef* which is separated from the coast by a lagoon, a *fringing reef* attached to, and extending seaward from, the land, and an *atoll* which is the annulus of reefs enclosing a central lagoon. The relationship of these forms to sinking volcanic foundations was discussed in Chapter 5 (see Fig. 5.12). Many Pacific islands have raised reefs which may date from the Last Interglacial

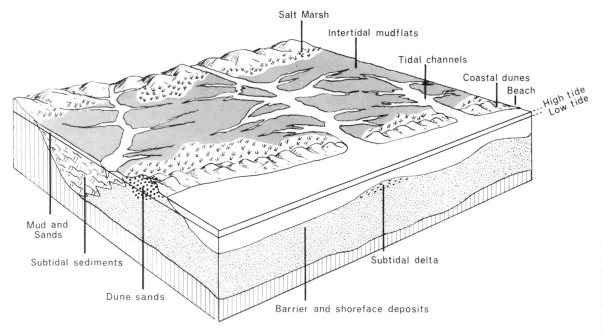

Fig. **13.16** Landforms and sediment of a barrier coast.

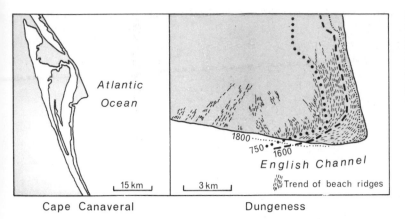

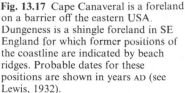

Fig. 13.17 Cape Canaveral is a foreland on a barrier off the eastern USA. Dungeness is a shingle foreland in SE England for which former positions of the coastline are indicated by beach ridges. Probable dates for these positions are shown in years AD (see Lewis, 1932).

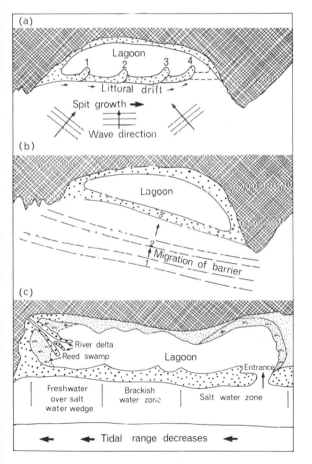

Fig. 13.18 (a) Enclosure of a lagoon by spit growth: either littoral drift or varying wave directions may be the cause of the growth and the hooked ends of the spits. (b) A lagoon formed by barrier migration: (c) Features of a tidal lagoon (the tidal range and zones refer only to areas inside the lagoon).

about 125 000 years ago and others have lagoons whose depths may date from karst erosion at the sea-level of glacial times. Whilst reefs are actively growing, and owe some of their form to modern processes acting at present sea-level, their inheritance from earlier periods with different sea-levels and tectonic events has also to be recognised. During glacial times many corals on the margins of the climatically suitable zone (Fig. 13.19) died and suffered severe erosion. The reefs of the Great Barrier of Australia are essentially a postglacial system built upon an eroded platform of older coral.

Deltas and estuaries

Deltas

A delta is a low, nearly flat area of land at the mouth of a river where sediment accumulates instead of being redistributed by the sea or by lakewater. The term was originally introduced by Herodotus, in the fifth century BC, for the tract of land at the mouth of the river Nile whose outline broadly resembles the Greek letter delta (Δ), with the apex pointing inland (Plate 13.11). As can be seen from Figs. 13.20–13.21, deltas may have many shapes with four common forms being widely recognized: *elongated* or *digitate*, like the Mississippi, where alluvium is abundant, the river can build into the sea, and wave action is limited; *cuspate*, like the Tiber, where wave erosion dominates the distribution of sediment away from the river mouth; *lobate*, like the Niger, where the river builds into the sea, but wave action is

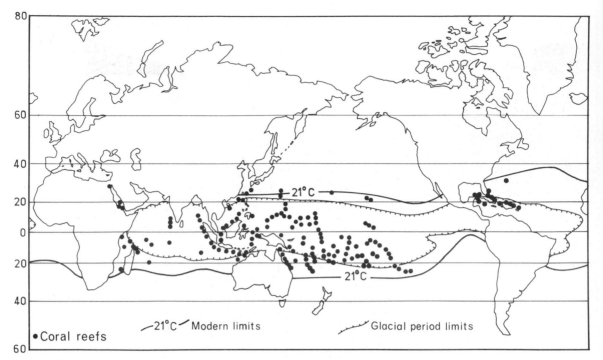

Fig. 13.19 Location of the main coral reefs. Not all of the 425 atolls in the world can be shown. It can be seen that, although the Great Barrier Reef of Australia was outside the limits of reef growth during glacial periods, most island reefs could continue lateral growth. The isotherm is for mean annual sea temperatures.

effective in redistributing sediment along coastal barriers; and *crenulate*, like the Mekong, where tidal currents produce numerous sandy islands separated by tidal channels along the delta front.

Much is known about delta sediments from the ancient deltas which have been studied in detail because of their economic significance. The major coal measures of the world are commonly within deltaic beds and many oil and gas reservoirs are in the sandy deposits of deltas.

At least six groups of variables may be identified as influencing delta formation. (1) Sediment input is controlled by the size, climate, and relief of the drainage basin, with large basins, high relief, and humid climate producing high discharges of water and sediment. (2)The nature of the delta deposits is strongly influenced by the density of sea or lake waters into which a delta is being built. (3) Waves, currents, and tides influence the redistribution of the sediment. (4) The depth of water into which the delta is advancing influences the rate at which the delta front can advance, and also the wave energy available for sediment transport. (5) The deposits

of the delta influence the compaction and sinking of the delta under its own weight. (6) The structure of the basin of deposition influences the stability and geological duration of the delta.

As a sediment-laden river enters a lake, or sea, mixing of the river with lake or sea water greatly reduces current velocities and deposition of bed-load may occur rapidly, while suspended sediment behaviour is largely controlled by the relative densities of the mixing fluids and by current and wave action. Fresh river water normally spreads out over salt water as a *plane jet* (Fig. 13.22A). Mixing and flocculation then takes place at the irregular interface between the two water masses and the flocculated sediment may be carried far into the sea before it can settle—as off the mouth of the Amazon.

Where rivers enter fresh-water lakes, mixing occurs readily and the flow is an axial jet of limited extent—often as little as four channel widths out into the lake (Fig. 13.22B). Deposition of sediment is essentially complete within this zone and the typical lacustrine delta (Plate 13.12) is generally

13.11 The Nile delta (satellite photo by NASA).

simple, with distinct bottom-set beds formed from former bedload and foreset beds partly of bedload and partly of suspended load materials, while topset beds may be overbank deposits and levees.

In the rarer conditions of a very cold river entering a warmer lake the flow is that of a plane jet, spreading out along the bottom, producing a dense turbidity current which carries material far into the lake depths and which may erode channels in delta-front and lake-bottom deposits (Fig. 13.22C).

A feature of many deltas of the elongate type is that they extend levees into lakes or seas. Such levees form because the river velocity is less at the sides than at the middle of the channel. Delta levees are usually below the heights reached by large floods and, therefore, they are readily

breached. The breach may then become occupied as the main channel, especially if it offers a much shorter, and hence steeper, route to the sea. Those breaches of levees which do not become main channels, but merely feed into a bay or a lake, are known as *crevasses*, and the deposits they form in the lake as *crevasse splays*. Such splay deposits are largely responsible for filling in the shallow zones between extending levees.

The areas between levees may also be gradually infilled by organic debris which forms peat. The peat lenses may have a considerable area and if the underlying sediments are compressing, or tectonically sinking, the peat can achieve a considerable thickness, it may then be buried and preserved as lignite or, eventually, coal (Fig. 13.23).

Most elongate deltas have numerous distribu-

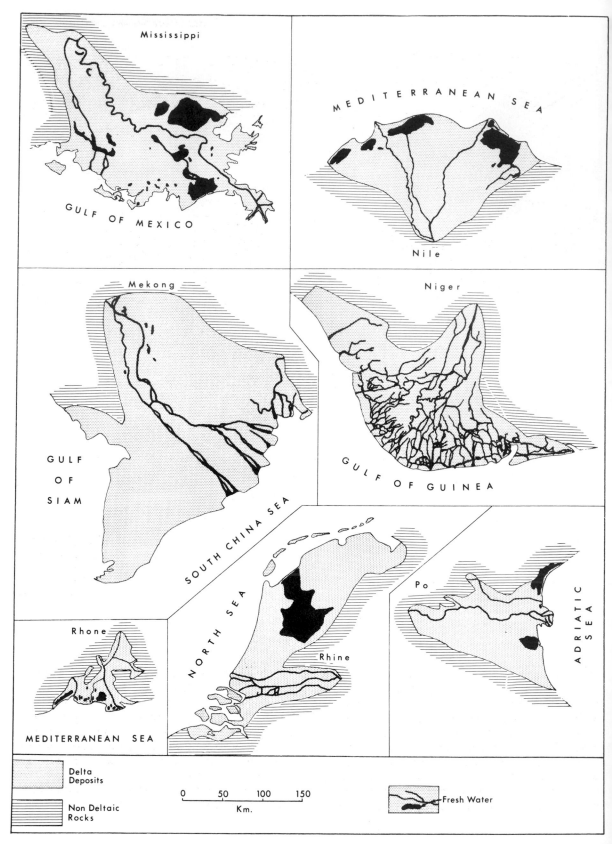

Fig. 13.20 Some major deltas.

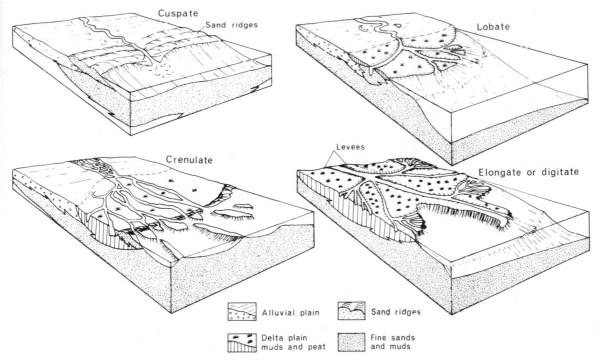

Fig. 13.21 Major classes of delta; slopes in the offshore areas are greatly exaggerated.

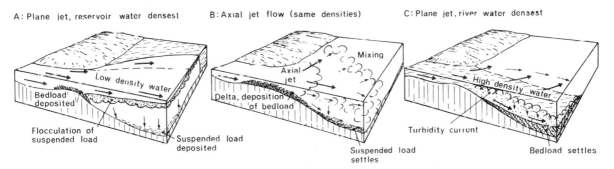

Fig. 13.22 Water density and sediment deposition of deltas and estuaries.

tary channels which further bifurcate as the delta extends seawards. Distributary channel formation is a result of the deposition of bedload material at a channel mouth. The sudden loss of velocity is most noticeable in the middle of a channel and a channel bar forms and diverts the jet around it. The diverging jet usually causes the mouth to widen seawards and mid-channel islands may eventually grow from the bar, but such channel obstructions reduce flow velocities and the river may overflow its levees immediately upstream of the bar. The new channel develops its own levees and bifurcation (or trifurcation) has occurred (Fig. 13.23). Multiple branching, elongate deltas are sometimes called *birdsfoot deltas*. Elongate deltas are most common in lakes or microtidal seas where littoral drift rates are low in comparison with sediment inputs.

In many deltas the deposition of sediment is greatly modified by marine processes, especially

13.12 Delta of the Tongariro River in Lake Taupo, New Zealand (photo and copyright: New Zealand Aerial Mapping Ltd., Hastings, N.Z.).

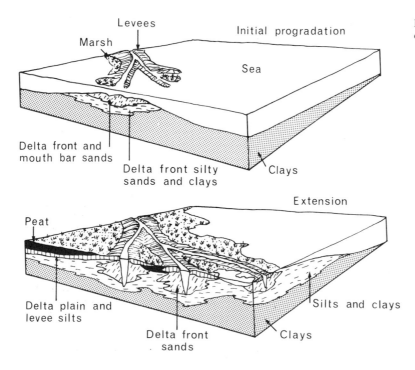

Fig. 13.23 The extension of elongate deltas and the resulting sediment types.

by waves and tides. In wave-dominated delta fronts much of the sediment is transported along the coast and beach and barrier ridges are built flanking the river mouths. Prograding shorelines can often be dated from the sets of beach ridges which mark the advancing coast. Sandy or shell beaches built upon muds, are called *cheniers* (in Louisiana French) after the belts of oak trees which grow along them in the Mississippi delta. The wave redistribution of sands commonly causes the delta form to be cuspate or arcuate. In tide-dominated deltas the fluvial sediments are reworked into bars and low sandy islands radiating from the river mouths. Muds and silts accumulate inland forming marshes and tidal flats. In warm climates these flats are mangrove swamps, as in the deltas of the Irrawady and Mekong where the tidal range exceeds 5 m.

The power of marine influences on a delta is, of course, related to the wave energy and the tidal range along a coast, but these are both strongly influenced by the depth of water off the coast. At one extreme deep waters off a very narrow continental shelf, or a submarine canyon off a river mouth, may inhibit delta formation unless the supply of debris is so great that a river can build out a submarine fan and build a delta upon that, as does the Ganges. Shallow waters inhibit wave energy and reduce wave effectiveness in redistributing fluvial sediments. Large rivers are able to modify and reduce the offshore profiles by creating large influxes of sediment and so reducing the control waves have on delta forms (see Fig. 13.3 for the distribution of main deltas).

The varying influences of fluvial and marine processes upon delta forms and sediments have been used as the basis for a delta classification (Table 13.1).

Many large deltas are of considerable antiquity as they occupy structural depressions and basins, e.g. the Ganges and Brahmaputra, the Tigris and Euphrates. The Mississippi has probably occupied its present alignment and been extending southwards since at least Jurassic times (see Figs 10.14, 10.31). In detail, however, the active lobes and channels of a delta may change very rapidly. The pattern of change during postglacial time is particularly well seen in the Mississippi delta where the various lobes have been dated (Fig. 13.24). Changing volumes of sediment being transported by rivers can greatly affect deltas. The Nile delta, for example, has been undergoing coastal erosion for much of this century as a result of the construction of a small dam at Aswan in 1902 and a much larger dam in 1964. The outstanding feature of all deltas, however, is the excess of sediment supply over the power of waves and currents to remove it. In postglacial time over 2800 km^3 of sediment has been deposited in the Mississippi delta and 1400 km^3 in the Niger delta.

Inland deltas are those forming in arid or semi-arid areas of internal drainage. Rivers like

Table 13.1 Delta Characteristics

Dominated by fluvial deposition		Dominated by waves or tides	
Examples	Mississippi—elongate; Nile, Niger—lobate.	Examples	Rhine, Po, Tiber—wave-dominated and cuspate; Mekong, Irrawaddy—tide dominated and crenulate.
Source	Continental interior, large drainage basin.	Source	Marginal or confined basins, small- to medium-area basins.
Sediment	Large volume and more or less continuous input; pronounced coastal progradation. Thick delta plain deposits with levees, marshes, and peats. In lobate deltas waves rework seaward edge to form barrier bars and lagoons. Thick muds deposited as pro-delta sediments.	Sediment	Moderate volume and often sporadic input (e.g. monsoon season). Slight or moderate progradation. Much marine reworking of sandy sediments with barrier bars and strandplain deposits. Thin pro-delta muds.

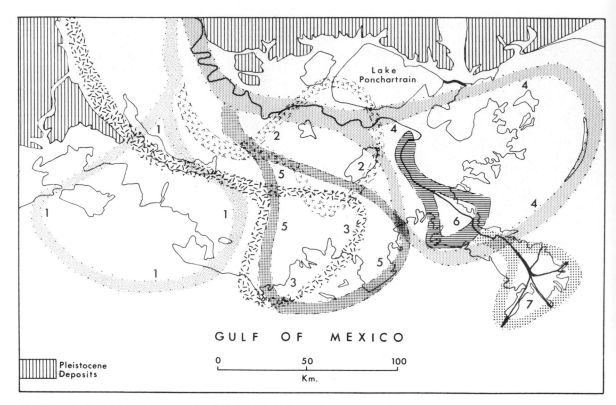

Fig. 13.24 Outline of the delta fronts of the Mississippi, during the last 4600 years. Delta 1 is dated at > 4600 BP; 2—4600–3500 BP; 3—3500–2800 BP; 4—2800–1000 BP; 5—1000–300 BP; 6—750–500 BP; 7—c 550 years (after Morgan, 1970).

the Cubango and Cuito, flowing into the Oka-vango Swamps of the Kalahari, gradually lose water by evaporation and seepage. In wet seasons their waters spread out, in the delta region, between many vegetated islands and form extensive shallow lakes, but in dry seasons the low islands stand above sandy, dry channels. During the peak of the last glacial, about 18 000 years ago, the upper Niger similarly terminated in an inland delta and these deposits now provide evidence of the past drier climate.

Estuaries

In glacial times rivers cut down to the prevailing low base level and subsequently the Flandrian Transgression has flooded the lower reaches of many valleys. The submerged sections of valleys have been given many names depending upon their origin and form.

The term *ria* was originally applied to drowned valleys, like those of SW Ireland, aligned with a geological strike normal to a coast. Even the rias of NW Spain, from which the term originated, do not follow the definition exactly, and it is now used as a general term for drowned river mouths. Well-known examples are those of Port Jackson (Sydney harbour), Carrick Roads (Cornwall), and the Marlborough Sounds (New Zealand); the latter, however, owe much of their submergence to tectonic downwarping (see Plate 13.3).

Inlets at the mouths of formerly glaciated valleys in mountains are called *fjords* (or fiords) (Plate 13.13). They are different from rias in that they have the features of glaciated valleys with straightened and overdeepened long and cross-profiles (see Chapter 15 and Plate 15.8) but also because the overdeepening by glacial erosion has been so severe; Scoresby Sound in E Greenland has depths of more than 1300 m and Sogne Fjord in Norway attains depths of 1244 m.

Inlets formed by submergence of valleys and

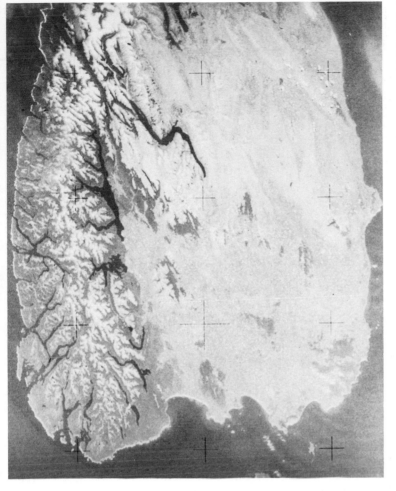

13.13 The fjords of SW New Zealand. Note also the shape of the bayhead beaches of the south coast which have adjusted to the approach of the swell waves from the south (satellite photo by NASA).

depressions in low-lying glaciated rocky terrain are known as *fjards* in SW Sweden and *förden* in Germany, and many of the *firths* of E Scotland are of the same class.

The term *estuary* is applied to the tidal and widening reaches of any river mouth. There is thus considerable overlap with other forms of submerged valley mouths even though many submerged embayments do not have major rivers feeding into them. The chief differences between estuaries and deltas is that the latter have an excess of sediment supply and deposition, but in estuaries tidal currents and river discharges keep channels open. In spite of this distinction there is considerable overlap between the two forms and many estuaries may eventually become deltas if sea-level

remains nearly constant for a few more thousand years, and some, like the Rhine, have features of both estuaries and deltas.

Well-known estuaries include the Thames, Humber, and Severn in Britain, the Elbe and other rivers of NW Germany, the Seine, Loire, and Gironde in France, the St Lawrence, Hudson, and Chesapeake Bay in N. America.

Most estuaries are areas of predominant sedimentation in which tidal channels meander between mudflats and channel bars of mud and sand. Such channels are seldom stable for long and where access to ports has to be maintained dredging is often essential. Deposition in many estuaries has accelerated in historical times as a result of forest clearance, agriculture, and induced soil erosion in hinterlands. Many small estuary ports have been abandoned in areas like the Mediterranean. The channels are strongly affected by tidal ebb and flood flows so that the meanders are adjusted to accommodate either a dominant flood or dominant ebb flow or to take both (Fig. 13.25). The processes of deposition, the cutting-off of channels by the growth of river mouth bars and barriers, and the formation of lagoons and coastal swamps have many features in common with those of deltas.

Erosional landforms of the coast

Erosion can predominate over deposition along coasts of any structure or relief as long as the transporting capacity of waves and currents exceeds the supply of sediment. The most easily recognized indicator of an erosional environment is the presence of a cliff, whether this is hundreds of metres or only a few metres high. Cliffs may plunge steeply into deep water or have an erosional shore platform stretching seawards from the cliff foot. Platforms may be a narrow ledge hanging over deep water or long gently sloping shelves. The controlling factors are usually structural, lithological, or tectonic.

Marine erosion takes place mainly during storms and is a result of wave activity. Both abrasive action of rock fragments held in suspension by waves, and the high pressures of air trapped by breakers in rock joints, are of significance in grinding and loosening cliff materials. Between storms rock is weakened by subaerial weathering processes, especially by repeated wetting and drying, salt weathering, and the activity of rock-browsing organisms such as chitons (Plate 13.17).

The simplest type of coast occurs where marine erosion has cut into the margin of a stable land mass of relatively resistant rocks and removed a wedge of material to leave a steep undercut cliff. There are many forms of cliff on which lithological and structural influences can be discerned, or past and present processes of slope erosion and deposition may be predominant influences on slope form. Lines of weakness are commonly picked out by erosion so that soft or closely jointed rock is eroded preferentially to leave a fretted cliff face, or a coastline of promontories of resistant rock separated by embayments or pocket beaches. At a local scale joints are exploited and opened to form caves, which may be extended horizontally to form *natural arches* (Plate 13.14) or vertically to form *blowholes* where water vapour and compressed air are expelled under the pressure of swells entering the cave below.

Retreat of coastal cliffs causes intersection of existing drainage lines to leave hanging valleys and breached valley walls (Fig. 13.26). The pattern of the offshore relief is usually controlled by the rock type and structure so that reefs, residual islands (called *stacks* where they are steep-sided), and headlands tend to be aligned in accordance with the geological strike.

Where rates of hillslope processes exceed rates of marine erosion landslides, gullying, or talus formation may control cliff form. Relict slope processes may also influence or control slopes as, for example, along much of the upland coast of Devon and Cornwall (Plate 13.15) where exposure to Atlantic storms is limited, in Britanny, southern Ireland, Washington, and Oregon. On all of these coasts there is a double slope form: an upper gentle slope with a mantle of weathering products which were produced, in most cases, under a severe periglacial climate during the Last Glacial, and a lower actively cliffed coast. On rocky sectors of some Arctic coasts physical weathering is still

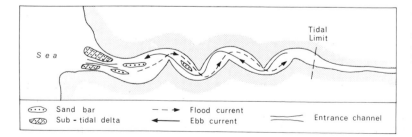

Fig. 13.25 The form of meanders in a tidal estuary resulting from flood and ebb flows. The pattern shows the Coriolis deflection for the northern hemisphere.

13.14 Cliffs, stacks, a natural arch, and offshore reefs cut in granite, near Lands End, SW England (photo by Geological Institute, Crown copyright reserved). (*Copyright*: British Geological Survey).

producing a hillslope more rapidly than marine erosion can form a steep cliff. Such slopes should not be confused with the dominant hillslopes, with virtually no marine cliffing, of rias (Plate 13.3).

Other examples of climatic influences on cliff form occur in parts of the humid tropics where deep weathering has decreased rock strength so that steep cliffs are rarer than might be expected.

Rates of cliff erosion are extremely variable. Where weak materials like unconsolidated glacial or fluvial deposits are being eroded, as on the Norfolk coast of England and the South Canterbury Coast of New Zealand, rates of retreat may be several metres per year and retreats of several kilometres in historical time are well documented. During one storm on the Suffolk coast in 1953, 300 000 tonnes of material was removed in 24 hours along a 1.5 km stretch of coast, causing 27 m of recession of cliffs 3 m high, and 12 m recession of cliffs 12 m high. Many examples exist of villages

and buildings lost over the last few centuries on coasts like those of Holderness where an average retreat of 120 m occurred between 1852 and 1952 along 61.5 km of coast. On coasts of weak rocks the rate of recession is nearly always controlled by wave energy.

The most stable cliffs are plunging cliffs in resistant rocks. These occur along faulted coasts such as those of northern Chile and inside Wellington Harbour, New Zealand, inside fjords, and on tectonically sinking upland coasts. They are stable because even where they are exposed to wave attack breakers do not hit them; waves are reflected and produce little transfer of energy or air compression in joints. Furthermore, there is no material available for abrasion.

On rising coasts, or those with a strong littoral drift, cliffs become fronted with beaches and are no longer subject to wave attack. They are then fossilized and subject only to slope processes

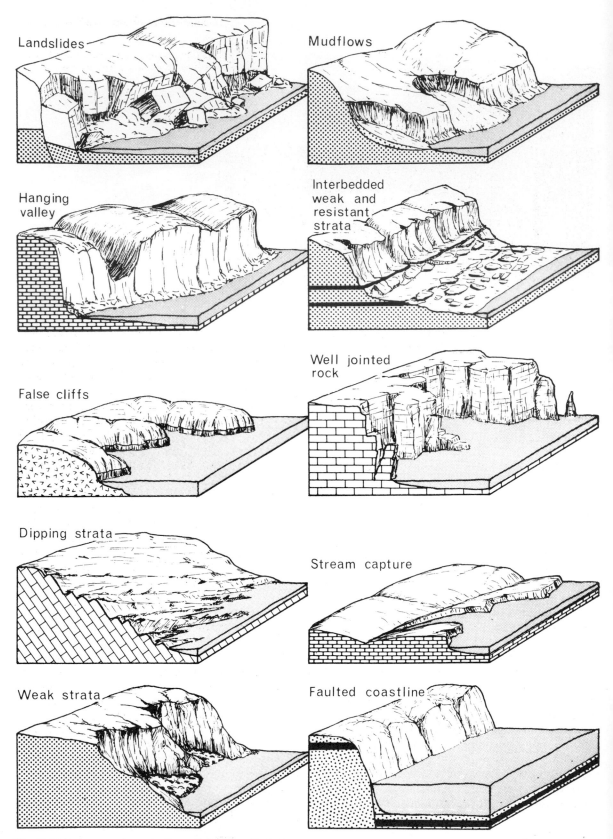

Fig. 13.26 Some controls on the form of cliffs (modified after Guilcher, 1958).

13.15 The upper relict slopes and lower cliffs of the south Cornwall coast.

which will eventually change their slope form as upper slopes are eroded and lower slope segments are buried by talus deposits.

Shore platforms

At the foot of most cliffs is a platform cut across rock. It has a gentle slope, commonly of about 1°, and extends seawards to terminate either at a steep drop-off into deep water, or continues at a regular slope into water 10 m deep or more. On the landward side most shore platforms terminate at the foot of a rock cliff which may have an erosional notch, or a cliff foot formed of talus (Plate 13.16).

Platforms may be single ramps dipping seawards, a single platform extending from about the level of mean high tide at the cliff foot to below mean low tide, a single platform at either high-tide or low-tide level, or stepped platforms with one at high-tide and one at low-tide level. Controls on the nature of the platforms include rock structure, tide range, and exposure to waves.

Shore platforms are in the surf zone and exposed to severe weathering by wetting and drying and salt weathering, to abrasion, wave quarrying of joint blocks subject to repeated fluctuations in water and air pressure, and to biological processes. Platforms are widened as cliffs retreat. On a tideless shore a platform with a low angle of slope could not be wider than about 500 m because its seaward edge would be below the wave base and not subject to abrasion. Where the tide range is 5 m it could still be no wider than

800 m. Shore platforms which are wider than this can be produced only by progressive submergence or have a very complex history of erosion by several processes. Even a small island could not be planed off to sea-level by abrasion alone.

Water-level weathering is thought to be a very important process in widening and flattening shore platforms. Intertidal pools (Plate 13.17) develop by weathering processes and are enlarged laterally, and deepened, by lithophagic organisms (rock-grinding animals) and weathering at rates of 1–10 mm/year. Such processes may produce ledges at a variety of levels around the pools so that breached pools have ledges at elevations ranging from several metres above mean high tide to several metres below mean low tide. It is not surprising, therefore, that attempts to relate particular ledges to assumed higher or lower-than-normal sea-levels are open to question.

On many limestone and coral coasts deep solution notches and overhangs are particularly common.

Shore platforms are often irregular forms on which geological structures impose an influence so that joints, dykes, bedding planes, outcrops of strata, and other features may be either zones of elevated or of selectively eroded rock.

One of the best-known, but least well-understood, shore platforms is the *strandflat* of the Norwegian coast (Plate 13.18), together with related platforms off Greenland and Spitzbergen. These platforms are up to 64 km wide and are on coasts which have been, and in many places still

13.16 Steep cliffs with a basal notch; shore platforms cut at mean high-tide level and a lower level which is just above low-tide level. The blocks standing above the platforms are undercut and have an overhanging visor. The cliff face has numerous tafoni resulting from selective salt weathering. The rock is ignimbrite, Coromandel Peninsula, New Zealand.

are, rising isostatically. They clearly cannot have been produced during the present postglacial period alone. Hypotheses seeking to account for them include abrasion by ice shelves, severe periglacial weathering and erosion followed by wave-removal of debris, glacial scouring, and selective erosion around resistant cores in weathered rock which form the steep-cliffed forms above the strandflat.

Most Norwegian geomorphologists seem to agree that frost weathering and wave action to scour the platforms and produce abrasion notches below the cliffs are, at least in part, responsible for their main features, and it is probable that they developed during many glacial periods of rising and falling sea-level.

Dominant influences on coastal landforms

Classification of coasts, their landforms, and dominant processes, was a major study of coastal geomorphologists in the first two-thirds of this century. Nearly all classifications were devised before the development of plate tectonic theory and before the present knowledge of climatic change, and resulting sea-level change, became available. Most early classifications need to be treated with caution or to be reformulated in the light of more recent information.

Coastal landforms may be classified according to: major structural influences; the relief, structure, and lithology of the hinterlands; the climatic influences upon coastal processes, and climatic influences upon the hinterland and the formerly exposed continental shelf; the current rate of erosion and sediment supply; the influence of sea-level change; and the relict influences of former climates or sea-levels. Each of these sets of influences could be the subject of a separate book and many of them have been mentioned already in

13.17 Part of the low-tide platform shown in Plate 13.16 with a cover of organisms, and pits produced by water-level weathering and bio-erosion. The inset (top right) is of a chiton, which is 2 cm long, in a pit it is enlarging.

this and earlier chapters. The following paragraphs are a set of notes which attempt to illustrate major influences only.

Major structural influences

Four categories of coast may be recognized from the application of plate tectonic theory. (1) The edges of actively diverging plates are rifted and block faulted and consequently tend to be relatively straight, with few embayments (unless a tilted block forms one) and a steeply sloping offshore zone. Examples are the Red Sea coasts of Africa and Arabia. (2) Actively converging plate boundaries are characterized by island arcs such as Indonesia, Japan, and other areas in the western Pacific, and also by rising mountains with narrow or no continental shelves, but deep trenches offshore. Such mountains usually have a structural trend parallel to the line of convergence and

produce gently curved mountainous arcs like those of South America and the Atlas Mountains. (3) Major transform faults produce coasts which are relatively straight and steep like those of southern California and the northern coast of Papua–New Guinea. (4) Continents, or sides of continents, which are embedded in crustal plates are far from sites of convergence, or transform motion. Many such coasts have shallow, broad continental shelves like those of the Atlantic coasts of the Americas which receive sediment from a number of large rivers. The east coast of Africa and the west coast of India have steep escarpments and narrow shelves because they receive few major rivers and there has been little sediment available for shelf deposition and construction.

This plate tectonic classification is reminiscent of an early classification by Suess (in 1904) into Atlantic-type coasts in which continental structures are truncated at the coast, and Pacific-type

13.18 A strandflat near Traena, N. Norway. The raised shore platform is settled by fishermen (photo by Widerøe's Flyveselskap A/S).

coasts with island arcs or mountain ranges parallel to the coast. The Atlantic type corresponds to the trailing edges of diverging plates and the Pacific type to leading edges with collision or transform motions. Associated with the modern classification is a consideration of rates of vertical movement along the coast, the amount of volcanic activity, and the alignment of structures.

Relief, structure, and lithology

Many coasts are influenced by variations in the resistance of rock to fluvial, glacial, and coastal processes and by a relief developed in response to ancient and inactive tectonic processes. They have inherited structural controls on their form. Areas with resistant rocks, or tectonically uplifted blocks, form uplands with cliffed coastal margins. Embayments, which are often exploited by local rivers and glaciers, form where weak rocks of closely jointed rocks outcrop and are bordered by headlands of resistance. Weakly resistant rocks form lowlands, perhaps with low islands offshore, and are likely to be areas of deposition with barriers and deltas, because rivers, exploiting

zones of weakness, and more effective coastal erosion cutting back the coast to leave a broad gently sloping offshore zone, produce an ample sediment supply.

Lines of structural weakness and strong lineations resulting from faults, major joints, and dykes are evident in many upland coasts like those of the Norwegian fjords. Where fold systems, or sedimentary rocks of varying resistance, control the landscape a drowned valley landscape with rias or, as in Yugoslavia, a Dalmatian-type coast of ridges and straits parallel to the coasts and interconnecting channels cutting across the ridges, may result.

Some upland coasts may become cut off from marine influences, as an alluvial or erosional plain is formed at a cliff foot. In western India the fault-line scarps now form the hinterland of narrow coastal plains.

Coastal plains of essentially erosional or depositional origin form coasts in much of the world. They are usually composed of, or have a cover of, relatively weak sedimentary rocks and the depositional type are often prograding. Barriers and lagoons are features of such coasts as those of the Gulf of Mexico. Alternating rises and falls of sea-level have had a major influence on such coasts and estuaries, deltas, and abandoned beach ridges provide evidence of the changes.

Climatic influences, present and past

The energy available for coastal processes is partly controlled by wave height and hence windspeed, direction and duration, although these are modified by the fetch. High-energy coasts tend to have severely eroded upland coasts and to have large deposits of coarse sediment if this is available. Low-energy coasts often have accumulations of sediment so that deltas, salt marshes, or mangrove swamps are common.

Available wave energy is controlled by zonal climate so that coasts between latitudes 45° and 60°, in both hemispheres, and east-facing coasts in the tradewind belts generally, have high storm-induced swells. Subtropical east-facing coasts in the Caribbean, around India, in N Australia, and the coast from Malaysia to south Japan are in the

tropical cyclone belt of hurricanes and typhoons that occur in most years. Equatorial coasts, especially in the doldrums, have low wave energies and those in the Arctic and Antarctic are sealed by ice for much of the year. Ice has the effect of protecting most of the coast from wave action although it can produce specific features such as ice-thrust ridges (Plate 13.19), frozen beach deposits, and scour features where drift ice is pushed over mud.

Climate also has an influence upon the distribution of vegetation and upon the rate of weathering. Mangroves and coral reefs are tropical and subtropical: the predominance of chemical weathering in the humid tropics limits sand supply to rivers, and hence beaches, so that sand dunes are seldom significant features of tropical coasts. Other limiting factors on tropical coastal dunes are low tidal ranges, the effect of tropical vegetation in stabilizing sand on prograding shorelines, and the nearly permanent dampness of some sands and the low wind speeds. Arid coasts tend to

13.19 Fast ice on the shore, and a tide crack with thrust ice blocks. Cape Chocolate, Ross Sea, Antarctica.

have salt flats and large dune systems which merge with desert dunes or supply them, as in the southern Namib.

Relict influences are particularly powerful in areas which have been glaciated. Not only fjords and fjards but also depositional coasts, with unusually large volumes of coarse debris, are the product of glaciation. In parts of the New England coast, for example, rounded hills called drumlins and morainic ridges left at the termini of glaciers now form important coastal features. The incomplete isostatic recovery of many coasts from glacial unloading inevitably produces many features which are imperfectly adjusted to prevailing wave directions and energies.

Current rates of erosion

The coastal zone is often one of very rapid change, with several thousand tonnes of sand being conveyed past a point on a beach in a single day and coasts advancing or retreating tens of metres a year. The general rise of sea-level of about 1 mm/year in many areas of tectonic stability, in the zones beyond those of glacio-isostatic adjustment, necessarily produces notable changes along many coasts and particularly pushes barriers towards the shore. Areas with large sediment supply from rivers commonly are advancing, although human interference with river loads and with littoral drift is modifying the natural rates of change.

Sea-level

The pattern of repeated glaciation followed by short intervals of warmer climate, like those of the present, during the last two million years, and particularly in the last one million years, means that coastal positions have changed almost constantly with only brief periods of relative stability, or still-stand, like the last 6000 to 4000 years. It is not surprising, therefore, that many areas of coast possess relics of past sea-levels in the form of raised terraces and submerged erosional platforms. Many warm interval (i.e. interglacial) periods had sea-levels very close to those of the present so that it is possible that some coasts have a compound history with both erosional and depositional landforms being added to, or further developed, for a few thousand years in every 100 000 years or so.

It may be appreciated from this discussion of the factors which may influence coastal landforms that many of them are influenced by more than one process, or material, so that simple exclusive classification is seldom possible.

Continental shelves and submarine canyons

Continental shelves

The discussion, provided in Chapter 4, of geoclines and geosynclines has demonstrated already that the surface of most continental shelves is formed of the upper beds of a sedimentary accumulation which may be largely undeformed on a passive continental margin, but have been tectonically deformed, and made very narrow, on an active margin. Only the inner coastal fringe of most shelves can be regarded as a predominantly erosional feature, and that may owe much of its development to fluctuating sea-levels in late Cenozoic times.

Submarine canyons

The term *submarine canyon* has, unfortunately, been used to cover a number of distinctly different types of valleys on the sea floor. The most important of these are winding V-shaped valleys, with many tributaries, which cross the outer shelves and extend down most continental slopes. Such canyons have many of the features of large valleys on land. Broad-floored and steep-walled troughs, like those of the Aleutian Island submarine slopes, may also fall within a classification of submarine canyons, but the gulleys etched into the submerged faces of deltas cannot be classed as canyons.

Many early accounts assumed that all submarine canyons have a similar form and origin. Studies carried out since about 1950, using echo sounding, seismic profiling, scuba diving, and submersibles have shown that the forms and origins of valleys beneath the sea are as varied as those above it.

Subaerial erosion has clearly played a part in the formation of many canyons, like those on the west side of Corsica and on the flanks of the Hawaiian Islands, because the submarine valleys extend into the estuaries of land canyons. The Hawaiian features were formed before those islands sank, isostatically under their own weight, but the Mediterranean features may date from the Messinian event of the late Miocene (10 My ago), when the Mediterranean was cut off from contact with the Atlantic at the Straits of Gibraltar, so that it became an evaporation basin with residual salt lakes in the deepest depressions. The base level for the rivers draining into the Mediterranean was consequently very low and the rivers incised. Many of the resulting canyons are now filled with late Cenozoic sediments, but it may be that some of the deep valleys on the southern side of the Alps were cut at this time and have been subsequently modified by ice, rather than wholly cut by glaciers, to produce troughs with floors below modern sea-level.

Many canyons along the Atlantic coasts appear to have originated in Cretaceous times, but to have been later buried by sediment and then re-excavated by turbidity currents. It is postulated, but not convincingly established, that many canyons on the New England coast and the Zaïre coast may have formed originally in the flanks of rift valleys formed at triple junctions at the sites of opening of the Atlantic. Sea-floor spreading is thus the original cause, but subsequent marine processes have produced the modern forms.

Individual coasts may have been variously upwarped and later depressed, or have emerged and then submerged with consequent incision of subaerial valleys and later drowning of their seaward ends during sea-level changes.

Faulting and glaciation are responsible for canyons in some areas. Faulting is most obvious where normal faults can be traced from the land into the continental shelves and the structural pattern is evident from seismic studies. Glacial troughs also are usually traceable from land onto shelves but, although cut at times of low sea-level, canyons with such an origin are limited in extent because the ice floats off the sea bed as the water deepens.

Submarine canyons are evidently produced by many processes. They are cut both into hard crystalline rocks and into soft sediments. Some have a history nearly as long as that of the ocean basins and have suffered phases of cutting, infill, and subsequent re-excavation by turbidity currents.

Further reading

A general review of coastal geomorphology is provided by Bird (1976) and in a number of other books such as King (1972), Zenkovitch (1967), and Guilcher (1958). Davies (1980) has presented a very useful guide to the variability of coastal processes and landforms. Beach development and processes are discussed by Komar (1976), at a level which is appropriate for senior undergraduates, and for the general reader by Manley (1968), Bascom (1964), and Steers (1962). Regional studies of the North American coasts are presented by Shepard and Wanless (1971), of England and Wales by Steers (1948), and of high latitudes by John and Sugden (1975). Zenkovitch (1967) uses many examples from the coasts of USSR and Bird (1976) from those of Australia. Engineering or shore protection works are most comprehensively treated in the *Shore Protection Manual* (1977) of the Coastal Engineering Research Centre. Knowledge of sea-level change has advanced so rapidly in recent years that textbook presentations are unsatisfactory. The following papers are particularly useful: Walcott (1972), Synge (1977), Clark, Farrell, and Peltier (1978), Donovan and Jones (1979). The effects of sea-level rise on beach erosion are discussed by Bruun (1962) and Dubois (1977). Submarine canyons are discussed by Shepard (1981). Isostasy and eustasy are examined in the book edited by Mörner (1980).

14 Snow and Frozen Ground: Processes and Landforms of the Periglacial Zone

The term *periglacial* was first proposed by a Polish geologist, Walery von Lozinski in 1909 to describe frost-weathering conditions associated with the production of rock rubble, or felsenmeer, in the Carpathian mountains. In 1910, he introduced the concept of a *periglacial zone* to describe the climatic and geomorphic conditions of areas peripheral to the ice sheets and glaciers of the Pleistocene. The zone was regarded as being that of the tundra, and extended equatorwards to the treeline. This definition is very restrictive and excludes many alpine areas and lowlands, such as those of eastern Siberia and interior Alaska, which owe their characteristics not to the presence of ice sheets but to low mean annual temperatures and perennially frozen ground, known as *permafrost*. A more useful definition is to use periglacial as a general term to include the processes and results of frost action, ground ice, and related features of cold climatic zones. The study of ice and snow on the Earth, and especially the study of permafrost, is also called *geocryology*. The prefixes 'cry-' for ice, and 'niv-' for snow, are in common use in terms for cold climate phenomena.

Nival processes

The actions of snow on the landscape may be called *nival processes*, they include *nivation*, that is the combined action of frost shattering, soil flow in association with ground ice (also called *gelifluction*), and slopewash processes occurring around snow patches; and abrasion by snow patches and avalanching.

Snow may form a protective cover on the ground surface or it may provide conditions in which physical weathering is promoted. Protection is most probable where the snow depth is greater than 1 m, the snow is fresh, and has a low density ($0.1 \, \mathrm{Mg/m^3}$ or less, compared with old compacted snow with a density of $0.3 \, \mathrm{Mg/m^3}$, or with ice $\simeq 0.85 \, \mathrm{Mg/m^3}$), and where there are few fluctuations of temperature across the freezing point. Thick snow keeps the ground temperature close to melting-point unless atmospheric temperatures drop well below zero for a prolonged period. Geothermal heat helps to maintain the temperature near zero and meltwater seeping beneath snow cover cannot be refrozen unless very low temperatures result from prolonged winter cold, or from the presence of permanent ground ice.

Nivation hollows (Fig. 14.1; Plate 14.1) occur on slopes where the snow cover is thin and uneven for part of the year. Around the margins of a snow patch snowmelt water is available to wet the soil, and during cold periods, and especially at night, some of this water will form ice and cause physical weathering (see Chapter 7, p. 190) of rock and some will separate and move soil grains. Rock and soil debris can then be transported by running water and by flows of saturated soil. Hollows have been observed to form beneath snow patches in a few seasons, and are most rapidly formed on bare ground, but more slowly on vegetated soil. The washing away of fine-grained soil may lead to the undermining and collapse of large blocks of debris in the back slopes of nivation hollows. Progressive enlargements of hollows is likely to occur at their margins only, as a protective snow cover inhibits physical weathering in the centre of the hollow. Large hollows have been called *thermocirques* and if these coalesce a terrace or gentle slope with a steep head may result. The presence of snow patches on one side of a valley and their absence on another can lead to valley-side asymmetry with

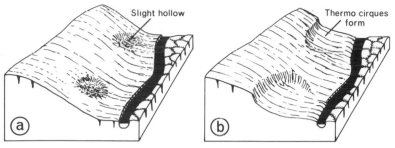

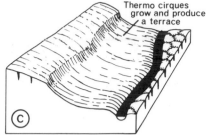

Fig. 14.1 Nivation hollows developing into thermocirques and then a terrace.

14.1 A nivation hollow with rills and fans of debris, Antarctica (photo by J. D. McCraw).

lower angled slopes on the side with the snow patches (Plate 14.2), but this is only one of the possible causes of asymmetry.

Snow sliding may be a cause of nivation hollows but the only measurements of this process were made in the Snowy Mountains of Australia, where steel rods fixed in bedrock were deformed, or broken, by the force of moving snow at stresses ranging from 160 to 3800 kPa, which is a stress range equivalent to that occurring at the base of valley glaciers. This sliding caused abrasion and scouring of the rock with the production of fine rock flour. In the experiment, however, movement occurred above a relatively smooth rock bed and may not be a common phenomenon on soil slopes.

Snow patches beneath cliffs may act as slide surfaces for debris falling from the cliff and a rampart of rock debris, called a *protalus rampart*, may be formed at the foot of the snow patch.

Avalanching of snow is a common phenomenon in mountains of the temperate zone and less common in polar regions. It is defined as the rapid mass movement of snow on slopes, and results from instability of the snow. An avalanche may be of dry or wet snow; it may carry rock and soil debris; it may slide over the ground, or it may be

14.2 The formation of a low-angle slope below a snow patch as a result of nivation. The valley slopes are consequently asymmetrical. Note the stream channel with little sign of erosion. Midsummer, Koettlitz Valley, Antarctica.

airborne. Instability may be caused by many factors. It is particularly common after very heavy snowfall and the rapid build-up of loose snow; it occurs also in old snow which contains refrozen layers, which provide smooth failure planes, or layers of snow in which the grains have become rounded by freezing on of meltwater or water vapour. Such old granular snow lacks cohesion and can be set in motion readily.

The two major types are powder-snow and slab avalanches. Clean *powder avalanches* move as a formless mass and have little geomorphic effect, although the wind blast associated with large ones may destroy buildings or trees. *Slab avalanches* usually travel over a well-defined sliding surface, starting as a large block which breaks up during movement. The slab may slide over old snow and do little geomorphic work, or over the ground where it can pick up soil and rock debris. Dirty avalanches are most likely to occur in the melt

season. They do considerable work, eroding their tracks, even into rock, and leave behind distinctive deposits beyond the foot of a slope (see Fig. 8.13). Many talus slopes and debris cones are initially composed of alternating layers of snow and rock debris and are clear evidence of the effectiveness of avalanches as transporting agents (Plate 14.3). *Slush avalanches* may also be very important agents in some mountains and are known to be able to transport boulders of up to 100 tonnes mass for 100 m or more on a slope of 5°. Slush flows along valley floors are a related phenomenon and are important during spring melting in the Arctic.

The geomorphic importance of snow has received very little attention by research workers. It is clear that it varies in its effectiveness across a landscape because it accumulates preferentially in hollows, and in the lee of obstacles, and because it is most effective where it is thin and melting. Its

14.3 Debris cones of snow and rock debris below an avalanche gully. Mount Selwyn, New Zealand.

contribution to erosion, transport, and deposition is greatest, therefore, during spring melt periods when it can add greatly to slope wash and stream processes. The annual variation in mean world seasonal snow and ice cover at present is between a low of 30 Gm² (7.5 per cent of the Earth's surface) and a high of 76 Gm² (19 per cent). This compares with an estimated annual mean of 60–70 Gm² during the Last Glacial. In the northern hemisphere most of this is snow cover, and in the southern hemisphere most is pack ice. The topic of snow as a geomorphic agent is clearly a neglected one.

Frozen ground

Rock and soil may be frozen either seasonally or perennially. Seasonally frozen ground remains frozen only during the winter. Ground freezing occurs over almost half of the land mass of the northern hemisphere; about one quarter of this land mass has seasonally frozen ground and the other quarter with freezing is underlain by permafrost. The area of frozen ground in the southern hemisphere is small because of the shape and latitude of the continents. There is still much to be learnt about the ground temperatures beneath the Antarctic ice sheets. The depth of seasonal ground freezing ranges from a few millimetres to about 3 m with the greatest depths occurring at higher latitudes. In a permafrost zone the theoretical depth of freezing also increases with latitude, but the depth of the summer thaw decreases towards the pole and this limits the depth of ground available for autumn freezing.

Permafrost

Permafrost is defined as the thickness of rock or soil in which a temperature below freezing has existed continually for at least two years. It should be noted that this definition is regardless of the water content and that permafrost may be 'dry' or contain ice.

It can be seen from Figs. 14.2 and 14.3 that the permafrost zone may be divided into two units: the zone of *continuous permafrost* and that of *discontinuous permafrost*. In both zones there is summer melting of the ground surface and this thawed layer is called the *active layer*; in northern Canada it is commonly only 0.3–0.5 m thick, but towards the southern limit of the continuous zone it is 1–2 m thick and in the discontinuous zone it is 2–3 m thick. Within the discontinuous zone an area of unfrozen ground below the active layer is called a *talik*. In the Arctic, and possibly in parts of the Antarctic, areas of the continental shelf are underlain by permafrost. This is possible because the sea water has temperatures of −1.5 to −1.8 °C, but much, or all, of this permafrost may be relict from a period of low glacial sea-levels when the continental shelf was dry land; since sea-level has risen the low sea temperature has maintained the fossil permafrost. Some writers recognize a zone of

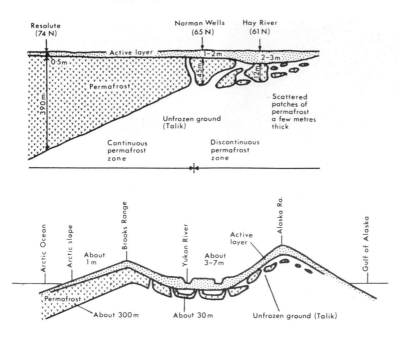

Fig. 14.2 Schematic distributions of permafrost from northern Alberta to Resolute on Cornwallis Island (north of Barrow Strait), and also from north to south across Alaska (after Brown, 1970).

sporadic permafrost in which small pockets of frozen ground occur; this division appears to be most useful in discussions of alpine phenomena.

About 80 per cent of Alaska and 50 per cent of Canada is underlain by permafrost. The volume of ground ice is greater in USSR than in North America, and reaches depths of at least 1450 m in Siberia compared with 1000 m on Ellesmere Island. Attempts to describe the limits of permafrost with air temperature isotherms are only partly successful, because the extent of frozen ground is governed by a number of factors including the presence of fossil permafrost, and the insulating effects of snow and vegetation cover. A cover of peat is often very effective in preserving patches of permafrost in the discontinuous zone (Fig. 14.4). As an approximate indicator, the southern boundary of continuous permafrost in North America coincides with the $-8.5\,°C$ isotherm, and with the $-7\,°C$ isotherm in Siberia. The equatorial boundaries of sporadic permafrost are largely controlled by vegetation, soil type, and altitude, with small patches surviving at 5000 m altitude on Mount Kenya on the equator.

The forms of ground ice include: ice which fills a few pores or fissures; ice filling all pores; ice coatings on particles; thin veins and lenses; and massive ice which may contain much soil or be nearly pure. The thickness and orientation of the ice tends to control the behaviour of permafrost on thawing and the development of distinctive periglacial landforms. Six main types of ice accumulation may be recognized:

(1) *Glacier ice* is formed above the ground surface and normally lies on it, but in alpine areas rockfalls and talus can cover it so that it is preserved as a form of ground ice.

(2) *Icings* (also called *aufeis* or *naleds*) are masses of ice formed during winter by the freezing of sheets of water from rivers or springs. They may be preserved by burial under sediment.

(3) *Pingo ice* is commonly a massive body of fresh-water ice created by the freezing of saturated sediment, or by the injection of groundwater under artesian pressure into permafrost.

(4) *Ice lenses* vary in thickness from a few millimetres to 40 m, and in horizontal extent from a few millimetres to hundreds of metres and up to 2 km. Most lenses are developed by the successive freezing-on of water to the layer so that it can grow and even force up the ground surface into hills.

(5) *Ice veins* may be oriented in any direction and up to 100 mm thick.

(6) *Ice wedges* develop from vertical ice veins and fill V-shaped cracks, in the ground, which are widest at the top.

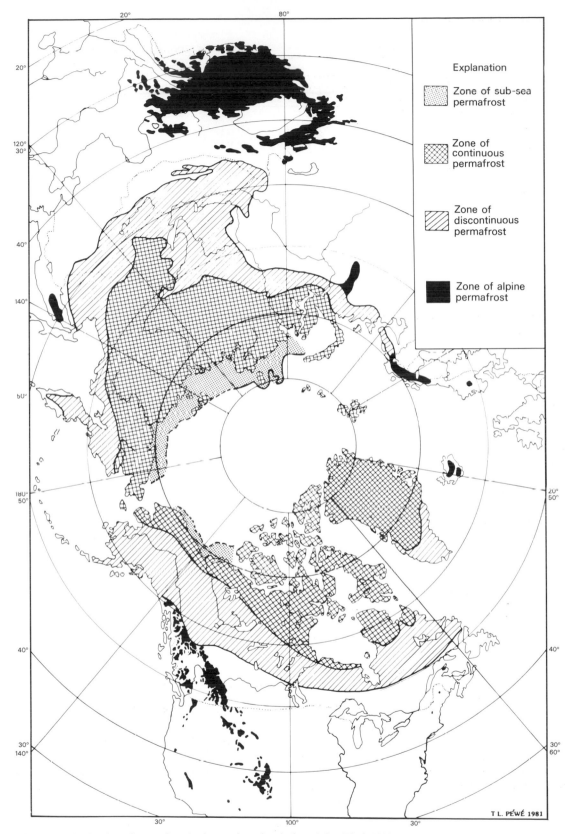

Fig. 14.3 Distribution of permafrost in the northern hemisphere (after Péwé, 1983).

Explanation

Zone of sub-sea permafrost

Zone of continuous permafrost

Zone of discontinuous permafrost

Zone of alpine permafrost

T L. PÉWÉ 1981

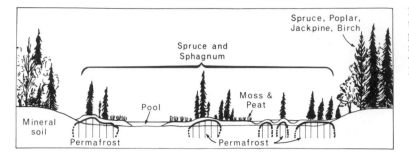

Fig. 14.4 A profile through peatland in the southern fringe of discontinuous permafrost in Canada, showing the relationship between permafrost, vegetation, and relief (modified from Brown, 1968).

The age of permafrost is often unknown, but it is evident that most permafrost in areas that have not been covered by glaciers during the Pleistocene is relict. The evidence for this is that (1) fossils of animals, such as the woolly mammoths which are now extinct, have been preserved in permafrost which can be no younger than the animal; (2) the upper boundary of some permafrost (known as the *permafrost table*) is considerably deeper than the depth of modern winter freezing; (3) in places the temperature of permafrost decreases with depth indicating residual cold; and (4) the thickest permafrost occurs in areas which were not covered by insulating Pleistocene glaciers. Some Siberian permafrost may have formed nearly 2 My ago. By contrast permafrost is now forming in some areas where Arctic glaciers are retreating. It should be recognized that the upper surface of the permafrost is in balance with the present climate, but that deeper permafrost may be either in equilibrium or out of equilibrium with modern climate.

Frost action

Frost-heaving and frost-thrusting

Frost-heaving is essentially upward, and frost-thrusting is a lateral movement of soil during freezing. Heaving usually predominates over thrusting because freezing commonly occurs from the ground surface downwards and ice crystals extend in this direction as a result.

The results of freezing and thawing of the ground are largely controlled by the amount of water available. Within a soil which is becoming frozen a boundary, called a *freezing front*, exists between the frozen and the unfrozen material. As freezing continues this front advances into the unfrozen soil. In dense clays of very low permeability the ground surface may be heaved upwards in proportion to the expansion of soil water as it turns to ice (a 9 per cent volumetric change). In more permeable soils the volumetric expansion is larger than this, implying that water has been drawn from deeper in the soil towards a freezing front: there it may be seen to have formed lenses of segregated ice parallel to the front, with the crystals in that ice developing their long axes normal to the freezing plane.

The migration of water during freezing is a result of the electrostatic charges and bonds between soil grains and water molecules (Fig. 14.5). The surfaces of most soil particles, and especially of clay particles, have negative electrical charges which attract the positive charges of the dipolar water molecules. The strength of the electrostatic field is greatest at the surface of the particle and declines, logarithmically, away from it. Only free, or gravitational, water has a random orientation of its molecules and can freeze at 0 °C. Strongly bonded, weakly bonded, and capillary water are all under the influence of the attractive forces of the molecules at the soil particle's surface. To freeze such water the surface attractive forces, and the molecular attraction of the water itself, have to be overcome. Thus strongly bonded water may remain unfrozen down to -78 °C, weakly bonded water will freeze completely only at -20 to -30 °C, and the freezing-point of capillary water is slightly less than 0 °C. In the soil the presence of salt and solid nuclei, for ice crystal formation, influence the temperature of freezing. It is evident that unfrozen bonded and capillary water can exist in a soil which has macropores filled with ice.

Water migrates within a freezing soil because ice crystals at the freezing front draw water from the hydraulic films of soil particles at the front. Where particles share hydraulic films (Fig. 14.5d(i)), that film nearest the ice crystal becomes thinner, by loss

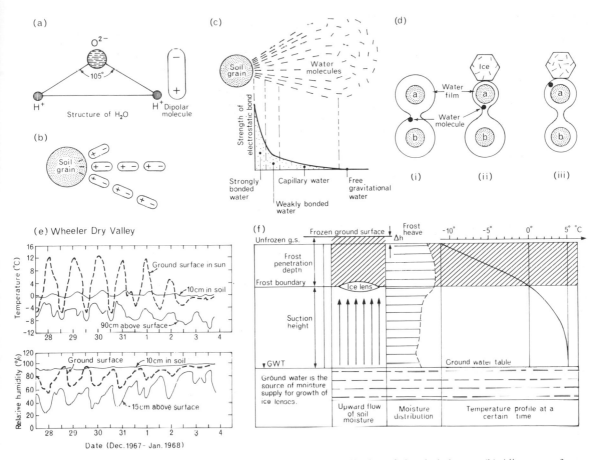

Fig. 14.5 (a) The structure of a water molecule, and the consequent distribution of electrical charges. (b) Alignment of water molecules around a clay or silt particle. (c) The strength of the electrostatic bond declines away from the grain, and the orientation of water molecules becomes increasingly random towards the freely draining water. (d)i. Water films in equilibrium around two soil particles; ii. Ice attracts molecules from out of the film around the nearest particle until; iii. a new quasi-equilibrium is established with thinner water films. (e) Air, ground surface and soil temperatures, and relative humidities in an ice-free valley in Antarctica. (f) A freezing model for a fine-grained soil with the formation of ice lenses (modified from Jumikis, 1956).

of water to the ice, and exerts a stronger attractive force on molecules of water which are equidistant between the soil particles (dii), thus drawing water into the thin film from the thick film until a state of equal thickness is attained. Thus, during freezing, there is a continuous migration of water towards the freezing front to form ice lenses. Soil heaving may be of large magnitude in permeable soils of silt and fine sand texture and high water tables, and the collapse on thawing will also be large; but in clays or soils with low water tables soils may become desiccated and heave will be less. In very permeable soils, of medium sand to gravel texture, capillary action is negligible and ice lenses do not form, thus limiting heave.

An important and widespread result of heaving is the transport of stones and boulders to the surface, at rates of 1–50 mm/y, from within the body of soil which has mixed grain sizes. The common explanations offered for this pheno-menon involve frost push and frost pull. *Frost push* occurs where an ice lens forms preferentially beneath larger stones which have a higher thermal conductivity than the mixed debris containing them, so that the base of the stone reaches the freezing-point before the containing soil freezes: the growth of the ice lens pushes up the stone. *Frost pull* develops where a freezing front pene-trates downwards into a soil, and as ice forms it freezes the upper part of a stone into the ice–soil mixture which is undergoing heave. Desiccation of soil around the lower part of the stone, or the growth of ice crystals around its upper part, may cause it to be held and pushed into a more nearly vertical plane while soil around it settles. Repeated push–pull movements progressively carry stones to the surface and, because voids are filled by growing ice lenses, which in thaws are replaced by soil, the process is not reversible.

The most spectacular results of heaving are the elongated blocks which are forced out of the ground and look like irregular tombstones (Plate 14.4a). Elongated, columnar, or tabular blocks are preferentially heaved to the surface—which is unfortunate if the columns happen to be the piles of buildings, of bridges, or even coffins in a burial ground! Heaved blocks are frequently tilted out of a vertical plane while they are within the active zone, but a preferred approximate vertical alignment of stones is a feature of periglacial zone soils.

More widespread than the heaving of large stones is the lifting of a soil surface by the growth of *needle ice*, called *pipkrake*. The ice needles are crystals which grow within the surface crust of a soil which is moist and losing heat rapidly by radiation to the sky. The needles are commonly 1–50 mm long, and grow daily while radiation continues and free water can migrate to the freezing front. Needle ice pushes up individual particles or the whole surface crust and so disaggregates it. Subsequently collapse of needles may cause soil to move downslope and after thawing the soil is more prone to rainwash and wind erosion. Pipkrake is particularly important in alpine environments.

The frequency of freeze-thaw cycles is an important control on the effectiveness of ground heaving and frost wedging, but the purely climatic factor of the number of times the air temperature passes through the freezing-point is not an adequate measure of effectiveness. It has been shown that although frost-wedging is common on the north-facing slopes of the European Alps, the number of freeze–thaw cycles per year is small—22 cycles on the Jungfrau. In contrast, on the south-facing slopes there are 196 cycles/y on the Jungfrau, with little frost effectiveness. The actual temperatures, their duration, penetration into the soil and rock, and the insulating effect of snow and vegetation all modify effectiveness. There are many examples of air temperatures having little relationship to ground temperatures (Fig. 14.5e) as a result of solar radiation warming dark surfaces to above melting-point even when air temperatures fall to as low as $-20\ °C$, as in Antarctica, or ground-freezing occurring at positive air temperature as a result of outgoing radiation, as in Iceland. In areas like Kerguelen Island, where there are many freeze–thaw cycles of small temperature range (200–240/y), soil frost is negligible. It was once thought that large numbers of freeze-thaw cycles per year were necessary for geomorphic effectiveness (i.e. the Icelandic type, see p. 191), but there is a growing belief by some investigators that deep penetration of fewer cycles (i.e. the Siberian type) may be more effective. Daily cycles cannot penetrate more than 10–15 cm into rock because of its low thermal conductivity.

Frost-wedging

Frost-wedging is the prying apart of materials, commonly rock, by the expansion of water upon freezing. Alternative terms are congelifraction, gelifraction, frost-riving, frost-shattering, and frost-splitting; because of uncertainties in many accounts the process is usually taken to include hydrofracturing. The prying force is not necessarily confined to the 9 per cent volume expansion which accompanies the freezing of water, but includes the directional growth of ice crystals and the extension of cracks by hydrofracturing. Some pore water may freeze and strengthen a rock, but where water can migrate and ice crystals can grow the tensile strength of rock is exceeded, and it splits. The result is that large accumulations of angular rock debris are characteristic of alpine and polar environments (Plates 7.1b, 8.11,) as are felsenmeer, talus deposits, and angular tors.

The shape and size of frost-shattered debris is largely influenced by the planes of weakness in the original rock. Thus schists, shales, and gneisses produce platy fragments. The production of such fragments reaches a maximum when there is an abundance of moisture, crossings of the freezing-point, and deep penetration of the freezing front. In many areas these conditions are most common in the spring, a conclusion supported by the frequency of rockfalls in mountains. Some cliffs in the Swiss Alps are thought to have retreated 10–25 m in the last 10 000 years by this process.

Frost-cracking

Frost-cracking is fracturing by thermal contraction at subfreezing temperatures; it can occur to a limited extent within the active layer or in soils which freeze only seasonally, but it is primarily a feature of permafrost. The phenomenon occurs because frozen ground behaves as a rigid solid which, unlike thawed soil, cannot respond to temperature changes by stretching: as temperature falls the permafrost contracts and the result is the formation of a crack. The first time cracking occurs in permafrost, it starts at the surface and extends downwards to a depth of 3 m or more. The pattern and spacing of cracks is controlled by temperature conditions and the behaviour of the ground. Crack spacing is commonly two to three times the crack depth, and cracks usually link up to form a four-, five-, or six-sided polygonal pattern (Plate 14.4b). In very cold arid environments, like those of Antarctica, the crack may stay open during the winter and summer, or it may partly close during the warmer summer period. Cracks may become partly filled with wind-blown debris (Plate 14.4c); where the ground movement is slight the surface will remain level (Plate 14.4d), but if the ground expands again the debris within the cracks will cause the permafrost on either side of the crack to heave upwards and form a low rampart on either side of the crack. In this way sand-wedge polygons are formed (Plate 14.4e).

In humid environments, like most of the Arctic, the cracks become filled with ice derived from the surface water, the ground water or water vapour. An ice-wedge starts as a vein which does not thaw in the next warm season and, because it is weaker than the permafrosted soil, cracks down its middle in the next cold season. In turn, the new crack is filled with ice and gradually an ice-wedge is formed (Plate 14.4f; Figure 14.6). The width added to the ice-wedge each year may be 1–20 mm. As a result of expansion of the ground and the ice during the warm season, the soil around the ice-wedge is bent upwards.

The largest known ice-wedges occur in Siberia where they extend to depths of 60 m and may extend laterally so that the volume of ice exceeds that of the soil. In most environments, however, ice-wedges rarely widen to more than 5–6 m.

Ice-wedges require certain temperature conditions for preservation and growth. In Alaska the southern limit for active wedges is a mean annual air temperature of −6 to −8 °C. Upon thawing of an ice-wedge its space is filled with soil. The wedge form is best preserved as a cast by sands and gravels because fine-grained soil flows into the space and destroys the wedge shape. Well-preserved wedge-casts are the best available evidence of former permafrost (see Chapter 18, plate 18.3).

Cryostatic pressure

Within the active layer, especially during the autumn, wet soil freezes at the ground surface so that a crust of rather rigid frozen soil overlies a nearly saturated soil. As the winter approaches, the freezing front penetrates downwards into the remaining wet soil and, because of the volumetric expansion of the water as it turns to ice, creates a high water pressure in the trapped water in the unfrozen soil pores. This high pore water pressure both causes the separation of the soil grains so that the soil becomes increasingly fluid and also the development of bulbous masses of wet soil. These tend to burst to the surface or create a local updoming of the ground as the frozen soil layer thins and weakens in the following spring. Where the bulb of high pressure and saturated soil reaches the surface it flows over the ground for a few centimetres and creates a mudboil (Fig. 14.7). Cryostatic pressure is a result of the transmission of freezing pressures to unfrozen saturated soil trapped between freezing fronts or between a freezing front and the impermeable permafrost table.

Patterned ground

Patterned ground is the collective term for the more or less symmetrical forms, such as circles, polygons, nets, steps, and stripes, that are characteristic of, but not necessarily confined to, the periglacial zones. They are, by their size, microrelief features and not major landforms but, because of their distinctive shapes, they have attracted much attention. The variety in patterned ground is caused by differences produced by a number of processes, variations in ground slope, soil texture, soil ice, soil moisture, and presence or lack of vegetation cover (Fig. 14.8). A further complicating factor is the uncertainty regarding the degree to which modern forms are relict from past processes or climatic regimes. All of the forms

14.4 Some features of frozen ground, (a) upthrust blocks aligned vertically but still frozen into the ground. (b) A network of rectangular sand-wedge polygons, Victoria Valley. (c) An open crack of a sand-wedge polygon (head of ice axe for scale).

mentioned above occur in both sorted and unsorted soils.

A discussion of the causes of patterned ground has to recognize that the origin of most forms is still uncertain; that many forms are polygenetic; that forms grade into one another; and that available, reliable data on soil conditions are inadequate to test many of the current hypotheses.

Types of patterned ground

Circles, polygons, and nets have the characteristic in common that they have nearly uniform dimensions in several directions, such as radii or length of side. They may be either small (< 1 m diameter) or large, and are always found in groups on nearly horizontal ground. On slopes (> 2–7°) their forms become elongated into garlands, lobes, or stripes. The forms of polygons and circles are well described by those terms: a net is a mesh whose components are made up of forms intermediate between polygons and circles.

Non-sorted circles, or mudboils, usually have a slightly dome-shaped centre, cracked into small polygons of fine-grained soil which may be free of larger fragments or have some stones with a nearly vertical orientation of the long axis. The sorted variety usually has a border of stones surrounding fine material (Plate 14.5a). A much rarer type is the so-called stone-pit with a centre of stones surrounded by fines. Debris-islands are sorted circles of fines amid blocks or boulders; they are unusual in that they can occur on slopes as steep as 30° (Plate 14.5b). In general all circles increase in size

(d) Fissures of a non-sorted sand-wedge polygon network (photo by J. D. McCraw). (e) Sand-wedge polygons with raised rims. These are part of the network shown in (b). (f) Ice wedges, Garry Island, North West Territories, Canada. The large wedge is 4 m wide at the top. Note the active layer, the lineation of the soil, and the ground ice below the wedges (photo by J. Ross Mackay). (g) Oriented pebbles in a soil. All photos, except f and g, were taken in the McMurdo region of Antarctica.

with the depth of frost action and with the size of the largest blocks. Tabular stones tend to be on edge where they are in a border or pit. Circles are best developed where there is permafrost, and because there are few large cracks or fissures associated with their margins they are usually assumed to owe most, or all, of their development to cryostatic pressures developed in fine-grained soils which may, or may not, have a vegetation cover. The stone-ring variety of sorted circles is thought to develop by both upwelling of fines and by the compression of soil beneath the load of stones around the central zone with fines (Fig. 14.7).

Polygonal forms may develop in the absence of soil-cracking, especially in soils of uniform fine texture (Plate 14.5c) and are then essentially closely fitting modifications of circles; but most polygons are associated with marginal cracks. Non-sorted polygons are those without a border of stones and they can develop both in soils of uniform texture and in those of varied grain sizes. Both sand-wedge and ice-wedge forms can be non-sorted and both may have either raised or depressed borders. A raised border is a rim formed on either side of a furrow which is immediately above the wedge (Plate 14.4e). Many depressed borders result from thawing of an ice-wedge. In low-lying areas in the Arctic the thaw season is often associated with ponding of water in furrows or in the centres of depressed polygons.

Small non-sorted polygons are often caused by desiccation of the soil during summer, but large forms are the result of thermal processes. In

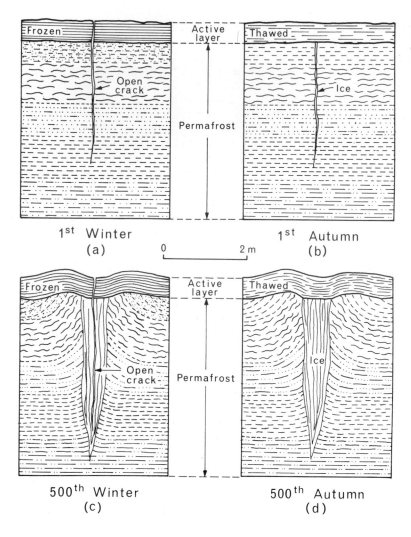

Fig. 14.6 Schematic evolution of an ice-wedge according to the contraction-crack theory (after Lachenbruch, 1962).

extreme polar conditions most polygons are of the ice-wedge kind, especially in the Arctic, and few are of the sand-wedge kind; both kinds are necessarily associated with permafrost. At the margin of the continuous permafrost zone, and thus in the areas of discontinuous or sporadic permafrost, seasonal frost-cracking may develop polygons which have the surface appearance of ice-wedge types.

Nets are not thought to be genetically different from circles or polygons and only two differences are commonly recognized. No nets are known to be as large as the largest non-sorted polygons and the features known as *hummocks* or *thufur* (Plate 14.5d) have no corresponding features in the circle or polygon variety. Hummocks are always covered by vegetation, especially grasses, mosses, and herbs, and have a core of either mineral soil, or humus, or a stone. They may be up to 0.5 m high and 1–2 m in diameter. They occur on slopes as steep as 15° but usually become elongated into lobes as slopes steepen. Because they form both in areas of permafrost and areas where permafrost has never occurred, they cannot be regarded as exclusively periglacial features (see below under 'palsas').

Steps are patterned ground with a terracette form and a downslope embankment of stones or vegetation. They occur on slopes with gradient of 3–20° and can often be seen to have developed from circles, nets, or polygons on flatter areas above them; they do, however, develop independently of circles, nets, and polygons. In the process of movement stones forming embankments tend to become aligned so that either predominantly vertical or horizontal alignments are present.

Stripes stretch directly down the slope, and

a. Cryostatic pressure

1.

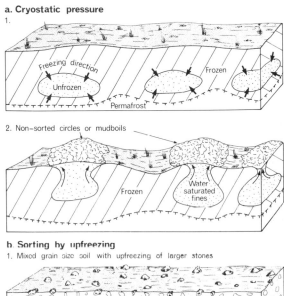

2. Non–sorted circles or mudboils

b. Sorting by upfreezing

1. Mixed grain size soil with upfreezing of larger stones

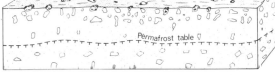

2. Upwelling of fines, sorting of stones, load deformation

3. Formation of sorted stone polygons or circles

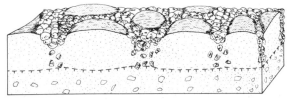

Fig. 14.7 (a) The development of non-sorted circles by cryostatic pressure on a bulb of unfrozen saturated sediment.
(b) The development of sorted stone circles by sorting of stones with uplift, then by upwelling of fines and further sorting of stones.

because they originate both in sorted and non-sorted circles, nets, and polygons, or develop from the start on slopes, they have a variety of sizes, forms, and origins. In both sorted and non-sorted stripes, movement is usually greatest at the depressed centre of the stripe and decreases towards the sides and with depth. Rates of movement vary depending upon gradient, moisture, texture, and vegetation, but may be up to 200 mm/y. In sorted patterns both the coarse and fine stripes move, but the stripe with fines moves fastest and can sweep coarser material aside so that it becomes concentrated in the coarse stripe. Many reports of stripes come from alpine areas where stripes are known to reform in a year or less, after being deliberately mixed, and where rilling, surface wash, and needle ice can cause sorting and selective downslope removal of certain particle sizes. Small stripes are not necessarily periglacial phenomena, but large ones originating as circles, nets, or polygons are usually associated with permafrost. The actual mechanisms of stripe formation originating on a slope are not well understood, but involve preferential growth of ice crystals in fines, and plastic behaviour of fine-grained soil in thaw seasons.

The only patterned ground phenomena which are unique to a periglacial environment are those formed by thermal cracking of the ground, since they indicate permafrost or extreme cold. Most other processes can occur in several climatic regimes as a result of variations in soil moisture. This is of considerable significance in attempts to interpret past environments from relict forms.

Non-patterned periglacial forms

Soil involutions, stone pavements, string bogs, palsas, and pingos are minor features which are commonly, but are not all exclusively, associated with periglacial environments.

Involutions

Within soils of the periglacial zone cryostatic pressures may break and displace soil horizons into irregular forms called *cryoturbation* structures, or into the more regular features known as *periglacial involutions*. Variations in vegetation cover, soil moisture, texture, and micro-relief may all influence the rate at which the freezing front moves downwards towards, or upwards from, the

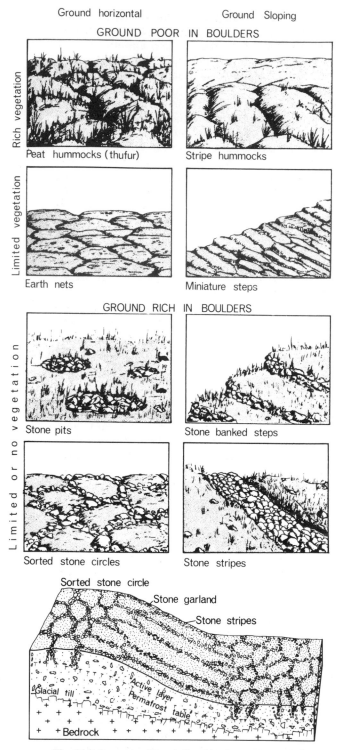

Fig. 14.8 Some possible relationships between types of patterned ground, vegetation, and ground slope (modified from Lundqvist, 1962).

permafrost and enclose bulbs of unfrozen soil. Where the cryostatic pressure bulbs break to the surface thufurs or boils may develop, but local differences in the density of saturated soil may cause some bulbs to sink and form nearly spherical, isolated masses or connected pods of displaced material. Alternatively veins or bulbs may be injected upwards into the soil. Many involutions and injection features are small, with vertical amplitudes of 0.25–2 m, but in soft rocks, such as lignite and unconsolidated sands and silts, with properties which vary vertically, the features may extend to depths of 10 m (Fig. 14.9).

Stone pavements

Stone pavements, or boulder pavements, usually occur in valleys where the ground is saturated. They differ from felsenmeer in that the stones may not be derived locally but be part of an alluvial, glacial, or colluvial deposit, and the underlying material is usually fine-grained. This implies that the large stones have been carried to the surface by upfreezing or that fines have been removed from the surface by wash or deflation. Overburdens of snow or ice may have caused high water pressures beneath some tabular blocks and caused them to be pushed to the surface where they 'float' in the dense saturated soil. Individual blocks, however, can be depressed by a load placed on them (Plate 14.6). Although they are common in Arctic areas it is not certain that they are unique to a periglacial environment.

String bogs

String bogs are areas of peatland, known as *muskeg* in Canada, characterized by ridges of peat and vegetation separated by depressions which frequently contain shallow ponds. The ridges are up to 2 m high and may be aligned across or along the slope which may be as great as 2°. The ridges consist of peat containing ice lenses for at least part of the year (Plate 14.7a).

The hypotheses of their origin involve elements of biotic, ice, and frost action. The habit of some plants, such as sphagnum moss, to grow in clumps and thus produce an irregular surface of humps and hollows is, presumably, a basic factor. Frost-heaving, and the seasonal growth of ice lenses, emphasizes the surface irregularities, but the formation of strings requires the gravitational movement of saturated peat and timber carried by floods. Many string bogs occur in the zone of discontinuous permafrost and some are probably the result of thawing permafrost in areas of poor drainage, but there is no invariable relationship with permafrost, nor is flood debris always available to be trapped and incorporated into ridges.

Palsas

Palsas (Plate 14.7b) are mounds of peat containing perennial ice lenses. They are often irregular in form, rounded, or elongated, with widths commonly in the range of 10–30 m and lengths of 15–150 m; heights range from 1–10 m.

Palsa surfaces are frequently cracked as a result of doming or desiccation. When cracks are deep, the ice inside may be exposed and it can then thaw and collapse, but as the ice lenses are seldom more than a few centimetres thick this does not result in the formation of a hollow. Palsas are essentially permafrosted peat mounds which grow because the better-drained part of a mound has a higher thermal conductivity than that of saturated peat, so that frost-heaving and ice lens formation are more effective in a mound than in its surrounds. Palsas and thufur may be genetically closely related, with palsas occurring in damper sites with deep peat, and hence being associated with bogs, whereas thufur can occur on well-drained uplands if the climate there is humid.

Pingos and other ice lenses

Pingos are large, perennial, ice-cored mounds ranging in height from 3–70 m and in diameter from about 30–600 m (Plate 14.7c). Most are more or less circular but a few are elongated. In outward appearance many look like palsas as they are covered by soil and vegetation, and have cracks over their surface. The essential difference is the ice-core, and the soil may be of peat or be of mineral material. The ice-core often makes up much of the volume of a pingo and may extend some metres below the ground surface. Deep cracking leads to decay of the pingo with the formation of a central crater (which may be breached or contain a lake), then the collapse of the walls so that a central depression is left with a surrounding rampart (Fig. 14.10). Pingo craters and ramparts have been recognized in western Europe as relict features from the Last Glacial.

Pingos are thought to grow in two main ways—by cryostatic pressure and by artesian pressure (Fig. 14.11 a,b).

(1) A lake which has no permafrost beneath it, but permafrost around its margin, becomes filled with sediment: As the water volume of the lake diminishes the permafrost encroaches and traps water and gas in the lake bed sediments. Freezing and expansion of the trapped water leads to the formation of an ice lens which grows as water under pressure rises beneath it. Once all the groundwater is frozen the growth of the ice lens, and the dome, ceases, unless freezing draws water from overlying soil and creates segregated ice. This type of closed-system pingo is very common in the delta of the Mackenzie River, northern Canada. Cryostatic pressures may also develop below oxbow lakes, abandoned river channels, along faults or uplifted portions of a former sea floor, and give rise to pingos.

(2) Artesian pressures are likely to occur where underground water is prevented from returning to the soil surface by impermeable soil or by seasonal freezing of the soil surface. For a pingo to form the water has to be under artesian pressures which are higher than those measured in other climatic zones. This suggests that a form of cryostatic pressure beneath a local permafrost lens is required before water will force its way through a talik to form an ice lens and dome up the surface. This type of pingo was first recognized in East Greenland and is known as the open-system type.

Maximum rates of vertical growth of pingos are about 0.5 m/y. All of those which have been dated are less than 10 000 years old and many small ones in the Arctic have formed in the last few hundred years. All pingos are associated with permafrost, but because they form only under very limited conditions they are far less common than the attention given to them in the literature might suggest. There are only a few thousand in the Arctic at present.

Many small ice lenses exist in lowland and badly drained areas of the Arctic. Many of these survive only for one or two seasons and are usually the result of the formation of segregated ice, or trapped water under cryostatic pressure below a frozen soil.

One feature of parts of the Mackenzie Delta and northern Siberia which is not understood is the presence of massive ice lenses and large volumes of segregated pore ice, as on Richards Island where

14.5 Patterned ground. (a) Sorted stone circles (photo by J. D. McCraw). (b) Sorted stone circles on a slope are also called debris-islands. (c) A non-sorted net, of nearly polygonal form, in a uniform silty sand. These features are probably caused by desiccation.

14 per cent of the upper 10 m of permafrost consists of ice. It follows that if this layer thawed the ground surface would subside 1.4 m. About 36 per cent of all excess ice is wedge ice and much of the remainder is pore ice. In parts of Siberia, and especially in Yakutia, ice exposures more than 70 m high form cliffs in some river valleys and offshore islands. Some islands have disappeared as a result of thawing of massive ice.

The origin of massive ice and icy sediment is still unclear; there are at least three hypotheses for their origin. (1) Buried snow, lake, river, or sea ice is one possible source. (2) Much of the ice may be exceptionally well-developed wedge-ice which has grown as alluvial deposits have been laid upon older floodplain surfaces. (3) The ice may be produced by the freezing of drawn-up pore water, which was trapped in lenses of sands and gravels beneath a permafrost zone which was deepening on an emerging coastal plain. For the Arctic areas of Canada this last hypothesis is generally favoured with the sea-level lowering at about 100 000 years ago, as the Last Glacial began, being a primary initiator of the permafrost growth in coastal areas. For Siberia there is evidence of massive wedge-ice formation over a large part of the last million years or so.

Hillslope processes and forms

Processes

Hillslope forms and processes in the periglacial zone are similar to those in other climatic zones, but the intensity of processes and the occurrence of particular landforms is often different from those in other environments.

The unique feature of periglacial slopes is the presence of phenomena which develop most effectively in areas with ground ice and snow patches. Ground ice, and especially permafrost, limits, or prevents, soil drainage during the thaw period, and melting soil ice adds water to the soil, so that

(d) Thufurs, at 1600 m a.s.l. in Central Otago, New Zealand. (e) Non-sorted steps of blocks giving way to stripes. (f) Unsorted stone stripes. (g) Sorted stone stripes about 30 cm apart. Note the rill in the fine material, New Zealand Alps. All features except (d) and (g) are from the McMurdo region of Antarctica.

in areas of fairly high winter snowfall, or summer rainfall the soil of an active layer becomes saturated: the pore water pressures rise rapidly and become exceptionally high because of poor drainage, and soil can then flow. Slow flowing of soil from higher to lower ground is called solifluction, but to distinguish solifluction in the presence of ground ice the term *gelifluction* is commonly used. Gelifluction movements are usually greatest in early summer and decline towards the autumn when drying winds and drainage have decreased soil moisture. Annual rates of movement are usually in the range of 5–40 mm but may be faster in lobes which are not restrained by a vegetation cover.

Gelifluction creates extensive sheets of rock or soil debris on slopes with angles as low as 1–3°. The sheets may be up to 1 m thick and they mask parts of the landscape and give it a smooth form (Plate 14.8a), both because of the presence of a moving sheet of debris and also because this sheet erodes protuberances of bedrock. Gelifluction sheets often incorporate large boulders which may be carried along at the same rate as the rest of the flow, or they may travel faster over a local slide plane as *ploughing blocks* which leave a track behind them. In areas devoid of vegetation gelifluction sheets are often called rubble sheets or *block sheets*, and *block streams* where they are of limited lateral extent. Such sheets may originate in felsenmeer or result from the accumulation of rock slope debris in shallow valleys, but they are only mobile where interstitial fine soil is available to hold moisture. The movement process often causes tabular blocks to lie nearly parallel to the slope, and taluses, or sheets and streams, with this common parallel layering of debris are said to be *imbricated*. Where plants are common the downslope limit of a sheet is often marked by a bench or riser. Lobes and steps develop on both low-angle and steep slopes of 3–20° and both on vegetated and block sheet slopes. They are fronted by turf

Injections

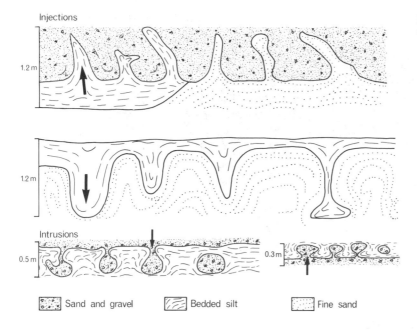

1.2 m

1.2 m

Intrusions

0.5 m

0.3 m

Sand and gravel   Bedded silt   Fine sand

Fig. 14.9 The formation of some types of periglacial involutions.

risers or stone walls up to 3 m high. Steps often develop below snow-patches and at the foot of slopes. Lobes may occur anywhere on a slope and often override one another.

Frost creep associated with the growth of needle ice and the heaving of freezing ground is a common slope process especially where debris is coarse, but it is usually regarded as being of less importance than gelifluction in slope downwearing and transport of debris. Too little evidence is available to test this assumption.

Slope wash and solution are slope processes which can occur in any environment, but in the periglacial zone wash may be more important than is usually recognized because the formation of soil ice disaggregates the soil, and on thawing individual mineral particles are readily transported. Layered slope wash deposits with imbrication of the larger fragments are often regarded as characteristic of periglacial environments and are commonly called by the French name *grèzes litées*. *Thermo-erosional wash* is regarded as being a major process in the flattening of periglacial slopes, to very low angles, by many Soviet workers. The evidence for this conclusion is not generally available, but the widespread formation of pipes and small collapse hollows by interflow and *suffosion* may be part of the process. Rates of

solution are probably low in the Arctic regions because of the limited duration of the thaw.

Rapid mass movement in saturated active layers is often associated with gelifluction and slush flows on steep slopes, and earth slides are common in areas of vegetation cover. The most dramatic failures, however, are discussed later and under the heading of thermokarst.

Slope forms and development

Periglacial slopes exist under a great range of temperature, moisture, and vegetation regimes with an equally large range of slope types and angles. In mountain areas all slopes between the treeline and the snowline can be regarded as part of the periglacial zone in the broad sense. At lower altitudes, in high latitudes, the abundance of free faces, talus slopes, and valley side tors are clear evidence of the importance of physical weathering, much of it having occurred since the retreat of the Last Glacial glaciers 8000 to 10 000 years ago.

The distinctive periglacial slope features are the extent of gelifluction sheets, the presence of stone stripes and related features, marked valley asymmetry in many areas, the formation of erosional cryoplanation terraces, and the formation of cryopediments.

(a)

(b)

14.6 (a) Stone pavement forming the floor of part of Wright Valley, Antarctica. (b) Individual stones 'float' in finer-grained sediment and can be depressed by a load.

Debis-mantled slopes, with angles up to 30°, result both from the effectiveness of physical weathering and gelifluction and also from the relative ineffectiveness of fluvial processes in many low-order drainage basins, because of low precipitation and hence low total runoff. Many such basins have smooth debris slopes extending to the valley centre with only limited stream channel development. Asymmetry develops because nivation is favoured by greater snow accumulation on one slope than another, by greater vegetation growth on wetter or sunnier slopes or those protected from drying winds, and because pole-facing slopes suffer less thawing than those facing towards the equator. Although there are many exceptions there is a general tendency for steeper slopes to be pole-facing in the high Arctic.

Cryopediments form at the base of hillslopes. In Siberia they are wash or gelifluction slopes with a thin veneer of soil and rock debris over eroded bedrock, and have slopes of 1–8° which extend for several kilometres from floodplains, or river terraces, to the foot of the hillslopes. They are essentially transportational slopes left by the parallel retreat of the hillslope which is undercut at its foot by nivation. They appear to be commonly recognized in the Siberian taiga (area of low, open larch forest). In Antarctica small pediments have formed at the base of bare rock slopes retreating as a result of physical weathering and the removal of the sand debris by wind.

On middle and upper slopes cut terraces in bedrock are known as *goletz*, *cryoplanation*, or *altiplanation* terraces. They are the equivalent in form, size, and origin to cryopediments of foot-slopes (Plate 14.8).

Slope development in the periglacial zone has been described as being distinctive as a result of progressive destruction of tors and free faces, the downwearing and smoothing action of debris sheets, and the filling in of hollows with that debris. It is unclear how rapidly periglacial destruction of a landscape occurs. The limited evidence available suggests that it is not an unusually

14.7 (a) String bogs in Labrador. (b) A palsa in Quebec Province. The palsa is composed of peat 1 m thick over a silty-sand soil. The permafrost table is at a depth of 0·6 m. There is no permafrost below the surrounding terrain (photos (a) and (b) by Division of Building Research, Canada). (c) A pingo of the open-system type in Eskerdalen, Spitsbergen (photo by E. A. FitzPatrick).

(a)

(b)

(c)

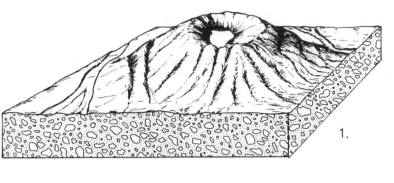

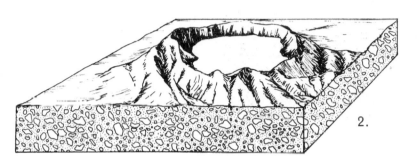

Fig. 14.10 The wasting of a pingo by exposure of the core and then melting.

rapid process and in arid areas like Antarctica is immeasurably slow. Many slopes in the periglacial zone have been largely 'inherited' from periods of glacial erosion so that it is not possible to recognize the results of slope evolution under a periglacial regime—only a superficial modification of a formerly glaciated one. In limited areas of eastern Siberia periglacial processes may have been dominant for two million years or more, but large areas are lowlands and the emphasis in research there has, so far, been upon studying processes rather than attempts to construct idealized sequences of landform development.

Fluvial processes

Fluvial action in the periglacial zone is different from that under other climatic regimes primarily because of the presence of icings and thick river ice in winter, and because of floods during the summer. Permafrost by itself does not increase or decrease river erosion.

In Alaska icings on floodplains can be up to 50 km long and 5 km wide, and in Siberia individual icings are reported to have areas of over 100 km². Icings may block channels during spring melts and cause greater bank erosion. The geomorphic effects are, however, relatively slight or little known.

The most distinctive fluvial phenomena of the periglacial zone are broad, braided channels of many small and medium-size rivers. Such channels result from an abundance of bedload, a lack of fine sediment, the availability of coarse debris derived from slopes, and of debris left by glaciers. Rivers are often widely braided because debris transport is very episodic. In areas of the polar deserts runoff is always low and confined to two or three months of summer with a spring peak flood. In the Arctic with higher snowfall the spring flood is more marked but it may be geomorphically ineffective if it occurs when channel floors are still frozen. The most effective regime is often that which derives much of its flow from glacier melting which reaches a peak in mid to late summer when stream beds have thawed. In the muskeg zone flows are more uniform because of the retarding and storage effect of the vegetation and soil. In the Arctic many rivers flow from south to north, consequently the thaw begins in the headwaters and moves as a wave down the channels, but the blocking action of ice floes causes widespread flooding.

Very large rivers like the Lena, Ob, and Mackenzie have many characteristics influenced by their sources on the southern margins of the periglacial zone. These rivers carry a large proportion of the sediment as suspension load and they are large enough to create a talik beneath their beds. As a result they can incise and maintain single channels. They braid only where they form deltas. Large rivers are also able to erode their banks by mechanical abrasion and by thermal erosion, or melting of the permafrost. In unconsolidated sediments this produces a *thermo-erosional niche*

14.8 Periglacial slope phenomena. (a) Block sheets on slopes, and a block stream in the valley floor, are evidence of gelifluction. Note the smoothing of the slopes. (b) Valley-side tors and talus slopes.

(a)

(b)

at the floodwater level. In parts of Siberia these niches may be 3 m high and 1–3 m deep. Such niches lead to the collapse of large blocks from the bank and to flows of the thawing block material (Fig. 14.12). Such processes may be significant in the widening of many channels and valley floors and may also promote the development of braided patterns.

There is considerable debate about the rate of incision of many small and medium rivers of the periglacial zone. The abundance of glacial and gelifluction-derived debris suggests that many rivers are overloaded and cannot transport all of the available debris, and consequently are not incising. By contrast some German workers, in particular (see Chapter 1, p. 18), believe that frost

(c)

(d)

(c) Cryopediments in sandstone covered with only a thin veneer of debris; note the nivation hollows and the undercut benches which may be a result of nivation and (or) salt weathering, Olympus Range. (d) Cryoplanation terraces at 1600 m in Central Otago, New Zealand (photos (a), (b), and (c) were taken in the McMurdo area, Antarctica).

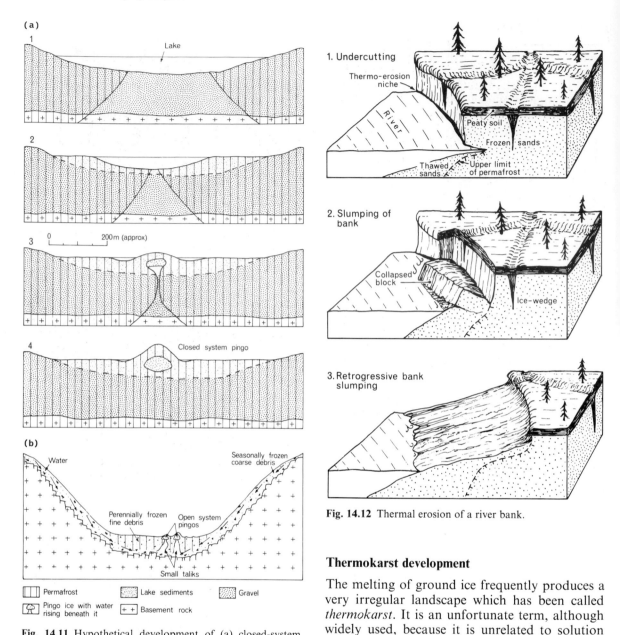

(a)

1 Lake

2

3   0   200m (approx)

4   Closed system pingo

(b)   Water   Seasonally frozen coarse debris

Perennially frozen fine debris   Open system pingos

Small taliks

☐ Permafrost   ▨ Lake sediments   ▨ Gravel

▨ Pingo ice with water rising beneath it   + + Basement rock

Fig. 14.11 Hypothetical development of (a) closed-system pingos and (b) open-system pingos.

1. Undercutting

Thermo-erosion niche

River

Peaty soil

Frozen sands

Thawed sands   Upper limit of permafrost

2. Slumping of bank

Collapsed block

Ice-wedge

3. Retrogressive bank slumping

Fig. 14.12 Thermal erosion of a river bank.

Thermokarst development

The melting of ground ice frequently produces a very irregular landscape which has been called *thermokarst*. It is an unfortunate term, although widely used, because it is unrelated to solution processes and few of its features have much resemblance to the landforms developed on limestones.

Thermokarst processes are believed to be among the most important of those shaping the periglacial landscape. They are most active in the western Arctic of North America and in northern Siberia where alluvial sediments with high ice contents are widespread. In alpine areas they are of limited extent and of little importance.

shattering of channel floors permits very rapid incision into bedrock. So few measurements of bedload transport have been carried out that no firm conclusions can be drawn.

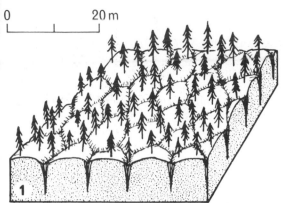

1

Lowland with ice wedge polygons and larch taiga, thawing begins

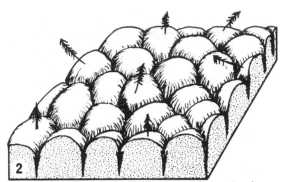

2

Polygon depressions deepen, conical baydzharakhs develop, vegetation cover is broken, erosion begins

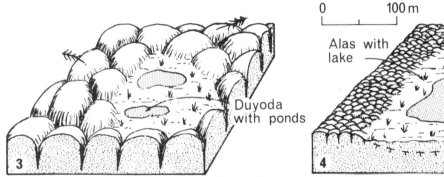

3

Duyoda with ponds

Duyoda forms with baydzharakhs and solifluction on the slopes

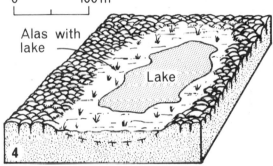

4

Alas with lake

Lake

Flat floored, grass covered, alas

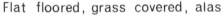

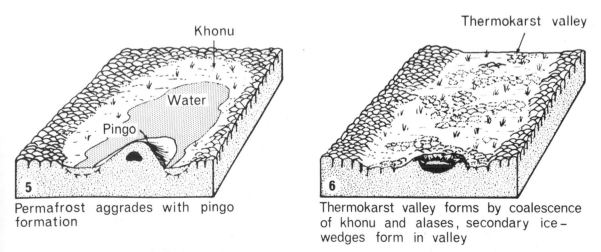

5

Khonu

Water

Pingo

Permafrost aggrades with pingo formation

6

Thermokarst valley

Thermokarst valley forms by coalescence of khonu and alases, secondary ice-wedges form in valley

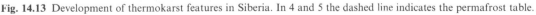

Fig. 14.13 Development of thermokarst features in Siberia. In 4 and 5 the dashed line indicates the permafrost table.

There are many possible causes of ground ice melting including: climatic change towards warmer and drier summers; changes in vegetation cover, especially thinning of that cover by natural or man-made fires; the laying bare of soil for construction, or by slumping and earthflows; and the development of standing water. On a large scale climatic change is probably the most important cause, but locally any loss of insulating cover of the ground ice can induce melting even in areas where mean annual temperatures are as low as − 5 °C. In parts of Siberia where the mean annual temperature is −9 °C the presence of summer thaw water along polygon cracks above ice-wedges can lead to further melting, and create self-induced thermokarst without a change in external conditions. Thermokarst phenomena include decaying pingos, meltwater gullies and pipes, and many forms of ground subsidence.

Ground subsidence

Where ground ice makes up a considerable proportion of the soil volume any large-scale melting will cause the ground to subside and may create irregular hollows and hills. In Yakutia a sequence of thermokarst landform development is recognized (Fig. 14.13). Conditions there are extreme because summer air temperatures are often high, reaching 30 °C in August, so that the active layer is 2 m or more in depth; also 60–80 per cent of the upper part of the permafrost is composed of ice.

Where the active layer is becoming deeper the ice-wedges begin to thaw and high-centred polygons with conical surface expression, called *baydzharakhs*, form. They may be 3–4 m high and 3–20 m wide. Collapse of decaying mounds leads to the formation of continuous depressions, with steep slopes, called *duyodas*, and these may be progressively enlarged and deepened until a thaw lake is developed at the bottom. Such a treeless feature is called an *alas*. Once an alas lake becomes deep enough to promote a talik beneath it the system may stabilize, drain to a lower alas or river valley, or it may become infilled with sediment from its margins and the permafrost may then rise again and form new patterned ground features on the old lake bed. Alternatively pingos may form in the lake, then the whole depression is called a *khonu*. Further widening of one alas, or joining up with other alasses, creates thermokarst valleys or depressions with an area of up to 25 km².

Large areas of Yakutia are formed of such thermokarst landforms: many are several thousand years old. At least 10 per cent of Yakutia has active thermokarst development.

In North America the full range of alas formation has not been identified but the early stages do occur. The limited range of summer temperatures may prevent more extensive development in Canada.

Backwearing thermokarst

Ground ice slumps and thaw lakes are widely reported from the Arctic. They form where ground very rich in ice, and with a slope of a few degrees, has a locally deep active layer. The saturated soil develops high pore water pressures and produces a low-angle flow backed by a headscarp (Plate 14.9a). Although such features are called slumps they are not rotational and lack many of the features of lower latitude slumps. The headscarp retreats as a result of exposure and thawing at rates of several metres per year. Most features have maximum dimensions of 200–300 m and stabilize themselves as soon as the headscarp is less than about 2 m high.

Perhaps the most common thermokarst features of lowland periglacial environments are shallow rounded depressions with semicircular, or circular, ponds or lakes in them (Plate 14.9b). The initial cause of *thaw lakes* may be the random melting of ground ice and the accumulation of water in the depression. Most are shallow (< 1 m deep), but they can attain depths of 4 m, and seldom have widths greater than 2 km. Most thaw lakes are temporary features which migrate laterally as banks thaw and the old lake bed emerges as a shallow depression, as silt and peat accumulate, or the lake drains to a stream. Most thaw lakes have a life span of less than 3000 years.

Much controversy exists over the causes of the near-circular or elliptical shapes and the regional orientation of many lakes. It is, presumably, related to the effect of the wind upon wave action and upon sediment transport and erosion by those waves, but the critical factor may be preferential thawing of the shore exposed to the wind, while lee shores are protected by beaches. Such lakes are different from the smaller beaded ponds which are the intersection points of thawing ice-wedge polygons linked by flooded troughs above the ice-wedges.

14.9 (a) An earth flow, in ice-rich sediment, with a head scarp which retreats at 2–6 m/y. The scarp is 8 m high, Garry Island, North West Territories, Canada (photo by J. Ross Mackay). (b) Thermokarst lakes, Mackenzie Delta (photo by J. Shaw). (c) Oriented thermokarst lakes on the flood plain of a freely meandering river. Note the old meander forms and ox-bow lakes. Old Crow River, Yukon, Canada (photo by Department of Mines and Technical Surveys, Canada).

Conclusion

The periglacial zone is extensive and of increasing economic importance because of mineral exploitation in lowlands and recreational use in mountains. Much research remains to be done so that details of processes may be understood, and so that the periglacial relicts left in the presently temperate zone may be recognized and their environment of formation estimated (see Chapter 18).

Further reading

The literature on the periglacial zone is voluminous not only in English but in Russian, French, German, Polish, and the Scandinavian languages. Valuable reviews have been published by Jahn (1971), Embleton and King (1975), French (1976), and Washburn (1979). An *Illustrated Glossary of Periglacial Phenomena* by Hamelin and Cook (1967) is a useful reference text but it contains few of the Russian terms applied to major features of Siberia. German work is reviewed by Karte (1981) and periglacial phenomena in China by the Research Institute (1981). Harris (1981) has reviewed periglacial mass wasting.

15   Glaciers and Glaciated Landforms

During much of geological time Earth has been free of major glaciers, but at intervals of about 150 million years ice ages occur: each of these lasts a few tens of millions of years. Each ice age is a period in which ice sheets, like those of Antarctica and Greenland, form somewhere on the continents. The Earth is thus in an ice age at present and about 10 per cent of the land surface is covered by glaciers. In the period centred on 18 000 years ago, ice was far more extensive and covered about 32 per cent of the land (see Fig. 16.24). Over the last million years, and to a lesser extent over the two million years before that, glaciers have alternately expanded to a maximum volume in periods known as *glacials* and contracted to a lesser volume, like that of the present, in periods known as *interglacials*. The present interglacial will certainly be followed by the next glacial within a few thousand years. The causes, effects, and patterns of ice ages and their climates are discussed more fully in Chapter 16. In this chapter the nature of glaciers and the work of erosion and deposition which they perform will be described.

Glaciers are important agents of geomorphic change not only because they cover substantial parts of the land surface, but because they have left their imprint upon much of the land they formerly covered, and meltwater draining from the glaciers has extended their influence far beyond the margins of the ice. A second feature which makes them important is that about 75 per cent of the world's fresh water is presently locked up in glaciers and ice sheets. Much of this water is inaccessible, but life is so dependent upon fresh water that serious investigations of the possibility of towing Antarctic icebergs to the desert coasts of the Arabian Peninsula and southern Africa are being carried out.

Glaciers

Ice forms and temperature

Glaciers are major bodies of ice which can move under the influence of gravity. They are of two main types: (1) *ice sheets* which have a flattened dome-like cross-section and are hundreds of kilometres in width (Fig. 15.1), or the similar shaped *ice caps* which have areas of less than about 50 000 km^2; (2) *valley glaciers* which occupy basins or valleys in upland areas (Plates 15.1, 15.2); a more detailed classification is given in Table 15.1.

In most environments glaciers are formed from snow which is gradually converted to ice. Snow crystals have an open feather-like appearance, but as the tips of the crystals are compacted, or melted, they turn to small globules and form a mass of loosely consolidated ice crystals, with interconnected air spaces between them, called *firn* or *névé*. Firn is usually developed in snow which has survived a summer melt season without completely melting. Further modification of firn leads to elimination of the air spaces and growth of the ice crystals until pure ice, with a density of about 0.9 Mg/m^3, is formed. Where temperatures fluctuate frequently about 0 °C, the transition from snow to firn and then to glacier ice is accomplished in 1–5 years within the upper few metres of the glacier. In a very cold climate, like that of central Antarctica (-30 to -80 °C), the transition takes several thousand years and may extend 100–200 m down from the surface. On ice sheets snow falls directly onto the glacier surface, but on valley glaciers much of it may accumulate on slopes and reach the glacier by avalanching. Other sources of glacier ice include *rime ice* produced where water droplets in the air freeze directly onto a glacier

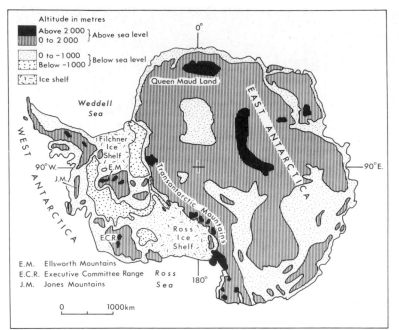

Fig. 15.1 (a) Bedrock relief in Antarctica indicating the area which is below sea-level, and the mountain ranges, all of which are buried by the ice sheet.

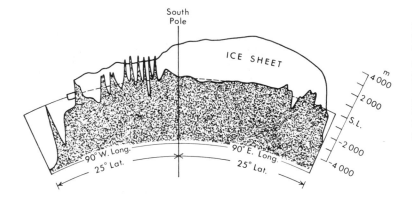

Fig. 15.1 (b) A profile along 90° longitude indicating that much of the West Antarctic Ice Sheet is grounded on bedrock below sea-level, and held in place by grounding on an archipelago. East Antarctica has the main ice dome and is separated from West Antarctica by the Transantarctic Mountains through which outlet glaciers drain into the Ross Ice Shelf.

surface, and streams and meltwater which refreeze and become incorporated in the glacier. Rime ice is of greater importance at high altitudes on polar ice sheets. In continental areas like northern Siberia and Canada, where snowfall is limited, refrozen meltwater can make up much of a glacier. The different types of ice are distinguishable by

their crystal structure and because firn accumulates in annual layers on the glacier surface.

The behaviour of ice is largely controlled by its temperature. Ice temperatures are influenced by heat derived from three sources: the atmosphere, the geothermal heat flow from rock below the ice, and from friction within the glacier and at its bed.

15.1 The East Antarctic Ice Sheet is in the distance and an ice stream drains from it to form an outlet glacier through the Transantarctic Mountains (right). Note the local ice fields on the mountains and the local glaciers flowing from the ice fields into buried cirques and then the outlet glacier (photo by NZ Department of Scientific and Industrial Research).

Atmospheric heat is controlled by direct solar radiation and air temperature, and also by the temperature of accumulating snow; latent heat is transferred as water freezes and snow is converted to glacier ice. Geothermal heat reaching rock surfaces from within the crust flows into glacier ice at an annual average of 60 mW/m² which is enough to melt 6 mm of ice at 0 °C per year. In volcanic regions, such as Iceland, the heat flow may exceed 100 W/m², but at the ground surface in cratons may be less than 40 mW/m². Frictional heat is produced as slipping occurs along the planes within ice crystals and as they move against one another or against rock surfaces; a rate of ice movement of 20 m/year within a glacier can produce as much heat as the average geothermal heat flow, and a movement of 100 m/year five times as much. As many large glaciers move at rates of 100–10 000 m/year frictional heat production has a major effect upon ice behaviour.

Two types of ice are recognized with respect to internal heat:

(1) *warm ice* has a temperature close to 0 °C and may contain water;

(2) *cold ice* has a temperature below its *pressure melting-point* (Fig. 15.3).

The term pressure melting point is used because the temperature at which water freezes is reduced

15.2 Névé fields, cirque glaciers, and a valley glacier (in the foreground) on the western side of the New Zealand Southern Alps. Note the ice falls and crevasses formed in these steep and fast-moving glaciers (photo by Mannering and Donaldson). (*Copyright*: Guy Mannering Collection: Copyright Reserved).

Table 15.1   Types of Glacier

Type	Description
Ice sheet	$> 50\,000$ km^2. Buries underlying relief. Flattened dome in shape.
Ice cap	Dome shaped glacier which buries the landscape. $< 50\,000$ km^2.
Ice dome	The central part of an ice sheet or ice cap.
Outlet glacier	An ice stream draining part of an ice sheet or ice cap. It may pass through confining mountains.
Ice shelf	A floating ice sheet of considerable thickness attached to a coast.
Ice field	A relatively flat and extensive mass of ice.
Valley glacier	A body of ice flowing down a rock valley and overlooked by cliffs.
Cirque glacier	A small ice body generally occupying an armchair-shaped hollow in bedrock.
Niche glacier	A small body of ice of an upland lying upon a sloping rock face or shallow hollow which it has modified only to a limited extent.
Diffluent glacier	A valley glacier which diverges from a trunk glacier and crosses a drainage divide.
Confluent glaciers	Two valley glaciers which converge.
Expanded foot glacier	A valley glacier which develops a widened snout where it emerges from confining rock walls.
Piedmont glacier	A large-scale widened glacier formed on a lowland or at a mountain foot. Piedmont glaciers have a flow system which is independent of the valley glaciers and ice falls which feed them.

Table 15.2   Numbers, Areas, and Volumes of Modern Glaciers

Type	Number	Area (km^2)	Volume (km^3)
Local	10 000	865 000	200 000
Greenland Ice Sheet	1	1 726 000	2 700 000
Antarctica	1	12 336 000	30 110 000

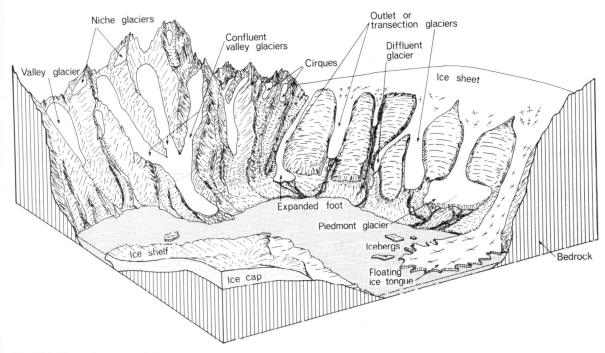

Fig. 15.2 The main classes of glacier.

under pressure at a rate of about 1 °C for every 14 MPa. The melting-point of ice at the base of parts of the Antarctic Ice Sheet is consequently about −1.6 °C, under a pressure of 20 MPa induced by the ice overburden. Cold ice forms at temperatures so low that there is little or no melting, and hence is characteristic of the surface of the Antarctic and Greenland ice sheets and also of many high altitude glaciers outside the tropics. Cold ice forms the surface few metres of most warm glaciers during winter. Warm ice forms wherever there is sufficient heat to raise ice temperatures to the pressure melting-point. One glacier may be composed partly of warm, and partly of cold, ice. At Byrd Station, Antarctica, the average surface temperature of the ice sheet is about −28 °C, but at the bottom of the 2164 m drill hole ice was found to be at the pressure melting-point. Because of possible ambiguity the old terms of 'polar', 'subpolar', and 'temperate' ice are now little used, as types of ice are not confined to set latitudes.

On many glaciers heat is transferred primarily by meltwater which is at, or above, the pressure melting-point; refreezing does not occur and latent heat is not released to cause further melting. If this were not so warm glaciers would soon disappear.

Ice sheet temperatures are of great importance because they control the behaviour of the ice and the processes occurring at its base. Six sets of influences may be recognized although the interaction of these influences is not simple. (1) Ice thickness controls the stress, and hence partly controls the temperature at the base of the ice sheet, but it also influences (2) the temperature of

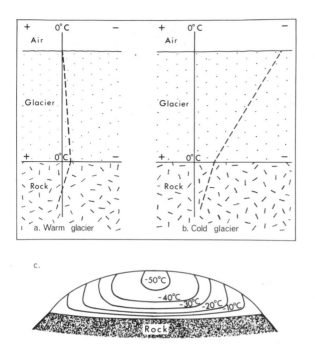

Fig. 15.3 (a) The dashed line indicates the temperature within a warm glacier to be close to 0 °C throughout. (b) The temperature in a cold glacier is well below 0 °C and rises towards the base because of frictional and geothermal heat. (c) In a cold ice sheet temperatures are lowest at the summit of the dome (which may be 3 km a.s.l.) and increase towards the base and the periphery.

the ice sheet surface because the surface temperature varies with altitude. In Antarctica surface temperature decreases at a lapse rate of about 0.01 °C/m rise in elevation; this means that for every 100 m rise in altitude the temperature falls by 1 °C. The thicker the ice and the warmer the surface temperature, the warmer the base of the ice tends to be. (3) The accumulation rate is important because it controls the rate at which accumulating firn is carried down into the ice. Basal temperatures tend to be lower where there is a high accumulation rate which carries cold down into the ice most effectively. Low accumulation rates favour higher basal temperatures. (4) The transport of cold ice to lower parts of the ice sheet reduces temperatures there and is influenced by ice velocities. (5) Geothermal heat and (6) frictional heat are both concentrated near the base of the ice sheet.

Mathematical models using these six character-istics (or surrogates derived from estimates of surface slope of the ice, velocity, vertical temperature gradient, basal shear stress, ice conductivity, accumulation rate, strain rate, warming rate, ice thickness, and base gradient) have been constructed and are used to calculate the temperature and availability of meltwater at the base of the ice. In East Antarctica it may take 100 000 years for ice to travel from the surface to the bed, so the rate of flow can only be estimated indirectly.

Ice movement

Glacier ice moves by three main processes: (1) internal plastic flow, or creep, of the ice; (2) alternate compression and extension of the ice mass in response to changes in the bedrock surface below the ice; and (3) by sliding of the ice over bedrock.

(1) Internal creep is a result of the weight of overlying ice and of gravity acting down the surface slope of a glacier. It can be shown experimentally that ice does not behave as a perfect plastic material (see Fig. 6.1g). Crystals of ice have no yield stress, but deform under any magnitude of applied stress, as does a viscous liquid. Provided that the stress level is below about 100 kPa the strain rate is nearly steady, but if the stress is increased to > 100 kPa the strain rate increases, and the behaviour approximates to that of a plastic material. This change in behaviour is thought to be associated with recrystallization of ice to form oriented crystals in which stacked platelets are nearly parallel to the direction of flow. Then displacement can occur by internal shearing of plates over one another. Continued plastic creep involves shearing, recrystallization, and the migration of crystal boundaries. The mode of behaviour is temperature-dependent: below about −8 °C deformation is dominated by internal shearing along crystal boundaries; between −8 °C and −1 °C creep is associated with loss of water molecules from one crystal and addition to a neighbouring crystal; at about 0 °C pressure melting and *regelation* (i.e. refreezing) are dominant.

The applied stress (τ) to strain rate ($\dot{\varepsilon}$) relationship for glacier ice is approximated by a power curve with the form of

$$\dot{\varepsilon} = A\tau^n \text{ (known as Glen's flow law)}$$

in which A is a temperature dependent constant,

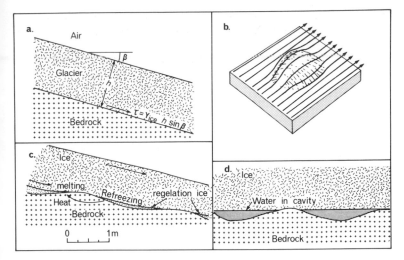

and n has values which range from about 1.3 to 4.5, with a mean of about 3.0. The value of n increases with confining pressure. Rock debris carried within a glacier affects its behaviour and may increase plasticity at the bed where much of the debris is carried.

If ice under stresses greater than 100 kPa is assumed to be a perfectly plastic material, which deforms in proportion to the stress of the overburden of ice, the basal shear stress of a glacier with parallel sides can be calculated from:

$$\tau_b = \gamma_{ice}.h.\sin\beta$$

where τ_b is the basal shear stress (kN/m^2),
γ_{ice} is the unit weight of ice (9.0 kN/m^3),
h is the ice thickness (m),
β is the slope of the glacier surface (degrees).

When τ_b is equal to the assumed plastic yield stress of ice (τ_c), sliding would begin at the glacier sole and ice thickness and basal shear stress could not increase further. It follows then that the thickness of ice can be calculated as a function of surface slope with:

$$h = \gamma_{ice}.\sin\beta/\tau_c.$$

This relationship is strictly valid only for an ice sheet, but it can be corrected and used to approximate conditions for a valley glacier.

(2) Plasticity thus implies that $h.\sin\beta$ is a constant and field measurements show that glaciers do approximate to this condition: where the bed steepens, the glacier thins, develops crevasses, and increases in velocity by sliding—processes collectively called *extending flow*; where the bed flattens, or becomes concave, the ice thickens, undergoes thrust faulting, and decreases in velocity—collectively known as *compressing flow* (Fig. 15.5b). Extending flow is often dominant near the head of a glacier where snow is accumulating, in steep sections, and below rock ridges. Compressing flow is usually dominant upslope of ridges and through the zone of ice loss, called the *ablation zone*, near the snout. A second line of evidence is that basal shear stress beneath ice sheets falls in the range of 50–150 kPa, which is a value close to the assumed plastic yield stress for ice of 100 kPa.

(3) At the base of a glacier where ice is at the pressure melting-point, the ice is separated from the bedrock by a film of water. This film reduces friction and the ice can slide. Melting of warm ice occurs because the ice melts when high pressures force it against the upstream side of bumps; the meltwater produced then flows around the bump to the downstream side where pressure is lower. Here it may refreeze, because its temperature is slightly lower than 0 °C, and form regelation ice. The refreezing releases heat and, if the bump is small enough (<1 m long), the heat may be transmitted through the rock to the upstream side where further pressure melting is encouraged (Fig. 15.4c).

Where water from melting or rainfall is present it may form a layer some millimetres thick, and fill all cavities (Fig. 15.4d), thus separating the ice from the rock. If the water is under high pressures

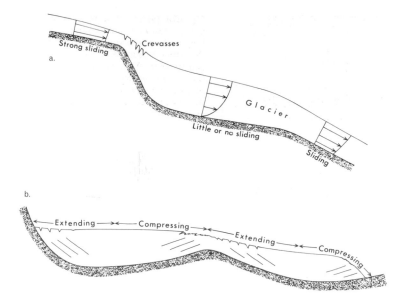

Fig. 15.5 Vertical velocity profiles within a glacier indicate the variability in the magnitude and location of the basal sliding. Creep is at a maximum a few metres above the bed and then nearly uniform to the glacier surface. (b) Location of extending and compressing flow along a glacier and the resulting surface features.

(p) the effective basal stress will be ($\gamma_{ice}.h - p$) and low effective stresses will give rise to enhanced sliding. Zero values of ($\gamma_{ice}.h - p$) represent the situation in which the ice is completely detached from the bed and 'floats'. Such conditions may occur when a glacier *surges* catastrophically, advancing its snout at rates of up to 100 m/day for limited periods.

Velocity profiles of glaciers indicate that most have highest rates of creep close to the bed, and that the rate is nearly constant above this basal zone in which the shearing may be distributed through the lower 10–20 metres of ice and debris. The contribution of basal sliding to total movement varies from zero in cold glaciers frozen to their beds to about 90 per cent in some warm glaciers, with 50 per cent being a more common value. The temperature regime of ice thus has a major effect upon the mode and rate of ice movement.

One variable which is extremely difficult to evaluate, in developing an understanding of ice movement, is that of bed roughness. Small protuberances probably enhance sliding by promoting pressure melting and regelation, but large protuberances may hinder movement, even though ice can deform around them and scour them to streamlined forms (Fig. 15.4b). A major effect of a very rough bed may be the creation of many water-filled cavities. It is unfortunate that the term 'cavitation' has been used in this context even

though it has nothing to do with implosion of air bubbles.

Flow patterns

In an ice sheet, or ice cap, ice movement is not confined laterally; but in a valley, friction with the valley walls reduces flow rates so that there is a zone of maximum flow rate in the centre of the glacier and at the surface, with a decline towards the walls and base. Lines of maximum flow rate may be offset by irregularities of the walls and bed so that velocity profiles may be asymmetric. In Fig. 15.6 common cross-profiles are shown. In (b) the rare blockschollen flow is indicated. This occurs most commonly in Himalayan glaciers and its cause is not well understood. There is a zone of shearing close to the margin of the ice and this may represent a boundary zone between ice which is partly, or periodically, frozen to the valley walls and that which can slide freely. The profile across the Athabasca Glacier (e) shows the distribution of velocity across a glacier which is undergoing movement by both creep and sliding and is in contrast to the theoretical profile (f) in a uniform valley.

Where stretching of the surface of a glacier, due to flow at depth, exceeds about 1 per cent per year *crevasses* are likely to form. Crevasses are vertical or wedge-shaped cracks in the ice, varying in width from a few centimetres to 10 metres or more, and

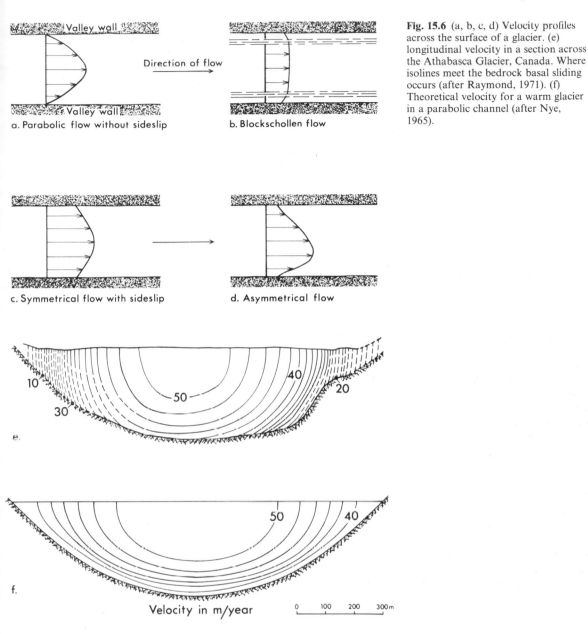

a. Parabolic flow without sideslip

Direction of flow

b. Blockschollen flow

c. Symmetrical flow with sideslip

d. Asymmetrical flow

e.

f.

Velocity in m/year

Fig. 15.6 (a, b, c, d) Velocity profiles across the surface of a glacier. (e) longitudinal velocity in a section across the Athabasca Glacier, Canada. Where isolines meet the bedrock basal sliding occurs (after Raymond, 1971). (f) Theoretical velocity for a warm glacier in a parabolic channel (after Nye, 1965).

in depth down to about 40 m. The largest crevasses occur in fast-moving very cold glaciers. Crevasse depth is limited because creep at depth within the glacier closes the fracture. Patterns of crevasses are shown in Fig. 15.7. Radial splaying occurs where ice spreads out and is common where valley glaciers emerge into open basins; splaying patterns are usually associated with compressing flow and transverse patterns with extending flow. Marginal crevasses, also called chevron crevasses, occur

because of the tension where the ice rubs against the valley wall. These alignments all occur because crevasses tend to open at about 45° to the direction of the maximum stress in the ice. The patterns of crevasses described here are very simple forms: most commonly the crevasses open in response to variations in the slope of the bed and walls so that complex and intersecting patterns are produced (Plate 15.2).

Within ice falls developed at steep sections of

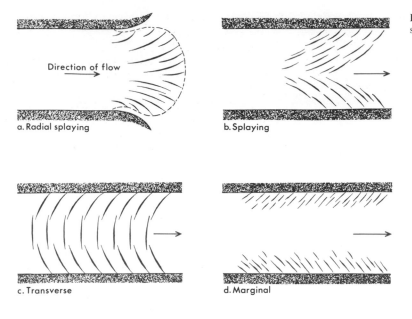

Fig. 15.7 Crevasse patterns on the surface of glaciers.

the glacier, crevasses may divide the ice into blocks or pinnacles, called *seracs*, and the transverse crevasses may extend at depth as normal faults in the ice so that the ice fall is a very irregular staircase. At the base of the ice fall the glacier reforms as the crevasses are closed. The reforming ice may have a wave-like form which bows down valley. The wave form, known as an *ogive*, is often produced from ice which has crossed the ice fall during the accumulation season and the depression between ogives corresponds to ice which has crossed the ice fall in the melt season (Plate 15.4).

Ogives should not be confused with alternating bands of 'bubbly' or 'milky' ice and dark bands of dirty regelation ice. The dark bands are often formed in summer and may correspond with the troughs between ogives but this is not always the case. Alternating bands of bubbly and dark ice are known as *Forbes bands*.

Rates of flow of glaciers may vary over short periods. Some glaciers increase velocity during warm days as meltwater volume increases. On a longer time-scale of months or years increased accumulation may lead to ice thickening to form a bulge which is transmitted down the glacier as a kinematic wave. Such a wave is like the hump produced by oscillating a string held in the hand: wave forms travel along it but the material of the string does not. The bulge of the glacier similarly moves down-valley at a velocity of about four times that of the remainder of the ice. Eventually it arrives at the snout which thickens in response to an accumulation event which may have occurred some years before.

A rare phenomenon is that of periodic surges in which the rate of ice movement may suddenly accelerate to 10–100 times its usual rate. Rates of ice advance of 5 m/hour have been recorded. Surging is usually associated with thickening in the upper reservoir part of the glacier until the ice is suddenly released, when catastrophic sliding is accompanied by severe crevassing. The snout may advance some kilometres or remain stationary, but be thicker after the surge (Fig. 15.8). Because the base of a surging glacier is inaccessible little is known about the mechanisms of surging. It may be that it is a form of kinematic wave associated with build up of meltwater at the bed of a glacier which suddenly becomes detached from the bedrock. Most reported surges have been in Alaska, the Yukon, and Iceland, but there is speculation that parts of the West Antarctic ice sheet could surge, delivering large volumes of ice into the ocean, raising world sea-levels, and even spreading floating ice masses into the southern ocean, thus creating a white surface which would reflect much solar radiation, and so cool the Earth's atmosphere. All this could happen in as little as 100 years

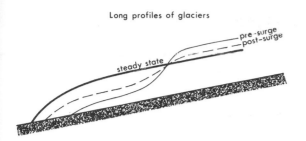

Fig. 15.8 Long profiles of a surging glacier compared with one in a steady state with a zero budget.

and have a major effect upon world climate. The idea of a surging ice sheet is a fascinating idea with potentially devastating consequences, but it remains largely speculative for it is not certain that large parts of an ice sheet will surge at one time.

Glacier budgets

The terms *budget* or *mass balance* are used to describe the input of snow into a glacier, its transfer as ice along the glacier, and its loss (or *ablation*) by melting, evaporation, or calving of icebergs. A positive balance implies that the glacier accumulation is greater than the ablation and a negative balance the reverse. A zero balance implies that accumulation and ablation are equal.

Somewhere along the glacier there is a change from net accumulation to net ablation: this position is known as the *equilibrium line*. It is the position on the glacier below which snow does not last through the melt season (usually the summer) (Fig. 15.9).

The rate at which ice is transferred from the accumulation zone to the ablation zone is controlled by the rate of supply and the flow rate. The rate of ice movement tends to be highest at the equilibrium line because a greater volume of ice must pass this line, than any other, in any period of time. Where supply is large transfer and ablation must be rapid. Glaciers with a high snowfall and warm climate are consequently much more active than those of low snowfall and low temperatures. There is a general trend of high to low activity from coastal temperate zones to those of continental and polar zones.

Variations in budgets are very common. Many glaciers are at present reducing in volume (Fig. 15.10), either because of greater summer melting or because of reduced snow accumulation over the last few hundred years. Ice sheets respond much more slowly to changes in budget with a lag of several thousand years. Because lag times of valley glaciers depend upon rate of ice movement and catchment area, a glacier in one valley may be diminishing while that in the next is growing.

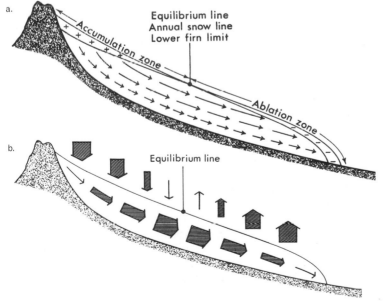

Fig. 15.9 (a) Long profile of a valley glacier showing the zones of accumulation and ablation separated by the equilibrium line, and the resulting flowlines of the ice. In (b) the magnitude of gains, losses, and transfers of ice along the glacier are indicated.

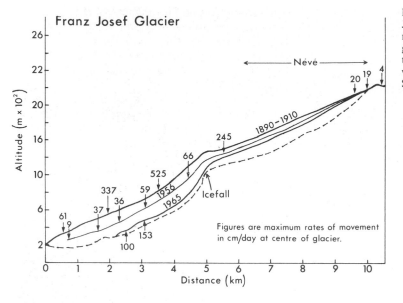

Fig. 15.10 Long profiles of the Franz Josef Glacier, New Zealand, showing rates of movement at the surface of the glacier at three periods, and the thinning and snout position retreat which has occurred this century (after Sara, 1974).

Form of glacier surfaces

A walk from the head of a glacier to the snout would enable an observer to recognize many of the essential characteristics of glacier forms.

At the head of a valley glacier there is commonly a narrow arc of snow and firn frozen to the rock wall, but below this a deep crevasse, known as the *bergschrund*, separates this snow from the active glacier. The bergschrund commonly follows the arc around the head of the glacier and is a distinctive feature which may be jumped by a walker heading down-valley, but may be difficult to cross in the up-valley direction because it also forms a step in the profile. To the side of the glacier avalanche cones and wind-blown deposits of snow give the firn basin a concave cross-profile.

As the walker travels down the glacier he will pass sites where rockfalls have carried debris onto the glacier surface and he will see how these piles are gradually spread out along the flanks of the ice to form *lateral moraines*, although much of the debris which is scattered beyond the flank will be buried by snow and carried down into the ice. It may be noticeable in the ablation zone that debris is melting out of the ice after travelling within it (Fig. 15.11; Plate 15.3).

At intervals a tributary glacier may flow into the trunk glacier and the lateral moraines of the two will be trapped as a *medial moraine*. Glaciers of nearly equal size will usually converge at about the same level so that the ice surface will remain nearly even and, in contrast to rivers, there will be no turbulent mixing of the two ice streams and each will retain its identity.

Where a tributary is much smaller than the trunk glacier it may 'hang' above the trunk ice stream and join it by flowing down an ice fall onto the main surface. Initially it may form a superimposed glacier with ogives spreading onto and gradually merging into the trunk ice. Because of its capacity to creep the ice will eventually form a uniformly level surface (Fig. 15.12; Plate 15.4).

Below the equilibrium line the glacier surface, in summer, will be one of ice rather than snow or firn and in cross-section it will develop an increasingly convex surface profile towards the snout. As a result of melting the lateral moraines may be left stranded above the ice surface if the glacier is diminishing in volume (i.e. it has an overall negative budget) or may remain level with the surface if the budget is positive. Rock debris which melted into the ice in the accumulation zone to become *englacial debris* will now melt out and become *supraglacial debris*. This may have such a large volume that the entire snout area is debris-covered. The wasting ice at the snout may be deeply crevassed if it is thin ice on a steep slope, but more probably it will be deeply incised by meltwater streams which cut meanders, potholes, and steep-sided channels into the ice. Where the pot-

15.3 The Tasman Glacier, New Zealand, is fed by niche glaciers, avalanches, and tributary cirque glaciers on the flanks of Mount Cook (3764 m) to the left, and Mount Tasman (3497 m) to the right. Note the large debris cones supplying rock debris to the glacier margins, the high lateral moraines indicating ice shrinkage, and the cover of supraglacial debris (photo and copyright: New Zealand Aerial Mapping Ltd., Hastings, N.Z.)

holes intersect a channel at depth in the ice the water will disappear: such deep sinkholes are called *moulins*.

At the snout itself the landscape may be one of chaotic jumbles of rock debris sliding down the ice faces to form irregular heaps before being washed away (this is the common situation on warm glaciers with a negative budget); but if the ice has a positive budget it may have a thick snout which bulldozes supraglacial debris, and other debris from beneath the glacier, into ridges. Meltwaters will carry the fine rock flour in suspension and the coarser debris as bedload.

Ice sheet surfaces are different from valley glaciers in that they are not confined by valley walls and ice can spread out in all directions. A large, very cold ice sheet like that of Antarctica has a surface that looks to be nearly flat over large distances, but it also has low domes and broad depressions, perhaps with crevasse fields, which reflect the underlying topography and areas of ice convergence and divergence. At high altitude most of the precipitation is rime ice which is fine and very granular. This ice becomes packed by the wind and then is cut into sharp angular ridges called *sastrugi*. Crevasse and sastrugi fields are both very difficult to cross.

Where an ice sheet drains towards the sea the

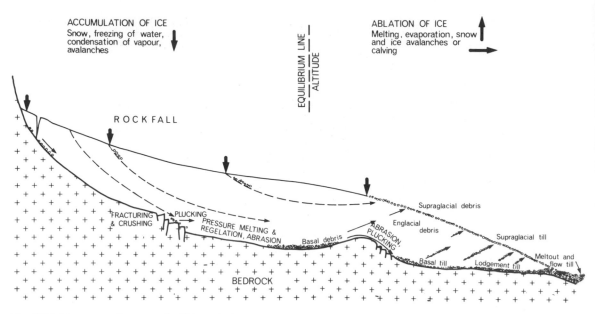

Fig. 15.11 Processes acting on and within a glacier.

traveller may notice little steepening of the ice surface until he is a few tens of kilometres from the margin, but he will probably come across crevasse fields, especially where these mark the edge of an ice stream which forms a main outlet from the ice sheet. Ice streams have much higher velocities than the main surface ice and occupy a broadly concave shallow trough in the ice margin; they may be some tens of kilometres wide.

The ice sheet may terminate on land as a steep ice face, 30–50 metres high, or reach the sea to form an ice shelf. Above the point on the sea bed where the ice floats (a point known as the *grounding line*) a deep crevasse known as the *strand crack* will form because the floating ice shelf is subject to tidal variations. For an ice shelf the equilibrium line is at the calving face and much of the shelf ice is actually formed from snow which falls onto it.

Where the ice sheet meets mountains it may diverge around them to leave peaks standing above the sheet as *nunataks*, or it may be blocked along much of this margin and overflow the mountain ridges into trough valleys cut in the mountains by outlet, or transection, glaciers. In these stretches the glaciers will have the characteristics of valley glaciers, but carry less debris because they gain it only from the trough walls and bed and not from a valley head. Outlet glaciers may reach the sea in fjords (i.e. glacial troughs

occupied by the sea) where icebergs will calve, they may flow into ice shelves, or terminate on land (see Fig. 15.13; plates 15.1, 15.5).

Glacial erosion

Glacial erosion can be a very effective process. It has been estimated to lower ground surfaces in bedrock at rates 10–20 times faster than fluvial processes, and average lowering has been estimated at between 0.05 and 3.0 mm/y. The distinctive landforms of glaciated areas largely reflect the greater cross-sectional area of glaciers compared with rivers and the much higher viscosity of the ice which prevents it changing direction around tight bends.

Processes

The most commonly recognized processes of glacial erosion are: abrasion, plastic moulding, fracture of fresh rock, plucking, friction cracking, meltwater erosion, and debris entrainment. At one site any combination of these processes may be active.

Abrasion is the process by which bedrock is scored and ground into small particles by contact with other rock particles held in the ice at the base of a glacier; ice itself is too soft to abrade rock. The

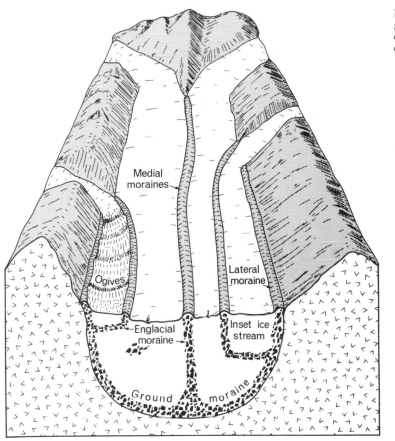

Fig. 15.12 Types and locations of valley glacier moraines. The volume and continuity of englacial debris is exaggerated.

15.4 A small cold glacier descends a 35°-slope to join the Canada Glacier (Antarctica). Ice falls on the Canada are replaced by ogives at their base. The hanging tributary is frozen to bedrock and is not eroding.

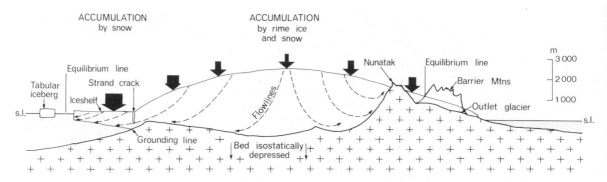

Fig. 15.13 Features of an ice sheet. Melting occurs where flowlines meet the bedrock. The equilibrium line of the ice shelf is located near its frontal position, indicating that the shelf gains much of its ice from local snowfall, and loses mass by basal melting and calving of icebergs.

15.5 The Byrd Glacier, an outlet from the East Antarctic Ice Sheet, flowing into the Ross Ice Shelf in the far distance. This glacier flows at about 2 m/day and is deeply crevassed. The valley walls are visible to either side.

grinding process has been observed directly and its two major products—*striated* rock (i.e. rock scratched or grooved, like wood by sand paper) and fine *rock flour* of silt-to-clay size removed during grinding—are widely observed (Plate 15.6). For abrasion to occur basal debris must be enclosed in sliding ice and held in contact with the bed. The process is accelerated by high sliding rates, soft bedrock, hard grinding rock in large fragments, melting at the base of the ice so that fresh angular debris can be carried down into contact with bedrock, by efficient removal of the

15.6 (a) Glacial striations across schist are aligned parallel with the hammer shaft.

15.6 (b) Striations, crescentic gouges, and plucked edges at a joint in schist. Ice moved from right to left parallel with the hammer head.

rock flour to leave fresh surfaces, and by an ice thickness large enough to produce high contact pressures between abrading particles and the bed. The frictional drag (F) between bedrock and debris is given by:

$$F = A_c.\mu(\gamma_{ice}.h - p)$$

in which A_c is the area of contact and μ is the coefficient of dynamic friction. Because ice deforms plastically, friction will increase with effective basal stresses only until the yield stress is exceeded, when the ice will flow over and around the particle rather than dragging it along the bed: in this condition deposition will occur. In cold ice abrasion may occur to a limited extent because of the plastic deformation at the bed, but its magni-

tude will be far less than at the base of warm ice moving by basal sliding. Plastic yielding of both warm and cold debris-laden ice is, probably, one possible cause of moulded and finely abraded streamlined rock hollows and ridges collectively called *p-forms*.

Fracture of bedrock can occur only where large blocks, held in the ice, are forced against smaller bedrock protuberances. Ice carrying small debris would merely flow around large obstructions. The triangular and crescentic grooves, called *chattermarks*, formed in bedrock surfaces result from pressure fracturing by the pointed corner of a large block held in the ice. It is the stress concentration resulting from ice pressure against a large block, and transmission of the entire stress into a small area, which permits fracturing of the bedrock. The chattermarks may be oriented up or down the flowpath of ice, being controlled by the angle of contact between the pointed block and the fracturing face.

Jointing and fractured rock is especially prone to glacial erosion. Any preglacial weathering or fracturing will obviously promote glacial erosion, but the problem is to understand the processes which permit the excavation of glacial troughs, like those of fjords, some 2000–3000 m deep. It may be reasoned that as a glacier invades a river valley it will exert an added pressure on the rock and hence close the joints, but as it abrades the rock and deepens the valley it will eventually remove high-density rock while filling the valley with ice which has a density of one-third that of rock (0.9 Mg/m³ compared with 2.8 Mg/m³). Abrasion will thus lead to a reduction of confining pressure and an opening of joints, especially joints which form nearly parallel with the valley floor and walls by stress release.

Whether hydrofracturing and freeze–thaw processes occur in joints below glaciers is still debated. It is possible that temperatures vary too little from 0 °C for this to be a significant process, except at glacier margins, even though melting and regelation occur beneath warm ice.

The incorporation of large rock fragments into a glacier bed by freezing on is called *plucking*. It seems most probable that this is entrainment of blocks already loosened by other processes, and not primarily erosional. It is unclear whether plucking is possible, or enhanced, beneath cold glaciers.

Meltwater erosion can be of considerable importance below ice sheets and valley glaciers. Water may be forced to flow at high pressure if it is confined in tunnels within the glacier, as a head of water of many tens of metres can be created within the glacier. Streams at the bed may consequently flow uphill for short distances and behave like water in a pipe. Subglacial gorges, potholes (see Plate 9.5) and large scallop-shaped features called *sichelwannen* are all possible results of meltwater action beneath ice.

Debris entrainment is essential for continued effective erosion of the bed. Some removal will be by subglacial meltwater from warm ice, but much entrainment of fine-grained particles doubtlessly occurs as relegation ice forms and sediment is frozen into the ice. Large volumes of sediment may

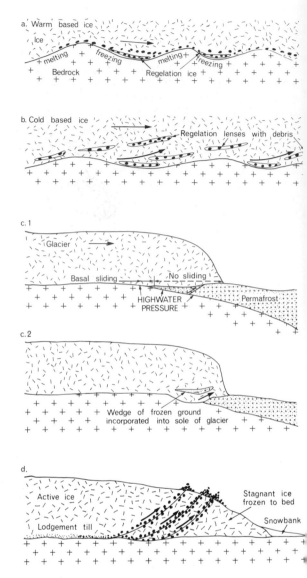

Fig. 15.14 Mechanisms of debris entrainment by glaciers. (a) Warm-based ice incorporates debris in regelation ice, but the debris is melted out again over the bumps so debris layers are few and thin. (b) Beneath cold ice near a melting zone successive regelation layers are incorporated into the glacier and layers are built up to considerable thicknesses. (c) In (c1) a wedge of permafrosted debris is frozen onto the bed of the glacier and high water pressure in the unfrozen substratum permits the wedge to be incorporated into the ice (c2). (d) A snout of static ice, or a rampart of snow, prevent ice advance by sliding. The glacier ice then shears along thrust planes and debris is transported upwards along these planes, and melts out at the surface.

be entrained beneath ice sheets which are warm-based in the interior and cold-based at the margin, for near the boundary of the two zones many layers of regelation ice will be built up (Fig. 15.14). The largest known mass of transported rock in Germany has dimensions of 4 km × 2 km × 120 m. It is probable that this mass was frozen into the base of a cold glacier at a time when the ground below the frozen mass was thawed and hence weaker than the ice-rockmass contact. Yet another process of entrainment may be that of thrusting along shear planes near a glacier snout. This process is probably most effective with fine-grained material but can also operate with large rock fragments (Fig. 15.14; Plate 15.7).

Landforms of glacial erosion

Glacial landforms resulting from erosion can be classified in several ways: by size, shape, or association with certain glacier types. Here they

will be discussed in relation to the two dominant glacier types—valley glaciers and ice sheets.

Glacial troughs are the outstanding features created by valley glaciers and form major features in high mountain landscapes collectively called *alpine topography*. Most glaciated valleys have a wide-open valley floor between steep walls. The whole cross-section is variously described as being U-shaped, parabolic, or as having a catenary curve (like that produced by a deeply sagging rope). The curve of the profile has a shape that is influence by the hardness and jointing of the bedrock and by the characteristics of the ice. Very hard rocks commonly support very steep valley walls (see Chapter 8) and slope angles may be close to 80°. Such cliffs are well known from the fjords of Norway, Greenland, New Zealand (Plate 15.8), and Alaska, and from valleys like the Yosemite, cut in igneous rocks. Valley walls cut in softer and more highly jointed rock are more readily abraded and plucked, and slope angles may be lower at

15.7 (a) Thick till at the base of the Taylor Glacier, Antarctica, and thin bands of debris in the snout face. In the absence of much melting (mean annual temperature is −20 °C) the face is nearly vertical and wastes by blockfalls of ice which forms an apron round the ice front.

15.7 (b) Dirt bands along shear planes in the snout of the Changri Nup Glacier, Nepal. The surface of the glacier is covered by ablation till.

15.7 (c) Flow-folds in the base of the Taylor glacier are evidence of plastic deformation.

15.8 Milford Sound, a fjord in south-west New Zealand. The steep valley walls are formed on strong igneous and metamorphic rocks, in contrast to the less resistant greywacke shown in Plate 15.3. Note the accordant summit levels of the peaks (photo and copyright: Whites Aviation).

30–40°; this is the case in the highly shattered greywacke of parts of the New Zealand Southern Alps. The open catenary form is, presumably, one well adjusted to transmitting viscous ice with minimum friction along the bed. The very wide, nearly flat-floored valleys in Antarctica, which are now ice-free, suggest that the higher viscosity of polar ice may have some influence on valley cross-sectional forms but, in the absence of detailed studies, this is speculative. One clear difference between polar and lower latitudes is that warm glaciers have much meltwater which may cause subglacial stream incision and narrow gorges. Narrowing of valleys also occurs in steep reaches, so it is possible for glaciated valleys to have sequences of wide basins separated by narrow gorges.

The cross-sectional profile is adjusted to the maximum ice volume, so large glaciers have the widest troughs. Outlet glaciers from ice sheets do not have a trough head but may have very large trough valleys. The Byrd Glacier, for example, is 30 km wide in places and ice in the trough is 3000 m thick with the valley sides rising another 3000 m above the glacier surface. If the Ross Ice Sheet were to disappear, the Byrd valley would be a fjord with walls 6000 m high from the bed (see Plate 15.5). Because glacial troughs are often cut into old fluvial valleys the former spurs have their ends cut off to produce ice-faceted faces, and high-level tributary valleys and basins are left as *hanging valleys* from which ice falls, or after deglaciation, waterfalls, cascade.

Long profiles of glacial troughs may begin at steep headwalls down which ice falls and avalanches descend, or they may begin in broad basins known as *cirques* (also called *corries* in Scotland and parts of Canada, and *cwms* in Wales). Cirques have a variety of forms but are commonly rather 'down-at-heel' with an over-deepened basin, and a

15.9 Cirques, inclined rock slabs which formerly carried niche glaciers, truncated spurs, and glacial troughs cut in hard rocks, Fiordland, New Zealand (photo and copyright: New Zealand Aerial Mapping Ltd., Hastings, N.Z.).

low bar across the mouth (Plate 15.9). The bar may be of solid rock but isolated cirques, or those no longer feeding a main valley glacier, may have the mouth enclosed by a bar of rock debris.

The forms of cirques are probably due in part to a rotational movement of the ice within the basin such that abrasion is most effective below the mid-point of the cirque glacier, while the trough headwalls are suffering severe physical weathering and supply abrasive debris which reaches the valley floor through the bergschrund and other crevasses as well as by being buried by snow and carried down in the ice. In some cirques the basin form may also be produced because at the mouth

ice becomes separated from the bed by debris and can no longer erode (Fig. 15.15).

An origin for cirques by rotational sliding and abrasion at the bed implies that the ice must be warm, otherwise there would be no meltwater available for basal sliding. Consequently it is probable that cirques in much of Antarctica are very old and were cut during Cenozoic times when the ice was warm, for it is evident that most alpine glaciers in Antarctica are composed of cold ice which is frozen to the bedrock and performing little or no erosion (Plate 15.10). Cirques at very high altitude in some mountains may have been initiated at lower altitudes and been uplifted since,

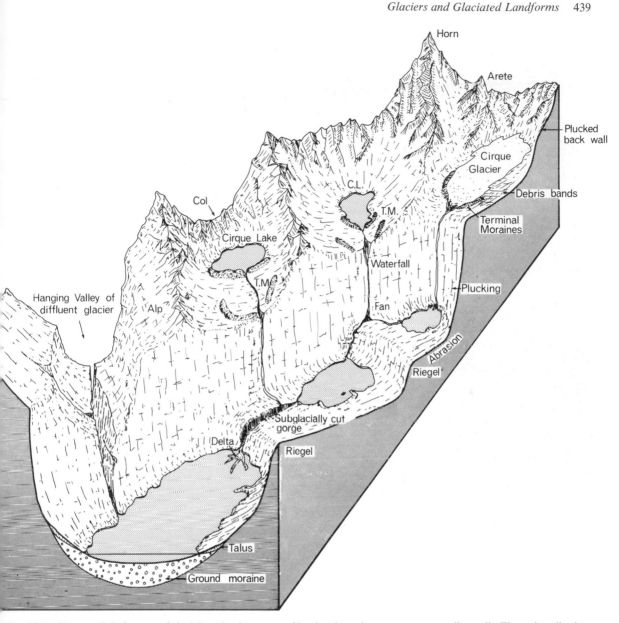

Fig. 15.15 Characteristic features of glacial erosion in an area of hard rocks and consequent steep valley walls. The main valley has few recessional moraines, but in the cirques minor cirque glacier expansion and retreat in the last 1000 years has left arcuate terminal moraines.

for they too may contain cold ice. The altitude of cirque floors may be nearly uniform in a mountain region and taken as an indicator of the summertime snowline, or the lowest altitude at which permanent snow and firn could accumulate when the cirque was being cut.

The long profile of many troughs is frequently irregular and composed of alternating rock bars and basins of varying height, depth, and length. It is unlikely that all such irregularities can be attributed to one cause and, because the evidence for the origin of such features is usually destroyed by subsequent erosion, there can seldom be definitive answers to questions of cause. Amongst the

15.10 Small slab-like glaciers occupying cirques in the Transantarctic Mountains. These glaciers are frozen to bedrock and are not eroding the basins they occupy.

15.11 Roches moutonnées with abraded upper surfaces and plucked faces. The ice flowed from right to left. Franz Josef Valley, New Zealand.

most general and plausible hypotheses are the following: basins are excavated in softer and more closely jointed rock while bars are formed on harder rock or rock with more widely spaced joints (this hypothesis is testable but few attempts have been made to do so); basins occur at sites of preglacial deep weathering and the formation of weak saprolite compared with shallow weathering over bars (usually untestable); basins occur below confluences with tributary glaciers which cause the ice surface to steepen and erosive power to increase (sometimes testable); basins occur below sites of accelerated basal ice sliding and hence deeper abrasion (hardly ever directly verifiable); basins occur where valleys narrow and erosive energy is increased by ice thickening (partly verifiable, but an association of basins with valley narrowing is often not the case).

A large rock bar creating a natural dam is often called a *riegel*, and strings of impounded lakes in the floor of a trough are called *paternoster lakes*, —an allusion to strings of prayer beads. As a result of abrasion on their upstream sides rock bars and outcrops becomes abraded, striated, and rounded, while downstream sides are plucked and jagged. These characteristically streamlined forms are commonly 5–50 m high and 5–100 m long and called *roches moutonnées* (Plate 15.11).

It is common for the central areas of glaciated valleys, which were formerly near the equilibrium line, to be considerably overdeepened and later occupied by a long lake or lakes. The equilibrium line is, of course, the zone of greatest ice flow and hence most vigorous erosion. The overdeepened form is particularly evident in *fjords*, which are glacial troughs now flooded by the sea. A glacier reaching the sea will eventually float off the bed so that erosion will cease somewhere near the snout, and the valley floor will rise towards the resulting

bars. The position of the bar is commonly near the outer edge of the mountains where glaciers are no longer pinned by the valley walls and can thin, spread, and calve icebergs.

In a deglaciated valley the floor becomes rapidly modified by reworking of old glacial debris by streams, so that lake basins are filled by encroaching deltas, medial and lateral moraine ridges are eroded and become less distinct, and valley floors may eventually be filled from side to side with reworked gravels and broad, braided stream channels. Valley walls will be modified at a rate largely controlled by rock resistance and will have talus

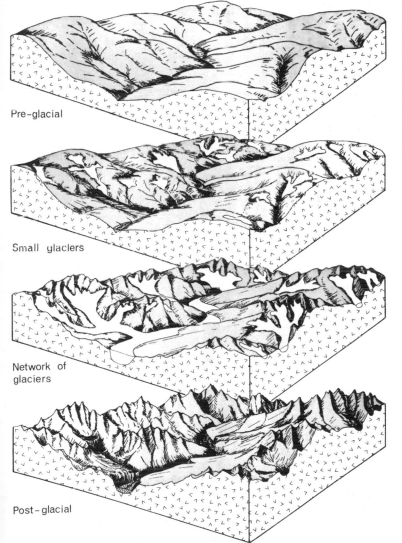

Pre-glacial

Small glaciers

Network of glaciers

Post-glacial

Fig. 15.16 The theoretical development of an alpine glaciated landscape from a fluvial preglacial landscape. Early glaciation is associated with cirque and small valley glaciers, and full glaciation with large valley glaciers incised below the valley wall benches (called 'alps' in Europe). The post glacial landscape is one of cirque and valley basins and troughs, and a steep upland rugged relief produced by severe physical weathering and slope processes (modified from Flint, 1971).

15.12 Mount Aspiring, a glacial horn with arêtes, produced by cirque glaciers cutting headwards (photo and copyright: New Zealand Aerial Mapping Ltd., Hastings, N.Z.).

slopes, vegetation, soil covers, or bare rock preserved depending upon the environment.

An alpine topography is commonly composed of steep angular jagged ridges called *arêtes* (Fig. 15.16) separating cirque headwalls. Where cirques grow they cut into the arêtes and may eventually produce high, steep *glacial horns*, as remnant mountains where three or four cirques cut back into one rock mass, and low *cols* where the arêtes are lowered (Plate 15.12). On valley walls a *trimline* may separate an upper slope segment controlled, by rock mass strength, preglacial valley forms, frost action, slab glaciers, avalanches, or protective snowpacks, from a lower U-shaped valley with profiles produced by the valley glacier (Fig. 15.16).

Ice sheets produce a variety of erosional forms because the extent of melting, freezing, and cold ice beneath them varies through the life of an ice sheet and with distance from its margin. The nature of bedrock erosion is directly related to the temperature of the ice.

In Fig. 15.17 an association between ice characteristics, availability of meltwater, and zones of erosion and deposition is implied. This model is based upon theoretical considerations and studies of the features recognized as being formed beneath the Laurentide Ice Sheet which covered central and eastern Canada during the peak of the Last Glacial (22 000–14 000 BP) and in earlier glacials. It emphasizes that the intensity of erosion varies beneath an ice sheet and implies that as an ice sheet thickens and thins the zones of varying intensity will change location. It is possible to recognize areas of little or no erosion, regimes of areal scouring, and also areas of selective linear erosion.

The lack of erosion beneath ice sheets is usually associated with cold basal ice and is consequently a feature usually associated with polar continental climates in which accumulation rates are low,

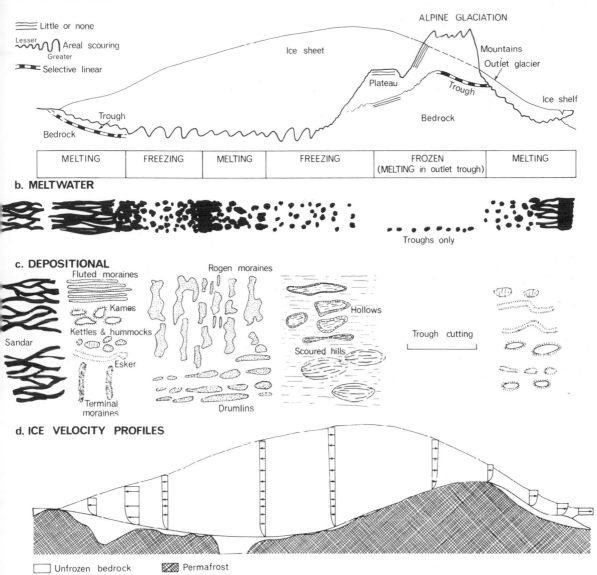

a. EROSION INTENSITY

ALPINE GLACIATION

≡ Little or none

Lesser ⌇⌇ Areal scouring
Greater

▬▬ Selective linear

Ice sheet

Mountains
Outlet glacier

Plateau

Trough

Ice shelf

Trough

Bedrock

Bedrock

| MELTING | FREEZING | MELTING | FREEZING | FROZEN (MELTING in outlet trough) | MELTING |

b. MELTWATER

Troughs only

c. DEPOSITIONAL

Fluted moraines

Rogen moraines

Kames

Hollows

Kettles & hummocks

Sandar

Scoured hills

Trough cutting

Esker

Terminal moraines

Drumlins

d. ICE VELOCITY PROFILES

☐ Unfrozen bedrock   ▨ Permafrost

Fig. 15.17 A schematic relationship of zones of erosion, production of meltwater, depositional and erosional landforms, ice velocity profiles, and location of permafrost beneath an hypothetical ice sheet.

surface ice temperatures are low and, therefore, ice has a low velocity. Cold basal ice at the present is found on the higher plateaux and interfluves in Antarctica, and at high altitudes on most polar mountains.

Cold basal ice is thought to have covered parts of the Queen Elizabeth Islands of northern Canada, parts of north-east Greenland, and smaller areas in parts of eastern Scotland and North America. The evidence for this conclusion is the presence of *glacial erratics*. These are large boulders of a rock which is different in lithology from the country rock and which must have been transported a considerable distance by a glacier (Plate 15.13). Erratics lying upon a deeply weathered regolith, near tors, or on a landscape which

15.13 A glacial erratic of pale-coloured sandstone on a ground moraine of dark dolerite, Wright Valley, Antarctica.

shows few or no signs of modification by ice are anomalous and imply that ice has covered the landscape without altering it. The lack of alteration could partly be because the regolith was permafrosted while below the ice, and partly because the ice was frozen to the bedrock. A second line of evidence is the presence of glacial deposits, derived from upstream of the uneroded bedrock, now found downstream of the uneroded area.

Regions of areal scouring are probably the most common landscape form created beneath ice sheets. The dominant features are sequences of basins, shallow grooves, and elongated swells and ridges which have all been scoured and streamlined by the passage of ice. Locally the influence of rock type and structural influences, such as the pattern of joints and faults, controls the details of

the landscape but the overall pattern is controlled by the direction of ice movement (Plate 15.14).

Glacially scoured basins and grooves may be tens of kilometres long, tens of metres deep, and tens to hundreds of metres wide. They are often overdeepened, and show clear evidence of striations. Structural influences may be evident in their form with elongation along fractures or in softer rocks. Since deglaciation rock basins have usually been occupied by lakes and a very disorganized drainage pattern links the lake basins, but has had too little time in which to develop a clear hierarchy of channels.

Higher rock outcrops show many of the features of roches moutonnées. Very large hills, some hundreds of metres high and many kilometres long and wide, show clear evidence of having been overridden and enveloped by ice. Such hills display

15.14 An area of northern Canada showing glacial scouring to produce lake basins and smoothed and rounded (mammillated) relief. The hummocky deposits are rogen moraines and to the left is an esker flanked by white sandy deposits. Some lakes in the distance are ice-covered. (Photo by Division of Mines and Technical Surveys, Canada).

the plastic sculpturing, smoothed rounded forms, and striations associated with scouring, and they may have either steep, plucked faces at their downstream ends or show smoothed faces all round. Joint patterns may influence surface forms with plucked faces developing in zones of close joint spacing so that a large hill has a surface composed largely of sets of roches moutonnées. In the lee of many features glacial deposits occur and, depending upon the degree of rounding and the volume and thickness of the deposits, such names as whaleback and crag-and-tail may be applied (Plate 15.15).

Selective linear erosion occurs where ice within a sheet is channelled either along a preglacial valley or forced to converge. Local thickening and/or steepening of the ice surface creates a

greater basal shear stress and hence higher basal ice temperature and erosion. It is also a self-perpetuating process with deeper scouring in a trough creating a channel for thicker ice flow later. The intervening upland or plateau surfaces, by contrast, suffer less or no erosion (Fig. 15.18). Selective linear erosion is certainly responsible for many of the deep trough valleys and fjords in the Queen Elizabeth Islands, Transantarctic Mountains (Plate 16.3), and western Norway and for the deepening of the Great Lakes and Finger Lakes of the USA–Canada border region. Virtually all uplands which have been inundated by ice sheets show some degree of selective linear erosion and in much of Scotland and Norway, for example, the preglacial drainage pattern has been selectively followed by the ice with troughs developed in the

15.15 Knock and lochan topography of regional ice-scouring, in northern Scotland. Note the lowered divides and lack of sharp ridges (photo by Institute of Geological Science, Crown copyright reserved). (*Copyright*: British Geological Survey).

old main valleys, areal scouring of lowlands (known in Scotland as knock and lochan topography), and virtual preservation of some uplands beneath thin cold ice. The thickness of ice is also attested to by the obvious enlargement of many high valleys which show clear signs that ice has flowed from one valley into the next and left a broad open col (Plates 15.15, 16.3).

Compound landscapes are the inevitable result of changing conditions during a glacial. As ice builds up during the onset of a glacial, first at high altitudes and then at progressively lower ones, or as ice thickens over a lowland the type of erosion must change. At one site linear erosion in a valley with cirque erosion in upland basins may be swamped by an ice sheet below which the cirques may be preserved and linear erosion be confined to the main outlet troughs. In the closing stages, as ice thins and warms, cirque erosion may again become effective and feed eroding ice into the low order valleys. Phases of erosion can often be seen in plateau areas where cirque erosion is active

along the edges of the plateau to create a deeply notched edge known as a *biscuit board topography* during the early and late stages of glaciation (Plate 15.16).

Local influences

The extent of rock and structural controls on relief, the influence of the preglacial form of the land surface, and the aspect of slopes, upon glacial landscapes is controlled largely at a local level. It must be recognized that there are exceptions to general statements but it may be said that rock controls are evident in both the depth of abrasion and the detailed alignment of features, with many glacial troughs following the lines of joints and weakness. Soft and impermeable rocks are scoured more readily than hard and permeable rocks. Glaciers preferentially erode along lines of preglacial basins and valleys because along them ice thickness is greatest. There are, however, many sites at which convergence and divergence of ice

Preglacial

1.

Glacial

2.

Postglacial

3.

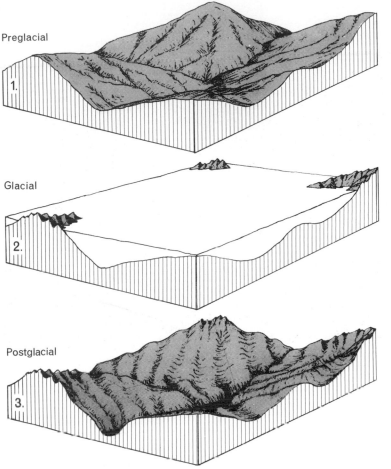

Fig. 15.18 Ice sheet inundation of an upland. Ice streams cut a wide and relatively shallow trough along the valley axis and protruding nunataks are shaped by periglacial processes, but much of the landscape is only slightly modified by glacial regional scouring (modified from Flint, 1971).

flows create hollows and swells which are unrelated to the preglacial form. Where thin ice is present, as in upland basins and cirques, freeze–thaw processes and local snow accumulation and survival can have a major influence on landforms. There is consequently a tendency for cirques to be more abundant and larger on north- and north-east-facing slopes in the northern hemisphere and on south- and south-east-facing slopes in the southern hemisphere (see niche glacier sites in Plate 15.9).

Glacial deposition

Materials and processes

The sediment transported and deposited by an ice sheet is derived from erosion and entrainment at the bed; that carried by a valley glacier is partly from the bed but commonly it is largely from the valley walls. Because it may be derived from freezing on to the base of the ice, by abrasion of bedrock, by transport of preglacially weathered soil and sediment, and by current slope processes, there is a large range of possible particle sizes and possible mineral and lithological compositions. In general, however, glacial deposits have the following diagnostic characteristics:

(1) they are poorly sorted with clasts (i.e. fragments) of many sizes from clay to very large boulders;
(2) they are usually massive and free from pronounced bedding;
(3) they have varied compositions and may contain particles from distant sources;

15.16 A view looking south into the Mackenzie Mountains of Canada. In the foreground the landscape has been buried by ice and scoured. The landscape becomes progressively more alpine to the south where cirques have been cut into a plateau, producing a biscuit-board relief (photo by Royal Canadian Airforce).

(4) many of the larger clasts are striated;
(5) many clasts are sub-angular with fractured or partly smoothed faces;
(6) elongated particles are commonly aligned in a dominant direction;
(7) the whole mass may be compacted by the weight of ice (Plate 15.17).

Some old terms are still in use to describe glacial deposits. In Britain the phrase *boulder clay* was once common and refers to an apparent composition; *drift* is still used for materials of uncertain or

mixed composition and origin. By far the most common and preferred terms are *till* for the glacially deposited sediment, and *moraine* for the landform developed in the sediment.

Till carried on, within, and at the base of ice is referred to as *supraglacial*, *englacial*, and *subglacial* till respectively (Fig. 15.11). Supraglacial material is derived from the slopes above a glacier or at the snout by the melting out of englacial material. Subglacial till is most commonly placed by *lodgement*. At its simplest this involves pressure melting of basal ice so that debris in it is released

15.17 Glacial till showing angular boulders in a fine-grained matrix.

and then 'lodged' or plastered onto the glacier bed. Lodgement may occur beneath stagnant ice, but it certainly occurs beneath actively moving ice and hence some of the till has the water squeezed out of it, and it may be sheared and made brittle by the pressure of the overriding ice. During this process alignment of the long axes of particles occurs. Where cavities occur at the glacier bed till may be squeezed, fall, or flow into available spaces. Lodgement tills can be built up to a thickness of many metres by progressive melting out from the glacier bed and by shearing of one layer of till over another.

At the snout of a glacier ablation processes cause supraglacial and englacial debris to accumulate on the ice surface. This *ablation till* is less consolidated than lodgement till and, because of the local slope on the surface of the wasting ice and the availability of meltwater, it is likely to suffer flowage and slumping (Plate 15.18).

The abundance of meltwater permits the development of fluvial processes so that the features of till are gradually lost and those of a fluvial regime or, if deposition occurs in a lake or sea, lacustrine or marine environment take over. Fluvial deposits are increasingly sorted, stratified, and rounded with the fines being carried away most rapidly in suspension and the coarse debris being involved in braided channel networks (Plate 15.19). In lakes, deltas trap the coarsest material and the fines are carried beyond the delta where they may form varves (Plate 15.20). *Varves* are

couplets of very fine and less fine silt-clay and silt to fine sand layers. It is usually thought that the silt to fine sand layers settle out in open water where wave action keeps clays and fine silts in suspension. When a winter ice cover prevents wave action then the very fine sediment settles, thus producing the couplet representing one year of accumulation. Varves may be produced by other processes also. Their importance for estimating ages since ice retreat was recognized early in this century by the Swedish geologist De Geer (see Chapter 16).

Where icebergs calve into lakes and the sea, bottom melting releases debris from the berg and it may fall to the bed either as a well-defined layered deposit or as an irregular pile, depending upon the area of the berg and its till content.

Many classifications of glacial deposits have been proposed. The most commonly used are based either on (1) the distinction between unstratified deposits derived from lodgement and ablation till and stratified deposits derived from sorting and bedding by meltwater, or (2) on the shape of the subglacial, endglacial, and outwash landform in relation to the nature of the sediment and the processes which created the landform. The second approach will be adopted here because many morainic landforms are composed partly of unstratified and partly of stratified material, and some have a partly solid rock core. A simple classification on sediment type is thus unsatisfactory.

15.18 Fine-grained flow till.

Landforms of glacial deposition

Most moraine forms can be classed into those formed (1) under the control of active ice, (2) from the uncontrolled chaotic deposition associated with stagnant ice, and (3) by fluvioglacial processes. In any attempt at classification, however, it has to be recognized that many deposits owe their original form to the support and confinement of ice, but once this is removed the form may be lost. It may also be lost as a result of wash, frost, and other non-glacial action. Large features such as lateral and endmoraines may be rapidly destroyed or reduced in size when ice melts.

Subglacial streamlined moraines parallel to ice flow include drumlins, fluted ground moraine, and many minor forms (Figs. 15.17, 15.19). Ridges up to 20 km long, 100 m wide, and 25 m high occur in Montana, for example, and probably represent an extreme for a single morainic feature, but drumlin fields may cover hundreds of square kilometres

and be composed of adjacent aligned features (Plate 15.21).

Drumlins are elliptical or ovoid hills, blunt on the up-glacier end and elongated down-glacier with a thinning tail. They usually occur in clusters with varying spacing between individuals. Sizes vary greatly but are commonly a few tens of metres high and wide and hundreds of metres long. They are usually formed beneath ice sheets but can occur beneath very broad valley glaciers. Drumlins are usually composed of lodgement till which is rich in clay, but some drumlins have a rock core and, more rarely, consist of a thin veneer of till on rock. Such features grade into wholly erosional whaleback or rock drumlin forms.

Many theories have been advanced for drumlin formation. It is certain that they develop beneath active ice and probable that high water pressures in saturated till cause it to be moulded so that it flows, with zones of high pressure beneath the ice

15.19 Three periods of development are shown in this cutting through outwash. (1) An older outwash to the right has been eroded and a terrace face cut in it along the diagonal contact. (2) Slope deposits rolled down this diagonal face. (3) Subsequently a younger outwash (to the left) was deposited against the older. Note the fluvial bedding.

15.20 Varved clay from the Ice Lake of central Jämtland, Sweden (photo by J. Lundqvist).

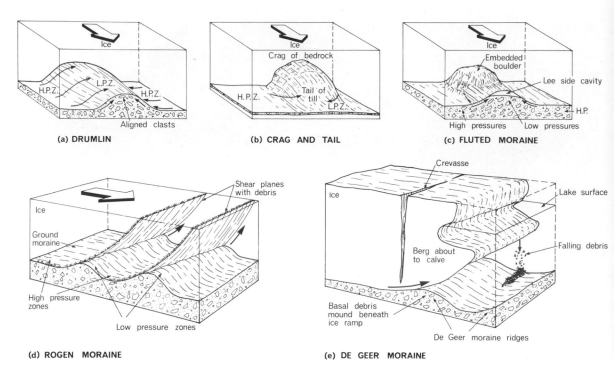

Fig. 15.19 Formation of glacial depositional landforms. (a) Drumlin. (b) Crag and tail, (c) Fluted moraine deposition in areas of low ice and water pressure at the bed of the glacier where pressure zones are parallel to the direction of ice movement. (d) One theory of rogen moraine is that it develops where shear stresses, transverse to direction of movement, are relieved. (e) De Geer moraines are thought to form in lakes where grounded ice calves and debris accumulates at the ice margin.

being scoured of till and zones of low pressure being sites of till concentration. Till will be moulded into the streamlined form as long as it is fluid or plastic, but if the water is squeezed out, or frozen, the till will become stiffer and lodge. The squeezing process may be enhanced by irregularities in the bed or by variations in the ice pressure. Such a process suggests that drumlins may have a form which is in equilibrium with the pressure of the ice and the strength of the till beneath a wet-based glacier. *Crag-and-tail* features may be the result of wet till being squeezed into the space downstream of a rock obstacle. *Fluted moraines* which are elongated ridges ranging in size from a few centimetres to many metres high and wide, and from tens to thousands of metres in length, may result from saturated till being squeezed into the cavity down-glacier of a boulder which is held on the glacier bed.

Subglacial ridges transverse to ice flow have often in the past been confused with endmoraine ridges, but the streamlining of their upper surfaces, or even 'drumlinizing' shows that they are

distinctive and formed beneath an ice sheet. In North America they are often called ribbed moraines but in Europe *rogen moraines*, from Lake Rogen in Sweden where they are well developed (Plate 15.14).

Individual rogen ridges tend to be 10–30 m high, over 1 km long, and have crests 100–300 m apart. Ridges usually have crescent-shaped parts with horns pointing down-glacier. They may be linked by cross-ridges and are often transitional in form to drumlins.

As with all subglacial features direct observation of the formation of a feature is either impossible or limited to meltwater tunnels and cavities near the ice margin, so there are more theories than facts concerning the origin of basal moraine ridges. The most probable theory of the origin of rogen moraines has similarities with that of the formation of drumlins and fluted moraines. It is probable that there are zones in the basal ice in which the stress varies transversely to the direction of ice flow. Such variations may result from compressing flow, or from the presence of large

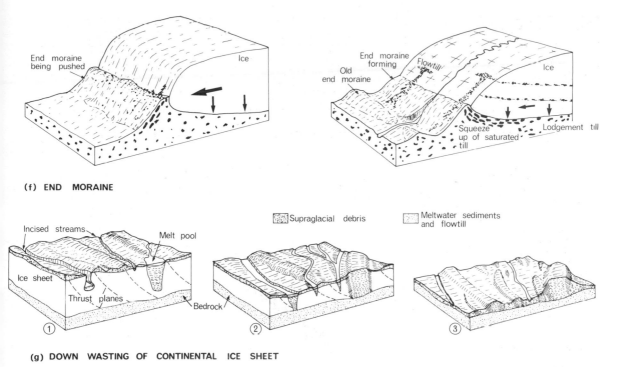

(f) END MORAINE

(g) DOWN WASTING OF CONTINENTAL ICE SHEET

(f) End moraines (or terminal moraines) are ridges produced by advancing ice or debris accumulating at the margin of a glacier. The dominant processes influence the dip of the long axis of elongated pebbles which may be forced to dip up or down glacier. (g) A wasting ice sheet with a cover of till (derived from thrust planes and englacial debris) has a complex of stream and melt-pond features. Fluvial and lake bed deposits consequently become increasingly important as the ice disintegrates, until the final deposit may be largely of fluvioglacial sediment. (c, d, are modified from Sugden and John, 1976 and g, from Marcussen, 1977).

bedrock obstructions, causing upward shearing of ice layers along thrust planes (Fig. 15.19d). This may squeeze saturated till upwards and cause lodgement in a ridge where stresses are lower. Once ridge building begins it may be self-perpetuating and grow as more till is plastered on the ridge face. The top of the ridge may be smoothed by contact with active ice. Near the margin of an ice sheet till may be squeezed up into subglacial cavities to produce similar, but usually smaller, features.

Subglacial transverse ridges of meltwater lakes are a separated form of moraine known as a De Geer moraine. Large ridges 5–30 m high may be formed but more commonly they are relatively small ridges composed of till, stratified sand, and lake deposits. De Geer moraines usually form in broad, shallow depressions in the landscape and are thought to develop at the margin of an ice sheet that is calving icebergs into the lake. Two possible mechanisms exist: (1) till at the bed is pushed into a ridge by ice just floating off the bed and (2) debris slides into crevasses or down the ice front to produce a submarine ridge.

De Geer moraines are widespread in Canada and Scandinavia and are probably more common than the better-known endmoraine ridges. They emphasize the importance of ice marginal lakes. In North America they are often called washboard moraines.

Concentration on regular features of moraines tends to underplay the abundance of nearly featureless till plains with occasional ridges or hummocks. Over much of Canada and Finland such plains now form the landscape and are presumed to result from the wasting of stagnant ice and letting down of basal debris onto a regular surface (Fig. 15.19g).

Icefront ridges develop at the margin of stagnant ice either during the recession of the ice margin or because it is reactivated and advances to form a push moraine (Plates 15.22, 15.23). Recessional

15.21 A drumlin field. The view is from Ribblehead towards the ESE, Yorkshire (photo by J.K.S. St Joseph. *Copyright*: Cambridge University).

moraines mark the former limits of ice and can form ridges some kilometres long and up to 200 m high, although they are usually lower. They are best known from the margins of glaciers which have readvanced in the last few thousand years, but the largest and most continuous survive from the major ice retreats between about 14 000 and 8 000 BP. They occur both on land and in shallow seas.

Processes of formation are varied and include: sliding of supraglacial debris down the ice face; flow of saturated till down ice; delivery of material to the snout along thrust planes; squeezing of saturated till from beneath the ice; and downwasting of ice to leave irregular piles of ablation till. Much of this till may be distorted by ice pushing and much, or all, of it may be further modified by meltwater (Plates 15.3, 15.24).

Dead ice forms a particularly irregular and chaotic topography of ridges, cones, pits and nearly flat areas. Some of the debris is till, but much is modified by water flow and deposition in meltwater ponds. Complex disintegration moraines lack recognizable patterns of landforms and may as a result not have been accorded the attention their abundance merits.

Meltwater

Meltwater is intimately associated with all warm glaciers and is produced in minor quantities by cold ice. Its importance increases from within the

15.22 An endmoraine left by a lobe of the Cordilleran Ice Sheet, Washington State, USA. Note the hummocky relief left by wasting ice behind the endmoraine (to the left) and the smoother outwash surface to the right (photo by John S. Shelton).

glacier towards the snout where it does more geomorphological work than the ice itself.

Meltwater may be produced both at the surface of a glacier and also at its base and internally. Surface melting is the most important source and is the main result of ablation. In maritime areas summer rainfall may join meltwater on the surface, and in valleys streams may flow onto the glacier margin. Melting at the surface is produced by heat from the atmosphere; within a glacier most heat probably comes from friction between moving water and the ice, for meltwater within a glacier has a temperature close to 0 °C; melting at the base results from basal friction and geothermal heat. The amount of meltwater produced varies with the seasons and with the environment. It is probably at a maximum in tropical mountains and in temperate latitudes near the sea, and at a minimum in high-latitude continental areas.

Surface streams on glaciers have channel forms and patterns which are similar to those in sediment, except that meanders are usually more regular. Internal streams cannot exist in cold ice, but in warm ice may form networks similar to those in a karst region, with water tables formed within the ice, and large hydrostatic pressures in some tunnels. At the base of a glacier meltwater may form tunnels or exist as a shallow lake. Radio-echo sounding suggests that there is a lake about 15 km wide in the rock depression at the base of the Lambert Glacier, Antarctica, and beneath the East Antarctic ice sheet there are at least twenty lakes with an area < 5 km^2, several with an area > 1000 km^2 and one with an area of 8000 km^2.

Meltwater production is irregular and dependent largely on the season and time of day, with largest discharges in summer and late afternoon. Sudden releases from cavities in, or below, the ice, or from lakes dammed by an ice barrier across a valley can produce catastrophic floods. In Iceland sudden floods are called *jökülhlaups*. They are releases of water formed subglacially above a volcano and may carry torrents of water

15.23 An arcuate terminal moraine left by a recent advance of a small cirque glacier. Britannia Range, Antarctica.

(25 000–100 000 m^3/s), and sandy debris with boulders, far from the ice front.

Meltwater erosion

The ready availability of sediment of all sizes ensures that meltwater streams carry large suspended and bedloads. Water under high hydrostatic pressures can abrade and cause cavitation erosion of bedrock. Channels are frequently cut below valley glaciers (Plate 15.25) and potholes cut in channel beds and walls. Many of the erosional bedforms have smooth, rounded, and scalloped forms which look as though they have been moulded from a plastic substance and are sometimes classed with *p-forms*. Because channels are influenced by the hydrostatic head in subglacial tunnels they can cut upslope and across ridges, and are thus clearly distinguishable from normal stream channels. Individual channels may rise up to 140 m in altitude although a rise of a few metres only is more common.

Many meltwater streams formed beneath a glacier flow under pressure towards the ice margin but, where the topography is suitable, streams may also flow for considerable distances along the margin. In warm glaciers the channels may be formed under the ice edge or outside it, but where ice is cold they must be formed outside. The form of such channels varies from a steep-sided gully to a bench cut in a hillside. In a few limited areas large channels and canyons are cut into bedrock and there is still speculation as to whether or not they are truly of meltwater origin. It is clear, however, that for large meltwater channels to be cut in bedrock the ice front must be stable for some time. Four major types of meltwater channel are commonly recognized: (1) tunnel valleys, (2) ice marginal channels, (3) spillways, and (4) coulees. Such distinctions are, however, not always possible.

Many meltwater channels have alignments which are not related to postglacial relief and are consequently anomalous. There is still argument, however, about whether channels were formed by

15.24 Hummocky ablation moraine left by a lobe from the Hatherton Glacier, Antarctica. The mounds are ice-cored and ablation rates are reduced by the cover of till.

overflow from lakes or subglacially. The debate about such well-known valleys as Newton Dale (Plate 15.26), which crosses the watershed of the North York Moors into the Vale of Pickering, still continues. It was regarded originally as a lake overflow channel, but more recently a subglacial origin has been suggested. Similar debate continues about many other valleys in Northern England.

(1) The importance of subglacial channels has been indicated from studies carried out in Denmark, where they are called *tunnel valleys*, and northern Germany, where they are called *rinnentaler*. These valleys are sometimes incised to a depth of 100 metres below the level of the surrounding landscape, they are commonly 1–2 km wide, and are sometimes traceable for distances of up to 75 km. They tend to have steep sides and rather flat floors, but with local irregularities in which small lakes have formed. It is probable that they were cut by streams flowing towards the ice margin and cutting laterally as

they did so. It is not realistic to suppose that a roof of ice spanned a 2 km wide valley beneath an ice sheet. Tunnel valleys usually end abruptly and it is supposed that this terminus is a site at which streams emerged from beneath the ice and created an alluvial fan. Large valleys beneath the North Sea may have a similar origin and smaller features in eastern England are regarded as being similar, although these are narrower, sometimes deeper with many enclosed depressions, and are infilled with sediment.

(2) *Ice marginal channels* are common and may originate from streams draining along the ice front. Particularly large channels, known as *urstromtäler*, occur in both northern Europe and North America where rivers have followed the ice edge. These channels may have originated largely because drainage was diverted from its original course northwards and flowed along the ice margin until it could reach the sea. The mouth of the Rhine was diverted westwards by the Scandinavian Ice Sheet, and the modern course of the Ohio

15.25 A hamlet of the Swiss village of Finhaut is perched on a bench above a gorge cut by subglacial streams.

River is believed to have been initiated around the edge of the Laurentide Ice Sheet (but see also Chapter 18).

(3) Channels cut by streams overflowing from ice marginal (i.e. *proglacial*) *lakes* are called *spill-ways*. For a channel to be clearly recognized as a lake overflow feature it is necessary for the old lake to be identified by its lake bed deposits, or from lake deltas, or from lake shorelines (such as the parallel roads of Glen Roy, Scotland). Even where such features are present it is not easy to understand how bedrock channels can be cut by the clear waters which normally flow from lakes, unless the channel bed and walls are of particularly weak rock such as mudstone. The Fenny Compton spillway from glacial Lake Harrison, in the English Midlands, is cut in clay to a depth of at least 7 m.

(4) By contrast with many supposed spillways the features known as *coulees* result from the instantaneous release of the entire contents of a large ice-dammed lake. In the State of Washington, north-western USA, glacial Lake Missoula was dammed against a lobe of the Cordilleran Ice Sheet. At its maximum the lake had a surface area of 7500 km^2 and a total volume in excess of 2000 km^3. At the dam controlling its outlet the lake was at least 700 m deep (this is all indicated by the old lake shorelines). It is believed that the dam broke suddenly, but the cause is unknown. A steep surface gradient developed across the lake, and at each submerged ridge gaps acted like the throats of flumes so that cavitation erosion was severe, and hollows up to 30 m deep were rapidly gouged from the rock. On the slopes below some of the ridges in the lake gigantic scalloped bedforms, 10 m high, are cut into the rock. Downstream from the ice dam water spilt across drainage divides and surged up tributary valleys carrying huge boulders.

About 80 km below the dam the flood waters fanned out across the Columbia plateau basalts, washing soils and loess from the surface and quarrying deep channels into the basalts. These steep-walled, irregularly floored canyons are the coulees. Individual canyons are 100 m deep and contain dry waterfalls with plunge-pools. The overall box-shape is largely controlled by the

15.26 Looking south down Newton Dale, Yorkshire, towards the Vale of Pickering. (Photo by J. K. S. St Joseph. *Copyright*: Cambridge University).

columnar jointing in the basalt. The anastomosing pattern of the coulees and stripping of soils has given rise to the area name of 'channelled scablands'.

The hypothesis of meltwater catastrophic flooding as the cause of the channelled scablands was first put forward by Bretz in 1932 and was met with incredulity, but is now generally accepted and the recently acquired evidence suggests that such events took place on several separate occasions. The most recent major flood took place about 20 000 BP and had a discharge of about 21 Mm³/s—one hundred times the mean flow of the Amazon.

A similar catastrophic origin has been proposed for the Labyrinth in Wright Valley, Antarctica (Plate 15.27). These canyons are similar in form to the Washington coulees but occupy a smaller area of about 30 km² and have a maximum vertical relief of 100 m. The evidence for a flood here is lacking, and alternative hypotheses of glacial erosion, subglacial erosion, and salt weathering are equally plausible.

Meltwater deposits

Meltwater deposits are recognizable from their internal stratification rather than their exterior forms which may be similar to those of glacial origin. Eskers, kames, kame terraces, lake beds, sandar, and valley trains are the main forms. The particles they contain have generally not been transported a great distance by meltwater, hence they are not so rounded as most fluvial deposits. They are also different in that kames and eskers are deposited in contact with ice and often have slumped margins produced as ice melts. Bedforms, however, are similar to those in fluvial deposits.

Eskers are long ridges which generally show a close correspondence with the most recent direction of ice advance. They may be sinuous or have a 'beaded' appearance and may be single features or have an anastomosing pattern (Plate 15.28). They vary in length from a few tens of metres to several hundred kilometres and in height from 2 to 100 metres. They follow irregular land surfaces up-and down-hill and are thought to be formed within subglacial tunnels. Because the esker is, in effect, a raised river bed it must be created when discharges are low enough for fluvial deposition, suggesting that late summer and early autumn are the most probable seasons for development. Esker deposits

have typical stream channel bedforms with ripples and dunes and the beaded type usually display a decrease in grain sizes in a downstream direction towards the bead, which has deltaic bedding. This suggests that the esker is formed subglacially and the bead is a remnant of a delta formed where the esker stream emerged from below the ice. Repeated marginal retreat gives rise to many ridge and bead forms. Nearly all eskers have bedding which has slumped towards the flanks of the esker ridge, suggesting that the flanks were modified as the confining ice tunnel walls melted and the flanks are now at the natural angle of rest of the sand and gravel.

Nowhere in the world are large eskers being visibly formed at present. This suggests either that eskers are formed beneath thick warm ice where they cannot be observed, or that modern glaciers lack the characteristics for large esker formation.

Kame terraces are remnants of stream beds that have formed along a valley wall at an ice margin. The stream channels are usually braided and the bedload has a large range of particle sizes. Because the source of sediment is the valley-wall talus, or the glacier, sorting is in its early stages and many pebbles and boulders are still rather angular and even striated. As ice wastes at the margin of the terrace slumping occurs and also erosion.

Kames are isolated hills of stratified drift produced in areas of dead wasting ice with an abundant cover of ablation till (Fig. 15.20). It is an environment of irregular melting, slumping, and fluvial modification of till with both ponding of meltwater and active surface and subsurface streamflow. Consequently kames are extremely varied in location, origin, size, and content of sands and gravels. It is common for one or more sides of a kame to slump after its supporting ice has melted and for the internal bedding to have mixed layers of bedded sands, gravels, and unsorted till (Plates 15.29, 15.30).

Many ponds in areas of stagnant, downwasting ice contain bedded silts and sands and also deltas so that *kame deltas* are widely recognized. Where blocks of ice are embedded in lake beds or a fluvial outwash deposit the ice may last until the deposit has become stable. When it melts it leaves a pond known as a *kettle lake* or *kettle hole* and large areas with ponds in a hummocky ground moraine are often called *kettle moraines*.

Outwash from an endmoraine is usually the product of deposition resulting from channels

15.27 The Labyrinth at the end of Wright Valley, Antarctica. Note the wide trough of the valley in the distance (photo by US Navy).

15.28 A group of anastomosing eskers near Nairn, Scotland (photo by J. K. S. St Joseph. *Copyright:* British Crown Copyright Reserved).

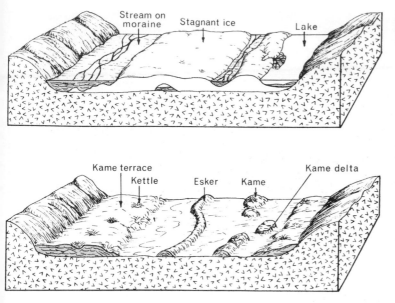

Fig. 15.20 Origin of ice-contact stratified drift and the resulting landforms (modified from Flint, 1971).

which braid and spread because the sediment supply is too great for the available stream flow and channel slope. Within valleys these braided channels are called *valley trains*, and on plains or wide lowlands the Icelandic term *sandur* is used for a single low-angle fan or sheet of outwash gravels (plural: *sandar* or *sandr*). Sandar and valley trains (Plates 15.31, 15.32) pass through endmoraines and zones of dead ice and are major zones of resorting of sediment. They may cover very large areas and pass downslope into deltas and lakes or into fluvial valleys cut beyond the ice margin. Outwash is a major source of loess.

Ice margins are seldom static for great lengths of time nor are rates of retreat uniform. Ice fronts may surge forward or advance slowly during periods of predominant retreat. Consequently fluvioglacial deposits may be eroded, and over-ridden, or pushed into fresh forms. Where cuttings through such deposits occur it is common to see a complex pattern and history (Plate 15.19).

Landscape patterns

As an ice margin retreats or downwastes there is developed a sequence of characteristic meltout, ablation, ice contact, and fluvioglacial forms, but many of the moraines formed of lodgement till may be associated with active ice and survive deglaciation so that zones of deposits can now be recognized and these may be associated with characteristics of the basal ice (see Fig 15.17). The zones are described below as they may be seen by a traveller who moves outwards from a zone which was immediately below the thickest part of a large ice dome.

Beneath a frozen zone erosion is at a minimum and even large glaciers perform little or no erosion or deposition, but in a zone of freezing where there is some meltwater basal velocities can be higher, erosion can occur and fluted and streamlined erosional forms are cut into the bedrock.

In a melting zone, with greater abundance of meltwater, tills will thicken although still have thicknesses generally less than a few metres, except where drumlinized forms or rogen moraines develop, and then true drumlin fields form as the tills thicken away from the centre of the ice sheet.

Drumlins may be replaced by rogen moraines where ice undergoes compressional flow close to its margin, and eskers may cross this zone indicating a period of deposition after the formation of Rogen moraines.

In the ice-wasting zone, various disintegration features occur with meltout tills, irregular ablation deposition, and kames. Endmoraine ridges with fluted moraines and fluvioglacial forms occur at the margin of active ice. At a lake margin De Geer moraines and lake beds may predominate.

Although there are many complicating factors it is possible to discern a close relationship between ice temperature and flow pattern and the erosional

15.29 (a) Kame deposits with ripple-bedded sands of fluvial origin and an unsorted till layer between them.
(b) A kame delta deposit with a till on top.

15.30 A meltwater channel, cut through hummocky endmoraines (top left), with well-formed terraces. Rakaia Valley, New Zealand.

15.31 A small sandur in front of the Garwood Glacier, Antarctica. The dark hummocky surfaces are an ice-cored ablation moraine.

15.32 A valley train in the Tasman Valley, New Zealand.

and depositional features formed at the base of the ice.

Conclusions

Glaciers, by erosion and deposition, have greatly modified the surface they have cut into and spread upon. This modification is particularly obvious in alpine areas, but the extent of the modification is much harder to assess where ice sheets have spread upon lowlands. Considerable controversy has arisen concerning the amount of erosion completed by the Laurentide Ice Sheets. It is evident that thickness of tills in northern USA, northern Germany, and Poland, is extremely varied, ranging from a few metres to over 400 m, but most of the glacial debris does not lie upon land but on the continental shelves and deep ocean floor. The best estimates, available in 1980, from studies of ocean-floor sediment thicknesses indicate that about 1 Gm^3 of sediment have been eroded from eastern North America and southern Greenland and deposited in the adjacent North Atlantic since the beginning of continental glaciation. This volume is a minimum estimate and does not include the sediment on the continental shelf or that carried into lower latitudes by icebergs or bottom currents. The volume of sediment represents an average eroded rock thickness of about 100 m. If to this is added the tills and fluvioglacial materials left on land, and from it is subtracted the sediment eroded by non-glacial processes, we can probably estimate that glacial erosion has lowered the land surface by an average of 60–100 metres during the last 2–3 million years.

Ice volumes throughout the world have repeatedly increased and decreased, and hence ice margins have advanced and retreated. The history of glaciation and some of its implications and evidence are discussed in the next chapter.

Further reading

The book by Sugden and John (1976) is strongly

recommended as the best general text in glacial geomorphology; that by Andrews (1975) provides useful additional reading. Embleton and King (1975) is a comprehensive review of the field. The papers of Sugden (1977 and 1978) on the Laurentide Ice Sheet appeared after any of these texts and indicate the value of modern approaches to relationships between ice sheets and landforms.

16   Ice Ages

An *ice age* is a period in the Earth's history, lasting some 20–100 My, during which ice sheets occur somewhere on land. Ice ages are separated by long intervals of 150 My, or more, during which there are no ice sheets and few alpine glaciers. The record of past ice ages is deciphered from the sedimentary rock record in which glacial deposits may be recognized. Because many glacial deposits are composed of unsorted, and often unbedded, silts and clays containing larger clasts of irregular shape and size they are not always readily distinguishable from deposits left by mudflows and landslides. The general class of unsorted and mixed grain-size deposits is called *mixtites*. These are only regarded as unequivocally being of glacial origin if they are found in association with glacially striated bedrock, crescentic gouges, or *dropstones* in varve-like strata. Dropstones are large clasts, released from floating ice, which fall into marine or lake beds.

Pre-Cenozoic ice ages

Evidence for Earth's earliest glaciation is best documented for an interval in the middle Precambrian at about 2300 My ago and the deposits are found in North America, Africa, and Australia. There were probably ice ages before this but the record is unclear. In the late Precambrian glacial deposits, called *tillites*, are found in Africa and Australia at intervals centred on about 900, 750, and 600 My ago. The exact ages, however, are still not well known and there is still debate upon the timing, duration, extent, and continuity of each ice age (Fig. 16.1).

Glacial deposits of late Ordovician to early Silurian age were discovered in the 1960s in north Africa and others have been reported from South America and the Soviet Union. The north African deposits are well exposed and displayed with long glacial grooves in bedrock, ice-scoured valleys, and sandy tillites occurring over such large areas that it is clear they must result from ice sheet erosion and not merely the action of local upland glaciers.

During this time the pole moved from West Africa into what is now the Zaïre rainforest and ice spread northwards from the central Sahara (Fig. 16.2). The ice mass was probably formed of several major domes and sheets and had an area perhaps twice as great as that of present Antarctica. Such a large ice mass occurring on the supercontinent of Gondwanaland could not readily be supplied with moisture, and it seems probable that the centre of glaciation migrated so that the evidence relates to successive ice inundations rather than to one event. Ordovician rocks are generally absent in central Africa but they occur in Namibia and South Africa. Near Cape Town, on the flanks of Table Mountain there is a 750 m thick sequence of littoral and lagoonal sandstones and shales resting on an erosion surface cut across Precambrian granite, and these coastal deposits are overlain by a tillite. These deposits may represent the southern limits of Ordovician ice. By early Silurian times the centre of glaciation had moved southwards as the pole also migrated (Fig. 16.3).

From middle Carboniferous to early Permian times an exceptionally long, cold, and extensive ice age caused major changes in the pattern of life and left a major imprint upon the geological record. At the beginning of the Carboniferous there were three major land masses, with climatic zones which were probably rather similar to those of today, but these land masses closed up so that by the mid-Permian there was one supercontinent, Pangaea, which survived until the Jurassic Period. On land, large areas were sparsely covered by plants, but in favourable areas there were large forests of tree ferns, scale trees, rushes, and seed

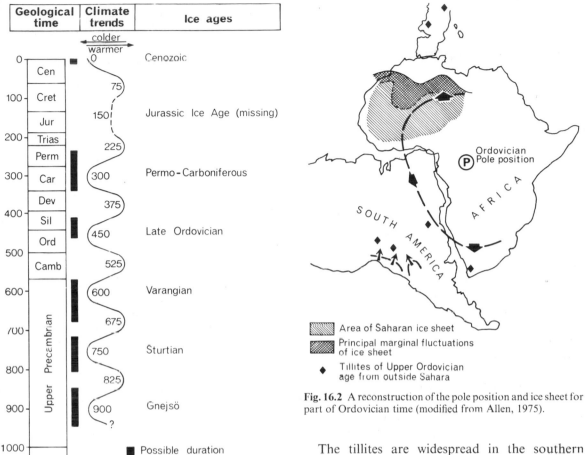

Fig. 16.1 An hypothetical sequence of ice ages during the last 1000 million years (modified from John, 1979).

Fig. 16.2 A reconstruction of the pole position and ice sheet for part of Ordovician time (modified from Allen, 1975).

ferns. There were also some pine-like conifers which were evolving rapidly so that by the end of Carboniferous time conifers were established, and many of the old species of seedless trees were replaced by the more advanced seed-bearing ferns and gymnosperms. Many of these plants grew in areas of impeded drainage and their remains have been preserved in peat bogs which were later buried and turned into coal beds. Some of the Permian coals are interbedded with tillites suggesting that the peats formed in climates similar to those of the boreal zone of Canada and Siberia today. The Carboniferous–Permian floras were characterized by the presence of *Glossopteris*, sometimes of *Gangamopteris* and some shales contain fossils of the aquatic reptiles *Mesosaurus* and *Lystrosaurus* (see Fig. 3.2).

The tillites are widespread in the southern continents and are associated with striated pavements, roches moutonnées, marine beds containing fossils and dropstones, and with far-travelled glacial erratics. From these many sources of evidence the patterns of ice movement and the changing centres of glaciation have been reconstructed. During a period of about 100 million years the centre of glaciation shifted gradually from South America and Namibia to South Africa, Madagascar, and India, and then to Antarctica (Plate 16.1) and Australia. There is little doubt that southern Africa was the focus of glaciation for the greatest length of time, and it is here, in the Dwyka tillites, which are up to 900 m thick, that we find the most convincing evidence for substantial glacial deposition. Of particular interest is the evidence that the huge ice sheets (probably exceeding twice the area of modern Antarctica) were at least partly wet-based, that they were partly marine with very large ice shelves, and were evidently affected by repeated surges

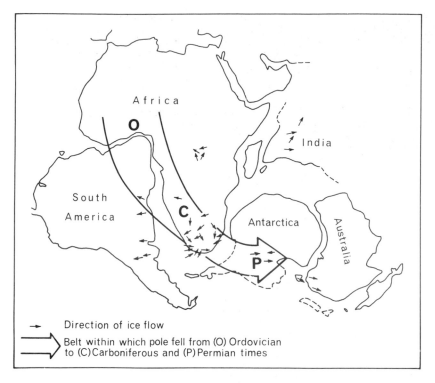

Fig. 16.3 Changes in the position of pole, and ice flow directions, for part Paleozoic time. Note that the position of the Ordovician pole is different from that of Fig. 16.2. This indicates the possible errors in reconstructions, and the possibility of pole movement during the Ordovician.

which caused very rapid rises of sea-level. This latter point is of concern to glaciologists at present studying the West Antarctic Ice Sheet and the rapid end of the last Arctic Ice Sheet about 14 000–8000 BP.

Beyond the Carboniferous–Permian ice sheet limits, permafrost and periglacial phenomena were widespread, and large glaciers existed to within 30° of the equator in Permian times. It is probable that this ice age was particularly cold; it certainly had the effect of drastically reducing the number of species of plants and marine invertebrates, and some 75 per cent of amphibian and 80 per cent of reptile families disappeared.

Mesozoic climates

Rocks of the Mesozoic Era provide ample evidence of widespread dry climates and equable temperatures seldom seen elsewhere in the geological record. Studies of evaporites, windblown sediments and deep-sea deposits, of oxygen-isotope work on paleotemperatures, and the paleontological record from fossil plants, pollen, and the newly arrived exothermic reptiles, all provide supporting evidence.

If a 150 My cycle of glacial climates truly exists then there should have been an ice age in Jurassic times, but there is no clear evidence for such an event and there is very strong evidence for climates much warmer than those of the present. The lack of any large land mass in near-polar latitudes may have been the cause for the failure of an ice age to develop at this time.

For Cretaceous times onwards there is abundant oxygen-isotope data on paleotemperatures and it is clear that the Cretaceous was a period of great warmth over the whole globe. Tropical conditions extended to latitudes of 45° N and S and the polar regions had temperate climates. Mean annual temperatures were some 10–15 °C warmer than those of today, and the temperature gradient from the equator to the poles was about half that of the present.

Throughout the Mesozoic the continents were clustered together in a single great land mass which stretched nearly from pole to pole. This configuration undoubtedly contributed to a high degree of efficiency in the transfer of heat from tropical to polar latitudes. It is probable that large single gyres (circular ocean current systems) occurred in the huge Pacific Ocean of both

16.1 An exposure of Permo-Carboniferous ice age sediment in Antarctica. The lower dark-coloured unit is of indurated muds containing pebbles and boulders. This mixtite is interpreted as being a glacial till. The upper pale-coloured unit is of fluvial sand containing lenses of angular gravel and is interpreted as being a glacial outwash deposit.

northern and southern hemispheres. As they circulated through a wide longitudinal expanse of the tropical ocean the water masses would be warmed to temperatures well above those of the present, and heat would be transferred to polar regions which consequently had temperatures of 8–9 °C. The history of global climates since the Cretaceous Period has been one of declining efficiency of heat transfer as more oceans, smaller gyres, and restricted circulations have resulted from sea-floor spreading and continental migrations.

Precambrian ice ages may have been caused by, or influenced by, decreases in received solar radiation, but the fundamental causes of Paleozoic and later Cenozoic ice ages may be found in the arrangement of continents and oceans.

Development of the Cenozoic ice age: the Tertiary record

At the close of the Mesozoic Era warm global climates had already started to deteriorate towards the glacial climates which characterize the later Cenozoic (Fig. 16.4). The Paleocene, Eocene,

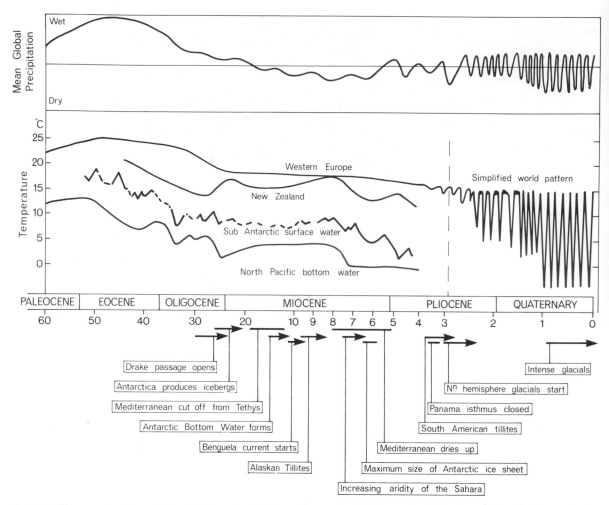

Fig. 16.4 The general pattern of climatic change for the Cenozoic as indicated by many lines of evidence. Note the changes in the time-scale. Major events are indicated.

and Oligocene Epochs experienced cool changes which were both more frequent and more intense than those of the Mesozoic. Over the interval from 65 to 22 My ago long episodes of relatively slight warming were interrupted by intervals of severe and often abrupt falls in temperature leading to successively cooler climates. The clearest evidence for this pattern comes from oxygen-isotope studies and from the study of plankton found in long cores taken from ocean-floor sediments. In parallel, the zones of abundant high precipitation in near-polar latitudes seem to have migrated towards the equator, so that by the late Cenozoic the mid-latitudes had become drier, the subtropical deserts had become well established and the

present ice age is clearly one of glacial and periglacial activity at high latitudes and altitudes, but aridity in the subtropics. Changes in the pattern of continents and ocean currents which accompanied the climatic change, and at least in part caused it, can be appreciated from a study of the Earth in the Eocene, Oligocene, and present (Fig. 16.5a,b,c).

Continents and ocean currents

About 50 My ago, near the beginning of the Eocene Epoch, there were no ice sheets anywhere on Earth; the Atlantic was still narrow; the Pacific was much wider than now, spanning about 210° of

OLIGOCENE

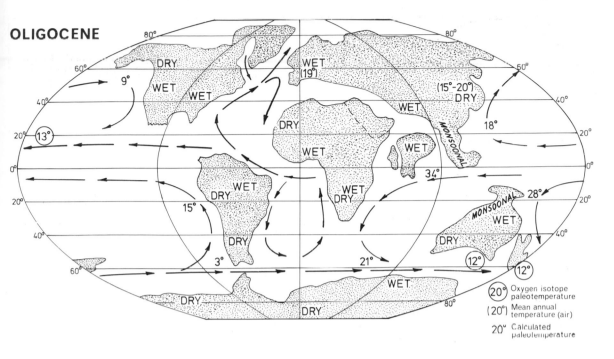

EOCENE

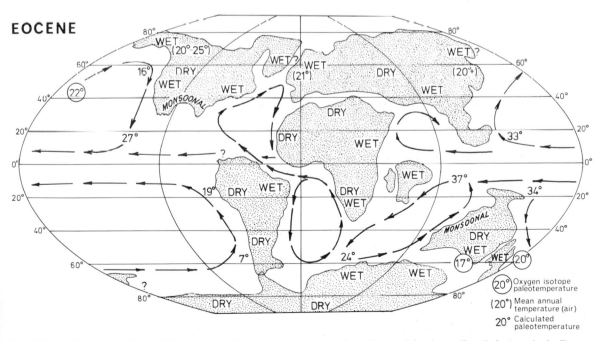

Fig. 16.5 (a,b) Reconstructions of the positions of the continents, oceanic circulation, and dominant climatic features in the Eocene and mid-Oligocene.

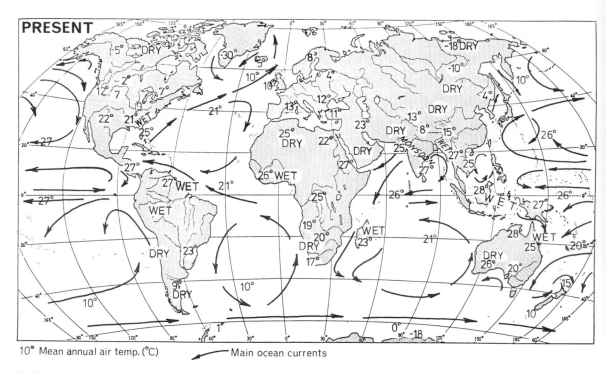

10° Mean annual air temp. (°C)   Main ocean currents

(c) the present distributions (a,b, after Frakes and Kemp, 1972).

longitude at the equator; North and South America were widely separated by a Panamanian Sea; South America was joined to Antarctica; Africa was separated from Eurasia by a shallow sea, the Tethys; and many of the great mountain systems of the world—the Andes, Himalayas, and Alpine chains—either did not exist or had limited elevation. As a result the transfer of heat from equatorial to polar regions was still very efficient. The tropical climatic belt occupied at least 50° of latitude, and the polar regions had temperate climates with mean annual temperatures of 8–10 °C with few or no frosts.

By Oligocene times, 38–25 My ago, a general cooling had occurred and the contrasts between the equatorial and polar regions had become much greater. The contrast became very strong after Drake Passage, between South America and Antarctica, opened 30–25 My ago. This opening permitted the formation of a circum-Antarctic current which restricted the transfer of heat southwards, and permitted Antarctica to become increasingly cold and the formation of its ice sheet.

During and since Oligocene times the Atlantic has opened wider, the Panamanian Straits and

Tethys have closed, and the large longitudinal sweep of tropical ocean currents has been reduced so that tropical seas have become cooler. The circum-Antarctic circulation has become wider and more dominant in the southern hemisphere as Australia and Antarctica have moved farther apart. The greater separation of Greenland and Scandinavia may have permitted the North Atlantic Drift to enter the Arctic Sea and so allow warm and moist air masses to carry precipitation onto northern Greenland and Siberia. As the southern continents of Africa–Arabia and India converged upon Eurasia, the Alpine–Himalayan mountains were uplifted. This has reduced the inflow of moist air into central Asia and increased the area of high-altitude glaciers. Similarly the uplift of the North and South American cordillera has influenced the climates of those continents.

The evidence for these reconstructions comes from many sources. From the residual magnetism in rocks the position of the paleomagnetic pole and the position of continents in relation to it, at the time of formation of those rocks, may be determined. The rate at which ocean basins have changed in shape is determined from the ages of sediments covering the ocean floor. Paleotemperatures are estimated from fossil plants and animals, and more precisely from the oxygen isotopes in carbonate deposits.

The role of Antarctica

The spread of glacial ice over 98 per cent of Antarctica has obscured most of the evidence for early Cenozoic climatic events on that continent, but a few fossil deposits on King George Island and Seymour Island north of the Antarctic Peninsula, and in the area of the Ross Sea, indicate that a varied flora of coniferous and deciduous trees existed in Eocene times and, in places, survived until the Miocene.

Core samples from ocean-floor sediments around Antarctica contain layers of exotic striated and fractured sands and pebbles which contrast in texture with the more usual ocean-floor muds. Their presence at an inferred paleolatitude of 54° S in upper Oligocene sediments, 26 My old, is taken as evidence that the sands and gravels had been carried in the sole of drifting icebergs which calved off glaciers from Antarctica.

The evidence for extensive ice cover on Antarctica by about 26 My ago is clear but the ice itself did not, apparently, cause severe cooling of the ocean around the continent; if it had there may have been a restriction upon moist air reaching the interior and so nourishing the developing ice sheet. The presence of trees at this time suggests that some of the Antarctic coasts may have been similar to those of British Columbia and Alaska today, where glaciers pass through forested lowlands to reach the sea.

During the Miocene, 24–6 My ago, ice on Antarctica continued to accumulate. In several parts of the continent glacial tills and striated pavements have been covered by basaltic lavas which can be dated (Plate 16.2) and it is evident that a full ice sheet had developed by 5–6 My ago. This sheet extended in the Ross Sea area some 200 km beyond its present limits and may have reached

16.2 A basalt cone on the valley wall, and a lava flow which has spread over tills on the valley floor. Koettlitz Valley, Antarctica.

the edge of the continental shelf. The Transantarctic Mountains prevent the spread of the main Antarctic ice sheet into the Ross Sea embayment but outlet glaciers have cut through the mountains and now feed the Ross Ice Shelf. Near McMurdo Sound the Taylor, Wright, and Victoria Valleys are now ice-free (Plate 16.3, Fig. 16.6) but periods of ice erosion are well dated as occurring before the formation of small basalt cones which erupted on valley floors after the last major retreat more than 4 My ago. This retreat probably resulted from both uplift of the mountains which then cut off the flow of ice from the Polar Plateau to the Ross Sea, and also from decreases in ice sheet volume as the interior of the continental ice sheet was starved of moisture by distance from the sea, and by altitude. It is probable that the East Antarctic Ice Sheet has had a volume close to that of the present throughout the last 4 My.

The growth of the Antarctic ice sheet to its maximum volume 5–6 My ago had a major effect upon the southern hemisphere, where severe cooling was recorded in the oxygen-isotope record and by changes in the faunas in coastal sediments around New Zealand. Its possible effect upon the west coast deserts of the subtropics will be discussed below.

Onset of glaciation in the rest of the world

Fossil plants and animals, both from marine and terrestrial sediments, indicate general but uneven cooling trends throughout the middle and late Cenozoic. The earliest mountain glaciations in Alaska are recorded in tillites interbedded with lavas dated at 9–10 My ago, but glaciers also reached sea-level, as indicated by ice-rafted dropstones in marine sediments along much of the Gulf of Alaska. These glacial events may be related to uplift rather than general cooling as widespread

16.3 Part of the McMurdo Dry Valleys, Antarctica, showing outlet glaciers from the East Antarctic Ice Sheet. The diffluent glacier forms the Ferrar which reaches the sea and the Taylor Glacier which terminates on land (photo by US Navy).

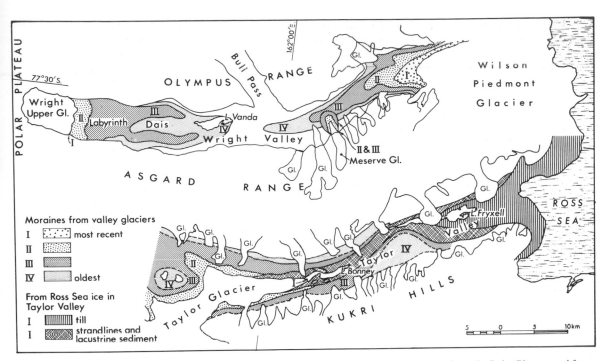

Fig. 16.6 Glacial deposits in the Wright and Taylor valleys, Antarctica. Ice has entered the valleys from the Polar Plateau and from the Ross Sea to leave a record of tills which have been dated. This evidence provides the best record of ice thickness on the Polar Plateau for the last 4 My (based on Denton, Armstrong, and Stuiver, 1971, and Calkin, Behling, and Bull, 1970).

ice-rafting in the north Pacific is not detectable until the middle Pliocene at about 4 My ago. There is, however, accumulating evidence of widespread cooling around 5–6 My ago, and limited evidence of Arctic Ocean ice by 5 My ago. This has led some geologists to postulate the existence of substantial lowland glaciers on southern South America and on northern North America as early as the late Miocene, 7 My ago. Glaciers of this age in the Pamirs and Gissar-Alai of southern central Asia are, almost certainly, caused by uplift. In North America the earliest well-dated deposits from the Sierra Nevada have ages of 3.1–2.7 My, and the earliest ice sheet formed on Canada and spreading to northern USA could have formed at least as early as 2.5 My ago. At low altitudes, at latitude 65°N, tillites interbedded with basalt lavas date the first glaciations of Iceland at 3.1 My ago. The Icelandic deposits record at least ten periods of ice advance with intervening warmer periods. The deposits are thought to indicate the formation of ice caps near sea-level and suggest that Greenland may also have had an ice sheet cover during the late Pliocene.

Cooling of the North Atlantic was well developed by 3 My ago, as is indicated by glacial detritus, presumably dropped by icebergs, in marine sediments. The Labrador current was formed at this time in response to Arctic cooling, and a characteristic cold water fauna appeared in northern oceans.

Accumulating evidence, therefore, is for a very large ice sheet on Antarctica by 5–6 My ago; for mountain glaciers to be common at this time; for lowland ice of unknown volume to be present in high latitudes, of both hemispheres, by 5 My ago; and for clearer evidence of lowland ice sheets in Iceland and Greenland by 3.0 My ago. Each piece of new evidence is for earlier onsets of glaciation than had been previously accepted.

Tropical deserts

There is still much uncertainty about the extent of arid climates in the early Cenozoic world. In general, smaller areas of land were arid than now, and this may have been the result of higher temperatures causing more evaporation from the sea and hence increasing the amount of water in circulation in the atmosphere. The distribution of the continents was different in the early Tertiary from that of the present with many consequences. Africa, for example, occupied more southerly latitudes but, because the equator has been close to its modern position since the Eocene, the remarkable dryness of southern and central Africa during the Eocene was probably due to a great expansion of the southern hemisphere subtropical high-pressure belt. In the Eocene much of what is now the Sahara had a humid equatorial climate, as is shown by fossil plants and extensive ferricretes and deep-weathering profiles. Throughout the Oligocene the Saharan region had a climate which supported a tropical savanna of open woodlands. This implies that moist air from the Atlantic and Gulf of Guinea covered much of the region and that moist air also penetrated southwards from the Tethys.

In Miocene times northern Africa suffered catastrophic changes. About 14 My ago uplift of the Atlas Mountains formed a climatic barrier between the Mediterranean and the Sahara, and the Mediterranean itself became nearly blocked from the Atlantic by narrowing of the Straits of Gibraltar and was cut off from the inland sea of the Paratethys which stretched from the Black Sea to the Sea of Aral. Deprived of these sources of fresh water, and sinks for salt water, the Mediterranean became a great evaporating basin for sea water flowing over the sill from the Atlantic. Flow from the Atlantic ceased about 6.5 My ago and the Mediterranean floor became a great desert basin with deep hollows filled with evaporites and brines. Rivers entering the basin cut down deep gorges to the new low base level so that the Nile entrenched itself 570 m into the bedrock gorge near Cairo, and was entrenched 170 m below present sea-level near Aswan. This period of great aridity—known as the Messinian Crisis—was certainly of major importance in drying up the Sahara, but it was not the only factor. It is probable that the general cooling of the atmosphere by the early Pliocene (see Fig. 16.4) also caused an expansion of the subtropical high-pressure systems and a retreat of the Intertropical Convergence Zone (ITCZ) to the south of the Sahara (Fig. 16.7) with a consequent increase of rainfall over central Africa and the Guinea Coast.

In the southern hemisphere the continents all have dry zones along their western margins—the Atacama of Chile and Peru, the Namib in Africa, and the Western Australia desert. Each of these deserts is largely caused by an upwelling of cold water off the coast to form the Peru (or Humboldt) Current, the Benguela Current, and the Western Australia Current. This cold water is mostly of Antarctic origin and the cold current did not exist until Antarctica was producing large quantities of cold meltwater and sea ice. Antarctic Bottom Water had formed by 15 My ago and the Benguela Current by 10 My ago. It is probable that the Namib Desert was established at about this time but the degree of aridity is still unclear. As will be discussed later, expansion of the Benguela Current carried dry conditions into equatorial Africa during glacial periods of the late Quaternary and had a major effect upon the climate of western Africa (Fig. 19.2).

The Tertiary record

Climatic changes during the Tertiary are mostly related to the development of a circum-Antarctic oceanic circulation and the resulting build-up of ice on that continent, a build-up which could not have occurred unless a large continent had been in a polar position. Antarctic influences spread by a general cooling of the southern hemisphere as ice accumulated, and through the establishment of the cold upwelling currents along the west coasts of southern land masses. Thus aridity was a consequence of cooling. The mechanism by which the subtropical anticyclones became established is, as yet, not well understood; but it is clear that by the early Pliocene the great deserts of the world had become established, both because of the pattern of general atmospheric circulation and because of the rise of the Cenozoic mountain systems with the creation of rain-shadow deserts on the continents, especially in Asia, northern Argentina, and south-western USA.

The opening of the Greenland Sea and the general cooling of the Earth, plus the rise of the North American cordillera, was followed by the establishment of glacial conditions in the northern hemisphere by 3 My ago.

This brief review of some of the better understood major events can be summarized in a diagram of a stylized continent. It indicates the relative area occupied by major vegetation types for latitudinal belts which have a width related to that available on the modern continents (Fig.

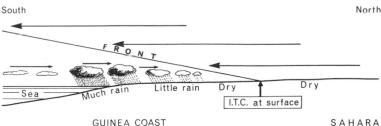

South   North

GUINEA COAST   SAHARA

Fig. 16.7 A schematic section through the atmosphere of West Africa from the Gulf of Guinea to the southern Sahara. When the Intertropical Convergence moves northwards wetter conditions prevail in the southern Sahara, and as it retreats southwards dry conditions dominate.

16.8). It emphasizes the relative decline of the forest zones, especially the tropical forests, the extension of grasslands and deserts, and the establishment of tundra.

The Quaternary

The term 'Quaternary' is defined in a number of different ways and its duration is still not well established. The generally accepted method of dividing geological time, since the Precambrian, into units is by the evolutionary change of plants and animals, so that each time unit has a characteristic assemblage of fossil organisms. This method breaks down when applied to the late Cenozoic. The Quaternary is simply not long enough in relation to the slow rate of evolutionary change to permit subdivision of strata according to the fossil record of animals and plants. Secondly, much of the interest in Quaternary geology is related to the history of climatic change, not to evolution of organisms. Isotopically measured temperatures, at times determined by radiometric and magnetic dating, are now the chief means of investigating the history of climatic events.

The age of the base of the Quaternary is decreed by the International Geological Union as being a time when cooling conditions were first recorded in marine sediments of southern Italy. These sediments have now been raised above sea-level by tectonism. Unfortunately the type site has turned out to be more complicated than was originally thought and the exact date is still not universally agreed. Many geologists accept a date of 1.8 My ago for the Pliocene–Quaternary boundary, others believe that 1.6 My is more appropriate and some workers regard 2.0 and 2.5 My ago as suitable dates. To avoid confusion 1.8 My is accepted in this book.

A second set of confusions relate to the use of the term 'Pleistocene'. This is regarded as extending from 1.8 My to 10 000 years ago and the time from 10 000 BP to the present is internationally called the Holocene or the Recent. The Holocene is actually the duration of the present interglacial, and some geologists regard the term Pleistocene as redundant as it covers virtually the whole of the Quaternary.

The Quaternary as revealed by ocean sediments

During most geological periods, history of the Earth is read from the deposits which may have been formed on land or in the sea, but are now exposed on land. For the Quaternary this is less satisfactory because few accessible sites on land have experienced continuous deposition—most have experienced severe breaks because of periods of erosion. Only sites like deltas and tectonically subsiding basins have long records of sedimentation and these can be interpreted only from long cores of rock recovered by deep drilling. Even then the problem of dating and correlating events, interpreted from one borehole with those from another borehole, is very severe. The most widely accepted record available at present is that from two cores recovered from the equatorial Pacific Ocean—cores V28-238 and V28-239 (Fig. 16.9).

Over most of the deep ocean-floor sediments consist partly of (1) brown muds derived from the continents (as windblown dust or suspended sediment or turbidity currents from coastal deposits); (2) calcium carbonate ($CaCO_3$) which has formed the hard skeletal parts (Plate 16.4) of animal plankton such as foraminifera and pteropods, and plant plankton such as coccolithophorida; and (3) opaline silica which formed the skeletons of diatoms (plants) and radiolarians (animals).

The calcium carbonate part of the ocean-floor sediment is of primary value for determining past climates. As the foraminifera die their hard parts sink to the ocean floor where this is not too deep. In very deep parts of the ocean $CaCO_3$ is dissolved

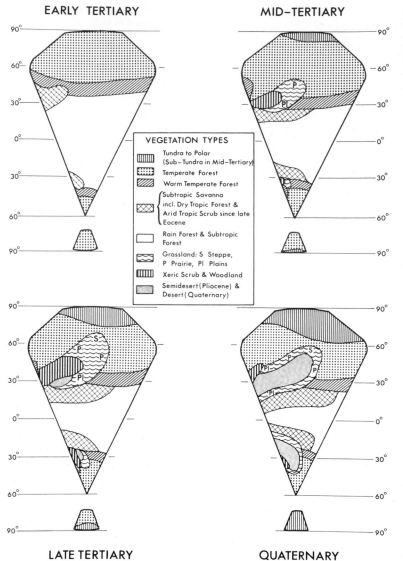

EARLY TERTIARY

MID-TERTIARY

LATE TERTIARY

QUATERNARY

VEGETATION TYPES

Tundra to Polar
(Sub–Tundra in Mid–Tertiary)
Temperate Forest
Warm Temperate Forest
Subtropic Savanna
incl. Dry Tropic Forest &
Arid Tropic Scrub since late
Eocene
Rain Forest & Subtropic
Forest
Grassland: S Steppe,
P Prairie, Pl Plains
Xeric Scrub & Woodland
Semidesert(Pliocene) &
Desert (Quaternary)

Fig. 16.8 The distribution of major vegetation types on a schematic continent, whose area in each latitude is related to that of the actual continents, for times during the Cenozoic (after Axelrod, 1952).

before it reaches the floor. In much of the ocean, sediment accumulates to form a layer 1–3 cm thick in 1000 years. A useful core is one in which sedimentation has been continuous and the sediment has not been disturbed by erosion or burrowing sea animals.

Determining the ages of layers of sediment in cores is done by a number of methods. Radiocarbon dating is effective back to 40 000 BP and the decay of uranium-238 to thorium-230 in the skeletal parts of foraminifera provides dates back to 400 000 BP. By far the most important method

used on long cores is by determination of the paleomagnetic record.

From studies carried out on numerous volcanoes whose lavas have preserved the magnetic record, it is known that during the late Cenozoic the direction of magnetic polarity has reversed at many times to give a record which is set out in Fig. 1.5. In core V28-239 the Brunhes–Matuyama boundary and the Jaramillo and Olduvai events are recognized.

The climatic record is interpreted in two ways: (1) from the oxygen isotopes preserved in the

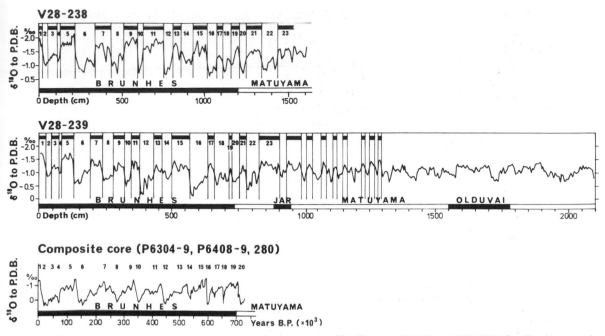

Fig. 16.9 The oxygen-isotope and paleomagnetic record in the equatorial Pacific cores V28-238 and V28-239 (after Shackleton and Opdyke, 1976). A very similar pattern is evident in the composite record from cores taken in the Caribbean (P 6304-9 etc.) (after Emiliani, 1966 © Univ. of Chicago). The record is given as a change in parts per thousand of ^{18}O from a standard -P.D.B.

calcium carbonate, and (2) from the assemblage of species of fossil plankton.

(1) During a period of glacial climate the light isotope oxygen-16 (^{16}O) is preferentially evaporated from the oceans and concentrated in the ice sheets while sea water becomes relatively enriched in the heavy isotope ^{18}O (Fig. 16.10). Organisms living in the sea absorb the oxygen in the isotopic proportion in which it is available, and when they die their skeletons retain this proportion in the $CaCO_3$ as a record of the relative abundance of the oxygen isotopes. When a core is taken from ocean sediment or from ice sheets, the change in relative abundance of ^{18}O (expressed as $\delta^{18}O$) can be used either as an indicator of temperature or of sea-level for a time in the past. The sea-level indicator is determined from ocean cores, on the theory that ^{16}O evaporated preferentially from the sea surface in glacials which were also periods of sea-level fall. The alternating periods of warm climate and high sea-level with colder climate and low sea-level are divided into oxygen isotope stages. The Holocene interglacial is stage 1 and the previous interglacial at 75 000–128 000 BP is stage 5. All cold periods have even numbers (Fig. 16.9). (Note that some

workers define the last interglacial interval as 115 000–128 000 BP).

(2) The assemblage of organisms living at the sea surface is controlled by water temperature, so by studying assemblages (i.e. communities) living now, the fossil assemblages can be statistically assessed for the temperature of the water in which they lived.

Core V28-239 from the Solomons Rise is 21 m long and represents sedimentation over the last 2.1 My. It provides by far the longest reliable record of the Quaternary and late Pliocene available from any source. The record shows that there have been seven glacials in the last 700 000 years, separated by much shorter interglacials, each of about 10 000 years length. Each glacial–interglacial cycle thus lasts about 100 000 years.

The period between 0.8 and 1.4 My was characterized by climatic flucations with a period of 40 000 years; sea-level was not lowered as much as during the period younger than 0.8 My, and ice build-up was more limited in volume and duration. Before about 1.4 My, warm periods were much longer than the later ones and sea-level falls not as great, but some did last for a long time.

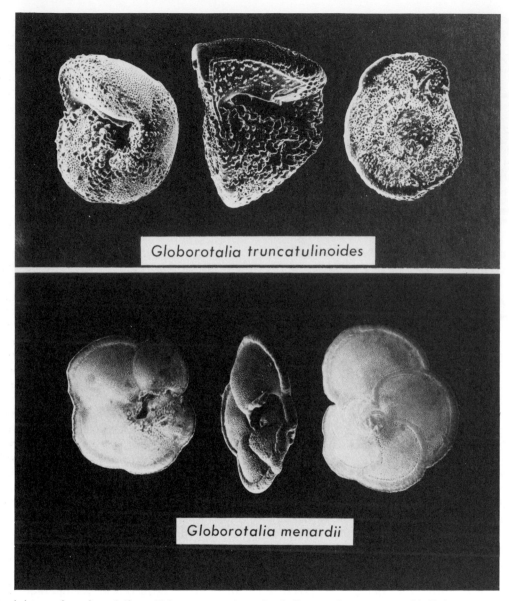

16.4 The skeletons of two foraminifera which are commonly used as indicators of water masses with distinct temperatures.

The dating of the core is thought to be too uncertain to extend the oxygen isotope stages beyond 0.9 My.

Recognition of stages

A generalized climatic curve (Fig. 16.11) can be divided into units of time such that a whole glacial, or a whole interglacial, is a stage represented by a body of sediment whether this is a set of glacial tills, a fluvial terrace deposit, or a marine or lake deposit. It is common for glacials to contain relatively warmer periods which are called inter-stadials and very cold periods called stadials. By convention the oxygen isotope stages have boundaries at the mid-point between a coldest and a warmest part of the glacial-interglacial cycle; but this is geologically unhelpful, for a sea-level or a

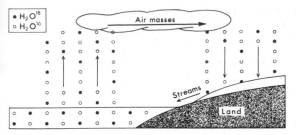

Fig. 16.10 An illustration of the principle behind oxygen-isotype analyses. Water containing the light iso-tope ^{16}O is preferentially evaporated from the ocean at a rate which is temperature-dependent. This water becomes locked up in ice sheets during glacials so that both ice sheets and oceans preserve a record of the ocean depeletion.

cold stage is usually represented on land by a deposit or erosional interval which formed at the coldest or warmest time, not at an indeterminate mid-point. Some confusion about definition of stage boundaries is consequently still with us.

The loess record

Studies of the oxygen isotope record began in the 1950s but reached a peak with publication of the V28-239 record in 1976. It was not realized at the time that, independently, studies of loess on outwash river terraces in Czechoslovakia and Austria were establishing a comparable record on land (Fig. 16.12).

The central European lowlands lie between the Scandinavian Ice Sheet and the ice cap of the Alps. Outwash gravels from these ice masses were carried into the lowlands by many rivers, but especially by the Rhine, Danube, and their tributaries. Alternating periods of outwash accumulation during glacials, and periods of downcutting in warmer intervals, left flights of terraces in the valleys (Fig. 16.13). As will be seen later in this chapter, many valleys contained four main groups of terraces and it was commonly thought that these represented outwash from four glacial events.

The terraces, however, are covered by layers of loess blown from the braided river beds during cold periods when vegetation was sparse. If each terrace represented the outwash of one glacial then the lowest terrace should have one layer of loess with a soil formed in the Holocene Interglacial developed in it. The next higher terrace should have the Last Glacial loess underlain by a soil of the Last Interglacial of 75 000–128 000 BP formed in a loess layer of the glacial which ended at about 128 000 BP. The highest terrace should have four loess layers each with a soil in it. Scientists working in Austria and Czechoslovakia on the terraces of north bank tributaries of the Danube during the 1960s and early 1970s realized, however, that each terrace contained many more layers of loess and more buried soils than this simple theory provided for (Fig. 16.13).

In sections exposed in brick pits at Krems and Brno, and in many boreholes, it could be seen that the deposits were composed of loess with various types of soil developed in it and that each soil had a type of weathering which was itself an indicator of climate, but also pollen, plant remains and snails were preserved (Fig. 16.14). Both vegetation and certain species of snail can only survive in limited

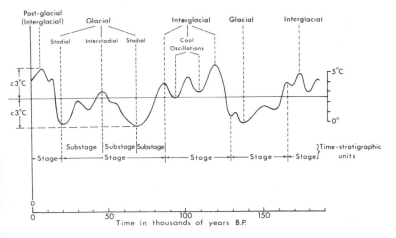

Fig. 16.11 A proposal for subdividing Quaternary records according to temperature, or surrogates for it. For geological purposes the boundaries are most easily recognized at high or low points, but oxygen-isotope stratigraphers use mid-points (modified from Suggate, 1965 © Univ. of Chicago).

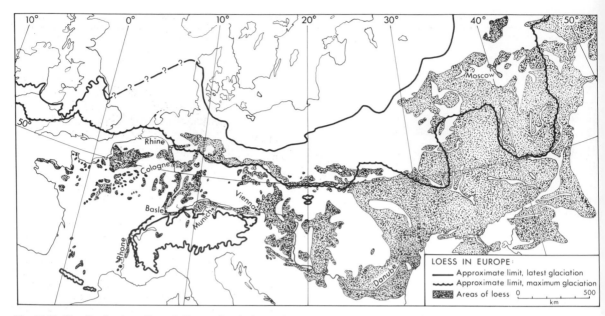

Fig. 16.12 The distribution of loess in Europe in relation to the Scandinavian and Alpine ice. The stratigraphically critical areas are near Vienna (modified from Flint, 1971).

climatic conditions, and so indicate past interstadials and interglacials. The soil types include
tundra soils developed on permafrost, rendzinas
which are characteristic of cool climate grasslands,
chernozems of warm summer grasslands, and
brown earths of forests. The soil–vegetation–snail
complexes are records of retreats and advances of
the ice and the associated climatic changes. From
these kinds of evidence, and that of the extent of
ice determined from moraines, the pattern of
change of a glacial can be reconstructed. A
reconstruction for central Europe at various dates
is shown in Fig. 16.15b.

The soils and the intervening unweathered loess
layers of the coldest glacial periods show that there
is, in total, an identical number of glacial and
interglacial periods represented on these terraces
to those revealed in the deep ocean cores, although
the dating and resolution is most reliable for the
last 0.95 My in which geomagnetic dating can be
applied. It is probable, however, that the Olduvai
Event is recognizable at Krems and if this is the
case the correlation can be extended back to 1.6
My BP (Fig. 16.15 a,b).

The classical record of glacial deposits

The European Alps was the birthplace of the first
theory of glaciation and an ice age, and especially
of the theories of its most widely acknowledged
exponent, Louis Agassiz, who published his
Études sur les Glaciers in 1840. The work which set
the pattern for glacial studies during the first half
of this century was the three-volume *Die Alpen im
Eiszeitalter* (1901–9) of Penck and Bruckner.

In their classical study Penck and Bruckner
postulated that the Alps had suffered four glacials—Günz, Mindel, Riss, and Würm glacials—
which they named after four streams in southern
Bavaria which are tributaries of the Danube. They
recognized these glacials from bodies of outwash
gravels which form alluvial terrace surfaces and
which grade into endmoraines on the plateaux
north of the Alps. The Würm terrace is the
youngest in each of the northern valleys and above
it is the Riss terrace. The Mindel gravel sheets are
remnants of sandar found at low altitudes on the
plateau-like interfluves, and the Günz gravels
form higher and more dissected sandar of the
plateaux.

The Würm and Riss terraces are clearly defined
and can be traced upvalley into endmoraines near
the frontal ranges of the Alps. The older sets of
gravels, which in German are called Deckenschot-
ter, occur mostly as discontinuous remnants. The
relative ages of the deposits are established by a

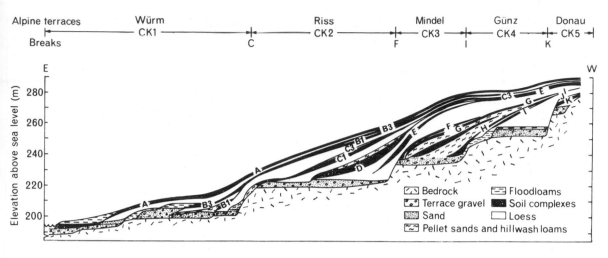

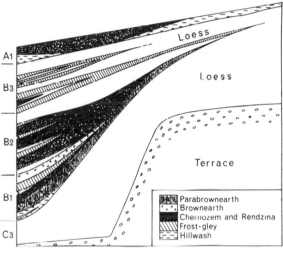

Fig. 16.13 (a) Diagrammatic section through the five terraces at Červený Kopec (CK) near Brno, Czechoslovakia, showing the loess and soils of Brunhes and Late Matuyama age on the bedrock. A to K are the individual glacial cycles (i.e. a glacial with the preceding interglacial): A is the present interglacial (after Kukla, 1975).

Fig. 16.13 (b) The soils of terrace CK1 shown in detail. B is the Last Glacial with the Last Interglacial. Note the strongly weathered forest soils—parabrownearth and brownearth of the full interglacial B1, and chernozems and rendzina of the interstadial B2. The very cold end of the glacial B3 has tundra soils and much loess deposition (based upon Kukla, 1975). In the FAO soil classification frost gleys are called gelic regosols, the rendzinas are calcaric regosols, brown earths are cambisols and parabrownearths are chromic luvisols.

number of criteria: (1) the relative altitude—the most important—with the highest deposits always being regarded as the oldest—at least in areas close to the mountains—and the youngest being the lowest; (2) weathering profiles of the gravels and the soils developed on them, so that the youngest deposits have suffered least weathering, and (3) the morphology of the deposits with the most eroded, discontinuous, and altered gravels being the oldest.

The system of gravels recognized by Penck and Bruckner can be illustrated from those found in the Isère Valley in the Western Alps (Fig. 16.16; Plate 16.5a,b). Here the Würm deposits form the low terrace which is only a few metres above the present floodplain. The Riss outwash surface is

10 m or so higher, and the Mindel deposits are higher still. Few remnants are left of the Günz terrace and the Plateau of Chambaran is capped by much older gravels which contain Pliocene pollen. It will be seen in Fig. 16.16b that the terraces converge downstream, indicating that the Alps have been uplifted during the period of Quaternary deposition, and also that in each successive glaciation the glaciers and their meltwater streams have cut deeper into the valley floors, so that each succeeding set of outwash is deposited at a lower elevation and within the valley walls cut in the gravels of the next older glaciation. The oldest gravels are carried to considerable elevations by the uplift which is continuing at a rate of 1–2 mm/year. Such rates imply

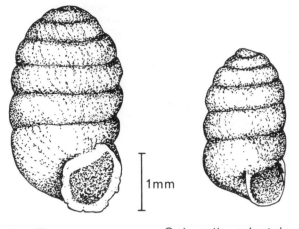

Pupilla muscorum
densegyrata   *Columella edentula*

Fig. 16.14 The shells of two snail species, which are key indicators of climate, from the central European loess (redrawn after Piechocki, 1977).

uplifts of 100 m or so during a single glacial. It will also be seen from these figures that many glacials have more than one glacial advance, and hence leave more than one endmoraine with its associated valley train (which becomes more than one terrace).

Influence and problems of the classical method

The Penck and Bruckner chronology for Alpine glaciations had such an impact upon other scientists that it became not only the standard method but its terminology was widely used. The evidence from such widely separated regions as central USA and the New Zealand Alps was interpreted to show that there had been four glacials there also, and it was believed that these glacials could be correlated, although it was recognized, from the 1930s onwards, that in places like Britain there was no evidence for more than three glaciations, but there were far more than three cycles of cold-to-warm climate.

As research continued during the twentieth century it became apparent that a fourfold pattern of glacials was not always indicated by the evidence. Where terraces or gravel bodies are a considerable distance apart, correlation on the basis of altitude alone is not reliable; even in one valley there is not always agreement on the designation of outwash bodies and soils in or on

them; at some sites the Mindel gravels are spread farthest from the mountains, elsewhere the Riss are more extensive. The Günz was usually similar in extent to the Mindel, but deposits are so eroded that this is not always certain. The Würm is nearly always less extensive than deposits of other glacials, but the distinction between a low Würm terrace and a Holocene floodplain is not always clear, because in parts of Bavaria the floodplain and lowest terrace are composed of gravels and silts deposited during the period of forest clearance and agricultural expansion of the Middle Ages. Elsewhere a Holocene valley is incised into the Würm surfaces. Other problems arose when high-level gravel bodies were recognized and interpreted as indicating glacials older than the Günz and called the Donau and Biber deposits.

The fundamental problem is that bodies of gravel can seldom be dated and the surface of a terrace, which may be composed of several outwash sheets of several separate glacials, indicates only the most recent event unless deep cuttings through the terraces are available (see Plate 16.6).

The studies of central European loess, on terraces which were thought to be related to the classical outwash sequence, demonstrate quite clearly that one terrace seldom results from one glacial alone. The so-called Mindel terrace at Brno, for example, has on it soils and loess which suggest that it developed over a span of two whole glacials and parts of two others.

The question now arises, regarding the use to be made of such terms as Günz, Mindel, Riss, Würm, and the North American terms Nebraskan, Kansan, Illinoian, and Wisconsinan (Fig. 16.17) which have been regarded as the North American equivalents of the classical European glacial sequence. The only useful method now seems to confine the terms to distinguishing bodies of till and outwash—each of which may represent more than one glacial event—and to apply the terms only to their type area, e.g. Würm to the Alps and Wisconsinan to North America.

The concept of four glacials only was so widely accepted, and for so long (70 years or so), that many of the ideas associated with it are still with us. Geological, geomorphological, biological, and archaeological theories have now to be adjusted to accommodate the evidence, derived from deep-sea cores, that there have been twenty-one cold periods in the last 2.1 My years. This evidence

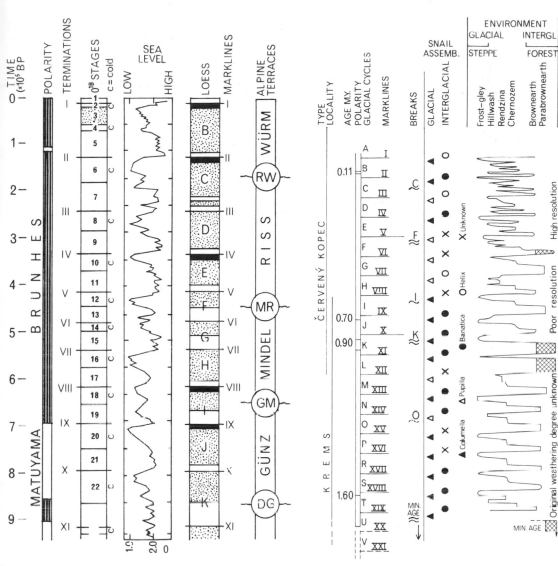

Fig. 16.15 (a) A comparison of the oxygen-isotope and loess records for the Middle and late Quaternary, and the alpine terraces with which the glacials correlate in Czechoslovakia. Terminations are the oxygen-isotope equivalent of loess record markline boundaries.

(b) Environment changes around Brno and Krems, Czechoslovakia, for the whole Quaternary, as reconstructed from the loess record, its soils and snail faunas. Marklines are the boundaries of glacial cycles and mark the end of a glacial (modified from Kukla, 1975).

implies that world sea-level was lowered and that ice sheets formed in polar regions. It does not mean that ice sheets formed everywhere to the thickness and extent they achieved in the last 700 000 years. There is, for example, no evidence yet discovered for more than three glacials with extensive ice sheets in Britain. Many of the early glacials of polar regions are represented, in Eur-

ope, by fossil periglacial deposits, tundra vegetation, and animals, including Man, and are separated from other similar cold periods by interglacials with temperate forests and warm climate faunas and floras.

The record of the twenty-one glacials in Western Europe is apparently not complete. In Britain, for example, there are two major gaps in the

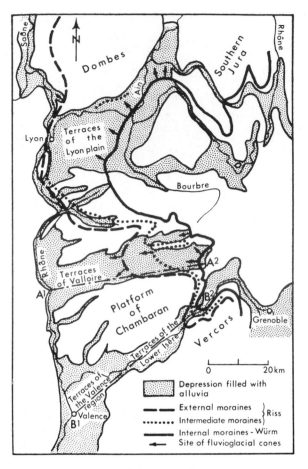

Fig. 16.16 (a) Location of the Isère Valley.

sedimentary record: from the base of the Brunhes Normal Epoch to the top of the Olduvai Event (a gap of one million years) there is an absence of evidence, and again in the Pliocene from about 2 My BP to 2.4 My BP there is virtually no evidence. It could be that these gaps are the result of erosion of the deposits laid down in the 'gap' period, or it could mean that the deposits have yet to be recognized and dated.

In many areas subjected to glaciation the evidence for changing climate and glacial processes is confined to that for the last event, usually that which has occurred in the last 20 000 years. Erosional evidence can seldom be dated and erosion inevitably destroys much of the evidence left by earlier events. The record of change therefore has to be sought away from the centres of ice accumulation and near the fringes of glaciation where glacial, fluvioglacial, lake, and fluvial deposits may retain a record of ice advances and retreats. Most glacial deposits are not directly datable. It is only where organic matter has been incorporated in deposits that dates can be obtained by ^{14}C-dating, and then only for the last 40 000 years or so. Older deposits have to be dated by geomagnetic methods or occasionally lake and marine shellbeds may be dated by ^{238}U-^{230}Th. These methods inevitably leave gaps in the well-dated record which have to be filled by extrapolation of the evidence and by correlation of deposits. Extrapolation and correlation, plus errors in dating, inevitably lead to uncertainty and debate.

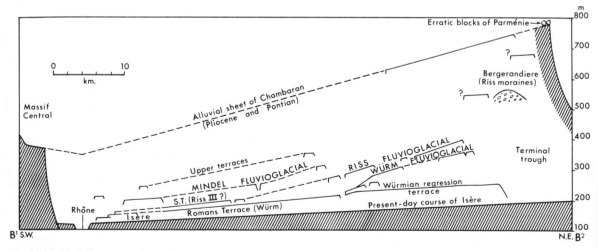

Fig. 16.16 (b) A diagrammatic section along the Isère Valley showing terrace surfaces.

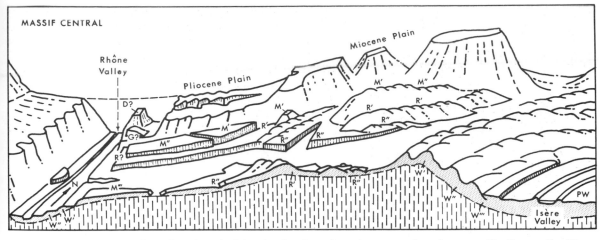

Fig. 16.16 (c) A diagrammatic section along the Isère showing the endmoraines, outwash, and terraces. D—Donau; G—Günz; M—Mindel; R—Riss; W—Würm; W^1— Würm one, or oldest Wurm (modified from Gignoux, 1960, and Bourdier, 1958, as reproduced in Alimen, 1967).

Methods of determining past environments

The most important methods used by students of the Quaternary have been mentioned already: glacial and periglacial deposits, loess, buried soils (called paleosols), organic deposits (especially pollen), fossil marine plankton, and oxygen isotope changes are among the most important.

Over large areas of the continents—especially in Africa, South America, India, and Australia—Quaternary climatic events were not primarily glacial but related to changes in rainfall or runoff. Because of lower temperatures during polar glacial periods there was a reduction in evaporation losses in tropical and subtropical areas, but the lower evaporation from oceans also resulted in reduced precipitation in many regions. There was a resulting change in the nature of geomorphic processes: sand dunes were more active and widespread, lakes dried up and in some areas became playas, and rivers suffered a decrease in discharges and hence deposited much of their load and built up their beds. This general statement should not be taken as being a universally valid description of events in areas outside those directly affected by glaciation, as local factors were of great importance in controlling climate and geomorphic responses to it. It should also be realized that during every glacial there were alternating stadials and interstadials and just as these were associated with thickening and advance of the ice and a colder climate during stadials, and thinning,

retreat, and warming in interstadials, so in the tropics and subtropics there were also temperature variations and variations in precipitation, lake levels, runoff, and river terrace formation.

In much of Africa, for example, bedrock lies close to the ground surface and Quaternary deposits are thin or non-existent. It is in limited areas that deposits can accumulate and preserve a record of past climates. Of greatest importance are the lake basins of the East African rifts from Omo and Turkana southwards, and the great basin of Lake Chad which has at various times expanded to become an inland sea the size of the present Caspian, and at other times dried up and had sand dunes blown across its floor. Dune fields with a nearly complete vegetation cover are widespread in Australia, the Kalahari, western USA, the Thar Desert of India, and parts of the southern Sahara—clear evidence of former dry and windy climates.

Many deposits cannot be dated directly but in volcanic regions such as East Africa, Japan, New Zealand, the Andes, Indonesia, and western North America volcanic ashes interbedded with stream, lake, coastal, marine, and peat deposits provide a method of precise dating and of wide-ranging correlation of events. The ashes can be dated directly by fission-track, $^{40}K:^{40}Ar$ or paleomagnetic methods.

In many areas of Africa and Europe deposits cannot be dated directly, but various imprecise methods are useful for correlating the deposits of

16.5 (a) The Isère Valley. In the foreground is the Würm terrace surface, in the valley floor is the Holocene floodplain. On the far side of the valley are the Würm terrace and the higher Riss terrace. The Plateau of Chambaran forms the surface of the upland in the distance.

16.5 (b) The incised surface of the Plateau of Chambaran which is underlain by Pliocene gravel sheets.

one area with those of another. Thus if one lake deposit contains fresh-water snails or pollen of similar assemblages to those of another deposit, not too far away, then they may be regarded as being related in age and environment of deposition. In some areas one of the most useful indicators for correlation is that of human tools, called artefacts.

As Man has evolved his culture over the last two million years he has improved his ability to make stone tools. The earliest tools are found in the Pliocene beds of Lake Turkana and consist of choppers or pebble tools formed by breaking off one end of a rounded natural pebble. This chopper, or Oldowan, technology was gradually improved upon and eventually gave rise to an Acheulian industry (named after Saint Acheul in France). The Acheulian handaxe is a tool flaked over the whole of both its main faces so that a working edge is formed round the greater part of its perimeter. Acheulian industries also developed cleavers, scrapers, and picks by the same methods (Figs. 16.18, 16.19). In the Sahara, Acheulian assemblages of artefacts are very useful indicators of the relative ages of many fluvial terraces.

From about 100 000 to 60 000 years ago the striking of flakes of stone from prepared cores was a distinguishing feature of a group of cultural industries, called Levallois when found around the Mediterranean, Mousterian in a wide band from the Atlantic to Inner Asia, the Fauresmith in southern Africa, and Sangoan in central Africa.

Between about 35 000 and 40 000 years ago a new industrial development was the punching of narrow blades from a prepared core to produce knives, chisels, points, borers, needles, and

16.6 Cliffs, on the south side of the Rakaia River, New Zealand, formed in outwash gravels. The gravels of the Last Glacial overlie those of the previous glacial which are indicated by the pale weathered band forming the lower half of the cliff.

Fig. 16.17 Limits of tills of the four stages recognized in central North America (after Flint, 1971).

scrapers. The new technology spread rapidly in Europe and the Middle East but developed many local special forms. Thus it is possible to recognize Aurignacian, Gravettian, Solutrean, and Magdalenean forms. In north Africa the local Aterian industry is characterized by tanged points which were probably used as arrow heads.

Human artefacts do not provide a precise method of correlation until about 80 000 years ago when cultural evolution became rapid (Fig. 16.20), but it is still useful in the absence of more precise methods.

A method which may be used around all coasts is that of sea-level change. The creation of beaches at various sea-level heights leaves a record which can be found round most coasts and can be traced for some distance upstream as river terraces. Because either ice has loaded the crust and

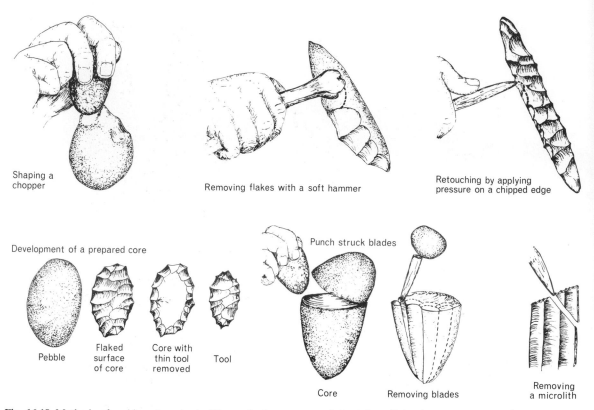

Shaping a chopper

Removing flakes with a soft hammer

Retouching by applying pressure on a chipped edge

Development of a prepared core

Pebble

Flaked surface of core

Core with thin tool removed

Tool

Punch struck blades

Core

Removing blades

Removing a microlith

Fig. 16.18 Methods of making stone tools. The methods are arranged, in a time of development sequence, from chopper to retouching to prepared core to microliths.

depressed it during glacials, but subsequent unloading has carried glacial-age beaches to great elevations, or because tectonic uplift has raised beaches, there is a record of old sea-level variations to be found round the coasts of most of the world. This record can be related to the climatic record from deep-sea cores and locally correlated to other features such as deltas, terraces, moraines, outwash, etc. As a method of long-distance correlation it is very valuable but, of course, of little use in continental interiors, and it is subject to varying isostatic adjustments (see p. 351).

Conclusion

It is now evident that application of many methods is showing up a broad pattern of Quaternary events. These are best known for the Last Glacial, for which evidence is most complete and dating is most secure. The correspondence of various forms of evidence is illustrated in Fig.

16.21. Migration of water masses in the North Atlantic, the central European loess record, the ocean sediment record, and the uplifted coastal terraces of Barbados, all show a pattern of events for the last two glacials which is independent of the direct glacial evidence and far more complete than it. The relatively rapid changes in land and sea environments are evident and indicate how frequently the dominant geomorphic processes must have changed also.

Terminology

The adjectives Upper, Middle, Lower, or Early, Middle, and Late are often used in conjunction with geological epoch or period names. As a general rule it is better to confine Upper and Lower to stratigraphic situations, that is, to layers of rock. Early and Late are used with reference to time. There is no universally accepted definition of Early or Late Quaternary but common practice

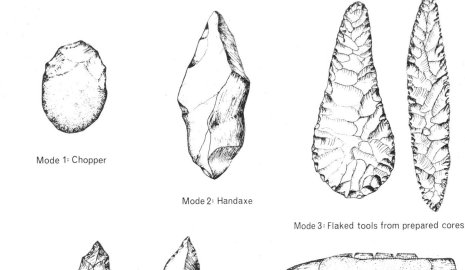

Mode 1: Chopper

Mode 2: Handaxe

Mode 3: Flaked tools from prepared cores

Mode 4: Blades with steep retouch

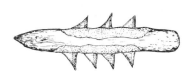

Mode 5: Microliths set in wood

Fig. 16.19 Developing modes of tool making from earliest (1) to latest modes (5).

Fig. 16.20 The span of Quaternary time during which specified cultures existed. Note the rapid diversification and development in the Upper Pleistocene record. The Maghreb is from northern Africa (data from Butzer and Isaac, 1975).

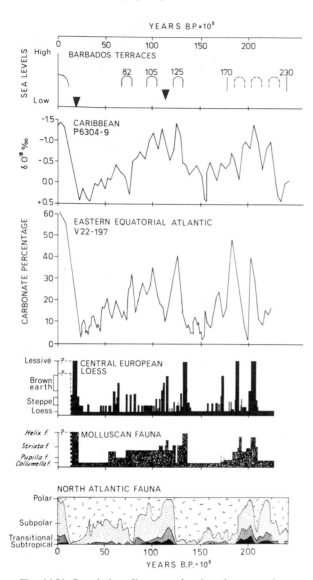

Fig. 16.21 Correlation diagrams showing the approximate coincidence of the climatic interpretation of data from numerous sources over the last two glacials and three interglacials (for sources see Hays and Perruza, 1972, *Quaternary Research*, 2, p. 360.

usually uses Early to refer to all of the Quaternary falling within the Matuyama Reversed Polarity Epoch 1.8–0.73 My BP; Middle Quaternary is 0.73 My until 128 000 BP; and Late Quaternary is 128 000 to the present. The Late Quaternary is thus the Last Interglacial, the Last Glacial, and the Holocene Interglacial. Note that in the previous sentence the defining words are capitalized

because they are being used as formal geological terms defining the events. In the absence of universally accepted stage names this is much better than attempting to use a term such as Würm outside its type area—the European Alps. Where the words 'early' and 'late' are being used in a general sense without precise definition then they are not capitalized.

Ice sheets of the Last Glacial

There is relatively little controversy concerning the extent of southern hemisphere ice sheets (except that of West Antarctica) during the Last Glacial (75 000–10 000 BP) but considerable controversy about the extent and timing of the maximum extent of the northern hemisphere ice sheets. It is a simplification to class the various views into two categories but differences of opinion can be represented as falling into support for either a maximum model or a minimum model (Fig. 16.22).

The minimum model represents the view of those who believe that there was restricted ice in polar regions and that this ice was largely formed on land. Frequently associated with this model is the view that glacier fluctuations in polar, and most temperate areas, did not always occur at the same time because of varying precipitation patterns.

The *maximum model* represents the views of those who believe that extensive ice sheets were based on continental shelves, as well as on land, and that large ice shelves formed in the polar seas. Associated with this model is the view that glacier fluctuations in polar areas were in-phase with ice sheet margins of temperate areas, and that the maximum ice cover occurred between 21 000 and 17 000 BP.

It is evident that the distribution of the Last Glacial ice sheets is not well known despite over a century of research. Our ignorance in this matter is of concern because without a thorough understanding of ice sheet sizes it is impossible to reconstruct the global glacial period climatic pattern. The uncertainties are caused partly by lack of evidence in crucial areas and partly by differing interpretations of that which is available.

Was there a Eurasian ice sheet?

The most controversial aspect of the two models of ice sheets concerns the extent of ice in Eurasia. The older and more firmly established reconstruc-

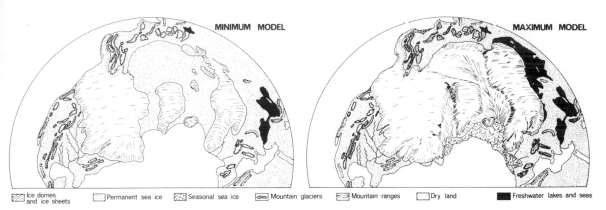

Fig. 16.22 a, b, Extent of ice in the northern hemisphere for maximum and minimum reconstructions.

tion is that in the northern hemisphere there were a number of disconnected ice caps and ice sheets with the European zone being dominated by the Scandinavian Ice Sheet which was confined by the edge of the Norwegian continental shelf, and which reached only to the mouth of the Mezen River in USSR. Smaller ice caps were centred over the mid-Siberian Plateau, over Svalbard, Franz Josef Land, Novaya Zemlya, and Severnaya Zemlya. The northern Urals, the Byrranga Range, and the Putorana Mountains were ice-dispersal centres. The continental shelves and the Arctic seas remained unglaciated. The newer maximum model contends that large portions of northern Eurasia, and particularly the Arctic continental shelf, were covered by an integrated ice sheet that included marine domes in the Barents and Kara Seas, as well as terrestrial domes over the British Isles, Scandinavia, and the Putorana Mountains.

The reasons for these two extremely different reconstructions fall into six main classes.

(1) The extent of ice cover on continental shelves is not well known. All of the evidence is covered by sea, and samples of mixtites and other sediment collected by dredge and drill core have to be interpreted as to origin. Organic matter, such as wood and sea shells thought to come from till sheets and moraines, may be contaminated with older carbon from coal, peat, or old shell beds formed before the Last Glacial and merely redistributed during it, consequently they give false radiocarbon dates.

(2) Many depositional features, of continental shelves, which may be moraines, and erosional features such as glacial troughs, have been identi-

fied only in very recent bathymetric surveys and their age is often unknown.

(3) Whilst depositional and erosional features of large size can be identified on the continental shelves, small but diagnostic features such as striations, roches moutonnées, and chattermarks cannot be identified. These three classes of evidence largely explain the problems of interpreting what happened in the Barents and Kara Seas.

(4) Where the ice extended beyond the present shorelines, endmoraines and other evidence of ice extent are now drowned. A major method of reconstruction is then dependent upon interpreting the amount of isostatic depression under ice load and the amount or rate of uplift from isostatic recovery since deglaciation. These calculations depend upon three factors which are imperfectly known: the maximum ice thickness; the viscosity and hence depression and recovery of the crust; and the ages of the uplifted shorelines (strandlines) which are the chief sources of evidence for rates of uplift. The ages are dependent upon radiocarbon dating of shells and wood and, like dredged material, this can be contaminated or redistributed older material. The best hope for resolving the controversy lies in a more critical re-examination and redating of many pieces of key evidence.

(5) On land the extent of ice and its thickness is usually judged by the distribution of both directly modified erosional and depositional phenomena, but also by the extent of soils, old weathering profiles, tors, and nunataks. The assumption by supporters of the minimum view is that such features could not survive the passage of ice over them, and hence that where they exist on inter-

fluves between glacial troughs—as they do in many areas near the coast of Norway, northern Canada, and north-eastern Greenland—then the ice must have been confined to the troughs and not overrun the interfluves during the last glacial maximum advance. It follows that the ice sheet surface slope and ice extent would have been limited. By contrast it is now believed by many glaciologists and glacial geologists that ice streams within the ice sheets converged into the troughs producing frictional basal melting, high velocities, and hence erosion, but on the interfluves the basal ice remained cold, non-erosive, and preserved the soils, tors, and other features (see Plate 16.7).

(6) In support of the maximum model is the relatively recent (1980) presentation by Russian scientists of a map of a series of huge inland lakes in Eurasia created when large north-flowing rivers were dammed by the Barents and Kara ice domes (Fig. 16.22). The extensive lacustrine deposits are not well dated but appear to support an age of about 20 000 BP for the maximum Eurasian Ice Sheet.

Other areas of controversy

In North America both the minimum and maximum reconstructions are in agreement along the

16.7 Outlet glaciers from the Greenland ice sheet near Bartolius Brae, Blosseville Kyst. Note the medial and lateral moraines of the glaciers and the inactive ice on the plateau between the glaciers (photo by the Geodetic Institute, Denmark).

southern margin of the Laurentide Ice Sheet; there is some disagreement about ice extent in New England and areas northwards along the Atlantic coast, but the greatest differences are over the ice cover in Arctic Canada. The minimum model restricts ice on Baffin Island, other islands, and involves an ice-free corridor between Greenland and the eastern Queen Elizabeth Islands. The maximum version depicts an extensive Innuitian Ice Sheet coalescing with the Greenland and Laurentide Ice Sheets. There is also conflict about the extent and duration of contact between Laurentide and Rocky Mountain Ice Sheets. These differences in view are possible largely because of interpretations of isostatic recovery and age of shells in strandlines. In the minimum model it is usually contended that as the ice advanced southwards the Arctic would have been cut off from its supply of moisture, and the lack of snow would cause a decrease in ice volume, so that the Arctic greatest ice extent would occur before that of the southern ice along the Canada–USA border. In the maximum model it is held that the Arctic was supplied with ice crystal precipitation (hoar frost or rime ice) from high altitude air masses, just as is present-day East Antarctica (Fig. 16.23; Table 16.1).

In Antarctica the minimum view holds that the present limits of ice are similar to those of the coldest part of the Last Glacial, while the maximum view is that the West Antarctic Ice Sheet expanded to the edge of the continental shelves and the East Antarctic Ice Sheet thickened. Recent data in support of the maximum view comes from dates of ice advances into the McMurdo Dry Valleys and cores from the Ross Sea.

Conclusion

It is not possible to support firmly one side of the argument because so many more data are needed; it must be said, however, that the maximum model looks a much more probable hypothesis than it did in the early 1970s. The importance of the differences can be assessed from the calculation that the models involve an ice volume at the coldest part of the Last Glacial and a fall of sea-level of:

maximum reconstruction—98 Gm^3 and 163 m,
minimum reconstruction—84 Gm^3 and 127 m.

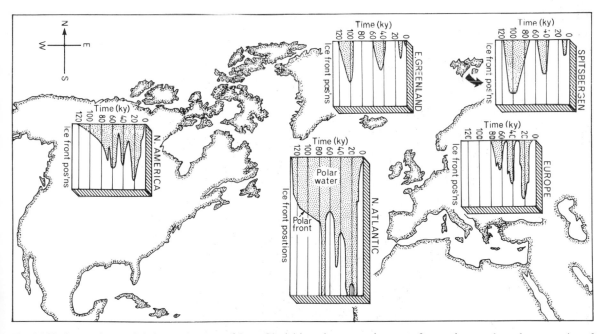

Fig. 16.23 Comparisons of timing and extent of Last Glacial ice advances and retreats for northern and southern margins of northern hemisphere ice sheets, together with the southern boundaries of sea ice in the North Atlantic. This reconstruction conforms with the evidence for a minimum model. Data are from Boulton *et al.* (1982).

Table 16.1   Hypotheses of Major Ice Fluctuations in the Northern Hemisphere according to a Minimum Model

Time (ky BP.)	Arctic	Mid-latitudes	Atmospheric circulation
Holocene, 10–0	Relatively warm and moist, reducing ice cover, but readvance in Little Ice Age	No ice except in high mountains, Little Ice Age period of cold winters	North Atlantic Drift and low-pressure systems reach Arctic Ocean
Stadial, 20–18	Cold arid climate, with retreat of glaciers	Maximum extension of the southern margins of the North American and European ice sheets	High-pressure systems in Arctic block moist air masses, but low-pressure systems nourish southern margins of ice sheets
Interstadial, *c.* 30	Cool, relatively moist, ice cover limited	Ice cover limited, cool	Arctic partly open to low-pressure systems
Stadial, *c.*40	Less extensive ice than at 75ky	Major ice advance, but less extensive than at 20–18ky	Arctic partly blocked by high-pressure systems
Interstadial, *c.*50	similar to *c.*30 ky		
Stadial, *c.*75	Major ice advances, glaciers fill west coast fjords	Small ice sheets form, but they are less extensive than during later stadials	Arctic open to low-pressure systems with cool, but moist, air masses: temperate areas cool, but dry, under high-pressure systems

Source: Boulton *et al.*, 1982.

Initial development of an ice sheet

Five main theories of the formation of ice sheet domes have been published.

(1) Many early workers thought that domes originated in place by the accumulation of snowfall.

(2) Others thought that very low winter temperatures in permafrost areas caused rivers and lakes to freeze to the bottom and valleys to fill with ice until freezing water and snow produced a dome.

(3) It was proposed by R. F. Flint, in 1943, that snowfield accumulation in the mountains of Scandinavia and north-western North America could lead to valley glaciers, then piedmont glaciers and ice domes as ice spread inland from the mountains.

(4) In 1957 J. D. Ives suggested that large areas of lowland Labrador could become snow-covered after a few winters of high snowfall followed by cold or cloudy summers with little melting, so that nearly instantaneous glacierization could occur from the albedo effect and accumulation.

(5) The most recent hypothesis is for the formation of marine ice shelves which grew rapidly into grounded ice domes. This hypothesis first put forward by Hughes, Denton, and Grosswald, in 1977, is part of the maximum Arctic Ice Sheet theory.

Before about 1975 it was unknown how long it took for the Last Glacial ice sheets of the northern hemisphere to form. When, however, the record of sea-level fluctuations derived from the oxygen-isotope record became available, it also became clear that between about 124 000 and 115 000 BP sea-level fell by more than 80 m. This large drop in only 9000 years, and the fall during the 13 000 years between 82 000 and 69 000 BP of 90 m, implies a very rapid growth in ice volume; yet at present precipitation rates the Canadian Ice Sheets would need at least 20 000, and possibly 30 000 years, to form. Thus theories (1), (2), (3), and (4) seem to be incapable of explaining the available data.

Theory (5) was developed from studies of the present West Antarctic Ice Sheet, which lies partly on an archipelago and partly on the continental shelf, and is stabilized by both islands and the grounded sections of the Ross Sea and Weddell Sea ice shelves, while being drained by ice streams which pass between the islands.

Calculations have shown that, assuming sea ice surface temperature was −1.8 °C, as it is now in the Arctic Ocean, sea ice could thicken by freezing until it became grounded on continental shelves of most of the Arctic within 1000 years. Once formed, shelves would continue to grow by the addition of snow and by the freezing of any meltwater draining off the land (something that could not occur to a great extent in mountains). Sea-level accumulation sites also have the advantage of high precipitation. For ice shelves to form and remain stable it is necessary for the ice to be protected from iceberg calving and wave action. This is particularly possible in the Kara and Barents Seas which are bounded by closely spaced islands on the edge of the continental shelf, and similarly the Queen Elizabeth Islands bound the straits and bays of the Innuitian area. The shallow waters of Hudson Bay and Gulf of Bothnia in the Baltic Sea, are naturally confined also. The ice domes, once grounded, would spread inland as lobes, similar in appearance to piedmont glaciers. Ice flowing towards the sea would advance like the front of the Ross Ice Shelf, but where it was confined by islands or ridges on the continental shelf floor ice streams would flow through the gaps to create floating ice tongues from which icebergs would calve. This hypothesis answers the puzzle presented by striations and nunataks on the northern Ural Mountains and Baffin Island which imply that ice had invaded those areas from the sea, not from the land direction.

The build-up of ice shelves and then domes would cause sea-level to fall, and icebergs calving from the tongues and shelf fronts would become grounded on shallow rises and bars like the exits from the St Lawrence River and Hudson Bay

where water depth is less than 200 m, and then in the straits between Baffin Island, Greenland, Iceland, the Faeroes, and Scotland where glacial-age depths would be less than 350 m. Thus the whole Arctic Basin would be blocked by grounded ice in former marine straits, and it would become possible for an Arctic Ice Sheet to form, with a volume of perhaps 54 Gm^3 (compared with 38 Gm^3 for modern Antarctica) (Fig. 16.24). The ice sheet would be nourished largely by ice crystals from high altitude air masses. It would be composed of many domes over uplands and continental shelves, together with ice shelves filling the Arctic Seas (and thus buttressing each other), and ice streams flowing rapidly through confining land or ice masses to calve, mostly in the North Atlantic. The southern margins of the Eurasian and North American ice would be composed of ice lobes.

It is important to emphasize that, although this maximum model and ice shelf origin is both plausible and explains much of the evidence for ice flow directions and rates of ice formation, it is still only a theory which needs further testing.

Arctic Ice Sheet disintegration

At its maximum the postulated Arctic Ice Sheet was composed of three main elements:

(1) land-based domes covering a frozen bed, and hence being stable;
(2) marine-based domes covering a thawed bed and hence being potentially unstable, but temporarily stabilized by confining land-based domes and ice shelves;
(3) ice shelves, several hundred metres thick, floating on the ocean and stabilized by confining ice domes, land masses, or bedrock rises.

These three elements form one integrated ice mass which terminated along its southern margin at nearly stable melting ice lobe fronts on land, and over the sea in calving bays in shelf ice. Draining the domes into the Arctic seas were ice streams which supplied much of the ice of the ice shelves (the rest came from bottom freezing and surface snowfall).

The Arctic Ice Sheet disintegrated during the 10 000 years between 16 000 and 6000 BP, when about 90 per cent of the ice melted (Greenland remains covered). It has been calculated that the total excess solar insolation available due to the

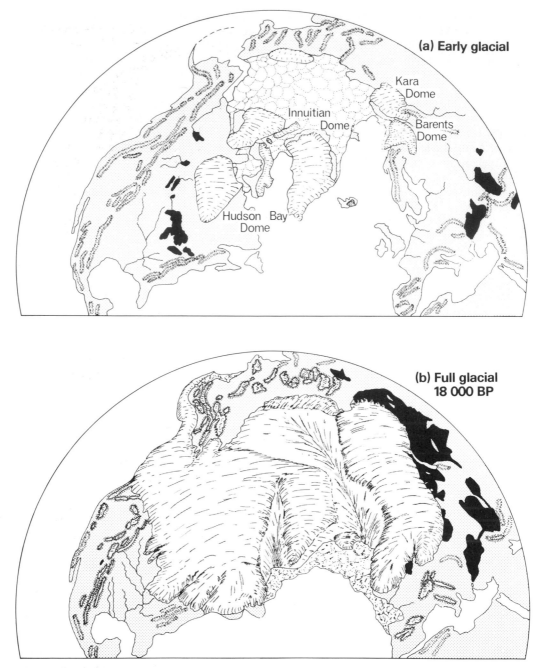

Fig. 16.24 a, b, c, d, Ice cover on the northern hemisphere at four stages of the Late Quaternary according to the maximum model. Note the extensive meltwater inland seas of Eurasia during the coldest part of the Last Glacial, and the calving bays with icebergs during the early stages of deglaciation—the calving bays are, inevitably, hypothetical.

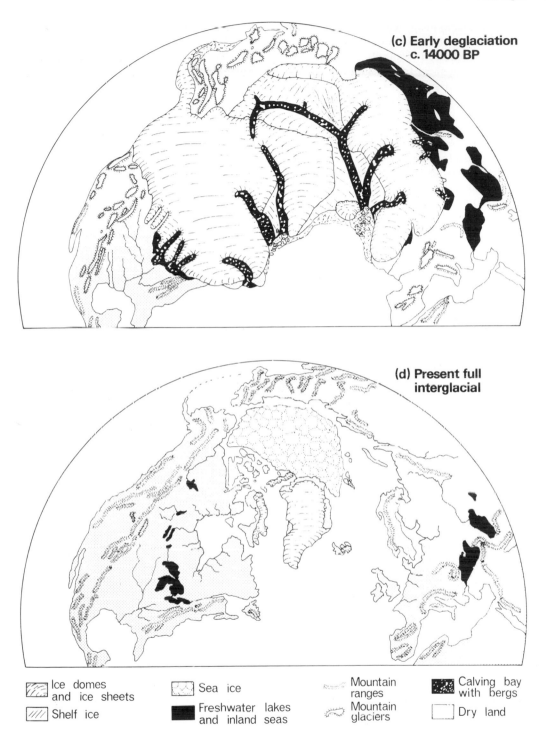

(c) Early deglaciation c. 14000 BP

(d) Present full interglacial

Ice domes and ice sheets	Sea ice	Mountain ranges	Calving bay with bergs
Shelf ice	Freshwater lakes and inland seas	Mountain glaciers	Dry land

Milankovitch mechanism was about 2.5×10^{24} Joules, while on the minimum model about 7.6×10^{23} J was required to supply the latent heat of fusion of the ice sheet for melting (see later in this chapter a discussion of Milankovitch radiation). The heat required for melting is thus clearly unavailable if the ice sheets were bigger than the minimum, and because the albedo of an ice sheet is 0.75–0.80 such a large proportion of incoming radiation is reflected away that not even the minimum ice sheets could be melted. Thus large land-based ice sheets are stable under any but very extreme climatic variations. Ice sheet wastage therefore requires a mechanism which is only partly dependent upon melting.

It has been proposed by Hughes, Denton, and Grosswald (in 1977) that rapid Arctic Ice Sheet disintegration can occur because portions of it are marine-based and potentially unstable. They envisaged a sequence of events which began when the seaward ends of some ice streams became detached from bedrock rises and the ice of the streams consequently floated, became deeply crevassed, and calved large tabular icebergs into the North Atlantic. Such calving occurs today from ice streams draining East and West Antarctica (Plate 16.8). The calving would continue in floating ice and rapidly proceed along the ice stream, because ice streams are located above troughs, or fjords, cut in the ocean floor by that ice stream,

16.8 Large tabular icebergs in the distance and sea ice in the foreground. Ross Sea, Antarctica (photo by NZ Department of Scientific and Industrial Research).

and hence are more likely to float than surrounding stranded shelf ice. Ice streams would thus be replaced by large iceberg-strewn channels, tens of kilometres wide, cutting back across the ice shelves into the edges of the marine ice domes. Where the head of the calving ice stream became stranded on a bedrock sill the process might stop, but in places like the deep straits between the Canadian Arctic Islands, the Skaggerak south of Norway, Baffin Bay, in Hudson Strait, and the Gulf of St Lawrence, ice would be drawn down into the calving bay and the equilibrium of the grounded ice sheet would be disturbed. The Laurentide Ice Dome, in particular, would be attacked from its heart. The icebergs would drift into the Atlantic and so be moved towards a heat source which would melt them (Fig. 16.25).

Once the process of disintegration started it would be self-reinforcing. The development of calving channels replacing ice streams removes lateral support from ice shelves and ice domes so that ice flow can increase or surge; melting of ice from the land raises sea-level and causes more ice shelves to float off bedrock. The process of ice sheet disintegration thus requires a trigger factor, such as a warming phase leading to melting and sea-level rise, but once this has started the calving channels tear out the heart of the ice sheet and the available heat factor is of less importance.

Disintegration of the Laurentide Ice Dome was aided not only by calving channels in Hudson Bay, by 10 000 BP, and Gulf of St Lawrence, but also by more limited calving into the fresh-water lakes round its southern margin — Lake Agassiz, Lake Algonquin, and other predecessors of the Great Lakes (Fig. 16.26). The Scandinavian Ice Dome lost icebergs into the Barents Sea, the Skaggerak, and by 9000 BP the Gulf of Bothnia. The Barents and Kara Sea Domes may have lost ice by calving into the huge meltwater lakes of the Dvina, Pechora, and Ob lowlands. The collapse passed a crucial point when the collapsing margin of the Arctic Ice Sheet lifted off the broad sills between Greenland, Iceland, the Faeroes, and Scotland, a little before 14 000 BP, and allowed the North Atlantic Drift to enter the Arctic Ocean with its warming influence.

In the northern hemisphere the Greenland Dome remains because its base is mostly above sea-level and confined by mountains. This situation is similar to that of the East Antarctic Ice Sheet, but the West Antarctic ice is less secure,

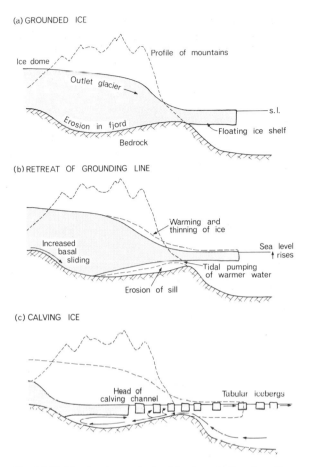

Fig. 16.25 The development of calving bays in the edge of an ice sheet showing the draw-down effect of an ice stream.

being still largely a marine-based dome. This ice mass survived the postglacial maximum warming, possibly because it is protected from warm ocean currents by the cold Antarctic Circumpolar Current.

Whereas a glacial begins with the build-up of ice in a marine dome on a continental shelf where abundant water is available, it finishes with remnant ice caps and glaciers at high altitude and on polar continents where low temperatures, cloudiness, and snowfall permit the ice to persist.

Evidence from ocean cores, moraines left during ice sheet margin recession on land, and the advance northwards of plants during the ice-front retreat has permitted a reconstruction of the ice front in the Atlantic region. By 7000 BP only remnants of ice survived in the Scandinavian Mountains, on Labrador, in a small area west of

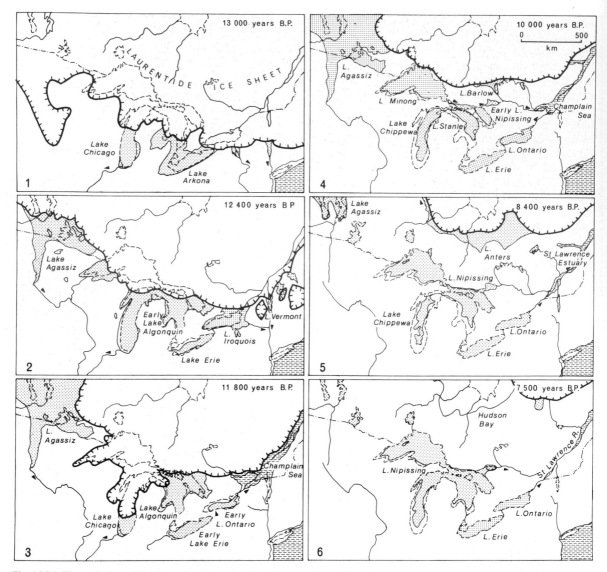

Fig. 16.26 The evolution of the Great Lakes at the end of the Last Glacial (data from Geological Survey of Canada, in Bird, 1972).

Hudson Bay, and of course on Greenland. Minor readvances during late glacial stadials are also widely recognized (see Fig. 16.27).

Climate at 18 000 BP

Detailed studies of surface temperature for periods during this ice age are being undertaken by a consortium of scientists working in the CLIMAP Project. As yet the only published result is that for July–August 18 000 BP. This represents a northern hemisphere summer and southern hemisphere winter for the coldest part of the Last Glacial.

In Fig. 16.28 the main features of the CLIMAP conclusion are shown. They result from an assembly of all of the available data on temperature indicated by fossil pollen, periglacial phenomena, and marine organisms. From these data a mathematical model of the Earth's atmosphere was built, and from this the pressure conditions of the

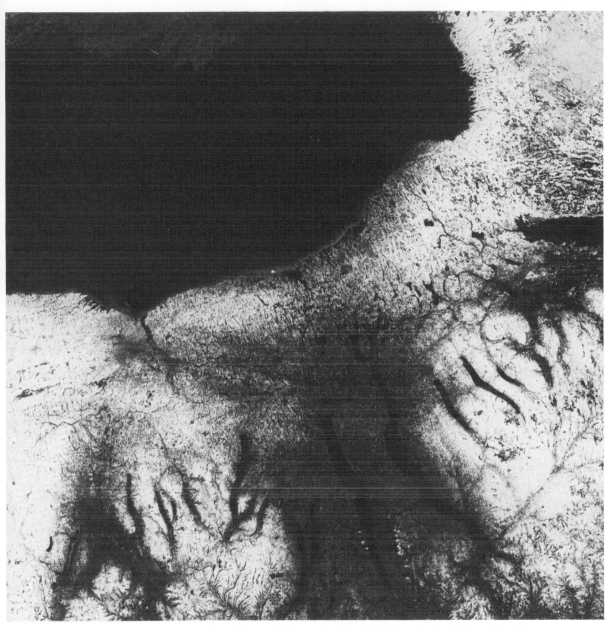

16.9 The Finger Lakes of central New York which now drain northwards to Lake Ontario (top). The valleys they occupy are thought to have been cut by ice streams within the southern lobes of the Laurentide Ice Sheet (satellite photo by NASA).

atmosphere were derived. The model could then be used to indicate probable temperatures in areas for which data are sparse or unavailable.

Permanent ice covered much of North America, the polar seas, and Eurasia, but ice-free areas existed in Alaska and much of Siberia. In the month of July there was probably little snow lying south of the northern hemisphere ice. In the southern hemisphere sea ice would have been far more extensive than it is today. Changes in land ice were small, compared with today, but locally were important in southern South America, South

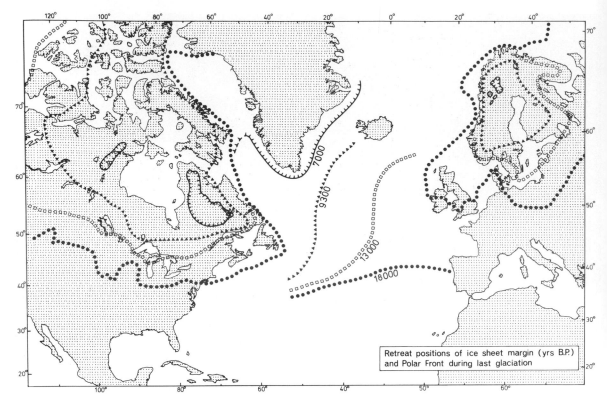

Fig. 16.27 Retreat positions of polar water in the North Atlantic and of the margins of continental ice sheets (data from Ruddiman and Glover, 1975).

Island of New Zealand, Tasmania, and the Snowy Mountains of Australia. Winter snow covered much of the land in these areas as well as in the south-eastern tip of Africa.

Continental shapes at 18 000 BP were significantly different from those of today, in a number of places, as a result of lowered sea-levels. The Bering Strait was closed, the Sunda Shelf of South-east Asia was exposed, the Sahul Shelf was dry and joined Australia to New Guinea, and the Japan Sea was enclosed. The fall in sea-level alone was probably responsible for drier climates in Japan and New Guinea.

The globally averaged difference between full glacial age sea-surface temperature and that of the present is 2.3 °C, but the differences are unevenly distributed. The largest local anomalies occur in the North Atlantic poleward of 40°N where the maximum change was − 17.2 °C. The large differences occur because, at the present time, the North Atlantic Drift covers this area, but in full glacial times it was forced southwards and crossed the

Atlantic to the shores of the southern Iberian Peninsula. The western Mediterranean was cooler than at present as Atlantic water entered it, but the eastern Mediterranean was little different.

In the North Pacific the large negative anomaly of − 6 to − 10 °C probably extended into the Sea of Japan. Lowered sea-level blocked the southern entrance to this sea and prevented the northwards movement of tropical waters.

In the Southern Ocean the polar front, forming a boundary between temperate and polar water masses, moved northwards 3–6° of latitude, but the temperature anomalies were not as large as those in the northern hemisphere being about − 4 °C, and as great as − 6 °C off the west coast of southern New Zealand. In equatorial areas of the eastern Pacific and the Atlantic there was a marked cooling which was probably caused by increased upwelling of cold waters to the surface. In the eastern equatorial Atlantic an enlarged Benguela Current caused temperature lowering of about 5 °C.

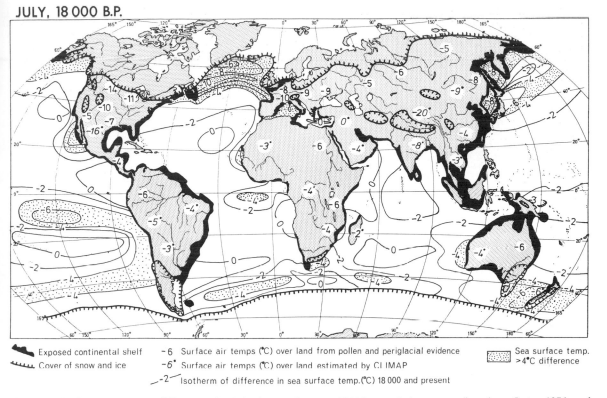

Fig. 16.28 Surface temperature differences, for July–August, between 18 000 BP and the present (based on Gates, 1976, and CLIMAP project members, 1976).

On land, temperature differences between the present and 18 000 BP were generally greater than they were over the sea, being on average 4–9 °C lower. Over much of South America, Africa, India, South-east Asia, and Australia the mean air temperatures were usually 4–6 °C lower than today. The largest anomalies occurred in the centre of Asia, especially north of Tibet, and in central North America where the anomaly is commonly − 10 °C or more.

A reconstruction of the major glacial climatic and vegetation zones at 18 000 BP is given in Fig. 16.29

The Holocene

Until quite recently the best record of climatic change in the last 10 000 years (the span of the Holocene) was obtained from the pollen record, particularly of north-western Europe. The pollen record is influenced by the rates of migration of plants and, in the last two or three thousand years, by human interference with the vegetation. The most sensitive areas are the marginal environments such as the borders of the forest and tundra zones or the sub-alpine belts in mountains. Also informative is the presence of specific warm climate indicator plants such as hazel in the boreal forests of Scandinavia, or holly, ivy, and mistletoe in the mixed forests of central Europe.

On the basis of pollen a general trend of climate was established (Fig. 16.30) which recognized cool periods at about 9000, 5000, 2000, and 300 years ago, and a warm period at about 10 000 years ago before a prolonged warm period in higher mid-latitudes *c.* 8200–5300 years ago. This long warm period is recognized in many parts of the world, although its dates vary because it was noticeable first in low latitudes and later in high latitudes; it is variously called the Climatic Optimum, Hypsithermal, Altithermal, and Postglacial Optimum. It was a warm period in which temperatures were

EARTH ENVIRONMENTS
18 000 B.P. (Maximum glaciation model)

Ice sheet (thickness in metres) and mountain glaciers

Ice shelves

Sea ice in summer

Tundra

Loess semidesert

Dry steppe shrubland and semidesert

Desert

Grassland savanna (some open woodland)

Forest

Lakes

Fig. 16.29 A reconstruction of ice and vegetation cover for the coldest part of the Last Glacial. Over most of the area that is now under a temperate climate some type of periglacial environment or cold semi-desert existed. The subtropical and tropical deserts were far more extensive, and extended into areas which are now covered by seasonally dry shrub and woodland. In South America and Africa a general intensifying of aridity caused the tropical rainforests to nearly disappear; the forests remained extensive only in Central America and South-east Asia. Note the great extent of the fresh-water inland shallow seas in Eurasia. (Reconstruction modified from one proposed by Bowen, 1979).

from 1–2.5 °C higher than they are now. The major cool oscillation of *c.*5300–4500 BP was 1–2 °C colder than now.

This general pattern is broadly confirmed by the Greenland Camp Century ice core, but there are some discrepancies because of insecure dating, latitudinal variations in the onset of a climatic phase, uncertainty in the effect of a complex

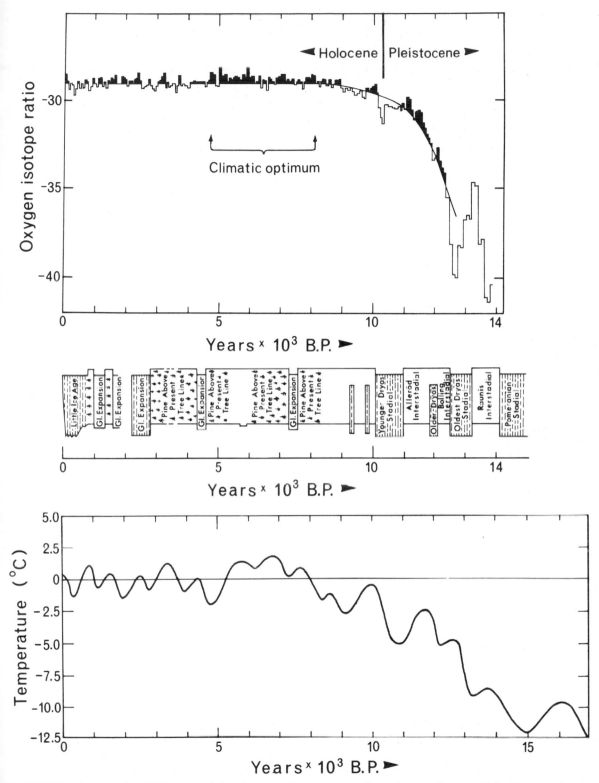

Fig. 16.30 Top: late glacial and Holocene variations in oxygen-isotope ratios at Camp Century, Greenland (after Dansgaard *et al.*, 1971). Pulses of glacier expansion and treeline variations in Scandinavia (after Karlén and Denton, 1976). The general trend in temperature in Europe (based on Butzer, 1974). All figures have the same time-scale.

climatic change on vegetation, the buffering effects of forests on their own composition, and the rate of migration of plants. These changes emphasize that an interglacial is just as subject to minor climatic variations in which glaciers advance and retreat, or treelines rise and fall, as is a glacial.

Northern hemisphere glacier and treeline fluctuations show a broad correlation with the Greenland record. There seems little doubt that the seventeenth to nineteenth centuries were generally cool, although the commonly used name for them of 'Little Ice Age' may contain an exaggeration. There was a general expansion of glaciers from 3300 to 2400 years ago and a minor advance at 1250–1050 years ago.

That the climatic variations, mentioned above, are of world-wide significance is shown by ocean cores in which the Climatic Optimum can be recognized and also the cold periods corresponding to glacial advances. Cores from the South Atlantic show the Optimum at 7000–5500 years BP and cold periods at about 4600, 2800, and 1000 years BP. Glacial advances in New Zealand occurred at about the same times.

Holocene precipitation changes are much more difficult to interpret from the record of mid-latitudes than are temperature changes, but in the semi-arid zones the record of precipitation is clearly seen in discharges of rivers and levels of lakes. From *c*.8000 to 4000 years ago Arizona and New Mexico had a relatively humid climate and in Africa the southern Sahara and East Africa experienced a relatively wetter climate from 9000 to 5600 years ago (see Chapter 19). In the humid tropics of Africa there was probably a small increase in temperature and possibly of precipitation during the Climatic Optimum but the record for most of the rainforest area is uncertain.

Causes of glacials and interglacials

The causes of the long period of climatic deterioration from the Eocene to the Pliocene have been discussed already. The rapid changes from glacial to interglacial climates cannot be explained by shifting of continents or development of ocean currents. Regular fluctuations require a mechanism which can explain the periodicity. Many theories have been put forward, ranging from collision of the Earth with comets to enveloping it in volcanic ash. The only mechanism which appears adequate to explain the periodicity is that of fluctuations in the amount of radiation received from the Sun.

The most influential, and currently most widely accepted, theory of climatic change is the one associated with the name of Milutin Milankovitch, a Serbian physicist whose main works were published between 1913 and 1938: it is variously called the astronomical theory, the solar radiation theory, or the Milankovitch theory. Three effects are recognized.

(1) As was established by Kepler, in the early seventeenth century, the Earth's orbit around the Sun is not a circle but an ellipse, but the shape of this ellipse varies with a period of about 100 000 years. This variation is called the *eccentricity of the orbit*.

(2) Earth spins on its axis as though it were a spinning top with an uneven load and 'wobbles'. The wobble causes a change in the position of the Earth along its orbit at times of the equinoxes and solstices. This *precession of the equinoxes* has a period of 21 000 years.

(3) The Earth behaves like a spinning top leaning over to one side. The amount of lean varies through about 2.4° and alters the latitude of the tropics over a period of 41 000 years. This is the *obliquity of the ecliptic* (Fig. 16.31).

The variations in solar radiation received at the top of the Earth's atmosphere, resulting from the three perturbations, were calculated as a function of season and latitude for the last million years by Milankovitch. In Fig. 16.32 the received radiation variations are compared with the $\delta^{18}O$ record from ocean core V28-238. The correspondence is remarkable: 14 out of 16 radiation maxima in the northern hemisphere average insolation curve are closely associated with high sea-levels, as indicated by the dashed lines. Statistical studies indicate that the 100 000-year cycle is dominant in the spacing of interglacials and has obviously been effective over the last 700 000 years (see Fig. 16.9). In this time also the shorter cycles are important for their controls of stadials and interstadials. In the period 0.8–1.4 My the 41 000-year cycle seems to have been dominant. The three effects, however, are not quite synchronized and this produces considerable variation within the general pattern.

Calculations also show that the Milankovitch mechanism is nearly quantitatively adequate to explain the build-up of ice sheets, and their destruction, once the effects of albedo, heat stored in the oceans, and ice sheet mechanisms are taken

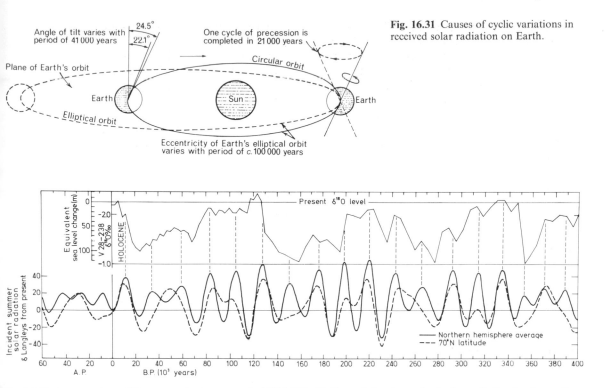

Fig. 16.31 Causes of cyclic variations in received solar radiation on Earth.

Fig. 16.32 A comparison of the oxygen-isotope record of V28-238 with the variation in received solar radiation for the northern hemisphere at latitude 70°N for the last 400 000 years (after Johnson and McClure, 1976).

into account. Albedo is important because both ice and snow cover reflect away radiation and so have a reinforcing effect during early glaciation, but the opposite effect when ice and snow cover starts to decline.

Milankovitch mechanisms occur whether or not Earth is in an ice age, but a continent must be in a polar position so that ice sheets can form, and interchange of heat between the equatorial and polar regions must be reduced before the radiation variations can be effective. The Earth will continue to suffer glacial–interglacial climates as long as the present configuration of land and oceans prevails.

Further reading

There are many books on ice ages, the Quaternary, and glaciation, but ideas have changed so rapidly that few of the older books are now of value except from an historical perspective. Earth's climatic history and the record of ice ages is discussed by Frakes (1979) and in the book edited by John

(1979). The Cenozoic climatic record for the Arctic is the topic of many papers in a special issue of the journal *Palaeogeography, Palaeoclimatology, Palaeoecology*, 3, no. 3/4 (1980). Antarctic glacial history is the theme of the volume edited by Van Zinderen Bakker (1978). A useful collection of essays on many aspects of climatic change appears in the volume edited by Gribbin (1978). Quaternary stratigraphy is critically reviewed by Bowen (1978). Climatic change in the tropics is regularly reviewed in the series *Palaeoecology of Africa*, most of which have been edited by Van Zinderen Bakker. The history of the Last Glacial and of ice sheet models is discussed in detail in the volume edited by Denton and Hughes (1981). The search for a cause of glacials and ice ages is described by Imbrie and Imbrie (1979). A well-illustrated series of twelve articles has been published in *Geographical Magazine* starting with Vol. 51, no. 2 (1978) and finishing with Vol. 52, no. 1 (1979). Each article is concerned with the Quaternary of a particular region. One of the best brief studies of Quaternary environments is that by Goudie (1977). Critical reviews of the maximum model, and statements of evidence in favour of the minimum model are provided by Andrews (1982), Boulton (1979), and Boulton *et al.* (1982). A controversial calculation of the effects of Milankovitch radiation variations, which suggests that this mechanism could form and melt the minimum ice sheets, is provided by Fong (1982).

17 Ultimate Planation

Extensive erosional plains of continental dimensions and erosional bevels on hill summits and flanks, together with stepped relief, have been regarded as evidence of a progressive lowering of relief towards ultimate base level. The way in which this lowering takes place, and the significance of bevels and extensive erosional plains and pediments of limited extent has been a subject of great contention. The arguments are perhaps best appreciated from a brief study of the cycle of erosion, of planation surfaces in general, and of pediments in particular.

The cycle of erosion

The phrase 'cycle of erosion' is intimately connected with the name of William Morris Davis (1850–1934), an American geologist, geographer, and teacher. The cycle of erosion as devised by him was essentially a teaching device used to set forms visible in the landscape into an historical sequence of development as part of his method of *explanatory description*. Features of Davis's concept of a cycle of 'normal' erosion are illustrated in Fig. 17.1: the cycle involves the uplift of a region and its reduction by erosion to a plain. The cycle can be divided into three main stages: youth, maturity, and old age. The uplifted highland is in youth when its rivers have excavated only narrow steep-sided valleys; in maturity it is dissected by many tributary valleys, the floors of the main valleys being wide, valley walls being soil-covered, and the available relief being at a maximum, with summit heights having lost little of their altitude before late maturity; old age is a stage reached when the landscape has low relief and in its late stages will be characterized by the presence of a *peneplain* (i.e. almost-a-plain) which will be hardly above the base level at its seaward margin, but will rise gradually inland towards residual hills with gentle slopes, called *monadnocks*. Davis's cycle may be interrupted by limited uplift which will cause rejuvenation of the streams with incision working headwards from the sea. Thus a cyclic landscape may have the imprint, left in it, of many phases of uplift and thus be polycyclic.

The idea of a cycle of erosion with almost invariable features at each stage has been severely criticized on many occasions, and it is not necessary to repeat such objections except to point out that the rapid uplift, which was envisaged as preceding a stage of youth, was tectonically unreal and the progressive lowering of relief did not take into account continuing isostatic adjustments or tectonic events. Davis himself realized that his scheme was extremely simplified—it was his aim that it should be so for teaching purposes—and that tectonic interruptions to the cycle were possible, or even probable. More fundamental objections arise from research carried out since Davis's death. It is now realized that new mountains are formed by accretion at plate margins and that events at such margins do not produce landforms of simple structure and rapid uplift followed by tectonic stability during which a cycle can progress through an ordered sequence of stages. Even more importantly, the site of mountain building is always marginal to the accreting plate so the centres of uplift are changing progressively (see the conclusion of Chapter 4). Equally fundamental are the objections that the idea of a 'normal' cycle (i.e. one dominated by a humid climate and fluvial processes) has been invalidated by recent knowledge of the numerous and rapid climatic changes through geological time, and especially during each ice age. It is possible that the only widespread events leading to polycyclic landscape formation are those like the splitting of Gondwanaland, during Triassic to Eocene times, which produced new base levels for the valleys which developed

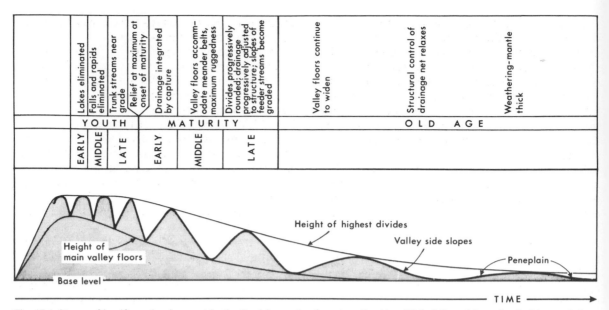

	YOUTH			MATURITY			OLD AGE
Lakes eliminated / Falls and rapids eliminated	Trunk streams near grade	Relief at maximum at onset of maturity	Drainage integrated by capture	Valley floors accommodate meander belts, maximum ruggedness	Divides progressively rounded; drainage progressively adjusted to structure; slopes of feeder streams become graded	Valley floors continue to widen · Structural control of drainage net relaxes · Weathering-mantle thick	
EARLY / MIDDLE	LATE		EARLY	MIDDLE	LATE		

Fig. 17.1 Stages of landform development in the Davisian cycle of erosion. Rapid uplift is followed by valley incision and then progressive stages leading to the formation of a peneplain.

along the margins of the separating continents. Gondwanaland had certainly been largely planated during Paleozoic and early Mesozoic times, and new cycles of erosion were initiated at each coast created by the splitting-off of new continents.

The cycle concept as outlined by W. M. Davis is part of the history of geomorphology; it has not been used as the basis of geomorphological research since Davis's death, but some of the terms used by Davis are still used and it is desirable that students should be wary of the implications in those terms.

While the importance of the cycle as a basis for research has often been exaggerated its influence on teaching in schools has been very great. In research the related concept of denudation chronology and the recognition of erosion surfaces has been more widely used and still has much significance, especially in the southern hemisphere where the remnants of Gondwanaland have many extensive erosional plains, or planation surfaces.

Denudation chronology and planation surfaces

Amongst English-speaking geomorphologists two books are commonly identified as having had a major influence on ideas of denudation chronolo-

gy—'*Structure, Surface and Drainage in Southeast England*', published in 1939 by Wooldridge and Linton, and L. C. King's '*Morphology of the Earth*' first published in 1962 and revised in 1967.

In most of Europe, and especially in Britain, the nearly horizontal or gently inclined surfaces, which are the remnants of, presumably, once-extensive surfaces cut across rock structures, are the focus of attention in denudation chronology. Although they may make up less than 10 per cent of the area of the whole landscape, and may be difficult for the uninitiated to recognize in the field, they are the only remnants of the former landscape whose history the researcher is attempting to reconstruct. The secondary aim is to interpret the history of the drainage pattern from the surviving terraces, river gravels, elbows of capture, channel segments, and air gaps.

In Africa, Australia, India, parts of South America, and interior Asia, the remnant surfaces are often very extensive and make up most of the landscape. It is perhaps not surprising that a modified form of denudation chronology should still be a significant part of geomorphology in southern Africa, India, and Australia.

The term planation surface is a general one encompassing nearly level surfaces, or valley-side benches, of whatever erosional origin. In a strict sense all parts of the land surfaces which are not directly of depositional origin are surfaces of erosion. Thus 'structural' surfaces aligned with the surface expression of a major feature, such as a bedding plane, are only visible because erosion has stripped all of the cover beds (Plate 17.1). The term 'planation surface' covers one with such structural influences whereas the term 'erosion surface' is often preserved for one cut across geological structures.

The existence of planation surface remnants is almost certainly due to two causes: (1) the long-term fall of sea-level since Cretaceous times which caused a lowering of base level during the Tertiary Period; and (2) broadscale uplift or upwarping of continental surfaces, with many such uplifts being related to sea-floor spreading mechanisms. The surfaces are consequently of Tertiary or Mesozoic age in origin and the valleys cut below them are predominantly of Quaternary age. The age of planation surfaces inevitably means that they have had a long and complex history of weathering and erosion processes so that the exact nature of the origin, as peneplains or by marine abrasion for example, may no longer be apparent. They are most likely to survive when formed on resistant rocks, and where they form the drainage divides and are far from the main drainage lines.

Identification and dating of planation surfaces

African surfaces, and especially those studied by L. C. King, offer good examples both of the features and the methods of research, as well as examples of the objections which may be raised against denudation chronology.

The surface of much of Africa is composed of extensive plains, some of which bear isolated hill masses and mountains as inselbergs (literally, island hills), and these plains are bounded by scarps between upper and lower surfaces (Plates 17.2, 17.3). Many of the surfaces are cut discordantly across geological structures (Plate 17.4), and in a few places their youngest possible age is indicated by the age of datable volcanic rocks which have been emplaced upon the surface (Fig. 17.2).

It has been contended by King that African erosion surfaces have been produced by the process of backwearing of scarps (Fig. 17.3), following rapid and large uplifts of continental blocks, to leave extensive erosion surfaces called *pediments* at the scarp foot. The coalescence of pediments creates a *pediplain* which is the planation surface. Repeated uplift is thus regarded as being responsible for a number of continent-wide erosion surfaces, or incisions, at successively lower altitudes with decreasing age.

Scarp retreat and the preservation of steep rock faces (or free faces) above a debris slope, over which material from the scarp is transported to the lower pediment which has been left by the retreat, is fundamental to King's theories and is essential for his explanation of the inselbergs and scarps of Africa. The hillslope form is maintained throughout the cycle of erosion so that a higher surface may continue to extend headwards at the same time as it is eaten away at its outer margin, as the

17.1 The Fish River canyon in southern Namibia. The skyline surface is regarded by King as part of the African cycle. The benches at lower elevations are structurally controlled and formed along bedding planes during Quaternary incision into this updomed part of the continental margin.

17.2 The African surface in southern Namibia with inselbergs, of which the summits are thought to be remnants of an older Gondwana surface.

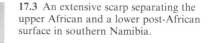

17.3 An extensive scarp separating the upper African and a lower post-African surface in southern Namibia.

17.4 A pediment in the Transvaal showing the erosion surface cutting across the geological structures. Note the virtual absence of regolith (some pediments have a much deeper regolith than this).

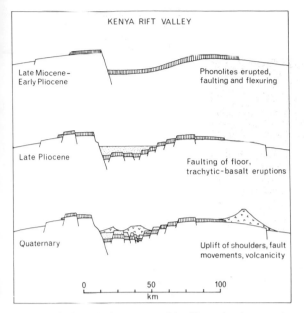

KENYA RIFT VALLEY

Late Miocene–
Early Pliocene

Phonolites erupted,
faulting and flexuring

Late Pliocene

Faulting of floor,
trachytic-basalt eruptions

Quaternary

Uplift of shoulders, fault
movements, volcanicity

0   50   100
km

Fig. 17.2 A simplified sequence of landform development in the area of the East African Rift Valley. Upwarping in Mid-Tertiary times was followed by faulting and flexuring of the African planation surface. Segments of the surface were preserved beneath volcanic rocks. Faulting has caused some segments to be at markedly different altitudes from other segments and some are tilted. Correlation on the basis of altitude alone would thus be hazardous (modified from Baker, 1965).

next lower surface extends laterally at its expense. This type of process is most clearly effective in semi-arid environments where rock is exposed on the cliff, and slope angles of the cliff are in equilibrium with the strength of the rocks. In more humid climates slopes become covered with talus, colluvium, and soil and may decline, in slope angle, with age.

In the early stages of a pediplanation cycle the streams are rejuvenated by uplift and start cutting headwards from the coast, thus creating a nick-point with a gorge below it. The stream on the raised surface is not affected by the uplift. If the uplift is a domed upwarping with the greatest amount of uplift inland, then the nickpoint will get higher as it cuts inland. The Victoria Falls on the Zambezi (Plate 10.2) and the Aughrabies Falls on the Orange River are regarded as two such nick-points. Scarps retreat away from the drainage lines. The Natal Drakensberg is one of the most magnificent of such scarps and according to King its wall-like face, over 1000 m high, originated in the Cretaceous and has retreated 150 km or more to the east (Plate 17.5). One of the factors most favouring the survival of such scarps is that they shall be large.

At the stage in which an erosion surface has removed virtually all remnants of a preceding landscape a continental pediplain consists of a complex assemblage of closely related surfaces, chiefly pediments, which slope down to local drainage lines and basins which are unified only in relation to a long-stable single base level. The detailed forms of the pediplain will be controlled according to bedrock, tectonic, and climatic factors. Pediments will vary in size; drainage density may be coarse or fine; inselbergs may occur or be absent; broad convexities may have developed on weak rocks; and local land movements may have produced local and minor nickpoints. Warping can produce basins of deposition and the depth of weathering or formation of duricrusts will be partly dependent upon climate.

Erosional land surfaces, by their nature, are not easy to date because they seldom bear deposits which are associated with their formation. Where

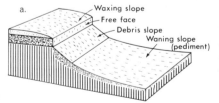

a. Waxing slope
    Free face
    Debris slope
    Waning slope
    (pediment)

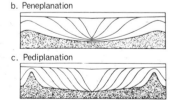

b. Peneplanation

c. Pediplanation

Fig. 17.3 (a) L. C. King's view of the major elements of hillslope profiles with the free face and pediment being the main persistent features.

(b) Under Davisian peneplanation the landscape is worn down so that the planation surface is evolving over all segments at all times.

(c) Under pediplanation the scarps are backwearing progressively so that the pediment is oldest near the drainage lines and youngest at the foot of the scarps.

17.5 Part of the Drakensberg escarpment, in Natal, showing the stepped remnants of the African surfaces below it.

fossils within cover beds are available a youngest possible age for the surface may be determined, but in their absence indirect methods which are dependent upon correlation have to be resorted to: elevation above sea-level; correlation with adjacent, already dated, surfaces; correlation with sediment deposition upon unconformities which are correlated with planation surfaces (Plate 17.6); the stage of development of landforms and weathering profiles in the erosion cycle; ^{40}K-^{40}Ar dating of volcanic rocks upon the surface; and paleomagnetic dating of ferricretes formed on surfaces.

These methods are all fraught with uncertainties. For example, if an eroded surface acquires a thin cover of, say, Cretaceous deposits the origin of that surface must be in Cretaceous or pre-Cretaceous times. But if the Cretaceous beds are removed by subsequent erosion and Oligocene beds are left on it, the only direct evidence for its age will indicate that it is of Oligocene or pre-Oligocene age. The removal of the Cretaceous beds may well imply that during their removal the original surface would also be modified. In phases of repeated deposition and erosion, a surface may thus be partly (say) Cretaceous, Oligocene, Pliocene, and Pleistocene. Thus to derive the fundamental age of a land surface, King argues that the landscape, viewed as a whole, was first planated in (say) Cretaceous times and acquired at this time its fundamental form. Subsequent modifications are of local significance only. The oldest deposits known to exist upon a cyclic surface are significant in establishing its age for comparison with other surfaces.

A second problem is that as surfaces extend inland by pediplanation, during succeeding geological periods, they become progressively younger inland. Thus a surface has a local or specific age at any point which will be different from that at its inland and coastal ends. However, in comparative studies of land surfaces which have been substantially formed over a few million years, surfaces older than the Quaternary can usually be ascribed to a geological period with little danger of serious misrepresentation.

Because of the paucity of datable deposits on major continental erosion surfaces, dating is often most secure when carried out on monoclinal coasts like those of Natal and Zululand. As a surface is initiated by uplift, rapid erosion inland will result in deposition offshore, and the extending erosion surface will thus have an age indicated by offshore fossiliferous deposits. Towards the end of the cycle, when the land surface is becoming stable, the sedimentary sequence will be broken by an unconformity and then the next uplift marked by renewed deposition. Thus each sedimentary sequence rests upon an older surface, which may be erosional or depositional, and the upper surface of that deposit will be contemporaneous with a related continental erosion surface. On coasts where the axis of upwarping shifts seawards a full sequence of offshore deposits may be exposed to give a series of dates for the inland surfaces (Fig. 17.4).

Within Africa matching of deposits and planation surfaces has given the following chronology:

Highest: Gondwana   (G)   – Jurassic
          post-Gondwana   (p-G) – Cretaceous

17.6 A major unconformity cut across metamorphic rocks and overlain by sedimentary rocks. This unconformity is known as the Kukri planation surface where it has been exhumed from beneath the sedimentary coverbeds and where it forms the modern ground surface in the Transantarctic Mountains. The cover beds indicate a youngest possible age for the erosion surface.

African	(Af)	–	early Cenozoic
post-African 1,2	(p-Af)	–	late Cenozoic (Miocene, Pliocene)
Lowest: Congo	(Q)	–	Quaternary

The Gondwana surface is found only on the highest divides. It is interpreted as remnants of the landscape of the original supercontinent which existed before Gondwanaland fragmented in the late Mesozoic. The remnants are today found only at high altitudes in such places as the Nyika Plateau of Malawi or in high Lesotho. In some places in the Transvaal and in Zimbabwe it appears to correspond with exhumed surfaces on which pre-Jurassic rocks have lain, and so it may be that the Gondwana surface was evolved over a greater span of time than the Jurassic Period.

The post-Gondwana surface is found only on high divide areas where its lineaments are present as systems of broad valley heads dissecting the Gondwana plateaus (e.g. the Sani Pass in Lesotho). In northern Africa, Gondwana base levels persisted into Cretaceous times and there a separate post-Gondwana cycle cannot be seen.

The most common high surface of Africa stands at elevations of 800–1200 m and was formed in the cycle that persisted from late Cretaceous to mid-Cenozoic times. As mid-Cenozoic deposits lie upon it in many places, this 'African' surface is often called the mid-Tertiary surface. It is sometimes covered with ferricretes in Uganda, parts of Nigeria, and in Natal. Where there are no remnants of older cyclic surfaces the African surface gives rise to broad even crestlines or wide plateaux (e.g. the Serengeti Plains of Tanzania). This planation is regarded by King as being the provider of the fundamental land surface of the southern continents.

Incised below the African surface are broad valley plains and rolling topographies which now

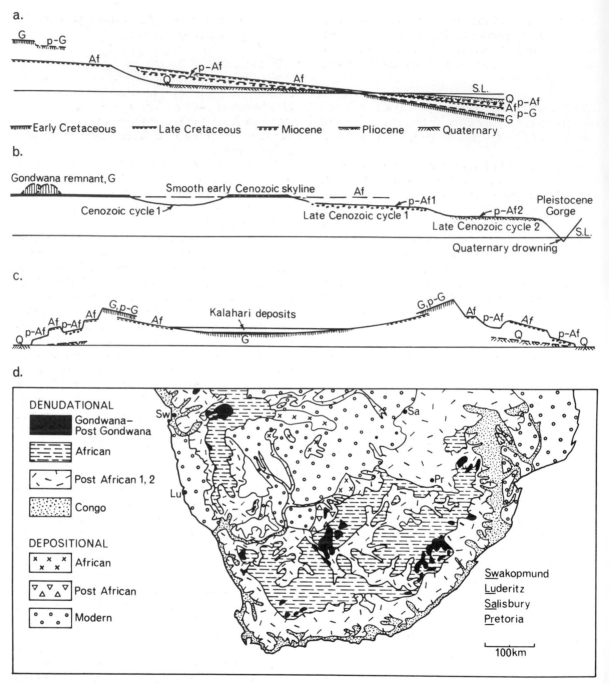

a.

⌁⌁⌁ Early Cretaceous   〰〰 Late Cretaceous   ⟋⟋ Miocene   〰 Pliocene   ⟋⟋⟋ Quaternary

Fig. 17.4 (a) In the Natal monocline planation surface remnants on land are dated by long-range correlation with sediments offshore. The Gondwana surface (G) is the major basal unconformity offshore and the highest remnant on land (after King, 1967).

(b) The relative altitude of planation surface remnants in southern Africa.

(c) Updoming of the continental margins has tilted planation surface remnants both towards the sea and towards the interior. The centre of southern Africa is a sedimentary basin which has received some of the eroded material from the marginal highlands. The African surface forms the internal unconformity.

(d) King's (1967) map of southern Africa surfaces.

make up much of the landscape. In many areas two cycles, or partial cycles, are evident—post-African 1 and post-African 2—of which the first was current during Miocene times and the second in Pliocene times.

About the close of Tertiary times a violent upwarping of earlier surfaces gave rise to a period of rapid and deep incision by the major rivers so that many now flow in deep gorges in their lower reaches.

At each period of uplift the nature of the earth movements in southern Africa was broadly the same; the Kalahari interior basin rose least and the peripheral highlands rose most in a broad uparch-ing which tilted both to the interior and to the sea with the result that the Kalahari and coastal areas received the sediment from the highlands (Fig. 17.4c). The maximum uplift thus took place in a horseshoe-shaped zone outside the Great Escarp-ment so that the greatest number of cycles are recorded there, especially in the Eastern High-lands and in Zimbabwe. The sequence of events for Natal is described in Fig. 17.5.

The idea of rapid intermittent periods of up-heaval separated by long periods of stability is inherent in the work of King. The difference between a vertical lowering of the landscape (peneplanation of W. M. Davis) and lateral denu-dation (pediplanation) should result in differences of isostatic adjustment. Under a Davisian type of cycle the continuous removal of debris from a large area should result in a virtually synchronous compensating uplift which could delay peneplana-tion almost indefinitely. Under pediplanation, isostatic uplift would occur only when a threshold amount of denudation had taken place from the rim of the continent, and it is thus necessarily intermittent.

It has been calculated that the threshold dis-tance of headward erosion to proceed inland before uplift would occur is about 480 km. Once this has been exceeded and uplift taken place, the actual amount of uplift is given by:

$$h = \frac{B}{A} \cdot r$$

where h = height of isostatic compensation (metres)
   B = unit weight of surface rocks removed (26 kN/m^3)
   A = unit weight of the sima inflow mater-ial (34 kN/m^3)

r = thickness of the surface layer removed (metres).

If the equation is applied to West African landscapes the African surface should be at about 350 m, the post-Gondwana at 800 m, and the Gondwana at 1400 m. The agreement with mea-sured heights is good, even though not perfect, and supports the concept of intermittent uplift and pediplanation.

The rates of denudation implied in the retreat of the Drakensberg scarps are about 1 m per 1000 years. This rate is high, but it is within the rates being estimated for modern erosion of high moun-tains with steep faces.

King has extended his ideas on African erosion surfaces to South America and Australia and even proposed a world-wide scheme applicable to all the components of Pangaea (Table 17.1). Whilst this represents a bold idea, and would form a framework for large-scale historical geomorpho-logy over much of the cratonic continental sur-faces, it has to be admitted that the scheme is open to question because of the lack of detailed map-ping, and especially of dating, in most areas of the southern continents. A bold attempt at indicating the ages and altitudes of recognized planation surfaces of widely separated areas of Africa and Asia is shown in Fig. 17.6. It should be realized that this figure does not represent the lateral extent of the surfaces.

In Britain there are no extensive erosional plains of the kind commonly recognized in the southern continents. Planation surfaces there are small, nearly horizontal, or gently inclined remnants which occur as plateaux, bevels on the summits of cuestas, or valley-side benches (Plate 17.7). Rather more questionable are the assumptions that the accordant summits (i.e. those with similar alti-tude) of hills and uplands are necessarily remnants of a surface. The origin and dating of these surfaces is based upon different methods from those appropriate to Africa.

Britain may, broadly, be regarded as an upland to the north and west composed of old and resistant rocks which have been eroded during much of Paleozoic and Mesozoic time, and the resulting deposits have been laid down in shallow seas to the east and south (Fig. 17.7). Thus an erosion surface cut across the uplands in early Cretaceous times yielded sands and clays depo-sited in a 'Wealden Lake' in the south-east. This

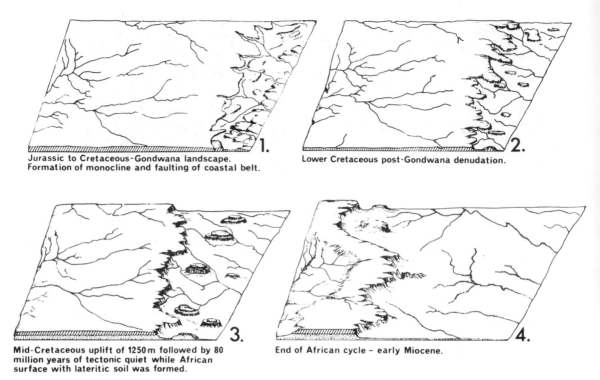

Fig. 17.5 Schematic blocks showing King's concept of the cyclic evolution of the Natal surfaces. Note the extensive Gondawana surface formed before continental separation (in 1). The Drakensberg is the major east-facing escarpment in stage 7. The cross-sectional blocks show the layered stratigraphy of the area which has prompted some critics to suggest that many of the planation surface remnants are actually structural benches.

erosion surface may have been buried by Cretaceous chalk during the great marine transgression of that time and so fossilized, but there are no outliers of chalk left in the north and west (although chalk does occur beneath basalts in Ireland) hence this hypothesis cannot be verified.

Furthermore the summits of the high remnant hills in Wales, like the Brecon Beacons, which have been regarded as the residual survivors of this surface, are of very limited area. The attempt to explain an accordance of summit levels in Wales with an extrapolated gently dipping unconformity

Table 17.1   World-wide Scheme of Erosion Surfaces, as Proposed by L. C. King (1976).

I	The 'Gondwana' Planation	of Jurassic age
II	The 'Kretacic' Planation	early-mid Cretaceous age
III	The 'Moorland' Planation	current from late Cretaceous till the mid-Cenozoic
IV	The 'Rolling' Landsurface	mostly of Miocene age
V	The 'Widespread' Landscape	the most widespread global cycle, but more often in basins and lowlands than uplifted by recent tectonics to form mountain tops; Pliocene in age
VI	The 'Youngest' Cycle	Modern (Quaternary in age) represented by the deep valleys and gorges of the mountainlands

5. Early Miocene uplift of 300m in inland areas.

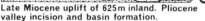

6. Late Miocene uplift of 625m inland. Pliocene valley incision and basin formation.

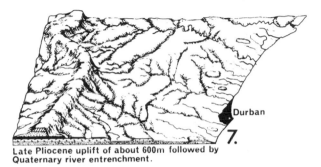

Durban

7. Late Pliocene uplift of about 600m followed by Quaternary river entrenchment.

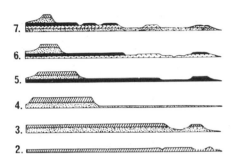

7.
6.
5.
4.
3.
2.
1.

at the base of the English chalk, thus requires an act of faith, and stretching of the imagination, which goes well beyond the 'hard' evidence (see Fig. 17.7b).

A similar argument has been used to explain the summit levels of the Welsh 'Tableland'. It has been postulated that the chalk sea transgressed across Britain and cut a marine surface on which the chalk was deposited. Updoming of Wales created a radial consequent drainage pattern on the chalk,

and during Tertiary times this drainage pattern would have become adjusted to the underlying structure and been modified by the earth movements which caused broad folding in the Jurassic and younger rocks of the south-east. As the chalk was removed by erosion in Eocene times the Tableland surfaces were cut and this episode is recorded as the unconformity underlying the resulting Eocene sediments in the south-east.

An alternative view is that the Welsh Tableland

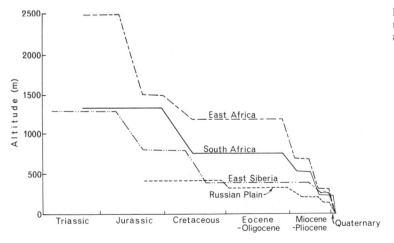

Fig. 17.6 A plot of the altitude of recognized planation surfaces of known ages in USSR and Africa.

17.7 Tertiary planation surface remnants in southern Devon with Quaternary age valleys cut into them.

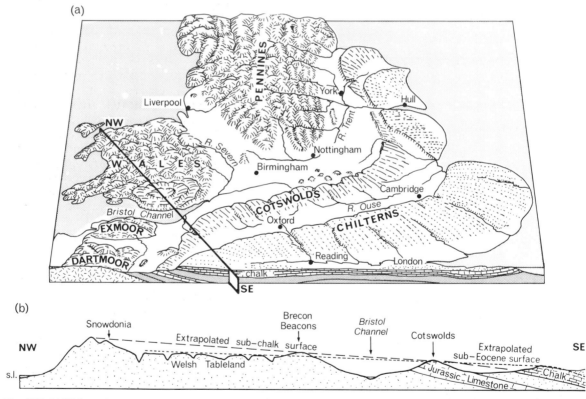

(a)

(b)

Fig. 17.7 (a) Wales and central England showing the uplands of predominantly Paleozoic rocks and the lowlands and cuestas to the south-east. The location of the line of section for (b) is indicated (the drawing is based upon Davis, 1912).

(b) A schematic section through Wales and south-eastern England indicating the extrapolations underlying assumed ages for the Welsh summit surfaces and the area of downwarping in the Bristol Channel which affected initiation of the Severn drainage. Both figures are diagrammatic, and do not attempt to indicate details of relief or altitude.

represents, not a sub-chalk surface, but a sub-Triassic surface which has been exhumed from beneath a cover of Mesozoic and Tertiary rocks.

From Eocene times onwards the geological history, at least in the south-east, becomes clearer because the formation of a gradually deepening syncline in the London Basin (Fig. 17.8) permits the deposition and survival of sediments which gradually lap onto the flanks of the chalk dipslopes so that, although these sediments are often removed from the chalk slopes by erosion, they have left behind relics of their presence which provide evidence of past events. Thus strongly weathered soils, deposits of clay-with-flints derived by weathering from the underlying chalk, and sarsen stones which are thought to be relics of

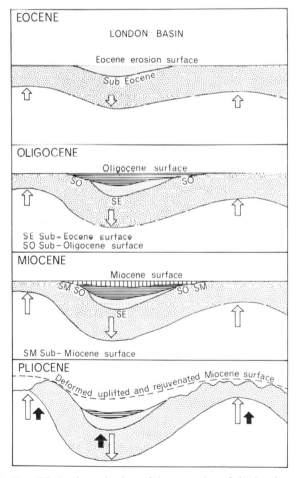

Fig. 17.8 A schematic view of downwarping of the London syncline and later updoming of the Weald (compare with Fig. 4.40) during the Tertiary Period (modified from Small, 1980).

an extensive silcrete formed in a warm Tertiary climate, are all widespread. In few cases, however, can they be firmly dated and debate still continues on the age of surfaces which can be broadly classed as being of early and late Tertiary age. Whatever the age, however, the remarkable continuity of summit heights and the polycyclic nature of the landscape cannot be denied.

Objections to denudation chronology come from many sources.

(1) Of primary importance is the necessity to extrapolate supposed summits over great distances in order to estimate ages of bevels and accordances of summit levels. This is seen clearly in the case of the sub-chalk and sub-Eocene surfaces in Britain. The same objection applies to King's use of the offshore unconformities in the sedimentary record in Natal to indicate possible ages of erosion surfaces near the foot of the Drakensberg.

(2) Tectonic tilting, folding and faulting dislocate landscapes so that projection of surfaces on the basis of altitude must be prone to error. As a compensation, however, it must be recognized that down-faulted sections of a surface may preserve segments of a planation surface, complete with a regolith cover, and so allow more information to survive, as it has done in Otago (Plate 17.8).

(3) Attempts to use remnants of ferricretes and silcretes on surfaces for correlation purposes are also fraught with difficulty as these duricrusts can form on upper and lower surfaces contemporaneously and so may not provide a useful guide to age.

(4) Narrow ridges and summit bevels may be at a nearly uniform altitude in a given area because that is the altitude at which there is an approximate balance between the rate of uplift and the rate of downwearing. This is particularly the case with alpine gipfelflur (see Plate 15.8 and Fig. 17.10).

(5) All surfaces are progressively reduced by weathering and erosion. The assumption that surfaces are all reduced to the same degree may be false, especially where extrapolations are over considerable distances.

(6) In the case of the African surfaces it is questionable whether scarp retreat is a process of such universal occurrence as King assumes, and the possibly complex origin of pediments (see below) illustrates this point. Africa is so large, and the changes of climate during the late Mesozoic

17.8 Remnants of a Cretaceous planation surface in central Otago, New Zealand. In the distance block faulting has lowered fragments of this surface which is now preserved beneath the valley-floor sediments. The planation surface has been warped and partly tilted by tectonic movements and incised by Quaternary channels to leave closely spaced, and nearly accordant, summit levels (photo and copyright: Whites Aviation).

and Cenozoic so great, that it is improbable that one mechanism would operate alone. This objection questions elements of King's explanation but not the existence of surfaces.

(7) The unbeliever may go further and declare that many summit bevels are the figments of a too vivid imagination. It must be countered, however, that recognition of planation surfaces, their dislocations, dating, and forms, is an important line of evidence in deciphering the geological history of many landscapes, and in many areas it is by far the most important method for developing an understanding of the Cenozoic contribution to modern landscapes.

(8) Studies of cyclic landform development are not readily compatible with modern process geomorphology and they take little note of modern work on climatic change and variable rates of denudation. An alternative approach which partly

synthesizes these methodologies is outlined in the last chapter of this book.

Pediments

Pediments are usually regarded as piedmont plains or extensive surfaces cut across bedrock and separated from a backing upland by an abrupt change of gradient. Hillslope angles commonly exceed 20° and the pediment at its head has slope angles of 2–10°. Many pediments are concave in longitudinal profile and may have slopes of less than 1° at the outer limits where they terminate at stream channels or are buried by sediment. As a general rule pediment slopes appear to be least, and to be straightest, in areas of extreme aridity. They may be bare bedrock or mantled by a veneer of sediment. Such a definition includes hill-foot plains of all types from gibber-strewn or alluvial sheets beneath silcrete-capped hills in Australia, to gravel-mantled glacis of erosion across soft rocks on the inland side of the Saharan Atlas, and the hill-foot bare, or gravel-covered, plains below mountains in SW Arizona.

The term 'pediment' was first used in the USA to distinguish the hill-foot erosion surfaces of the Basin and Range Provinces. Such basins are nearly always graben, or fault-angle depressions, and it is usually assumed that the pediments have been cut as the hillslope retreated, leaving basins of internal drainage filled with alluvial and lake sediments, surmounted by mountains which dec-

lined in size until inselbergs were left, with extensive pediment passes between them (Fig. 17.9).

Many attempts have been made to relate the pediment slope and lithology to the calibre of particles being transported across its surface, the grading of those particles along the pediment, and to the water depth in channels or in sheet floods. As a general rule granitic rocks and sandstones weather to a fine grus or sand, but strongly silicified or closely jointed rocks produce irregular gravel. Most pediments have coarse debris near the head and progressively finer debris towards the foot. The assumption underlying such studies is

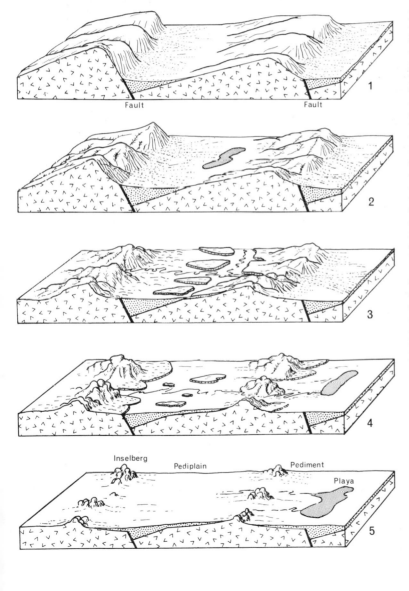

Fig. 17.9 Diagrammatic evolution of a pediment in a fault-angle depression. (1) Hillslopes retreat by weathering, wash and mass wasting. Fans may form in the piedmont if slope erosion provides debris at a faster rate than it can be removed. (2) A pediment forms at the mountain foot as debris supply diminishes. The pediment extends into the hill base because of preferential weathering at sites receiving runoff from hillslopes. Fans may be regraded, as supply of debris diminishes, by wash and resorting of the fan and alluvial material. Playas may form in low points of enclosed basins. (3) Pediments extend headwards and the alluvial infill of basins may be dissected and terraced if basin drainage is captured and diverted. (4) Pediment passes form between diminishing residual hills. (5) Isolated inselbergs surmount peripheral exposed rock-floor pediments which extend around all hills, and have a common pediplain surface extending across the infills. This set of processes and developments was commonly hypothesized for the Basin and Range provinces of USA.

that modern processes are responsible for the pediment. This may well be a case where the origin of the whole form is being confused with the process which is merely retouching the present surface. Modern process studies can tell us something about the mechanics of the modern processes and the sedimentary deposits they produce; their relevance to the overall origin of pediments is far more questionable. The evidence for an exhumed origin for many bedrock surfaces, which originated as weathering fronts beneath a soil, seems indisputable. Such an origin is far less clear where relict weathering profiles do not survive. Origins by the erosive work of anastomosing channels, sheetwash, removal of pre-weathered detritus, stream channel migration, and as the residual surface left by slope retreat, have to be considered. The lack of sediment accumulations at the head of pediments implies that inselberg, or scarp slope, retreat is at a sufficiently slow rate for debris removal from the slope foot to keep pace with debris supply. The weakly incised rills at the head of pediments imply that wash processes may be more important than channel flow, and the lack of streams in many deserts indicates that channel migration has little part in pediment formation at present. Frequently, however, it is impossible to distinguish ancient from modern effects, or the causative effects of a process (e.g. wash) from the situation in which it is the slope and debris of the pediment which permits that process to occur, or the condition in which process and form develop together.

Pediments vary enormously in area from a few tens of square metres to hundreds of square kilometres. Their age also varies, and hence the range of climatic environments to which they have been subjected. It is improbable that one style or origin is appropriate to all pediments, or that one set of processes have shaped a pediment throughout its history. Regrading and periodic dissection and stripping of older surfaces, as a result of climatic, tectonic, or base level changes, may well lower pediments over time and, in so doing, destroy evidence of earlier processes of evolution. One pediment may have suffered all of the commonly recognized controls: initiation by tectonism, extension by fluvial incision and channel migration, development of piedmont lowlands as hillslopes retreat and maintain strength-equilibrium angles of slope, deep weathering and double planation followed by exhumation as a stripped

etchplain, and finally reshaping by sheetwash and fluvial processes.

Gipfelfluren

One of the most contentious items of evidence in reconstructions of erosional history relates to the broad accordance in altitude of many mountain summits within some ranges. Such an accordance (see Plate 15.8) has been taken as evidence that the mountain range has been derived by erosion from a formerly planated landscape which may have suffered uniform uplift or doming. When some ranges are seen from a distance, a summit accordance (i.e. a *gipfelflur*) may be an obvious feature and it can be demonstrated from high-quality maps (give or take a 100 m or so), but the cause of the accordance is seldom demonstrable from unequivocal evidence. Some of the problems are illustrated in Fig. 17.10.

Where the rate of downwearing is uniform across a mountain mass the summits may maintain their elevation relative to one another, even where drainage density varies, provided that adjustments to lithological and climatic controls result in change of slope angles (a,b,c,e). Where the rate of uplift varies across a mountain mass the summit surface may be a reflection of this rate with the highest peaks being above the zone of greatest uplift (d); such an adjustment may be accompanied by progressive increases in slope steepness if the drainage density remains constant (f) or by a decrease in drainage density if the slope angles remain constant (g). These relationships exist in a fluvially eroded landscape because slope angles are related to the relief available for dissection and to the spacing of the drainage channels. The height of summit surfaces can consequently be seen as a possible consequence of (1) survival of an original surface which has been uplifted, but this is improbable because of erosion as uplift occurs; (2) the uniform downwearing from an original summit surface (Fig. 17.10a,b); (3) the relative rate of uplift of the mountains lying below it (Fig. 17.10f); (4) the balance between uplift and denudation of the mountains lying below it (Fig. 17.10d); (5) the relative resistance of underlying rocks in relation to the regional climate, but this is unlikely to be uniformly varying across a landscape; (6) relatively uniform drainage density within a large area of a mountain mass.

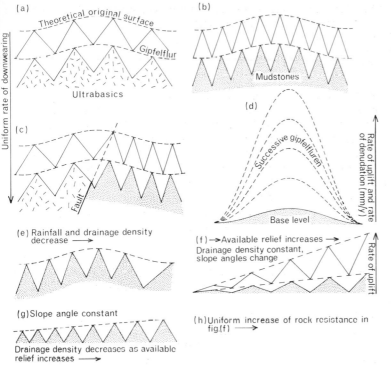

(a) Theoretical original surface
Gipfelflur
Uniform rate of downwearing
Ultrabasics

(b) Mudstones

(c) Fault

(d) Successive gipfelfluren
Base level
Rate of uplift and rate of denudation (mm/y)

(e) Rainfall and drainage density decrease ⟶

(f) ⟶ Available relief increases ⟶
Drainage density constant, slope angles change
Rate of uplift

(g) Slope angle constant
Drainage density decreases as available relief increases ⟶

(h) Uniform increase of rock resistance in fig.(f) ⟶

Fig. 17.10 Theories of the origin of gipfelfluren.

Further reading

An appreciation of the methods and teaching of W. M. Davis can be obtained from the compilation of P. B. King and S. A. Schumm (1980). The methods of denudation chronology are well displayed by Wooldridge and Linton (1939, revised 1955). L. C. King's major work is best appreciated from his study published in 1967. Features of mountains and gepfelfluren are discussed in the following texts: Osburn and Wright (1968), Ives and Barry (1974), and Slaymaker and McPherson (1972).

18 Landforms of the Cold and Temperate Zones

The cold zones

The Earth's cold climatic zones may be divided into two main units: a glacial zone where discharge of water is in the form of ice; and a periglacial zone where there is seasonal discharge of liquid runoff, but where the formation of ground ice plays a predominant role in the development of interfluves.

The limits of the glacial zone are clear: they coincide with the margins of glaciers. They are not strictly climatic boundaries, for ice flow extends below, or beyond, the regional snowline, except in Antarctica, where the limits are the calving line of icebergs. Within the glacial zone are periglacial enclaves on nunataks, mountain peaks and, in Antarctica, a few ice-free valleys where snowfall is too limited, and ablation too great, for permanent ice to form, and mountain ridges are too high for glaciers to flow in from neighbouring areas.

Much of the present periglacial zone was formerly covered by ice of the Last Glacial. Consequently the time during which periglacial influences have been effective is relatively brief and the landforms of glacial origin may be modified only slightly. Equatorwards of the ice margins of the Last Glacial, however, and especially in eastern USSR, there were areas of extensive permafrost which still survive and some are out of equilibrium with modern climate.

Many attempts have been made to define the periglacial zone and periglacial climates. Most definitions seek to distinguish the areas of severe polar influences, the varying importance of maritime and continental influences, altitudinal influences, and relict features (Fig. 18.1). For many purposes it is satisfactory to define the periglacial zone as including all areas where the mean annual temperature is less than +3 °C, with intense frost action, ground which is snow-free for part of the year, and where perennially frozen ground, but few or no trees, can survive. Within such a broad definition it is possible to recognize five distinct types.

(1) *High Arctic and Antarctic areas* usually have extremely weak daily temperature ranges of 5–10 °C, but very strong seasonal patterns with midsummer temperatures close to 0 °C but midwinter temperatures of −40 to −50 °C. Summer thaw depths are limited to 0.1–0.3 m into the ground. Precipitation is low at around 100 mm, hence polar deserts and Arctic barrens have limited melt or ground ice phenomena, but strong eolian activity and sand-wedge polygons may predominate over other forms of patterned ground.

(2) *Continental periglacial areas* of subpolar climate have weak daily temperature ranges, but strong (40–60 °C) seasonal ranges. Summer temperatures of +15–20 °C in Yukon and Siberia, for example, permit a full vegetation cover of tundra herbs and grasses, or low trees. Annual precipitation may be 250–600 mm with much falling as summer rain. The active layer may be 2–3 m thick and patterned ground and thaw phenomena may be very common, except in arid areas like the Gobi where eolian activity is significant.

(3) *Alpine areas* above the treeline have relatively high precipitation and high summer temperatures. Physical weathering is of importance, but the broad range of latitudes in which mountains lie creates a great range of intensities and phenomena.

(4) *Subpolar oceanic islands* of both hemispheres experience strong maritime influences so that mean annual temperature ranges may be only 10 °C, but precipitation may be high at 1000–2000 mm. Such climates prevent the formation of permafrost and limit the penetration into the ground of the many, but brief, freeze–thaw cycles,

Rock	Ground ice		
Basement rock	Permafrost	T Talik	Active layer (seasonal freezing)
Sediments			

Patterned ground

Ice wedge polygons	Sand wedge polygons	Stone pits	Earth nets and mudboils
Cryoturbation, involutions	Stone circles	Frost crack polygons (seasonal freezing)	

Gelifluction

Earth steps	Stone banked steps	Earth lobes	Block sheets and non-sorted stripes
Sorted stripes			

Organic patterned ground

Peat hummocks	Palsas	String bogs

Pingos

Active ∘ open system c closed system	Decaying

Thermokarst

Earthflows	Alasses	Oriented lakes

Erosion

Rills and gullies	Avalanche tracks	Nivation hollows	Cryoplanation terraces and pediments

Eolian

Loess cloud	Deflation	Sand dunes

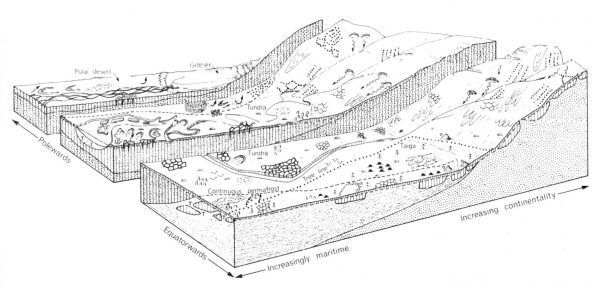

Fig. 18.1 A schematic representation of the relationship of Holocene periglacial phenomena with latitude, continentality and altitude in the northern hemisphere (this diagram was inspired by a similar one reproduced by Eismann, 1981; after Karte, 1979).

and so promote gelifluction and needle ice activity.

(5) *Relict permafrost* may have survived through the whole of the Quaternary in parts of Yakutia. Its limits there are well known, but in Canada and Alaska there is still much uncertainty about its distribution. The most obvious relict permafrost exists beneath boreal forest and parts of the taiga. Gelifluction there is probably limited by the rapidity of thawing in spring, and beneath a continuous and undisturbed forest the ground is mostly stable. Natural, or human-induced, fires, deforestation, and ground disturbance are the chief causes of degradation of the permafrost and may cause alasses and similar features to form.

Reference should be made to Chapters 14, 15, and 16 for details of cold climate phenomena.

The mid-latitude forested zones

The climatic zones of late Holocene times, with mid-latitude forests as their natural vegetation (Fig. 1.7), are zones of great geomorphic complexity. This complexity is due partly to the varying intensity of glacial and, or, periglacial processes superimposed upon older landforms which still bear relics of events of Tertiary time, and partly to varying degrees of continental and maritime climatic influence during the Quaternary.

The glacial and periglacial processes and landforms of the present are the key to an understanding of many fossil features which survive in areas which now have temperate climates. The analogies between Pleistocene and Holocene phenomena must, however, be used with caution, for the periglacial zones of Pleistocene times cannot be assumed to have had the same characteristics as those of the present. During the coldest parts of the Middle and Late Pleistocene glacials, between one-third and one-half of the land surface experienced air temperatures which regularly fell below freezing-point, compared with one-quarter at present. The periglacial zone, however, was not merely moved towards the equator, for at the southern boundaries of the ice sheets conditions were quite different from those of the modern polar regions. The primary difference was that the length of night and day was the same as it is now. Thus the long, dark, extremely cold winters of the present tundras and polar deserts may have been characteristic of unglaciated Siberia during glacials, but in the mid-latitudes winters were only a little longer than those of the present, and sum-

mers probably fairly warm. Areas of true tundra vegetation were probably very limited in extent and available soil moisture may have had more influence on vegetation cover than did temperature.

The northern hemisphere temperate zones, because of predominantly continental influences, and in western Europe because of the diversion of the North Atlantic Drift to low latitudes in glacial times, experienced far more extreme climatic changes than the much smaller areas in the southern hemisphere, which were glaciated only in alpine areas and which, because of their maritime locations, suffered little or no periglacial influences. These southern hemisphere areas, consequently, have the imprint of Tertiary climatic influences, or tectonic and volcanic events of the Quaternary, clearly preserved in their landscapes.

Geomorphic inheritance from the Tertiary

The extensive erosion surfaces of the southern hemisphere continents are, perhaps, the most impressive examples of Tertiary landscapes which have survived, with only minor modifications, through the Quaternary. They have been subjected to repeated climatic changes, and particularly to variations in humidity. The area of such surfaces which can be considered as being in the modern mid-latitude climatic belts is small, but the inheritance of deep-weathering profiles upon extensive plains around the southern fringe of Africa and south-eastern Australia, and the preservation of these profiles upon raised and domed uplands in Australia, is a major feature of these modern landscapes. In South America tectonic movements and volcanism continued throughout Tertiary times in the Andean cordillera, and had a marked effect upon the cratonic areas to the east through the deposition of terrestrial deposits by rivers draining from the Andes, and also by deposition of airfall tephra. In New Zealand tectonism and volcanism is so young that most of the landscape is Quaternary and much is Late Quaternary in its detailed form.

It has been claimed by Büdel (see Fig. 1.12) that considerable parts of the European landscape are essentially Tertiary in their main outline, with Quaternary events being responsible only for detailed modifications. In considering such views we should remind ourselves of some of the main geological events of Cenozoic times so that these

Table 18.1   Cenozoic Epochs, Time and Summary of Major Geological Events

Era	Period	Epoch	Radiometric age (My)	Major geological events
CENOZOIC	QUATERNARY	Holocene	10 000y	Postglacial
		Pleistocene	1.8	Low-altitude glaciations take place in NW Europe, Greenland, and N. America. Snowline in equatorial regions is lowered.
	TERTIARY — NEOGENE	Pliocene	5	Hominids begin to make tools in Africa. Main uplift of the Jura in Europe; deposition of the Crags in East Anglia.
		Miocene	24.0	Messinian evaporites form in the Mediterranean. Main folding in S. Alps and widespread regression in NW Europe. Reactivation of basement faults causes drape-folding of later cover rocks in S. Britain. Glaciation in Antarctica.
	TERTIARY — PALEOGENE	Oligocene	37.5	Thick molasse sequences are emplaced in the central Alps.
		Eocene	53.5	Main metamorphism and deformation occur in the E. Alps. Deep-water sedimentation takes place in the North Sea where rapid subsidence occurs. Periodic transgressions from the North Sea inundate the adjacent sub-tropical lowlands. Volcanism ceases in Britain and becomes centred on the Atlantic Oceanic ridge.
		Paleocene		Thermal doming associated with rifting between Greenland and Rockall is reflected by regional regression and emplacement of turbiditic sands in North Sea. Intense igneous activity occurs in Greenland, Faroes, and NW Britain.
				Rifting between Greenland and Rockall culminates with the emplacement of new oceanic crust (*c.* 60 My) and the N. Atlantic Ocean is born.
				Mammals rapidly diversify after the extinction of the dinosaurs.

Source: Modified from Anderton *et al.* (1979).

views can be put into perspective (Table 18.1). The opening of the Atlantic was associated with widespread volcanism which has left its imprint in E Greenland and northern Britain. Uplift of northern and western Britain, and subsidence in the south-east, began in the early Tertiary, and this style of deformation has continued since, with the largest-scale movements in the Pliocene. The uplift was not simple: in NW England, for example, the Irish sea was downwarped and the Lake District gently domed, while the Pennine blocks were raised by movements occurring along reactivated faults which may have originated in the Triassic.

The rise of NW Britain provided sediment for deposition in the North Sea and to the SE of Britain. The most widespread marine transgression occurred in the Eocene and covered the whole of E Britain, the Low Countries, N France, and NW Germany. The monotonous mudstone known as London Clay dates from this time, and it grades upwards into sandier beds. The sedimentary basins around the southern end of the North Sea experienced many transgressive–regressive cycles of marine deposits, including red-mottled kaolinitic clays which probably derived from emergent uplands in W Britain, which were experiencing weathering and erosion under a tropical climate. It is, then, in SW Britain and in related uplands that any landforms relict from early Tertiary times must be sought.

By the Miocene, broad uplifts and a general marine regression in many parts of Europe had resulted in a paleogeography similar to that of today. Miocene earth movements were the culmination of the Alpine orogeny, although deformation and metamorphism had occurred in the late Eocene and early Oligocene in the E Alps. In the Jura fold belt the main movements took place in the Pliocene. In Britain Miocene movements may have magnified effects which were related to the early Tertiary opening of the Atlantic, with further eastward tilting and the production of a drainage pattern of which such rivers as the Thames, Humber, Tyne, and Forth are descendants. In SE England and NW France broad open folding of Cretaceous and early Tertiary sedimentary rocks was largely controlled by movements along deep seated faults.

Tectonism has resulted in doming, faulting, and folding of many land surfaces with resulting differential erosion of higher and steeper areas, and the creation of new drainage lines and modification of old ones. Marine transgressions have covered lowlands, and after succeeding regressions left new surfaces exposed. Terrestrial deposits have infilled basins with river deposits, or molasse north of the Alps, and with organic deposits in Germany and Poland. Tertiary landscapes are consequently likely to be tilted, fragmented upland relicts, sometimes exhumed from cover beds, but lowlands will have a cover only of the latest sedimentary phase.

Examples of upland relict surfaces with pockets of residual tropical soils with kaolinized pallid zones beneath ferricrete remnants, or tropical red earth soils, are fairly common in the unglaciated uplands of Europe, especially in the Variscan blocks such as the plateau of east Devon (see Plate 17.7) in England. Silcrete remnants, called *sarsen stones* in England, occur widely in Europe and North America, although they are often redistributed by periglacial processes, and are regarded as evidence of former tropical climates, although the degree of humidity required for their formation is unknown.

Examples of remnants of tropical etchplains, with hills of inselberg type standing above deep basins are described from the Sudetes Mountains of Poland, where deep weathering and double planation are regarded as being responsible for their original form. It is probable that such features may have experienced 70 My of tropical climatic regimes and as little as 2 My of stripping and remodelling in temperate and periglacial climates.

The Pleistocene glaciations

The Pleistocene is commonly regarded as being synonymous with the present ice age. The inadequacy of this view has been discussed in Chapter 16, but it is confirmed by the recognition that, although the earliest continental glaciation of North America may have taken place in the late Pliocene, in parts of Europe the first signs of intense (glacial) cold did not occur until after 0.7 My ago.

In England this cold phase is represented by sands and gravels, at Kesgrave (near Ipswich), left by an early course of the River Thames. These sands and gravels may be either glacial outwash or periglacial deposits. In England there are only three sets of tills separated by interglacial deposits, so it is unknown if highland Britain has experienced more than three glacial episodes. The first, or Anglian Glaciation, left tills as far south as Essex, St Albans, and Finchley, north London (Figs. 18.2, 18.3). The extent of the next, Wolstonian, Glaciation is unclear, although it is roughly along a line from Bristol to London, but in the process of ice advancing towards the Cotswold escarpment water was trapped to form glacial Lake Harrison covering much of the triangle of the Midlands, with Birmingham, Leicester, and Evesham at its apices.

The Last Glaciation, known as the Devensian in Britain, was cold with a tundra vegetation, but with periods of interstadial warmth. The Devensian ice sheet built up very quickly, perhaps in about 10 000 years in Britain, but not until very late in the Glacial. At many sites there is clear evidence of a long and climatically complex early glacial period without ice cover. In continental Europe ice sheets built up first in Scandinavia and only after about 35 000 BP did ice spread across the Baltic into N Germany and Poland. During the peak of the Weichselian (the Last Glacial of Europe), periglacial processes were active over all of Europe north of the Alps and spread south into N Spain under the influence of cold maritime conditions, as winter sea ice and icebergs reached latitudes which are now subtropical (Fig. 18.4, 18.5). A belt of polar desert with cover sands and outwash stretched from The Netherlands around the margin of the ice sheet; across central Europe there was a belt of loess steppe and tundra with riverine gallery woodlands along the lower Danube, and forest species were limited to small refugia in sheltered basins of the Mediterranean uplands.

As an example of the geomorphic activity of a relatively brief episode of ice sheet formation in the Last Glacial the British zone may be considered in modest detail. Reconstructions of the British ice sheet indicate that, at its maximum extent, there were probably extensive areas of basal freezing with, presumably, little or no erosion or till formation; yet many areas of northern Scotland, the Lake District, and Snowdonia, like Norway and the Finger Lakes of New York, show

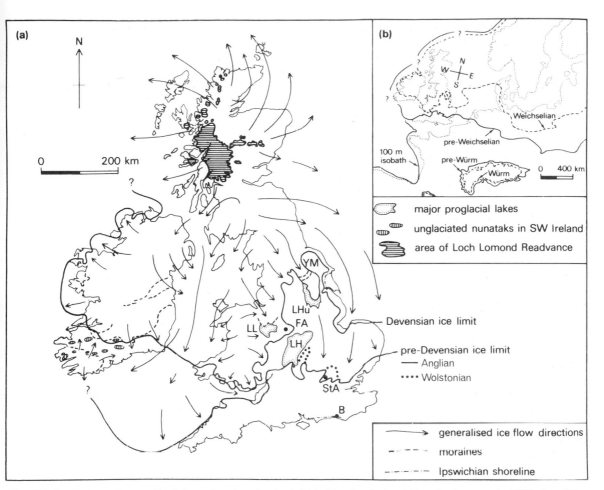

Fig. 18.2 Ice limits and generalized flow directions in the British Isles (ice limits follow Boulton *et al.*, 1977). Localities: B—Brighton; FA—Four Ashes; LH—Lake Harrison; LHu—Lake Humber; LL—Lake Lapworth; StA—St Albans; YM—York Moraines.

clear evidence of deep scouring and glacial trough formation. The probable explanation, for the lack of correspondence between ice sheet characteristics at the glacial maxima and erosional and depositional bedforms, is that most of the erosion, and hence most of the till formation, occurred during the growth, and again during the wasting, of the ice sheet. Because the last event leaves its imprint most clearly we assume that the detailed landforms we see today were modelled to their present form during the final stages of the Last Glacial: this would have been during the period 20 000–14 000 BP.

The most intense erosion has occurred in the highland areas where ice scoured the divides and

ice streams cut open troughs. Several areas of lowland near the margins of the ice-sheet have also been extensively modified and eroded. Such areas lay beneath ice at the pressure melting-point and on soft rock, consequently large blocks of chalk and clay were eroded and transported, and tunnel valleys and other glaciofluvial forms were produced. The volume of till and associated gravels has been estimated at about 2000 km³. In eastern England the volume of the Chalky Boulder Clay alone is about 300 km³. Much of this is chalk, with a matrix of Jurassic clay and thus locally derived, so it seems most probable that the clay vales and chalk cuestas of the Midlands and East Anglia have been extensively modified by the ice sheet. It

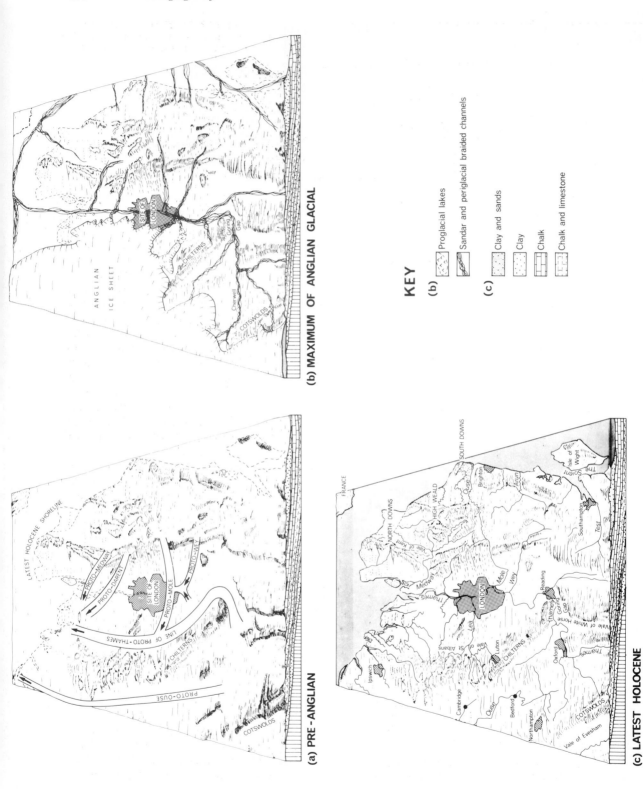

(a) PRE-ANGLIAN

(b) MAXIMUM OF ANGLIAN GLACIAL

(c) LATEST HOLOCENE

KEY

(b)
Proglacial lakes

Sandar and periglacial braided channels

(c)
Clay and sands

Clay

Chalk

Chalk and limestone

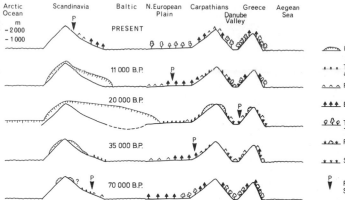

Fig. 18.4 Ice volumes and vegetation cover in north-west Europe during the Last Glacial (data from Starkel, 1977).

is thought that the cuesta south and east of Cambridge (see Fig. 17.7) has been lowered by 100 m or more by ice scouring and the scarp has been cut back by 3 km. The resulting deposits cover a belt 10–20 km wide on the dipslopes of the chalk cuesta.

The extreme highland erosion may be explained by intense abrasion as the ice thinned, and became warmer, at the end of the Last Glacial; the ice-marginal zones of erosion could have been active at the maximum of the Last Glacial. Less easily understood is the limited abrasion over such

areas as the Southern Uplands of Scotland and the Pennines (Plate 18.1a,b). It is postulated that such areas have a relief which is too low to support local ice caps and glaciers when a large British ice sheet does not exist, and during full ice sheet maxima the ice was cold at its base over these uplands and erosion was minimal. The lack of erosion in the lowlands of Buchan, NE Scotland, similarly may be due to a location beneath the centre of an ice sheet during glacial maxima and to a lack of local ice during periods of less extensive ice cover. Such conditions could account for the deeply weathered soils which survived beneath parts of the Scottish ice, and for the lack of glacial erosion over much of the Irish lowlands.

Deglaciation was well advanced by 14 000 BP but a major readvance, at about 11 000 BP, caused a new ice cap to form in Scotland as the Loch Lomond Readvance; this is broadly contemporaneous with the Ra moraines of eastern Norway, the Central Swedish endmoraines, and the Salpausselkä moraines of Finland. The Loch Lomond Readvance limits are shown on Fig. 18.2a and the Fennoscandian belt lies roughly along a line from Oslo through Stockholm to just north of Helsinki. The period lasted for about 800 years and was a time of severe periglacial activity and many talus deposits, ice-wedges, and solifluction deposits were formed during it. Over most of Britain a tundra environment was re-established.

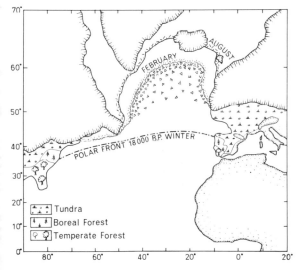

Fig. 18.5 Positions of the ice front in the North Atlantic for summer and winter at the glacial maximum. Continental ice sheets are shown with hachured borders. Pack ice is shown with crenulate borders, and loose winter ice floes with triangles. The position of the polar front indicates the southern edge of cold water masses and coincides closely with the northern limit of boreal vegetation on land (data from McIntyre *et al.*, 1976).

Periglacial influences

At, and beyond, the ice sheet margins the dominant deposits are those carried by glacial meltwaters. They are necessarily confined to lowlands and valleys, but there is good reason for

(a)

(b)

18.1 (a) Slopes in the Southern Uplands, Scotland, showing a lack of glacial trough development or scouring, but smoothing of the slope form by gelifluction debris. (b) The Pennines near Malham showing little evidence of severe glacial erosion.

contending that they, and the loess derived from them, are the most important of all periglacial phenomena. These deposits are stratified with rapid vertical changes in bedding and texture as a result of seasonally variable discharges of meltwater. The outwash streams braid and anastomose so that there is wide variability in sediments, with gravels forming bars and silts and clays being flushed away. Inclusions of till, and other complexities, are common near the ice margin. Away from the ice margin the channel and sediment patterns become less distinctly glacial. In changes of climate, sediment supply and discharge will certainly leave some record in the deposits of river valleys, but whether a clear record is left, as flights of terraces, depends upon many factors, especially upon rates of uplift or depression, and upon the

inland limit of sea-level change influences (see Fig. 10.14). Where uplift rates are high, terraces can form during both glacial and interglacial periods as a result of tilting and channel erosion; it may then be evident, from terrace-forming deposits, which climatic regime was dominant during their deposition. Because Middle and Late Quaternary glacials have been five to eight times as long as interglacials there is a high probability that most terraces will be of glacial age. In an area of downwarping, like the Rhine delta (Fig. 18.6) and Hungarian Plain, terraces will not form because of continuous sedimentation. In areas of little or no vertical movement the situation may be complex with some terraces being of climatic origin and others due to channel migration. To assume, as has sometimes been done in the past, that a terrace

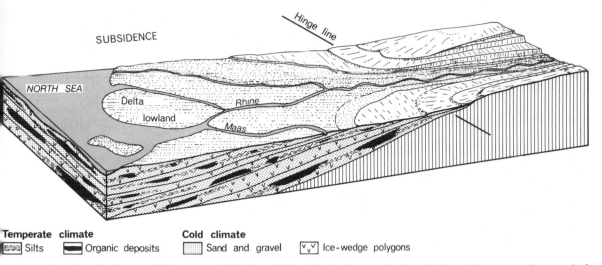

Temperate climate | Cold climate
[Silts] Silts   [Organic deposits] Organic deposits   [Sand and gravel] Sand and gravel   [Ice-wedge polygons] Ice-wedge polygons

Fig. 18.6 A schematic diagram of the terraces of the lower Rhine showing how continuing subsidence has preserved a record of climate and geomorphic processes in river deposits.

necessarily results from a glacial climatic stage is unwarranted.

In Britain, as in most formerly glaciated areas, flights of terraces are best developed in regions beyond the ice margins. Rivers like the Thames have terrace sequences developed over much of the last half million years. Within the areas which were glaciated complex terraces may exist, but these are almost entirely related to the wasting of the last ice sheet and they cannot be correlated from one basin to another.

Even in well-studied valleys, terrace recognition, dating, and correlation may be difficult. The middle and upper Thames terraces around Oxford, for example, are not satisfactorily correlated with those further downstream because in the Goring Gap (Figs. 18.7, 18.3), between the London and Oxford areas, there are few terrace remnants. In the London Basin high-level gravels, or 'Plateau Gravels', include late Tertiary deposits and outwash deposited when the Thames was diverted from its old course, through the Vale of St Albans, into its present valley by ice of the Anglian advance. Several of the terraces exposed in excavations within London contain interglacial faunas and floras, as do some around Oxford, but the same terraces may also contain glacial gravels and periglacial structures indicating that they are, in part, of cold climate origin. It is probable that many terraces are composed partly of cold climate gravels, with warm climate organic deposits and silts filling old channels, oxbow lakes, and forming river-bank deposits.

Drainage diversion is a characteristic of the Quaternary. Some of the best examples occur at the margins of Scandinavian ice in Poland and Germany. Channels here may be formed by meltwater streams which flowed towards the North Sea, and, at the latest stages, towards the Baltic, but they may also be solely, or in part, due to thermoerosion and bank collapse, as ice-wedges and permafrost were undercut by channel flow. Many of the depressions, called *pradolinas* in Poland and *urstromtäler* in Germany, are broad depressions in places 20–25 km wide. They have evidence of lateral channel migration across their floors and may be linked to each other by low passes. They cannot be called ice-marginal channels in any strict sense as they may be many kilometres from former endmoraines. Parts of many of these pradolinas are now followed by Holocene rivers, like the Elbe, and they have been used as routes for canals (Fig. 18.8). In USA much of the Ohio River course has a similar origin.

In addition to variations in type of load, and channel pattern, rivers of the periglacial zone have experienced many variations in discharge. One cause of this variation is the reduction of total precipitation in glacial times, but a greater warm season runoff as both snow and ice melted. The extreme cases of underfit channels in the English Cotswolds (Plate 18.2a,b and see Fig. 10.16)

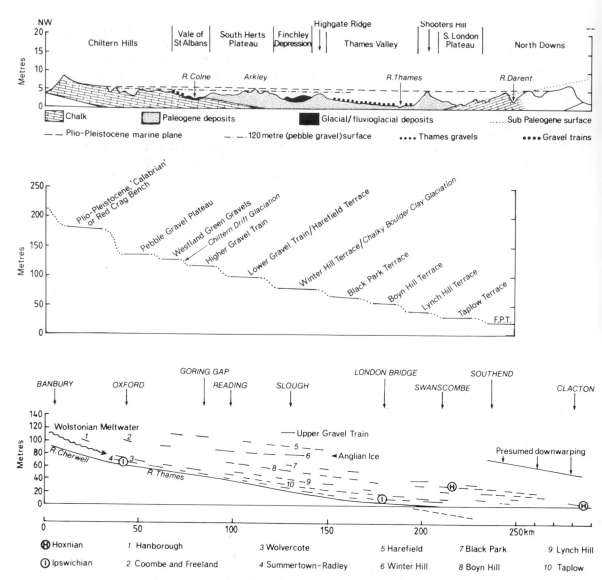

Fig. 18.7 (a) A cross-section of the Thames Valley showing erosion surfaces, location of gravel trains (see Fig. 18.3), and terrace deposits. (b) A theoretical complete sequence of terrace levels in the Middle Thames. (c) Longitudinal profile of the Thames and its terraces (a, b based on Wooldrige and Linton, 1955, with modifications by Jones, 1981; c, based on Clayton, 1977). The Hoxnian and Ipswichian are interglacials.

suggest that discharges must have been at least 50 per cent greater at various times in the past than they are today. An example of an underfit channel in the Black Forest, Germany, shows evidence of severe solifluction and permafrost phenomena and is thought to be a predominantly periglacial form. Gelifluction on the slopes, thermoerosion at the base of the slopes where streams impinge against

them, and abrasion by ice floes, may all have contributed to exceptional valley widening and contributed to the formation of periglacially modified valleys. The underfit nature of modern channels may thus be a result of meltwater floods in valley floors and potent periglacial processes on the valley-side slopes.

A second effect in the periglacial zone is that due

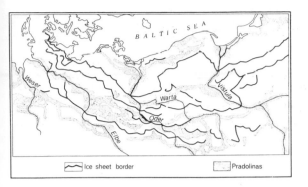

Fig. 18.8 Pradolinas (urstromtäler) in the Central European Plain (data from Woldstedt, 1950).

to permafrosted and hence impermeable soil. It is most probable that many, now dry, dipslope and scarp valleys in the French and English chalklands were cut during periods of permafrost with surface flow.

Relicts of periglacial processes

Relict, or fossil, periglacial features are widespread in Western Europe. They are most common in a zone which includes all of southern England, France north of the Seine, the Lowlands, the whole of central Europe north of the Alps and Black Sea, and most of western USSR, but much

18.2 (a) An underfit stream channel occupying a broad valley, with large meander wavelengths, in the Cotswolds.

(b) an underfit stream channel in a broad trough-like valley with gelifluction infill, in the Black forest, southern Germany.

18.3 (a, b) Ice-wedge casts in Scotland
(photos by E. A. FitzPatrick).

less of eastern USSR. In North America fossil periglacial features occur northwards of the southern margin of the Last Glacial ice sheets, but, with the exception of loess and sand deposits, few distinctively periglacial phenomena occur south of the ice margin, except in uplands. In the southern hemisphere there are few relict periglacial phenomena, apart from loess and some gelifluction deposits in New Zealand and Argentina, and some gelifluction deposits also in Tasmania and on the Snowy Mountains.

The presence of numerous ice-wedges and polygons in western Europe is clearly indicated by the numerous fossil cracks which are now filled by debris which has slipped into the space left by melting ice-wedges (Plate 18.3a,b,c). The presence of ice-wedge casts is the best indicator of former permafrost and it can be seen from Fig. 18.9 that, as the ice of the Last Glacial wasted, permafrost spread into residual tills and sediments. Many of

these ice-wedges are thus dated at 14 000 to about 10 000 BP. Permafrost necessarily implies a negative mean annual temperature, and according to some authors this must be at least $-2\,°C$. Ice- and sand-wedge polygons may require $-5\,°C$, open-system pingos $-2\,°C$, and closed-system pingos $-6\,°C$.

By far the most obvious result of periglacial processes is the smoothing of lower slopes and deposition of blankets, and fans, of gelifluction material at the base of slopes. This smoothing effect is very obvious in hill areas like the Southern Uplands of Scotland (Plate 18.1a). The dominance of loess and gelifluction deposits over all other relict periglacial forms is understandable because these two sets of phenomena are not necessarily associated with permafrost. The distribution of gelifluction-like material (also known as *head* in England) is so widespread that it cannot unequivocally be ascribed to a periglacial origin unless it

(c) Crop patterns in Suffolk, England, indicating the presence of fossil periglacial polygons in the soil below (photo by J. K. S. St Joseph. *Copyright*: Cambridge University).

is associated with features which are uniquely periglacial.

Other slope forms and processes relict from periglacial environments are asymmetrical valley-side slopes, and landslides produced within an active layer on slopes which are now at too low an angle for sliding, but which once suffered very high pore water pressures during summer thaws.

Temperate areas beyond the periglacial zone

In much of Europe mean annual temperatures during the coldest part of the Last Glacial were 10–18 °C below those of the present. This extreme difference accounts for the severe change in environment there, but in southern hemisphere areas, and in south-eastern USA, the temperature contrast is much smaller. In New Zealand, for example, the lowlands of the South Island had annual temperatures only 5–6 °C below those of the present, and in the northern North Island the difference may have been as little as 2 °C. As a result there is no evidence of former permafrost in the South Island and the only periglacial features are loess, gelifluction, and fluvial deposits. The southern part of the North Island lost its forest cover and was subjected to severe slope and channel erosion, with resulting periods of fan building and channel infilling and instability. The northern part of the North Island retained its tree cover over the lower hills and lowlands. These forests changed their species composition in response to climate, and the treeline was lowered. The slopes may have evolved throughout the Quaternary without losing their vegetation cover for climatic reasons. As a result, weathering profiles on interfluves are still deep, often up to 20 m on sandstones and argillites, and well-weath-

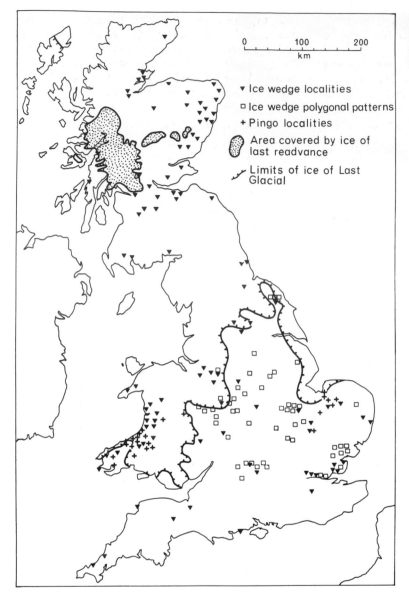

Fig. 18.9 Selected periglacial features in Britain, the limits of the Loch Lomond Readvance and of the Devensian ice (data selected from Williams, 1975, Watson, 1977, and Hutchinson, 1980).

▼ Ice wedge localities
□ Ice wedge polygonal patterns
+ Pingo localities
Area covered by ice of last readvance
Limits of ice of Last Glacial

ered volcanic ash beds have survived from middle, or even early, Pleistocene times on slopes of less than 12°. It is evident that forest removal under human influences has had more extreme effects upon slope erosion than appears to have occurred during the Last Glacial.

Holocene landform changes

Retreat of the Weichselian ice (= Wisconsinan in North America) from its most advanced position was well underway by 14 000 BP, but full deglaciation was not accomplished in Labrador and some other limited areas until after 7000 BP (see Fig. 16.27). Minor interstadial readvances, like the Loch Lomond, interrupted deglaciation, but by 10 000 BP most parts of the mid-latitude zone had a climate which was much the same as that of modern times. There was a time lag in the re-establishment of vegetation and soil cover, which was variable in length because of local drainage conditions and the variable rate of migration of plant species.

Large areas of the mid-latitude zone became forested as climate improved. The predominant geomorphic processes were consequently fluvial, and any fluctuations in the intensity of those processes should be recorded in the lowest terraces and floodplains of Holocene rivers. In Europe there is clear evidence of several periods of overbank deposition in many valleys, with flood-plain silts being deposited upon the soils of earlier

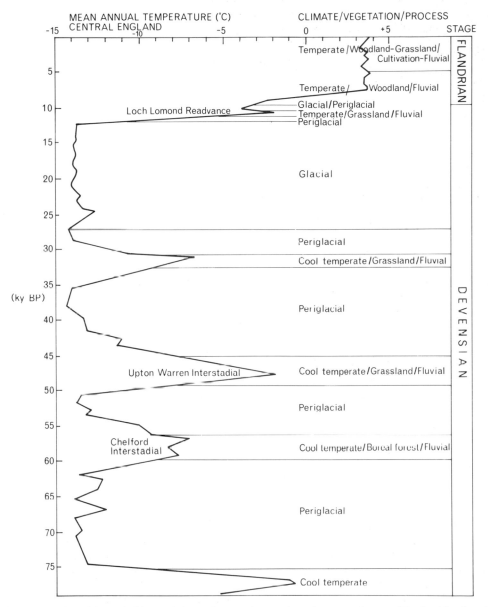

Fig. 18.10 A reconstruction of the probable set of alternating morphogenetic environments in central England for the colder part of the Devensian and for the Flandrian. The temperature curve is derived from oceanographic data presented by Sancetta *et al.* (1973), and the temperature scale is based upon data in Watson (1978). The temperature curve may imply that conditions in the early Devensian were colder than they actually were. Boreal forests might imply mean annual temperatures 3 °C colder than the present, and peak glacial times were at least 10 °C colder than the present.

floodplains. Many of these periods of deposition have been associated with periods of slope erosion under human activity, especially with the spread first of Bronze Age and then of Iron Age cultures.

Some of the clearest evidence for naturally induced phases of erosion and deposition in valleys comes from North America, where human interference with the landscape was limited until the eighteenth and nineteenth centuries. The best available evidence suggests that there were major periods of fluvial erosion at the beginning of each of the cold periods of Little Ice Age type.

Conclusion

It was believed by W. M. Davis that the temperate mid-latitude environments are the major zones of fluvially dominated landscapes, and that they provide the best examples of the 'normal' or fluvial cycle of erosion. The record of severe climatic changes, which has been unravelled mostly since Davis's death, shows that the mid-latitude zones have had temperate climates for only a small part of Cenozoic time and that the imprint of first tropical and subtropical and then glacial and periglacial processes is very strong in most mid-latitude areas. The influence of climatic change and associated changes in morphogenetic environments may be illustrated by the pattern of events interpreted for places like central England lowlands (Fig. 18.10) for the Late Pleistocene only.

Whilst we can now recognize the severity of morphogenetic changes, and the rapidity with which they occurred, we are still unable to evaluate the magnitude of landform change produced by a single suite of processes. Especially, we cannot distinguish the effects of successive glaciations from each other, nor can we evaluate the importance of the numerous and lengthy periglacial episodes. It does appear, however, that the least effective morphogenetic episodes are those associated with the warm climate intervals.

Further reading

The readings suggested for Chapters 14, 15, and 16 are relevant to this chapter. In addition the papers in Brown and Waters (1974), Shotton (1977), and Jones (1980) provide many useful detailed studies. Tertiary relicts are discussed by Büdel (1979), by Isaac (1981), and the Sudetes Mountains by Jahn (1980). The volume by Jones (1981) gives a particularly useful review of the geomorphology and Quaternary history of south-eastern England. Australian landforms are discussed in Jennings and Mabbutt (1967) and Davies and Williams (1978). New Zealand landforms are described in Soons and Selby (1982). Holocene valley changes are discussed by Brakenridge (1980). The distribution of permafrost in certain mountain areas is described by Gorbunov (1978) and by Harris and Brown (1981).

19 Landforms of the Subtropics and Tropics

The semi-arid, savanna, and tropical forest environments are usually treated as separate entities in geomorphological studies because the dominant processes in their modern (Holocene) evolution are distinctive and of different intensity in each region. Evidence obtained in the last 20 years shows that these regions have shared many major fluctuations of climate, and hence in suites of geomorphic processes, during the Cenozoic, consequently they have features and relict forms which are best understood from a unified study.

Climate history

The Tertiary record

Evidence from studies of oxygen isotopes (Chapter 16) and independent geomorphological and biological data show that the rainforests were far more extensive during much of Tertiary time than they are now. The geomorphological evidence for relict weathering profiles, ferricretes and alcretes on extensive planation surfaces, especially of the Gondwanaland remnants, is widespread even in areas which are now arid. Many examples are not datable but in a few places like the Mojave Desert, California, Miocene basalts (8–9 My old) (Fig. 19.1) lie upon deep-weathering profiles formed in granitic rocks. The arid climate, under which has occurred erosion and removal of part of the basalt and the weathering profile, must have developed later than by 8 My BP. In this case the change to aridity may be dating uplift of the Sierra Nevada, San Gabriel, and San Bernardino Mountains. Similarly uplift of the Himalayas is probably responsible for increasing aridity in the mid-latitude deserts of central Asia and uplift of the Andes, also in late Miocene to Quaternary times, may account for aridity in northern Argentina.

Aridity may have developed in the central Sahara in Oligocene times, or even earlier, but it seems most probable that the first major dry phase in the northern region was associated with the Messinian Event of the late Miocene. The Mediterranean at present is an area in which, overall, evaporation exceeds precipitation, and the Sea is maintained by an inflow from the Atlantic. The closing of the link with the ocean, through the Straits of Gibraltar, allowed the Mediterranean to dry up and arid conditions, with saline lakes in depressions, to become widespread. It is reasonable to assume that the extensive arid zone of the Mediterranean region spread its influence into the northern Sahara. After the Mediterranean basin had again become marine, it is probable that more humid conditions influenced north Africa, but the uplift of the Atlas Mountains and declining sea temperatures during the Pliocene probably caused semi-arid conditions to become dominant. The best available evidence for the southern Sahara also indicates that the Pliocene was a time of gradually increasing aridity, with severe aridity becoming widespread during the Quaternary.

In the southern hemisphere, cooling of Antarctica was accompanied by an expansion of the westerly wind belt and more vigorous atmospheric circulation, but dry conditions are not recorded in Australia until mid to late Miocene times when evaporites formed in lake beds, and pollen in lignite beds show deterioration of forest covers which had been of a type now found in rainforests of New Guinea and New Caledonia. The Pliocene in Australia is not well known, but it seems to have been distinguished by wide swings of climate, but overall trends to drier and cooler climates.

The west coast deserts of the southern hemisphere have all been influenced by upwelling of cold water of the Peruvian Current (also called the Humboldt Current) off the Atacama, the Benguela Current off the Namib, and the West

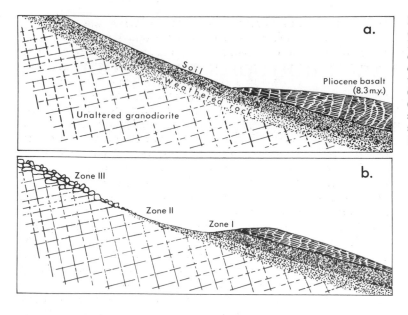

Fig. 19.1 Diagrammatic representation of conditions at one site in the Mojave Desert: a. Immediately after emplacement of the basalt. b. Present conditions, showing the relict weathering preserved beneath the basalt (Zone I); a partly stripped surface (Zone II), and exposed relict core stones forming the modern boulder surface (Zone III) (after Oberlander, 1972 © Univ. of Chicago).

Australian Current off Western Australia. From these areas, only the Namib shelf deposits have been studied in detail.

The Benguela Current was probably initiated in mid Miocene times (10—14 My ago) as the East Antarctic ice sheet developed, and there is evidence in the sediments off the Namib coast of increasing organic carbon production in the upwelling waters by late Miocene times. Desiccation of the Namib and increasing dryness in the Kalahari region (Fig 19.2) are features of the Pliocene. In the Kalahari the lower terrestrial sedimentary beds (see Fig. 17.4c) all have evidence of sub-humid to semi-arid climates, and only the surface Kalahari sands indicate truly arid climates. In the Namib there are no residual deep-weathering profiles or ferricretes and the fluvial deposits were laid down by rivers rising on the interior plateaux, so direct evidence of former morphogenetic environments and climatic change is lacking.

The Pleistocene inheritance

It has been recognized for most of this century that the arid zones have experienced many variations in humidity. The evidence for wet periods is provided by lake beds, evaporites, fluvial deposits, fans, tufa, shells, bones, wood and charcoal, paleosols and pollen in terrestrial deposits, especially in areas now too arid to support animal and plant life

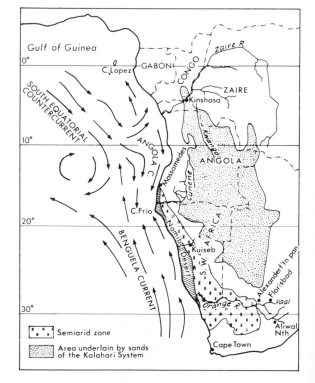

Fig. 19.2 The extent of desert, and deposits of desert orgin, in southern Africa. Note that the northward extent of the Namib is controlled by the extent of the Benguela Current, and that the Kalahari Sands reach across the Zaïre River although they are now covered by forest and savanna woodlands in the northern area (after Selby *et al.*, 1979).

of the kind represented by the fossils. Arid phases are most obviously represented by active dunes and evaporites and where dune and humid climate deposits are interbedded the pattern of variations can be recognized.

Until radiometric dating became available, correlating between scattered and isolated deposits was highly conjectural and caused much controversy, with the result that many interpretations of the relationships between glacial climates in mid-latitudes and arid or pluvial climates in the subtropics may be found in the literature. In rare sites the dating of basalts by K:Ar and of lake beds and mollusca by U:Th ratios has provided ages of deposits, but most ages are derived from ^{14}C dates and, as these are seldom reliable for materials older than 40 000 years, it is only the deposits of the Holocene and the second half of the Last Glacial period which can be correlated widely from region to region, or even locally. The longest datable records of detailed climatic changes have been obtained, by inference, from ocean-floor muds recovered in cores taken from off the Saharan coast in the eastern equatorial Atlantic (Fig. 19.3).

In sediment from cores it was found that the abundance of the clay mineral illite is nearly uniform throughout the sediment and, as it has no climatic significance, illite can be used to estimate the variations in quartz deposition. The quartz is from dust blown off the Sahara and carried over West Africa by the Harmattan winds (NE Trades). An increase in dust, as indicated by the quartz variations, suggests that the glacial periods were either windier, or drier, than the present (or both). Whichever interpretation is adopted it can be seen, from a comparison with the record of sea water temperature derived from studies of ocean fauna in the North Atlantic, that quartz dust is abundant in cold periods and least abundant in periods of warm climate. There is thus a direct correlation of climate between a sub-tropical desert and northern hemisphere water masses for the last five glacial cycles. Similar results have been obtained from cores taken from the Caribbean.

Terrestrial evidence for former more arid climates is widely available in Australia, the Kalahari, West Africa savannas, and north-western India. Fossil dunes (i.e. those no longer active) (Plate 19.1a,b) are commonly vegetated and stable, they may be gullied, have soils formed on them, and their lee slopes degraded to angles

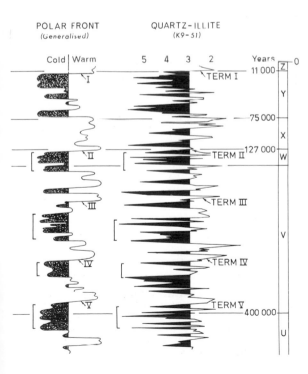

Fig. 19.3 Comparison of quartz/illite ratios, in core K9-51 from the equatorial Atlantic, with generalized oceanic polar front curves, indicating water temperatures, for the North Atlantic. Shading is arbitrary. Correlations are indicated by brackets. The letters in the right-hand margin refer to foraminiferal zones and are used for dating ocean deposits by correlation. 'Term' is an abbreviation for 'Termination' of a glacial cycle (after Bowles, 1975).

(a)

(b)

19.1 (a) Linear dunes in the Kalahari, with river channels crossing the now-vegetated dunes and small 'pans' showing as rounded white patches. The alignment of the dunes is evident (ERTS-1 imagery by NASA). (b) A longitudinal dune in the Simpson Desert, Australia, with plants on its flanks but with an active crest. In the foreground is the bed of an intermittent stream and an exposed interdune surface with a 'gibber' lag.

below those of the angle of rest of sand—commonly 32—33°. In many sites interdune areas have been flooded by lakes, and hence have lacustrine clays in the depressions, and shorelines cut into the dune flanks. In parts of Lake Chad many dunes are now islands in the shallow lake waters. Vegetation is usually effective in restricting dune move-ment where annual precipitation exceeds about 100–300 mm but fossil dunes occur in West Africa and Western Australia where annual rainfall exceeds 800 mm.

In several parts of Africa fossil dune systems cross large river channels; in the Sudan fixed dunes, known locally as *Goz* cross the Nile;

Kalahari sands, now silicified and covered by forest, extend north of the mouth of the Zaïre (=Congo) River (Fig. 19.2); similarly the middle reaches of the Niger River now cross a dune field. All of these examples imply that the rivers dried up during a period of active dune migration. This period was consequently one of lower rainfall, and possibly great windiness.

Exceptionally high lake levels may result from higher rainfall, lower temperatures, and hence less evaporation, or, rarely, from some interference with the outlet of a lake. Repeated changes in lake levels are usually climatic phenomena. In Africa south of the equator, many depressions in the Kalahari were once filled by lake waters. In Botswana, Lake Ngami formerly had an area of between 500 and 1000 km² and paleo-Lake Makarikari had an area of 34 000 km². Under present

climatic conditions the annual discharge of the Zambezi River, or an input of 40 km³, would be needed to sustain such a lake. In east Africa many lakes occupy the floor of the East African Rift between the Danakil Depression in Ethiopia and Lake Malawi in the south. These lakes, and Lake Victoria, have provided a record of fluctuating levels for the Late Quaternary which is among the best indicators of paleoclimates (Fig. 19.4).

The importance to people of the high lake level periods, which were certainly also high rainfall periods, and hence are called *pluvials*, is that: (1) expanded lake beds and wide floodplains have provided extensive arable soils, especially where these can be irrigated; and (2) they have provided large bodies of groundwater in areas now lacking surface flow. It has been estimated that in the Sahara–Arabia desert zone there are some 65 000

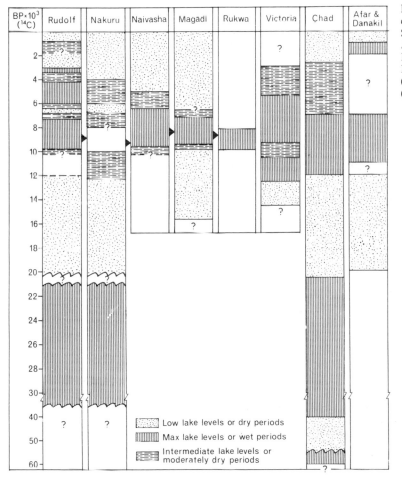

Fig. 19.4 Fluctuations in lake levels in east Africa and the southern side of the Sahara. Data: for the smaller east African lakes from Butzer *et al.* (1972); Lake Victoria from Kendall (1969); Lake Chad from Servant and Servant (1970); and Danakil from Gasse *et al.* (1974).

km³ of groundwater. Over much of this area infiltration is impossible, or limited to sites where floodwater from exotic rivers, like the Nile, can infiltrate permeable sedimentary rocks. Locally extreme rainfalls or floods in *wadis* (i.e. valleys with intermittent streamflow) may recharge groundwater. Numerous ¹⁴C dates of groundwater have confirmed that most Saharan water is 'fossil' and fell as precipitation in the period 40 000–25 000 BP. The period 20 000–9000 BP was hyper-arid with no recharge, but the period 8000–5000 BP produced considerable recharge, especially where streams from uplands supplied water which infiltrated in lowland basins.

At present about 10 per cent of the land area between 30 °N and 30 °S is covered by active sand deserts. It is evident from recent research that in the period 20 000 to 17 000 BP nearly 50 per cent of the land in this belt had active sand blowing. For the Climatic Optimum around 6000 BP the evidence is less direct but it is possible that only the Thar and Namib had extensive active dune fields (Fig. 19.5). The duration of the major dune-forming period at the end of the Last Glacial was variable. It may have lasted only 2000–3000 years in the northern Sahara and some 6000–8000 years in the southern Sahara. The relationship between cold glacial conditions in subpolar latitudes and aridity in the subtropics around 20 000 BP is very clear. The modern period is arid but less so than that at the end of the Last Glacial.

The subtropical pattern of climatic change indicated from lake levels is the opposite of that in mid-latitude mountain-basin deserts, like those of south-western USA. There the relationship between lake benches and moraines of mountain glaciers, and deltas of meltwater streams, shows that the lake levels were high during glacial advances. This pattern is the reverse of that of the subtropics. The lakes of the Great Basin were originally called 'pluvial lakes', implying that they were caused by higher rainfall. Recent studies suggest that snowmelt and glaciermelt, together with lower temperatures, and hence reduced evaporation, are responsible for the lakes and account for their coincidence with glacial advances.

Similar high glacial period levels are known in Eurasia for an enlarged Caspian Sea, which extended 1300 km up the Volga River from its present mouth, and united with an expanded Black Sea through the Mantych Depression. In the Middle East, 'Lisan Lake' extended along the

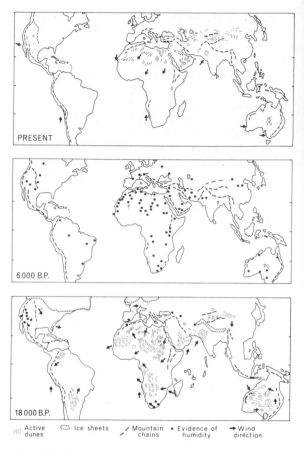

Fig. 19.5 The extent and alignment of major ergs at three periods. The evidence for great eolian activity at 18 000 BP is overwhelming, and most major ergs, fossil and active, have alignments created at that time. The Climatic Optimum at about 6000 BP was a time of greater humidity and vegetation cover, especially in the tropics and subtropics, and eolian activity was suppressed. Continental outlines at 18 000 BP represent sea-level lowering of 120 m (based on Sarnthein, 1978).

Dead Sea Rift Valley linking Lake Tiberias (= the Sea of Galilee) and the existing Dead Sea, with a water level at −180 m compared with the present −400 m of the Dead Sea, at 18 000–12 000 BP. Snowmelt and reduced evaporation probably explain these high levels.

Evidence for Quaternary climatic change in the humid tropics is almost exclusively confined to that for the last 40 000 years. For periods older than this, there is a strong implication that climatic change has occurred with the periodicity which is now being deciphered from the evidence in arid and semi-arid areas. The evidence is of two kinds—biological and geomorphic.

The biological evidence is best known from the Amazon basin and from the rainforests of West and Central Africa. In Africa and South America there are a number of areas of the rainforest which are particularly rich in species of plants, birds, butterflies, lizards, and small mammals with limited range. These areas include eastern Liberia, the western Ivory Coast, Gabon, Cameroun, an area of the Ruwenzori (Fig. 19.6), and a considerable number of localities around the margins of the Amazon Basin. It is postulated that these areas were refuges in which forest species survived during periods of drier climate when savannas and woodlands extended into what are now forest zones. This type of biological evidence is not directly datable and may be interpreted as an indication of one major event or as being the last in a sequence of drier and wet climatic periods.

The geomorphological evidence is of two kinds:

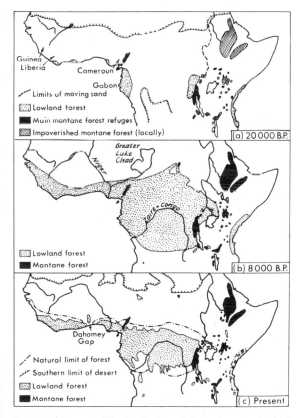

Fig. 19.6 The probable distribution of forest in central Africa at 20 000 BP and 8000 BP compared with the present (after Hamilton, 1976). Note that the existence of Greater Lake Chad at 8000 BP is now questioned.

indirect and direct. The indirect evidence is that the Sahara and Kalahari extended their influence into the areas now occupied by humid tropical climates. Similar evidence exists for semi-arid climates in what are now savannas in the Orinoco lowlands of Venezuela, where grasses and woodlands grow on relict sand dunes.

Direct geological evidence for South America comes from ocean cores and some from landforms. Cores from off the north-east coast, in the areas affected by sedimentation from the Amazon and Orinoco Rivers, show that Holocene sediments are predominantly rich in gibbsite and kaolinite—two clay minerals which form under conditions of humid tropical chemical weathering—and the content of feldspars is less than 20 per cent. Below the Holocene boundary in the cores, the sediments are rich in feldspars—up to 60 per cent—and low in the two clay minerals. The feldspars are thought to be derived from rocks of the Brazilian Shield and to have been produced in semi-arid climates with predominantly physical weathering. The feldspars could not have been derived from the Andes because feldspars would be physically broken down during such long transport. Because of the high rate of sedimentation off the Amazon mouth, the longest cores do not extend far back into the Last Glacial, and evidence for earlier climates is lacking.

In southern and south-eastern Brazil many of the valley-side slopes have stepped profiles. Cuttings through the steps have revealed that the bedrock may have a deep regolith formed in place by weathering; and that this saprolite is commonly buried by alternating layers of bedded and unbedded material, with stone lines and paleosols separating many of the beds. These step surfaces are interpreted as valley-side pediments formed in semi-arid climatic conditions. The flights of steps are interpreted as being indicative of alternating periods of pediment formation, and periods of channel incision and linear stream erosion under humid climates in which forest cover caused most slopes to be relatively immune to mechanical erosion (Fig. 19.7).

During pediment formation under semi-arid climates with thin—possibly shrub savanna—vegetation, landslides and mudflows occurred on steep back-slopes and left beds of unsorted colluvium, with weathered pebbles and corestones in a matrix of strongly weathered saprolite, on flatter footslopes. The colluvium was reworked by sheet-

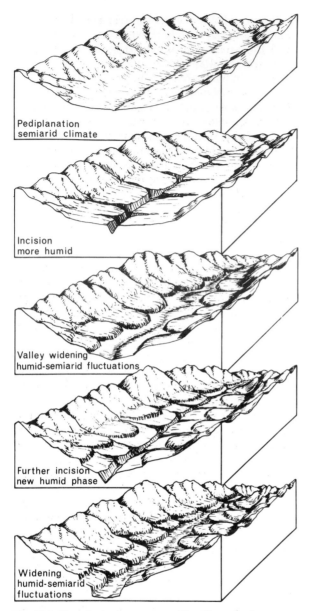

Pediplanation
semiarid climate

Incision
more humid

Valley widening
humid-semiarid fluctuations

Further incision
new humid phase

Widening
humid-semiarid
fluctuations

Fig. 19.7 Hypothesized sequences of landform development, in south-eastern Brazil, resulting from climatic changes during the Quaternary (redrawn and modified from Bigarella, 1969).

and rill-wash leaving layers of bedded silts and sands and with pebble lines marking the old rills. Some areas have stone lines of quartz pebbles indicating that surface wash has left an armoured surface after selectively removing the fine sediment (Plate 8.10). These processes are responsible for the formation of low-angle pediments with sur-

faces of erosion cut across weathered rock, and infills of colluvium and wash deposits in the hollows and at the lower ends of the pediments.

The phases of humid climate are thought to be periods of deep chemical weathering, soil formation across the pediment deposits and older saprolite, channel incision along the formerly shallow drainage lines, and the formation of floodplain deposits in channel floors, the latter deposits being the best surviving evidence of the humid environments. Tectonic uplift has doubtlessly played a part in the creation of this stepped landscape. Support for the correlation of semi-arid pediment formation with glacial periods is provided by evidence that the lowest pediments, and their covering sediment, dip below the Holocene interglacial high sea-level. The lack of firm dating of the pediments, however, prevents exact correlations being made and some workers have suggested that the dry periods are of late Tertiary age.

Evidence from Sierra Leone, West Africa, indicates similar patterns of landform change in the monsoon rainforest. In central Sierra Leone, however, abundant wood and organic matter for radiocarbon dating has been exposed in cuts made into valley floors in searches for alluvial diamonds. It is evident that lowland rainforest covered much of the area during the middle part of the Last Glacial to about 20 000 BP, when active slopewash occurred beneath a semi-arid or dry savanna vegetation; forest was re-established during the Holocene Climatic Optimum and then disturbed in the period of human occupance since about 3500 BP. Similar patterns of climatic change are interpreted from sites in northern Angola where alternating periods of sand movement and stable soil development under humid climates are dated.

In south-east Asia climatic changes, in the presently humid tropics, may not have been as severe as those in Africa and South America. Some evidence suggests that the rainforests may have not merely persisted but spread across the newly exposed continental shelves during the glacial maximum, and so had a much greater extent than they have now. There is evidence, in the terrace sediments of many river valleys, of considerable bedload movement alternating with periods of deposition of reworked saprolite. Much of this evidence is undated and interpretations from it of climatic change, and especially of forest–savanna transitions, may not be well founded. Alternative hypotheses may include tectonism and volcanism as the causes of some changes in depositional environments. Much research is still required.

Summary

It is evident that most modern arid and semi-arid areas originated in Late Miocene to early Pliocene times. The implication of this discovery is that nearly all of the deserts have landforms which owe their main hill, plateau, and plain forms to events and humid climate processes of the early and mid-Cenozoic. Ergs and small landforms like fans, alluvial flats, terraces, playas, and wadis are likely to have formed during the late Pliocene and Quaternary and have surfaces which have attained their present form in the last 40 000 years. It is improbable that a study of modern desert processes alone will tell us how the present landscape evolved. In the humid tropics climatic deterioration has probably led to phases of severe erosion followed by stability under re-established forest. Such alterations have certainly had an effect upon valley-floor sedimentation and may well have contributed to etchplain stripping.

Classes of landform

Morphostructure and present-day climate are the two most commonly used criteria for defining classes of landforms in the subtropics and tropics. The most extensive landform units occur on the ancient erosion surfaces of the Gondwanaland relicts, and more limited units occur in hill and mountain country where steep slopes, and climatic variations with altitude, contrast with the valleys and basins. Climatic criteria have the underlying assumption that there is a close relationship between modern landforms and modern climates. We have seen that this is not entirely valid, but it remains a useful distinction as long as the significance of relict, or paleoforms, is recognized.

Rainfall amounts, intensities, and periodicities provide useful classification criteria where they are related to present vegetation, depth of weathering, and dominant processes. Such relationships are, however, often complex or unclear, and in the case of many savannas there is evidence that natural and man-made fires may be of more importance in maintaining distinctive vegetation types than is climate alone.

The World Meteorological Organization climatic aridity index is an example of a bioclimatic classification for dry climates. It uses the ratio of annual precipitation to evapotranspiration (P/ETP): for the hyper-arid zone of virtually no vegetation P/ETP < 0.03; for the arid zone of sparse vegetation P/ETP ranges from 0.03 to 0.20; for the semi-arid zone of steppe and shrubland P/ETP ranges from 0.02 to 0.50; for the sub-humid zone of savannas and steppes P/ETP ranges from 0.50 to 0.75. In the humid tropics monthly variations of rainfall are more important and the formula P/4T (where P is mean monthly precipitation in mm, and T is monthly mean temperature in °C) may be used to distinguish between wet and dry months; in wet months P/4T > 1; in dry months it lies between 1 and 0.5; in very dry months it is < 0.5.

The equatorial forest zone has an annual rainfall of at least 1200–1500 mm; it has four seasons with rainy periods about the equinoxes, a short relatively dry season of a month or so, and a longer 2–3 month relatively dry season. These dry seasons are not long enough to restrict plant growth, and rainfall during them may exceed 80 mm per month, but they do cause a decline in stream discharges, a decline in groundwater tables, and may permit some precipitation of solutes in the soil profile.

Away from the equatorial zone the dry seasons are commonly longer, and at least one of them may be long enough for small streams to dry up and the water table to fall to such an extent that some plant species are excluded. With increasing seasonality of rainfall, water tables, stream discharges, precipitation of colloids in soil profiles, and production of kaolinitic clays all become more variable during a year. The transition to semi-arid conditions is recognizable from geomorphic evidence by the disappearance of kaolinitic weathering products and red soils, and by the appearance of desert varnish at the same places as savanna vegetation gives way to discontinuous grasses, shrubs, and succulent plants.

Many of the features of arid zone rainfall can be identified in a study of that of the Sahara (Fig. 19.8). Along the northern fringe of the desert is a narrow belt of Saharan steppe with low shrubs and much bare stony ground. This steppe has an annual rainfall of about 80–300 mm. The rainfall distribution is typical of the Mediterranean climatic zone (e.g. Béchar and Tripoli) with winter rain and dry summers. Along the southern margin of the Sahara is the Sahel where rainfall has a

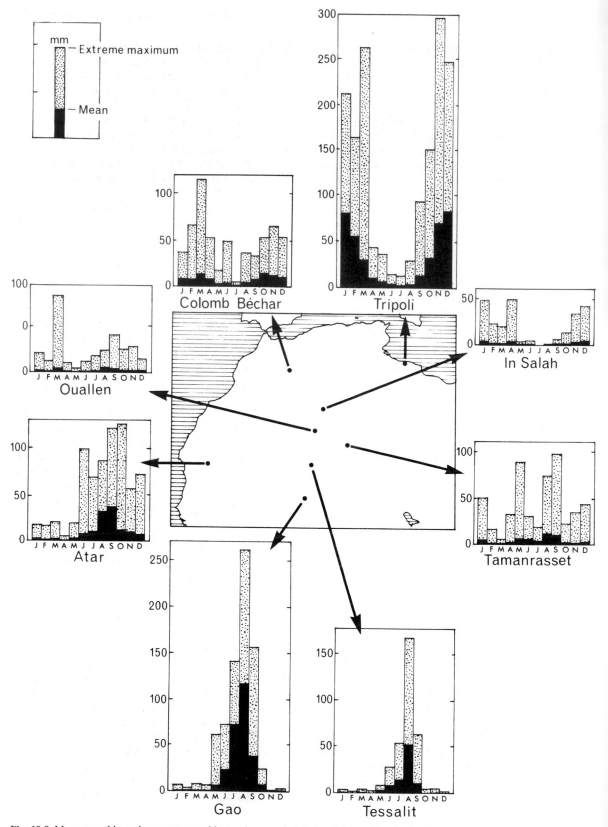

Fig. 19.8 Mean monthly and extreme monthly maximum rainfalls for eight stations in the Sahara.

summer maximum, caused by annual northwards migration of the Intertropical Convergence Zone (see Fig. 16.7), and an annual total of 80–500 mm. The natural vegetation is grasses and tall thorn shrubs with little bare ground. Between the northern steppes and the Sahel the Sahara has low rainfall, commonly less than 20 mm, and little or no vegetation. Over most of this extremely arid region cyclonic rain may fall at any season under Mediterranean or Gulf of Guinea air masses. The outstanding feature of all semi-arid and arid zone rainfall, however, is its irregularity and the wide departure of the extreme rainfalls and droughts from mean values. Figure 19.8 shows that in many months extreme rainfalls depart from mean values by 100 per cent and, rarely, by up to 300 per cent. Total droughts in the Sahara have been known to last for 8 years, and in the Atacama for 18 years. Extreme rainfalls are accompanied by very large runoff totals, severe floods, and rapid geomorphic change.

Deserts like the Namib and Atacama have coastal belts of prevailing fog and low cloud 5–30 km wide. These fog belts, and the chilled air from off the ocean, bring unusually cool conditions and heavy dew on most days in the year. Most continental deserts, however, have high sunshine hours, often 80 per cent of the theoretical maximum, and low relative humidities in the range of 5–40 per cent. Potential evaporation is, therefore, very high and up to 5 m in the central Sahara. Intense solar radiation and clear skies cause rock, pebble, and sand surfaces to become 10–15 °C warmer than the air temperature in the afternoon, but several degrees colder at night.

Vegetation, soils, and weathering mantles

Even within an area with a presumed uniform climate desert surfaces have very variable distributions of plants. Vegetation is usually most abundant where soil moisture is available and sites are stable. The flanks of inactive dunes, and the sediment bars of wadi floors, may consequently have a close plant cover, but neighbouring rock pavements have few plants. Seasonal variation in plant, root, and litter cover of soils may be extreme. Sparse covers expose soil to rainsplash and deflation, and rock or large clasts to extremes of temperature change and eolian abrasion.

Where fluvial activity has been dominant a surface of rounded pebbles lying upon fluvially bedded sands and gravels forms the ground surface. (In the Sahara, surfaces with rounded pebbles of alluvial origins are called *serir*, and those with more angular pebbles, which are derived by weathering from local rock, are called *reg* (Plate 19.2a, b); alternatively 'serir' is used for any gravel and pebble surface of the central Sahara, and 'reg' for similar surfaces elsewhere.

Soil profiles are often lacking in organic matter and clay minerals, and characterized by weak infiltration of water so that deposition of soluble salts occurs in the upper part of the profile. The accumulation of salts may, however, be a very slow process, as is illustrated by the study of calcrete profiles which have been approximately dated by the estimated ages of terrace surfaces in which they have formed (Fig. 19.9). During climatic fluctuations soluble salts may be translocated to greater depths in the profile under greater infiltration, or the surface friable soil deflated in dry periods to leave an impervious crust which armours the ground against other processes.

In some areas patterned ground can form if salts with high volume changes from dehydrated to hydrated forms, such as bassanite ($CaSO_4 \cdot \frac{1}{2} H_2O$) to gypsum ($CaSO_4 \cdot 2H_2O$), are present. Such surfaces are not common but are usually associated with the margins of sebkhas, or salty sediments in wadi floors (Plate 19.3). Where salt-rich deposits are in contact with fresh rock, salt weathering processes may be very effective in producing undercut slopes, tafoni, and pedestal, or mushroom-shaped, rocks.

In semi-arid climates, with seasonal variations of soil moisture, montmorillonite-rich soils may undergo extreme swelling and shrinking with wetting and drying. This may lead to the formation of raised mounds in a regular form—called *gilgai* in Australia. Mounds are usually 1–10 m wide and can be regular or irregular. On slopes they may form a terracette pattern.

Rock surfaces of deserts are commonly seen to be pitted and to be suffering granular disintegration, flaking, and spalling of various-sized fragments. On the flat surfaces of sedimentary strata of the Sahara tesselated *desert pavements* or *hamadas* develop loosened blocks, and the ground around outcrops of igneous rocks is often littered with fresh-looking rock fragments or a grus of single crystals. The abundance of cavernous hollows, overhanging visors, pits, pans, and rills has often been taken as evidence for rapid weathering

(a)

(b)

19.2 (a) Serir of the Plaine du Tidikelt, with the scarp of the Plateau du Tademaït beyond, Algerian Sahara. (b) rounded gravels forming the serir of the Plaine du Tidikelt.

processes, and numerous theories of the effectiveness of mechanical and chemical weathering processes have been advocated. In spite of such work there are few unequivocal conclusions which can be drawn about the effectiveness, magnitude, or frequency of dominant weathering processes. One of the most telling pieces of evidence for a very slow rate of weathering is the survival of cave and wall drawings in the Sahara and Australia for several thousand years, and of detailed carved surfaces on monuments in Upper Egypt. There is not even a clear consensus on whether *desert varnish* (the red-to-black coating on many rock surfaces, which is formed of iron and manganese oxides with clay minerals such as goethite) pro-

tects and case-hardens rock, or whether it is merely a surface discoloration.

Salt weathering, hydration, frost, thermal expansion and contraction are all possible processes of mechanical weathering in deserts, but their incidence and their varying intensity are usually unknown. The decomposition of feldspars in the heart of hyper-arid Sahara presents a characteristic problem—is the decomposition still occurring or did it cease after the last pluvial period?

The erosion surfaces of Western Australia, and parts of the Sahara, have preserved on them clear evidence of an origin under warm humid climates, as ferricretes are found overlying deep-weathering

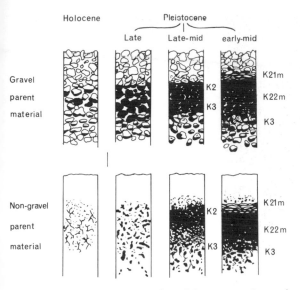

Holocene   Pleistocene

Late   Late-mid   early-mid

Gravel parent material

Non-gravel parent material

K21m
K2
K3
K22m
K3

K21m
K2
K22m
K3
K3

Fig. 19.9 A schematic representation of the nature and rate of formation of calcium carbonate horizons (K) in New Mexico, 'm' denotes massive formations (after Gile, 1975).

profiles, often with pallid zones. Stripping of these regoliths and duricrusts leaves an irregular basal surface with low rocky outcrops standing above a broad etchplain in which the hollows have been filled with eolian and fluvial sediments (Plate 19.4).

In virtually all lowland landscapes in the humid tropics nearly continuous forests and woodlands cover deep, chemically weathered, soil profiles and the few exposures of bare rock are either steep, cliff-bordered, plateaux of quartzite, rounded domes of granitic rocks, or limited outcrops of bedrock in valley floors. A further outstanding feature is the common lack of relationship between the form of the bedrock surface, below the weathering mantle, and the form of the ground surface. It is a common, but largely untested, assumption that these features have developed over much of Cenozoic time.

In areas of uniform lithology, weathering mantles are usually thickest on plains, plateaux, and hilltops where outcrops of fresh rock are most scarce. The mantle thins downslope on valley sides, and it is normally in valley bottoms, where streams have incised, that fresh rock outcrops. The rare exceptions to these generalizations occur in areas of very steep relief where landslides may expose bare rock on slopes, and colluvium may infill valley floors. Such situations have been widely reported in Brazil.

The dynamic processes which form slopes on the topographic surface do not affect bedrock surfaces which develop under mechanically static conditions by chemical weathering alone. Hence inclinations of weathering fronts may be greater or less than those of the ground surface but, because of the greater thickness of mantles on hilltops than in valleys, the bedrock surface usually has less total relief than the ground surface. Consequently planation processes may expose bedrock reliefs of domes, basins and ridges which are unsuspected from inspections of the ground surface.

Hillslopes

In arid areas the rate of weathering is usually slower than the potential rate at which debris can be removed from a slope by wash or wind action. The slopes are, therefore, weathering-limited. As rainfall increases towards the humid tropics the thickness of soil profiles increases and rock units become covered by a soil. Slope development rates are then transport-limited.

Hillslopes of the arid and sub-humid regions fall into six classes.

(1) *Slopes with forms dominated by structure* are most evident in arid climates because of the lack of a concealing plant and soil cover. These forms include some bornhardts (see Plate 5.24) and various features such as volcanic necks, dykes, and sills, and folded sedimentary rocks. They are prominent forms either because they are extrusive, constructional, forms or because they are more resistant than the surrounding rocks.

(2) *Strength-equilibrium slopes* are widespread because of the lack of vegetation and soil cover. Massive hard rocks with few joints can support nearly vertical slopes and each major change in jointing pattern and lithology produces a change in slope angle. Slopes, consequently, express the characteristics of the rocks faithfully and precisely.

(3) *Debris-mantled slopes* are especially common in arid areas with relict weathering profiles, such as Western Australia, the Mojave Desert, and the southern Sahara. These slopes are now surface exposures of a former basal weathering front overlain by corestones from which the surrounding matrix of fine weathering products has been stripped (Fig. 19.1). Such slopes have been de-

19.3 (a) Large rectangular sand-wedge polygons in the central Namib Desert.

(b) The surface of a small salt flat, in the central Namib, showing the curling and lifting edges of a salty crust, which is drying.

scribed in the older literature as 'boulder-controlled'; but in many cases the slope angle controls the stability of the boulders.

Corestones will not remain in place on bedrock slopes which are steeper than the friction angle of the contact between the boulder and bedrock (usually 32–39 °) and they will break down by weathering, especially where the corestones have already been weakened by chemical decomposition within the regolith. Steep bare bedrock slopes may then be exposed. Such bare slopes are particularly common on granitic rocks where boulders have suffered granular disintegration and the resulting grus has been washed on to the piedmont slope. Exposed bedrock will gradually be converted to a strength-equilibrium angle, but this process is dependent upon the opening of joints

and may take so long that slope segments may remain out of equilibrium for long periods.

(4) *Armoured mantles*, or regoliths, are most common in semi-arid areas which have suffered severe swings of climate from sub-humid to arid conditions, and in areas of low precipitation but relatively high humidity such as the coastal areas of the Atacama of northern Chile (see Plate 19.5). A pit dug into such mantles reveals chemically weathered saprolite with secondary clay minerals and, sometimes, strong coloration by iron oxides. Within this regolith large clasts survive. Surface washing by Hortonian overland flow, or deflation by wind, may gradually remove the fine clays, silt, and even sands from the mantle and leave a surface of coarse sub-angular debris which effectively armours the ground against further denuda-

(a)

(b)

(c)

19.4 (a) The head of Oued Tamanrasset in the Hoggar Mountains. The oued has a wide sandy floor and low terraces which date back to the last major wet period. (b) Part of the Namib Desert plain inland from Walvis Bay. The bedrock is schist, which dips at a steep angle indicating that this is an erosional surface. The regolith is merely a veneer of fractured schist fragments. (c) In the background is a low scarp cut into an ancient, deeply weathered profile. The cliff is a case-hardened remnant of a lower pallid zone. Induration of the cliff probably occurred in an arid phase during late Cenozoic erosional episodes. The glacis in the foreground is cut across a granitic grus. Western Australia (photo by C. W. Finkl).

(a)

(b)

19.5 (a) Hills of the Atacama Desert, inland from Antofagasta, Chile. Note the rills on the slope faces, the piedmont fan, and the gravels of the lowland. This area is now extremely arid and there is no vegetation. (b) The ground surface, of the hills shown in (a), is an armour of coarse angular gravel and boulders overlying a chemically weathered profile containing many angular clasts. The head of the valley is rilled.

tion, unless severe rill and gully erosion can cut through the armour during either extreme rainstorms or wet periods. In mountainous areas the armour may also be produced if coarse fragments are pushed to the surface by ice lenses or swelling and shrinking of clays (see Fig. 14.7): in such cases the regolith becomes progressively depleted of larger clasts.

(5) *Talus slopes* are rare in warm arid areas because the rates of slope recession are slow and high scarps, which are necessary as sources of talus debris, are only common where streams have incised steep gorges into the bedrock (Plate 19.6), or large escarpments have retreated over long periods. As taluses commonly have angles of 28–34°, they have angles which are markedly lower than the bedrock scarps above them. Several conclusions can be drawn from this juxtaposition.

(a) Where cliffs retreat to maintain strength equilibrium angles, at very slow rates controlled by weathering rates, talus debris may be broken down relatively quickly and washed or blown away. Rock slopes will therefore not retain a complete debris cover—this is the usual condition in the Namib and many other deserts.

(b) Where cliff retreat is relatively rapid, or cliff height (and hence the supply face) is large, talus will accumulate. In the situation shown in plate 19.6a, the talus may form a thin veneer over bedrock which will have an angle controlled by the talus angle. Such slopes are called *Richter denudation slopes*. They develop as particles fall from a cliff on to the top of a talus. If the newly fallen debris just covers a little of the base of the cliff the next fall will be over the new talus and hence the base of the cliff will now be higher, so the cliff will recede by a series of minute steps at the angle of the talus. The series of minutes steps forms on essentially straight slope cut across the bedrock below the cliff. The bedrock slope may be progressively buried by debris or it may be revealed if the talus is later removed. Richter denudation slopes are more common in semi-arid areas and in mountains than is usually recognized.

(c) Where talus production is very rapid, as it often is in mountain and basin deserts where cold winters promote severe mechanical weathering, debris may cover the lower slopes and be incorporated into large debris cones and alluvial fans. Some thick taluses are certainly relict from cold climates of the Last Glacial.

(6) *Badlands* (Plate 19.6b) and glacis of erosion

are not common in most deserts, but the spectacular landscapes of the former have attracted considerable attention. Weak rocks such as marine mudstones, but more commonly terrestrial beds such as folded and uplifted lake beds, playa deposits, loess, and alluvium are readily rilled and gullied by concentrated runoff. Their erosional forms are most readily developed in semi-arid areas with a sufficiently high rainfall to produce runoff, but with a rainfall too low to support a complete soil and vegetation cover.

Glacis may be produced in several ways. In arid areas they may be strength equilibrium slopes formed on rock with such low mass strength that the bedrock slope angle is 5–15 °C. Such glacis most commonly form below a protective caprock, but if the caprock is eventually worn away the whole hill has this low angle. More commonly, perhaps, glacis may be wash slopes which are periodically inundated by sheetwash and shallow channels: in this case they are essentially pediments cut across weak rocks.

Slope processes in semi-arid areas may be relatively rapid where regolith and weak rock are exposed. Surface wash on upper slopes may form a nearly continuous sheet of water which is disturbed only by the impact of falling raindrops. Such sheets form zones, on watersheds, of little or no erosion. Further down slopes water films thicken, become turbulent, channelled, and erosive; exposed soil and sediment can be readily transported to the piedmont and the slope may develop a very high density of drainage channels—as have badlands. Considerable experimentation with natural and simulated rainfall has been undertaken on badland slopes but it is doubtful if the processes recognized there are truly representative of arid or humid environment conditions.

Rockfalls are important processes on steep cliff faces—even though they may be rare events—but other forms of mass wasting are not. The lack of clay minerals in debris mantles, the lack of prolonged infiltration and high pore water pressures, and the shallowness of most slope regoliths, preclude most forms of landsliding, unless local geological conditions such as weak rocks and earthquakes promote failure. Some large landslides are recognized from desert areas, but most are probably relict from former wetter climates.

On hard rocks channel incision is rare but the lack of infiltration into permeable bedrock may produce extreme overland flow during occasional,

(a)

(b)

19.6 (a) Cliffs on the south site of the Orange River near its mouth. The upper steep cliff is a strength-equilibrium slope, and the lower part is a Richter denudation slope with a thin veneer of talus. (b) Badlands cut in weak sediments near the Salar de Atacama, Chile.

but high-intensity rainstorms. In such cases mud-flows and debris flows may occur on slopes or in wadis. They often leave levees of coarse debris at the flanks of the flow where water seeped away, and debris lobes at the foot of the slope where water was lost by infiltration and the flow ceased to be a viscous fluid, then lost plasticity and eventually became a solid. Because it is usually impossible to date debris flow materials, there is often doubt about whether debris lobes and levees of arid zones are relict from rare storms or from former pluvial periods. In the semi-arid zones, however, they may occur in nearly every wet season and be the major process of debris trans-port.

Hillslope processes in the humid tropics are the same as those which occur in temperate forest and grassland areas, but the controls on their distribution and intensity are related to distinctive tropical climatic and regolith features. The lack of long-term instrumented observations in the humid tropics, and the uncertainty of the representativeness of available data, raise doubts about the validity of many published generalizations.

Splash and surface wash have often been discounted as effective processes in the humid tropics, but there is clear evidence from the presence of miniature earth pillars, accumulations of soil upslope of fallen logs, and miniature rills that they can be effective processes. Splash is probably most effective where there are openings in the tree canopy, where plant litters are thin or non-existent, and where water can drip from the tips of leaves in high canopies onto bare soil. Surface wash may be produced along paths, animal tracks, or where there is nearly continuous soil exposed between plants in forests or savannas—this happens frequently after seasonal fires and can produce lag gravels and armoured surfaces. In forests surface wash is supplied not only by throughfall and drip but also by stemflow.

Piping and eluviation are probably far more important than is usually supposed. Piping develops where water is concentrated in the regolith and can flow along fissures left by decaying roots, along relict joints in the saprolite, or other lines of high permeability. The process of piping washes fine-grained material out of the regolith and this may be deposited as a small fan or patch of silt and clay where the pipe reaches the ground surface. Extension of pipes can cause rill and gully erosion.

It is probable that soil creep is the most important form of mass wasting but few data are available on this topic. Severe chemical leaching may cause volume changes in the soil and promote creep, as may the presence of clays which are wet enough to be close to the plastic limit throughout the wet season. Landsliding is widely reported in areas of steep slopes ($> 30°$ or so). Landsliding is promoted by deep clay-rich, saprolites, by permeable oxide-rich soils, high soil moisture contents, and by sharp contacts, between the saprolite and bedrock, which become slide planes; but sliding is inhibited by the lack of soil cracking and the presence of kaolinitic clays with their low-volume change characteristics and high liquid limits, which indicate that they have to absorb large volumes of water before exhibiting flow behaviour. Landsliding is reported as being a major process in slope development in the uplands of New Guinea (where earthquakes are also a trigger factor), on the basalts of Pacific island volcanoes, on steep slopes with deep saprolites in Malaysia, and on the flanks of many emergent granitic domes, such as those around Rio de Janeiro. Repeated shallow landsliding causes parallel slope retreat and the formation of narrow ridge crests. Such steep slopes with long transportational mid-slopes are widely reported in the hill country of Burma and New Guinea.

Erosion on slopes implies deposition on lowlands. In very deep valleys, as in parts of the Amazon headwaters, debris may be delivered directly to stream channels, but in wider valleys colluvium may fill valley floors and be retained there, especially if a rock bar forms a local base level for the valley. In New Guinea many wide, flat valley floors have deep infills which have been dated, as being several tens of thousands of years old, by the presence of recognizable volcanic ash beds in the colluvium. In many areas of Africa footslopes and shallow basins, called *dambos*, have layers of sands and of stone lines in the colluvium. Stone lines are often regarded as evidence of pedimentation and the fine earth material above the stone line either as a later surface wash deposit derived from fresh slope erosion, or as soil carried upwards through the stone line by termites (Plate 8.10). Such deposits seem to indicate alternating periods of land surface stability and instability which may be related to climatic change, to fires, or to human activity.

Many attempts have been made to classify slopes, in humid tropical areas, according to their form, but it is doubtful if enough slope profiles have been surveyed for this to be possible. As in all morphogenetic environments some slopes are prominent features of the landscape because they are features of geological resistance: the escarpments of the Roraima Plateau, Venezuela, and the western margin of the Jos Plateau, Nigeria, are good examples. The flanks of bornhardts and tors formed in massive granites are common examples of geological resistance (Plate 19.7). Where weathering mantles are thick two general features of hillslopes are commonly recognized. (1) Upper slopes are often broadly convex and lower slopes gently concave with few abrupt changes of slope angle unless a slope has been undercut. Whether or not such forms are indicative of a dominance of creep on the convexities and of wash and colluvial accumulations on lower concavities, as is often asserted, is unclear. (2) Upper slopes with ferricrete caps may have short steep escarpments in the duricrust, if this is stable, or convexities if the ferricrete is breaking away and blocks are being transported, by creep, to lower parts of the slope. Eluviation and piping commonly occur on such slopes, and the cappings may be undermined and sag or collapse. In areas of rapid uplift and incision valley-side slopes may be long and straight with narrow crestal convexities and short basal concavities. Such slopes are common in New Guinea and the Amazon headwaters.

Where bedrock is exposed, as it often is along escarpments formed of quartzite and at the margins of bornhardts, koppies, and tors, coarse talus accumulations are common, even in the rainforest. The debris may be angular, especially where it is derived from bedded rocks, or it may be formed of rounded corestones. The angular debris may be derived from the fracturing of massive and bedded rock as stress release occurs (with or without the formation of curved sheets of rock), or it may result from physical weathering where bare rock is exposed to direct insolation and heated to 70 °C or more and then 'quenched' in rain showers. In savannas severe surface heating, to 200 °C or more, during fires may also contribute to physical disintegration.

Plains and piedmonts

Some plains and plateaux are clearly of structural origin: this is the case with the hamadas of the northern Sahara. Many plains, however, truncate basement structures and are of erosional origin. There are three major theories of the origin of tropical plains (see Chapter 1, p. 19 and Chapter 17). (1) The theory of L. C. King (1967) emphasizes scarp retreat to leave pediments, cut across thinly veneered bedrock in semi-arid and savanna climates by overland flow and weak stream channel incision, and their lateral merging to form pediplains. (2) By contrast E. J. Wayland (1934) described the formation of etchplains, in humid climates, as chemical weathering processes 'etch' the bedrock to produce an irregular basal weathering front which reflects rock resistance, and which is covered by deep saprolite suffering gradual

19.7 Outcrops of granite just south-east of Lake Victoria, Tanzania. The large, roughly cuboid boulders which appear to have developed from the disintegration of a large rock mass are often called 'koppies' in Africa; the more general term 'tor' may also be applied to such features.

removal at the ground surface. (3) The concept of etchplanation was elaborated by J. Büdel, who stressed the separation of the bedrock surface from the ground surface in his exposition of the concept of double surfaces of levelling. For surfaces of levelling to occur either stream channels must migrate laterally and progressively remove topsoil and saprolite, or surface wash must occur on the plains and carry soil into the channels. In the view of many observers, surface wash over exposed soil is a feature of semi-arid climates and of savannas in the dry season, whereas surface wash is not of major significance under closed forest. It is usually concluded, therefore, that either channel migration is of great importance in forests or that repeated climatic changes have caused alternating humid and sub-humid conditions in areas which are now forested, and that surface wash and pediment formation have been particularly effective in the drier periods.

Presently available information does not allow a full analysis of these views, but some comments can be made.

(1) The history of arid plains has to be interpreted from the few remnants, such as duricrusts and weathering profiles, which survive on them, although it cannot be known whether the weathering profile formed in a pre-existing plain, or was lowered into an irregular surface. Such remnants may represent long periods of action of a single set of morphogenetic processes, or relics of many shorter duration, and varying, processes. Such evidence as there is indicates that much of Western Australia has a deeply weathered mantle which is being gradually stripped and that it is an etchplain (Plate 19.4c), or multilevel set of etchplains. Many of the plains of the southern Sahara may have a similar origin.

(2) Surfaces like those of the Namib Desert (Plate 19.4b) are far more difficult to interpret. They appear to be erosional plains left as scarps and inselberg slopes retreated (i.e. pediments in L. C. King's sense). This hypothesis is supported by the fact that all of the inselberg and scarp slopes measured so far are strength-equilibrium slopes, suggesting that the slopes are retreating in such a fashion that strength equilibrium is maintained. If they were exhumed slopes from beneath a regolith, they would not all be expected to be in equilibrium. Against this hypothesis is the apparent lack of morphogenetic activity at present, in an arid, or semi-arid, climate which appears to have been dominant since the late Miocene. Furthermore, much of the inland Namib plain has a thin crust of calcrete suggesting a stable surface formed in a semi-arid climate. An alternative hypothesis is that the Namib is an etchplain, but there is no evidence for this: nowhere are there remnants of chemically weathered profiles (and hundreds of kilometres of shallow trenches have been bull-dozed by mineral prospectors, thus giving far better exposure than is usually available). Thus for the Namib, as for many other areas, we are left with one possible, but apparently rather inactive process of development, and a second possible but totally unsupported process. It may well be that the lack of critical surviving evidence will permit little improvement on this unsatisfactory situation.

(3) Studies of footslopes in humid areas confirm that they are commonly deeply weathered and some even have well-defined *scarp-foot depressions* which are subject to greater wash and gully erosion—from water draining off the scarp—than the lower part of the piedmont. Deeply weathered footslopes appear to be fundamental features of many landscapes and obviate the possibility of scarp retreat across bedrock as the dominant process in many humid areas.

(4) Surface stripping may not be a feature of stable closed forest or savanna ecosystems but a characteristic of disturbances or transitions from one to the other during periods of climatic change.

(5) Etchplanation may, and probably does, involve lateral, as well as vertical, penetration of the weathering front and hence may be largely responsible for the recession of the foot of scarps. Such recession must obviously involve removal of the saprolite by solution, wash, or streams.

(6) Domes and other rises in the bedrock at the weathering front may become perpetuated in the landscape because of initial superior resistance to weathering and then upon exposure, as the saprolite is removed, by resistance to surface denudation. By shedding drainage water to their margins, such features may grow as their surrounds are further etched and lowered: thus some scarp protrusions and inselbergs may originate at the weathering front, but owe their present form to later differential erosion.

(7) The time-scale of planation surface formation is tens of millions of years, and the modern processes may be merely modifiers of features which are fundamentally very old.

(8) Differential weathering and erosion may be more significant in producing surfaces than is periodic uplift.

Pediments of arid and semi-arid footslopes may be morphologically and genetically different from major erosional plains (see Chapter 17).

Basins of deposition

Alluvial fans of desert regions are nearly always accumulated at the foot of mountains which have a sufficiently large supply of debris, and either seasonal or episodic runoff which is great enough to transport that debris to the piedmont. They are usually associated with high tectonic relief in mountain and basin deserts; for example, they flank the growing Andes, the ranges of central Asia, and they are major features of the landscape of fault-bounded ranges and tectonic basins, or *bolsons*, of SW United States (Fig. 17.9). In shield and platform deserts they occur along the margins of rift valleys such as the Jordan—Dead Sea Rift and along the edge of marginal uplands such as the Saharan Atlas and Flinders Ranges of Australia. The forms and processes operating on alluvial fans are discussed in Chapter 10 (p. 290).

Bajadas are coalesced fans which have converged and formed a depositional piedmont plain. The longitudinal slopes of bajadas, or those of the constituent fans, are usually of the order of 1–5° although they may steepen at the head to nearer 10°. The transverse profile along the mountain front is undulating due to the general convexity of the fans and the smooth curve of the profile may be broken by rills and gullies, particularly at the head and toe of a fan; mid-fan areas tend to be associated with sheetwash and are less incised.

Large bajadas may have accumulated over the whole of Pliocene and Quaternary time with vertical thickening of a few cm to 1 m per thousand years. In enclosed basins bajadas will grow progressively until they occupy nearly all of the bolson, and may lessen their slope angle as the supply of debris from the diminishing upland declines and weathering causes particle sizes on fan surfaces to decline. Sheet floods and rills will then transport the debris and infill the lowest part of the bolson. Alternatively fans may be progressively steepened if the rate of mountain uplift exceeds the rate of bolson infilling. The sediment accumulations of bolsons may also become dissected if climate changes, or if the internal drainage of a basin is captured and the local base level is lowered (see Plate 4.24b).

In spite of the foregoing comments it has to be recognized that there is no general tendency for piedmont aggradation under arid climates. On stable shields, upland valleys and piedmonts develop graded profiles over long periods of time and pediments are widespread. The aggradational surfaces of upland basins owe their existence to tectonism and, to some extent, to the lack of external drainage systems.

Alluvial basins occur within the catchments of many tropical rivers. The central Zaïre basin is a very large example of a tropical drainage basin with dominant incision in the headwaters, alluvial infilling in the wide central region of the basin, and an outlet from this basin through a confining upland. In the Zaïre basin this pattern is a result of tectonism. In some other cases the alluvial basin is an inland delta—such as the Sudd of the Nile, the Okavango Swamps of the Kalahari, and the central Niger basin—which may be caused by tectonic warping or subsidence, a declining discharge away from the headwaters, or by a river debouching into a very flat piedmont. Subsidence of the Ganges, Mekong, and Irrawady basins has produced extensive alluvial plains, and rivers like the Fly River of New Guinea have large, flat lowlands produced by enormous volumes of sediment from rising mountains in their headwaters. In contrast to the basins of arid regions, those of humid areas have outlets to the ocean.

Drainage forms and processes

The characteristic topographic break between desert uplands and lowlands is reflected in the respective hydrological regimes. In the hills are connected, close-branching drainage channels cut in rock, descending to gorges with alternating rock bars and gravel fills. At the hill margin, the valleys either remain incised or open into flat-floored re-entrant valleys which widen towards the plains and which are floored by sand fills bearing the imprint of the waning stages of the last flood (see Plate 19.4a).

The lack of infiltration on the bare or bouldery slopes causes very rapid rises in flood discharges, peak flows which have nearly the same duration as the rainfall, and a flood recession which is very rapid because of the lack of storage in the limited depressions, soil, vegetation, or sediment. Boulder and mud torrents of desert uplands are well known for their rapid rise but short duration. They are sufficiently common for the wise desert traveller never to camp in a wadi bed. The periodicity of such floods is varied; they may occur several times a year, or as little as once in ten years. Uplands with snow cover are more likely to have flows with a slower rise and longer duration.

In major floods of small catchments, total storm runoff may exceed 50 per cent of the precipitation, and yields may reach 10–30 $(m^3/s)/km^2$ with peak flow velocities of 7–12 m/s. Such yields and velocities are usually confined to small areas because of the limited areas covered by a single severe storm.

Flash floods in uplands can convey very high sediment concentrations of 20 000–150 000 p.p.m. These values are more than an order of magnitude greater than sediment concentrations in humid environments where sediment is retained by plant roots and the concentration is diluted downstream by increases in water from tributaries. The loss of discharge by evaporation and channel infiltration downstream, in deserts, results in a progressive downstream increase in deposition in the lower sectors of desert drainage channels. The coarse bedload, which may be a very large proportion of the total sediment load, is dropped first and the fine sediment carried farthest. Very high drainage densities are common in deserts, where weak rock is exposed, and this is also a contributing factor to rapid runoff.

In the humid tropics runoff and sediment discharge are relatively reduced by the vegetation cover, and flow regimes reflect the seasonal rainfall distribution.

Rivers

Desert rivers fall into three groups according to their sources. (1) Exogenous rivers rise in humid uplands outside a desert area and either cross it or terminate within it. The Nile and Orange rivers cross the Egyptian Sahara and southern Kalahari respectively, but the rivers draining into the Aral Sea terminate there and are part of an interior drainage system. (2) Another group of rivers rise in uplands within the desert. Where these are fed by large springs or snowmelt, like the Jordan, they may be perennial, but they are generally intermittent and usually terminate in interior basins or on plains. (3) By far the most common streams are

those with ephemeral flow which is produced by rainstorms. The flows of this third class of rivers have many characteristics in common with those of the uplands, except that their hydrograph patterns are likely to have numerous peaks where water from each tributary adds its component.

Unless a river has a long history of channel formation in pluvials, and is incised into bedrock—like sections of the Nile—a desert river is likely to have a relatively broad, shallow channel in the loose uncohesive deposits of desert plains. Floods, therefore, may spread widely, invading interdune areas and covering old lake beds. The spreading of floodwater reduces river transporting capacity, encourages evaporation, and limits the duration of the inundation, although this may last for several years in some extreme conditions. Deep pools in meander bends or rock basins (*billabongs* in Australia, and *gueltas* in the Sahara) may hold water for several years after a flood and be of major importance for people and animals.

The form of channels is related to high sediment concentrations, high bedloads, and lack of bank vegetation. Sand-bed channels widen until the resistance to flow across the bed and at the banks is approximately uniform: the sand bed is readily deformed and so are the weak banks. Width/depth ratios may be as high as 70. Wide sand beds are also likely to produce braid bars and the overall wide channel therefore has low sinuosity. Responses of channel form to climatic change were discussed in Chapter 10 with reference to the Murrumbidgee (p. 279).

Floodplains of sand-bed channels tend to be composed of sand and to be frequently incorporated into the active channel during extreme floods. This incorporation is not merely overbank flooding but involves channel processes so that the floodplain may be formed of alternate longitudinal bars and flood furrows.

As a stream reaches farther into a desert plain, and away from its main source of sediment, the proportion of the load composed of silt and clay increases. The channel bed and banks consequently become increasingly cohesive as a result of deposition and the channel form changes. Width/depth ratios decrease, meandering develops, sinuosity increases, and levees form along the stable banks. In floods levees may be broken and new channels form so that an anastomosing channel pattern is created. Some of the world's best examples are those of the Channel Country in

SW Queensland, Australia. This riverine clay plain is mostly drained, by the Diamantina River and by Coopers Creek, towards Lake Eyre. Only the largest floods reach Lake Eyre, at intervals of ten years or so. The Lake consequently is more aptly described as a playa which is occasionally inundated. Total inundation occurs about twice in each century (Plate 19.8).

In the lowermost reaches of internal drainage channels, it is common for channels to have very low angles of slope and for distributaries to branch out into digitate deltas which may stand higher than the surrounding *flood basin*. Flood basins may terminate against ergs; alluvial bars may be created as water is lost in lakes or in sebkhas (=playas). The clay flats of flood basins often develop hummocky gilgai relief.

Many controversies exist about the nature of the rivers of the humid tropics. One concerns the cause of the drainage network and pattern of incised stream channels found in many areas of low hills in Africa and South America. On one side it is argued that a forest cover is so stable that significant fluvial incision and channel lateral migration are prevented, and consequently that channel and valley forms must have been inherited from savanna-type environments during periods of drier climate. The alternative view is that valley

(a)

19.8 (a) A stream channel with cohesive banks of clay and silt, and (b) a sand bed channel. Both channels are part of the Great Artesian Basin drainage towards Lake Eyre.

(b)

formation by linear incision and lateral migration are both occurring under forest covers.

Quantitative data derived from long-term measurements are seldom available and arguments can not be supported or denied by them. The controversy can be put in perspective by considering the types of surfaces crossed by streams and the nature of the resulting channels.

Long profiles of many tropical rivers crossing planation surfaces and uplands are remarkably less regular than those of temperate zone rivers. Profiles commonly consist of reaches with low gradients broken by rapids or waterfalls. The channels are seldom deeply incised, and even at the edge of a quartzite escarpment a gorge below a waterfall may be very short and weakly developed, but channels do exhibit marked controls on their alignment by bedrock structures, especially by major joints, faults, and lithological boundaries.

The reasons for these features are twofold.

(1) Many rivers crossing partly stripped etchplains have bedloads consisting of quartz sand derived from old sandstones or from the chemically resistant quartz from igneous and metamorphic rocks. Unless they have emerged from mountains, where physical weathering and rockfalls have provided coarse debris, such rivers have no source of coarse debris and can do little abrasion of rockbars and outcrops in their channels. Most such rivers carry a large proportion of their load in suspension because the major source of their load is the clays forming deep saprolites. The only exception to this is the presence of nodular iron oxides which may have formed part of a ferricrete.

(2) The form of the weathering front beneath a saprolite is irregular. A channel which is being incised into a saprolitic cover will consequently be lowered onto bedrock rises along its course, while upstream and downstream of the rise the stream is still flowing over saprolite. If the channel cannot migrate laterally off the rise then it has a permanent rockbar across the channel and this becomes a rapid. Many rivers, even those as large as the Zaïre, have their long profiles interrupted by such bars. Channels are then likely to consist of sections with freely developing alluvial forms alternating with shorter bedrock-controlled reaches.

The Zaïre, and its tributaries, within the central part of its basin has extensive sections of both straight channels with braid bars and meandering channels with stable island forms. The main

channels and their larger tributaries have the same mean depth of about 6 m at flood, implying that deepening of the main channel ceases when the annual discharge reaches about 200 m^3/s. Large rivers then expand laterally by multiple braiding and the formation of several active talwegs. The braid bars are mostly elongated with sharp or triangular ends facing up- and down-stream. The bars are composed of sand and are nearly stable in the forest zone, because of their permanent tree cover, although they do extend upstream by 'capturing' migrating dunes. In savanna zones the long dry season inhibits permanent forest cover and the bars are subject to erosion and destruction. It is probable that the large source of sand has been created by soil and gully erosion associated with cultivation and forest clearance by people. It may be, therefore, that some of the channel forms are the result of human activity.

In both the Zaïre basin and much of the central and lower Amazon basin, levees and jetties of sediment have been built by the main channels across the junctions of their tributaries with the main stream. The tributaries are therefore ponded and their valleys are partly drowned and turned into branching narrow lakes which are gradually infilled with sediment and vegetation. In the Amazon the central channel is a drowned trough, or ria, produced by the Fladrian Transgression. This very wide trough has great depth and is only partly filled with alluvial bars. Aggradation has produced many spits, forelands, barriers, islands, and other depositional features normally associated with coastlines, as well as the more typically fluvial meander scrolls, oxbow lakes, point bars, and islands. Alternating channel erosion during periods of low glacial sea-level and of infilling during high interglacial sea-levels has left a complex pattern of abandoned floodplains and depositional forms along the length of the Amazon trough below Manaus. The presence of sand dunes and channel bars in those tributaries draining the Andes, and of 'black waters', with few bed dunes, draining the forest lowlands of the Brazil–Venezuela frontier zone, point to the probability that channel forms are most closely related to supply of bedload and not just related to forest cover.

Vegetation cover does, however, have a marked effect upon many channel forms. This is most evident from a comparison of savanna zone channels with those of the forest zone (Plates

19.9a,b). In savanna areas the seasonal drying up of streams inhibits the formation of graded profiles and even of permanent channels. Many valley bottoms become seasonally flooded basins through which many irregular streams drain in wet seasons, and over which depressions and vegetation control ponding and drainage ways. Such basins may become areas of deposition during the onset of the wet season and well-defined and incised channels may develop by the

19.9 (a) Dry channel beds amid grasslands, central Namibia.

19.9 (b) River channels near Kisangani, Zaïre, blocked by floating mats of water hyacinth, and a close-up of these plants (photos by K. Thompson).

end of the wet season. In the following dry season wind action, vegetation, animal trampling, and fire may combine to break down the well-defined levees once more.

A modern effect upon many channels is that caused by the introduction of plants, such as the water hyacinth (*Eichhornia crassipes*, Plate 19.9b), into many areas by people. This plant grows so quickly, and forms such dense floating masses, that it may block channels and cause local flooding. Some tributaries of the Zaïre have been virtually turned into plant-covered lakes during this century.

Lakes and playas

Enclosed basins, in which drainage waters accumulate, occur for a number of reasons in deserts. They may be tectonic, as in the East African Rift, and Death Valley, California, where the dislocation is by faulting; or by warping as in the Lake Eyre, Lake Chad, and Chott Djerid, Tunisia, basins. Deflation by wind can produce many small shallow basins with downwind dunes, as in the lunette-bordered basins of SE Australia, or it can produce large basins cut in sedimentary rocks as in the Qattara Depression, Egypt. More common in plains, are the evaporation sites in flood basins where alluvium or eolian deposits may impound water.

In playas subject to deposition from surface floods, the surface is generally 'dry' and formed of silt and clay deposited from the floodwaters, which become increasingly saline as the flood evaporates. In basins supplied by groundwater the proportion of salt to sediment increases and the surface is usually 'moist'. A single large playa may have sediment-dominated and salt-dominated parts.

The thickness of the playa deposits varies from a few metres to hundreds of metres, depending upon the supply of material, the depth of the basin, and the duration of the accumulation. Many playa deposits have layers of lacustrine sediment from pluvial periods and major floods, separated by beds of saline sediment, or pure evaporites, from arid phases of climate.

Lakes and playas near the sea derive much of their salt from sea spray carried inland by wind, and the lakes therefore have a solute composition which reflects that of sea water. Farther inland this 'cyclic salt' origin is increasingly supplemented from other sources such as weathering products,

old evaporites, and second-cycle salt which has passed through soil and sediment. Compared with coastal lakes, inland ones will usually be deficient in chlorides, but enriched in calcium. Many groundwater-fed playas are particularly rich in gypsum ($CaSO_4$) which is found as large crystals in the grey muds and white evaporite beds.

Playa surfaces have a detailed morphology produced by polygon formation as the surface dries, the curling up of slabs of salt and silt (Plate 19.3b), extrusions of salt mounds along cracks or above springs, gas rings and pits formed by escaping air or methane, sinkholes where a buried body of halite has gone into solution, and deflation hollows. Soft ancient lake beds may be cut into yardangs by wind action. Eolian activity is limited, however, at the water table.

Ergs

Eolian sands form the dominant feature of only 2 per cent of the North American Deserts, only 15 per cent of the Sahara, but 50 per cent of the Australian arid zone. The forms of these deposits have been discussed in Chapter 12.

The main source of sand is alluvium, as is shown by the common occurrence of sand desert in lowland basins of interior drainage. This relationship is particularly clear in parts of the Sahara and the Takla Makan. An alluvial origin may be recognizable in the composition of the erg sands, with very red sands of Western Australia and the Simpson Desert being derived from old lateritic terrains, and ferromagnesian minerals being more important in areas with volcanic rocks. Most sands are predominantly of quartz grains with some feldspars, but local influences may be important as in Chad and New Mexico where evaporites, notably gypsum, derived from old lake beds make important contributions to the dunes.

Some eolian sand originates in place from weathering of bedrock, as in the residual sand sheets on Nubian Sandstone in the eastern Sahara, and in parts of Egypt.

Local influences on erg composition presumably diminish with distance of travel of the grains from their source. Some dunefields are clearly rather immobile, but studies of satellite imagery indicate that in the Sahara many dunes are aligned along the trade winds for thousands of kilometres.

It is rare for there to be clearly datable deposits, in ergs, which can be used to indicate the age of origin of the erg as a whole, or of individual large dunes. Inferences may be made from some calculated rates of sand movements. It is estimated, for example, that most large Saharan ergs are at least Late Pleistocene in age because the supply of sand to lowland basins is, at present, far too small for the ergs to be of Holocene age. It also seems probable that the sand accumulation requires a period of active stream transport and deposition in sebkhas, followed by periods of drier climate with predominant eolian activity for sorting sand from fluvial and lake deposits, and then sand transport and accumulation in ergs. These requirements suggest that the wet period in the Sahara at *c*. 40 000 BP, followed by the arid period at *c*. 20 000 BP, may have been of considerable importance in erg history.

The repeated cycles of dust accumulation in the Atlantic, however, indicate that periods of eolian activity are rather regular features of Saharan ergs and many of them may be much older than Late Pleistocene and have passed through numerous phases of inactivity in wet periods and been reactivated in drier periods. The coincidence of many dune directions of advance with resultant wind drift directions, may be taken as evidence either of long-term stability of atmospheric pressure systems or of the readjustment of erg dunes to modern wind directions over a large part of the Holocene.

Conclusion

The range of landform types and processes found in deserts is very large. In the central Sahara, for example, they include: regions of coalescing fans and glacis in the north grading into depressions filled by ergs, or into fluvio-eolian sandplains; modified stripped etchplains with ferricrete residuals and higher inselbergs; around the Tibesti, yardang landscapes shaped by trade winds merge south-westwards into active dune fields and the fixed dunefields of the Chad basin; in uplands such as the Hoggar massif periglacial relict felsenmeer occur at high altitudes and at lower altitudes are areas of active gorge-cutting and wadi terrace formation. Deserts are consequently a complex of ancient and modern forms, often with the modern forms etched into the remnants of the ancient ones. The inheritance from Tertiary times has left extensive landform units out of equilibrium with modern processes; but in other areas, such as the great ergs, the modern climate and the landform are in equilibrium even though the erg may have a mid-Quaternary, or earlier, origin.

In the humid tropics we can distinguish greater depths of weathering and permanent streams as providing features of the rainforest zone, which contrast with the thinner weathering mantles, more extensive duricrusts, sandier soils, and seasonal channels of savannas. The paleoforms and polygenetic features are less evident in the rainforest than in savanna or arid areas, but this may be due to a lack of investigation rather than a fundamental feature. There is still doubt, however, about whether these differences have resulted in truly, and fundamentally, unique forest and savanna zone landforms. Uncertainty about the relative duration and latitudinal extent of the vegetation covers during the late Cenozoic makes it impossible to be definite on such questions. There is also doubt about the relative importance of rapid, compared with gradual, transitions of climate, so that we are in doubt about the significance of periods of instability of vegetation in erosional history.

What is clear, however is that people are probably having a more profound and rapid effect upon vegetation cover and dominant geomorphic processes than climatic change has ever done. The present rate of rainforest destruction exceeds one per cent per year and may well give rise to unique processes, with rainforest zone rainfalls impacting directly onto exposed soils which have no vegetation buffer. Such processes could cause major changes in soil depth, gully erosion, channel alluviation, and rock exposure in a brief period.

Further reading

The two most accessible modern reviews of desert landforms are Mabbutt (1977) and Cooke and Warren (1973). Mabbutt provides a more comprehensive review of landforms and Cooke and Warren discuss modern processes in more detail. The bibliography in Mabbutt's book includes many of the major works in French and German. A review of the evidence for climatic change is provided by Goudie (1977) and specific examples of evidence and interpretation may be obtained from Bowler (1976), Bowler *et al.* (1976), and Rognon and Williams (1977). Many papers on Quaternary events of the Sahara are included in the volume edited by Williams and Faure (1980).

A number of papers on fossil water in the Middle East are contained in *The Quarterly Journal of Engineering Geology*, 15, 2, (1982). Desertification is discussed in Paylore and Hanley (1976). Warren and Maziels (1976), and Grove (1977). A comprehensive review of desert dust is provided in the volume edited by Péwé (1981). Recent evidence from the Chad area is discussed by Durand (1982).

Landforms of the humid tropics are discussed at length in books by Tricart (1972) and Thomas (1974), and many examples are given in the volumes by Garner (1974) and Douglas (1977). Examples of evidence of climatic change are included in papers by Bigarella, Mousinho, and Da Silva (1969), Damuth and Fairbridge (1970), Fairbridge (1976), Thomas and Thorp (1980), and Klammer (1981). Feininger (1971) has given detailed descriptions of deep-weathering profiles, and fluvial processes and landforms are described by Blake and Ollier (1971), Savat (1975), and Tricart (1975). Infilled valleys are discussed by Blong and Pain (1976) and Löffler (1977b) has discussed concepts of morphogenetic stability of rainforest. Reviews of modern research are provided by Douglas (1978), Thomas (1978), and Bremer (1980).

20   The Tempo and Scale of Geomorphic Change

In many discussions of the rates of landform change emphasis is placed upon 'average rates' of denudation, uplift, sea-floor spreading, and other processes. Such figures are valuable because they can give an impression of the rate of long-term change of the Earth's surface. Average rates, however, can also obscure recognition of the fact that many natural processes are episodic in operation, and also that land surfaces may evolve in a series of leaps, with periods of stability followed by brief periods of severe erosion. Variations in rates of change through time are further complicated by variations in space. The plains and plateaux of Western Australia, for example, are thought to be some of the most stable land surfaces on Earth, with low rates of erosion, and few large variations in those rates, because of the low relief, semi-arid to arid climate and the rarity of extreme storm events. Some 2000 km to the north-east the mountains of Papua–New Guinea are rising rapidly; active intrusion of gneiss domes, periodic severe earthquakes, volcanism, and a warm humid climate in an area of very steep relief, all promote extreme instability, with relatively frequent episodes of severe landsliding and flooding alternating with less extreme, but nevertheless high, rates of erosion. In Western Australia much of the landscape is ancient with some paleoforms relatively little changed in the last 100 My. In Papua–New Guinea most of the major landforms are less than a few hundreds of thousands of years old, and most of the land surface is only a few hundred years old.

Landscape stability depends upon the ratio of resisting, or stabilizing, conditions to the forces promoting change. (1) Where these are in balance rates of change are steady; where resistance exceeds the forces of denudation the landscape is nearly stable with weathering permitting the deepening of soil profiles and a store of weathering products is created. (2) Where the forces of denudation are greater than the resistance, erosion carries away soil and weathering products as quickly as they are formed, and the topography is likely to have a cover of thin soils, or to be characterized by rock outcrops. These states are respectively: (1) *transport-limited*, and (2) *weathering-limited*.

Resistance to denudation is a function of rock, soil, and vegetation strength; it is promoted by low angles of slope and limited available relief above the base level for erosion; Resistance is also promoted by a capacity to absorb applied energy. Thus water may be stored in vegetation, deep-weathering profiles, lakes, floodplains, or temporarily as ice. Denudation, by contrast, is most effective where materials are weak, slopes are steep, available relief is high, vegetation cover is thin, climatic energy levels are high (especially rainfall), and periodically extremely high. To this catalogue must be added the observation that erosion rates depend partly upon the availability of transportable soil and sediment. As soil can only accumulate to considerable depths under stable conditions, the most extreme erosion occurs where, and when, there is a sudden change or break in an equilibrium. Thus a catastrophic process crosses a threshold from stable to unstable conditions. Extremely high rates of denudation cannot last for long because they remove the supply of readily eroded soil and sediment (see Fig. 1.14). Landscape sensitivity to change is, consequently, as effective in controlling the short-term denudation rate as is the energy of the processes of erosion and transport. In the long-term (or nearly average) condition, the storage of removable sediment is probably less important and, over an area of continental dimensions, rock and soil resistance is averaged so that climate, acting through the vegetation and available relief, is likely to be the dominant control on denudation rate.

Methods of estimating local denudation rates

Local denudation rates are measured for two main purposes: either to derive an estimate of the effectiveness of a single geomorphic process, or to obtain an estimate of the rate of erosion under a given class of vegetation or landuse.

Rates of single geomorphic processes

Rates of slow processes are most conveniently measured in mm/year, but faster processes in km/y or even m/s. For slow single processes, and average rates of denudational and tectonic processes, even 1 mm/y may be fast and some writers prefer to use the Bubnoff unit (B): $1\ B = 1$ m/My = 1 mm/ky = 1 μm/y.

Rates of movement of many slow, but nearly continuous processes, such as migration of sand dunes, flow of glaciers, or gelifluction, can be assessed by positioning stakes in or around the feature and then regularly resurveying the feature to determine its location. Very slow processes, such as soil creep, are studied by placing rods, pins, blocks, or other markers, with the same density as soil, in the moving material and then relocating the marker after an interval.

Rates of local processes such as gully, rill, and stream bank erosion are also estimated by using fixed pegs and repeated survey. Sheetwash and ground lowering may be estimated from pegs, from the exposure of the foundations of buildings whose age is known, and from exposure of tree roots where the age of the tree can be obtained by counting the growth rings represented in a core taken from the trunk. Typical rates derived from such measurements are indicated in Fig. 20.1.

Denudation and land use

The rate of soil formation by weathering processes, and the natural rate of erosion, are both controlled by slope, climate, nature of bedrock, and the vegetation cover. Over a broad range of mid-latitude environments between 20 and 200 t/km^2 (i.e a thickness of 10–100 B) of new soil is formed each year. On gentle slopes with a complete forest or grass cover the natural rate of loss by erosion is either less than, or approximately

equal to, the mass of soil formed. Natural erosion is, of course, an essential process to the maintenance of soil fertility, for without it freshly weathered minerals, which are a source of plant nutrients, would not be available in the root zone, and soils would become deeper but, at the surface, would consist largely of strongly leached clays.

Disturbance of vegetation by people, by natural fires, severe storms, or long-term climatic changes, can lead to *accelerated erosion*, but by far the most widespread cause of accelerated erosion is human interference and, at a local scale, land use usually controls rates of denudation. Even at continental scales human influences are often dominant. To distinguish naturally accelerated erosion from that caused by people the term *induced erosion* is often used for the latter.

The most accurate estimates of local denudation rates are derived from large plots which have raised borders, of steel or concrete, and collecting troughs at the downslope end, to receive runoff and sediment. In USA particularly, controlled experiments to determine rates have been carried out since about 1930. In Table 20.1 some representative rates are given for various land uses, on gentle slopes and moderately humid climates. These figures indicate the extreme effects of human interference with the vegetation, and also show clearly that many forms of land use cause short-term losses of soil which far exceed the rate of renewal by weathering, and hence cause a loss of an essential natural resource. The representative rate of erosion for cropland exceeds the rate of soil formation for most humid climate soils.

World denudation rates

Three studies of world denudation rates have been published: by Corbel, Fournier, and Strakhov. Each of these studies uses as its basic data the load of suspended sediment carried by major rivers. This procedure has the advantage that sediment yield can be averaged over the catchment to give a figure in tonnes/km^2 per year, and for many catchments the yields can be related to world climatic zones so that world maps of sediment yield can be prepared. The method has five major disadvantages: (1) bedload and dissolved load are not always taken into account; (2) distinctions between areas of different relief types are difficult, or impossible; (3) the effect of land use upon sediment yield is subsumed in the generalized data;

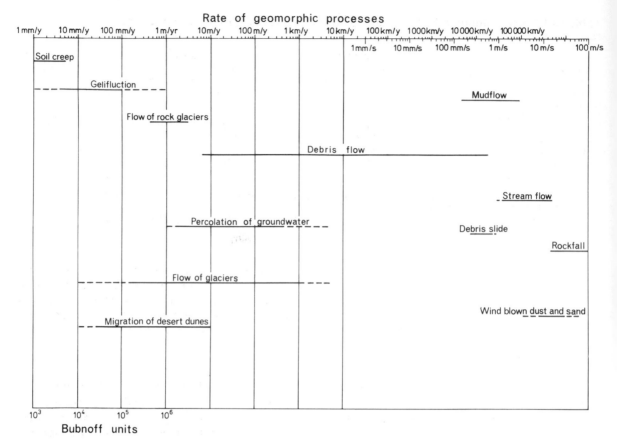

Fig. 20.1 A plot of the rates at which selected geomorphic processes are known to operate.

(4) it is assumed that all sediment eroded from the catchment is carried to the sea in a relatively brief period; and (5) the results of rare, but very large magnitude events, may be missed.

Problems of determining average rates

(1) The significance of bedload and dissolved load in total denudation is extremely variable from one catchment to another and also within one catchment. Bedload in a high mountain area, with a humid climate, may contribute 90 per cent of the total load, but by the time the river has reached the sea the bedload material has been so comminuted that 80–90 per cent of the solid load is carried in suspension. This fact justifies using suspended sediment yield as an index of total denudation. Solution load may constitute over 90 per cent of total load in catchments upon limestones, but as

little as 5 per cent in areas of quartzite or siliceous sandstones: it is common in humid climatic zones for it to exceed 40 per cent of total load. As few reliable sets of long-term data, on solute load of large rivers, are available the errors introduced by ignoring dissolved load, or applying correction factors to suspended sediment data, are unknown.

(2) Most available data indicate clearly that areas of high relief and steep slopes have far greater denudation rates than areas of low relief. Mean rates of ground lowering of 10–50 B are commonly reported for lowlands and rates for mountains range from 50 to 10 000 B. There is still much debate about the characteristic rates for mountainous regions, with available evidence for the Canadian Cordillera, for example, suggesting that characteristic rates are 60–120 B, but mountain regions elsewhere appear to have considerably higher rates. The Nanga Parbat region of the

Table 20.1   Representative Rates of Erosion from Various Land Uses in USA

Land use	Tonnes/km² per year	Average ground lowering (B)	Relative to forest = 1
Forest	8.5	4.2	1
Grassland	85.0	42.5	10
Abandoned surface mines	850.0	425.0	100
Cropland	1700.0	850.0	200
Harvested forest	4250.0	2125.0	500
Active surface mines	17000.0	8500.0	2000
Construction	17000.0	8500.0	2000

Note: 1 B of ground lowering = removal of 2.0 to 3.0 t/km² per year of bedrock, or 1.8 t of silty gravel, or about 1.8–2.5 t of soil, depending upon its bulk density. Also 1 B = removal of 1m³/km² per year.
Source: Methods for Identifying and Evaluating the Nature and Extent of Nonpoint Sources of Pollutants, Washington, DC, US. Environmental Protective Agency, Oct. 1973.

Himalayas, for example, has been eroded and uplifted at neary 10 000 B for the whole of the Pleistocene.

(3) Human interference with natural processes and use of the land have, over large areas of the continents, increased denudation rates by factors of 1.2 to 3.0, and locally and temporarily by larger factors.

(4) In small catchments, with steep valley walls, landslide and other debris is delivered directly to the stream channels and becomes part of the transported debris output almost immediately (Plate 20.1a,b). In valleys, with extensive terraces and floodplains, slope debris is trapped upon the flat surfaces and may be stored there for thousands, or even millions, of years. Denudation load from such areas is then largely, or wholly, derived from channel bank collapse, and may not represent the movement of material within the catchment (see Fig. 20.7).

(5) The effects of rare, but very large events, upon denudation rates of large areas are usually unknown. On lowlands the effects of very large floods may increase sediment yields by a factor of 2–10 for a few days or weeks, but as vegetation re-establishes itself on floodplains the sediment in transport declines: the effect on denudation rates averaged over a century is probably small. Very large magnitude events in mountains, especially very large landslides, may have an effect which can last for tens to hundreds of years. It has been suggested that in the New Zealand Alps, for example, infrequent but large landslides account for half of the total denudation. One rockfall in a large valley may easily represent the equivalent of 100 years of 'normal erosion'. The debris from a rare event is retained in a valley until it can be evacuated by stream or glacier action. It is consequently a rapid modifier of the landscape, but becomes part of the denudation load far more slowly.

Climatically controlled patterns of denudation

Of the three studies of world denudation rates, Fournier's indicates highest rates, those of Strakhov and Corbel are lower. Data are not yet available for making a detailed comparative study and comment is confined, here, to the work of Fournier. He studied suspended sediment yields in seventy-eight drainage basins and derived a statistical relationship between sediment yield E(t/km² per year), and a climatic parameter p^2/P, where p is the rainfall of the month with the greatest rainfall, P is the mean annual rainfall (mm), $\bar{H}$ is the mean height within the catchment, and α is the mean slope of the drainage basin:

$$\log E = 2.65\log(p^2/P) + 0.46\log\bar{H}\tan\alpha - 1.56.$$

Using this relationship Fournier mapped the world distribution of erosion rates (Fig. 20.2). By combining Fournier's world data with that of Schumm and Langbein for USA, a general pattern may be discerned in the relationship between rainfall which is effective in causing denudation and the sediment yield (Fig. 20.3). Maximum denudation rates occur in the seasonally humid tropics, especially in the monsoon regions, and in the semi-arid zone where vegetation cover is limited, lower rates occur in the humid tropics

20.1 (a) A large landslide which has slumped into a stream channel impounding a lake and providing sediment to the stream channel. (b) The upper reaches of the lake formed in (a) is being infilled by a delta. Repeated surveys of the volume of sediment being deposited indicate that annual erosion rate in the headwaters of this stream is about 4000 B. Hawkes Bay, New Zealand.

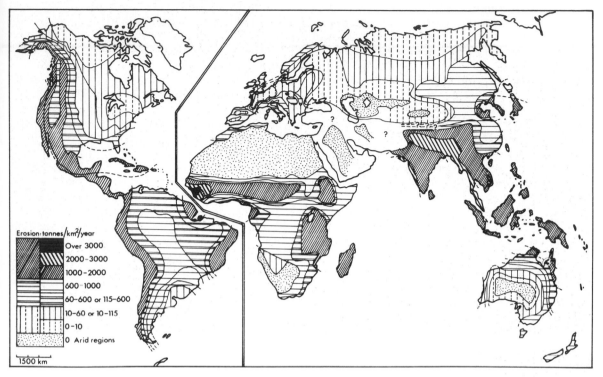

Fig. 20.2 World denudation rates (after Fournier, 1960).

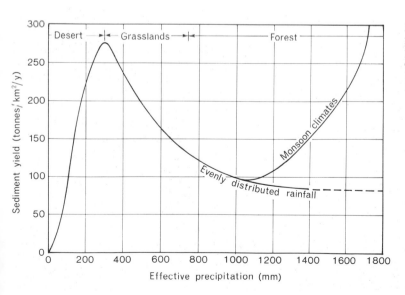

Fig. 20.3 Variation of sediment yield with effective precipitation (total precipitation minus losses by evapotranspiration) (data from Langbein and Schumm, 1958, and Fournier, 1960).

where the seasonal effect is lacking and the vegetation cover more complete. The lowest rates occur in the arid regions where fluvial transport of sediment is virtually nil. The rate of erosion rises in the seasonally wet Mediterranean lands, but in temperate and cold climates it is low. In all climatic zones steep and high relief increases rates.

Rates of tectonic processes

Rates of uplift or subsidence vary greatly in time and space, and over continental areas they may be reversed as periods of broad downwarping are followed by periods of doming. Rates may be estimated in the short term by repeated accurate surveying, and in the medium term of thousands of years by determining the elevation of features, of known age, formed at sea-level (see Figs. 1.4 and 4.38). Much longer trends may be estimated from the elevation of sedimentary rocks: in the European Alps Upper Cretaceous rocks are found at 2500 m indicating an uplift rate of 30 B, and similar-aged rocks on Kanchenjunga, at 8400 m, a rate of 90 B. These are, clearly, minimal rates and are much lower than most unplift rates measured over short periods.

Rates measured by various methods give such a range of rates that generalizations can be very misleading, but some which are thought to be representative are listed below. The available data suggest that rates of tectonic uplift may average 1000 B in orogenic belts, but localized short-term rates may reach 76 000 B in parts of Japan. Subsidence rates are generally lower than uplift rates and in the range of 10–100 B, but they can reach several thousand B for short periods.

Horizontal motions along large transform and transcurrent faults can be very large, with displacements of 5000–20 000 B being long term averages, and instantaneous movements, such as the 21 m on the San Andreas fault during the 1906 earthquake, being extreme.

Many tectonic rates are consistent with the sea-floor speading rates of 10 000–100 000 B (Fig. 20.4).

Rates of landscape change

Average rates (B) of landsurface change approximate to the following values:

downwearing of continental erosion surfaces	5
downwearing of lowlands	50
backwearing of low hills	200
downwearing of major uplands	300
backwearing of continental escarpments	1000
downwearing of major mountain ranges	1000
uplift rates	1000
horizontal displacements along large faults	10 000
convergence at a subduction site	100 000

These rates suggest that hillslopes are unlikely to have been greatly modified during the last 10 000 years, so studies of slope forms must take into account the results of late glacial and Holocene climatic changes. Minor valleys 20–50 m deep are reasonably attributed to the Late Pleistocene, but intermediate-sized features have to be considered in relation to the whole of Pleistocene time. The origin of erosion surfaces of continental extent may reasonably be attributed to Tertiary or even late Mesozoic time.

The average rates imply that it is possible to erode all mountains down to sea-level in about 25 My if renewed uplift and isostatic compensation do not intervene. It is obvious, of course, that the continents are not removed by erosion and that, with the exception of much of Australia, they are not generally reduced to plains close to ultimate base level. If continental isostatic uplift merely permitted the land surface to keep pace with erosion, then the total thickness of siallic crust would eventually be consumed, and sima would be exposed. This has, clearly, not happened, even though continental surfaces have been lowered into the igneous and metamorphic roots of ancient orogens. It seems most probable that the long-term denudation rates, of extensive continental erosion surfaces, are low enough for these surfaces to survive through major episodes of continental fracturing, dividing, and recombination, of the kind which produced assembly of the components of Laurasia and Gondwanaland in the Precambrian and their separation in the Mesozoic. Rates of subduction at continental margins are great enough to maintain the thickness of continental crust. Continents maintain their shape and fit at separating, passive margins but grow at, active, subducting margins so that their overall shape,

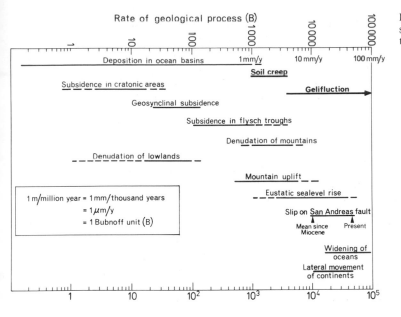

Fig. 20.4 A plot of the rates at which selected geological processes are known to operate.

size, and fit changes on a time scale of 100 My or so.

Very high rates of uplift explain the survival of high mountains at collision or accreting margins of crustal plates. These areas are also the zones of major plutonic intrusion and high rates of horizontal movement. The mobility of the Earth is much greater in the horizontal plane than in a vertical direction, and large crustal blocks can move laterally at a much faster rate than mountain chains can be raised.

Varying tempos of change

In a study of the Earth's changing surface a geomorphologist has to think within a time-frame which is appropriate to the size of the features he is considering. At the largest size, and longest time-span, continental erosion surfaces develop over tens, and even hundreds, of millions of years. The time over which a mountain range is raised during an orogeny and worn down to a residual plain is of similar duration. This period of orogenesis and progressive downwearing is still often referred to as *cyclic time* even though any implications of repeated folding and uplift on one site have to be rejected because continents grow by marginal accretion of mountain chains (Chapter 4), and any smooth progression is likely to be interrupted by

isostatic adjustments, broad tectonic warping and, perhaps climatic changes (Fig. 20.5).

Superimposed upon tectonic and 'cyclic' erosional changes are major swings in climate between ice ages and periods without ice ages. Whether or not a regular periodicity exists in ice age occurrences is still unclear (see Fig. 16.1) but, if regular periodicity does exist, these major climatic cycles may have an interval of about 150 My, with the possibility of a cycle being missed, as in Jurassic time. If the theory that ice age development is related to the position of a continent in a polar latitude is correct, then the minimum duration of an ice age is controlled by the maximum rate of sea-floor spreading.

Within each ice age regular and repeated fluctuations between glacial and interglacial conditions in polar and temperate latitudes, with alternating dry and humid conditions in the tropics and subtropics, can occur when overall world temperatures are decreased to a level at which Milankovitch mechanisms can be effective. In the Early Quaternary the 20 000- and 40 000-year cycles dominated the alternation of morphogenetic environments (see Fig. 16.9), but in the Middle and Late Quaternary the 100 000-year cycle dominated glacial to interglacial changes. Within the longer cycle the 20 000, 40 000, and perhaps shorter, cycles interacted to produce many changes of

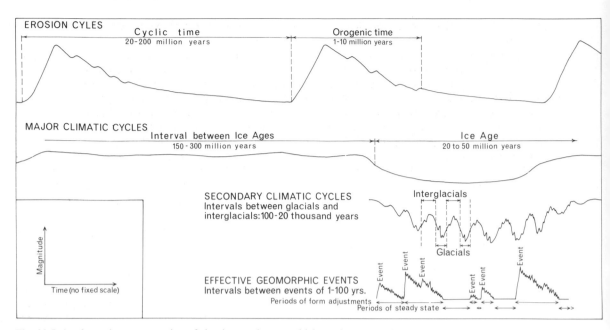

Fig. 20.5 A schematic representation of the time-scale over which erosional and climatic events operate.

environment (Fig. 20.6). A feature of the Quaternary record is the rapidity with which major swings from full interglacial to nearly full glacial conditions can occur (see, for example, the changes at 115 000 and 75–80 000 BP in Fig. 20.6).

During the Quaternary, alternating morphogenetic environments have dominated rates of change of landforms, and episodes of change from one system to another have been, frequently, periods of severe erosion (see Fig. 1.14) which are followed by longer periods of relative stability.

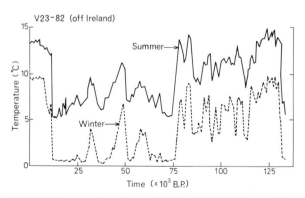

Fig. 20.6 Temperature changes for the last 130 000 years for winter and summer as interpreted from microfaunas within North Atlantic core V23-82, taken off Ireland (after Sancetta *et al.*, 1973).

Within the time-span of a human life we cannot see directly the effects of a climatic or tectonic cycle, but we can see the effect of major catastrophic events. When these have a return period longer than a few hundred years, they become incorporated into the sequence of secondary climatic cycle events, but the shorter return period events of tens to hundreds of years can be regarded as the effective events in many landscapes. A major geomorphic event can be regarded as one which has sufficient energy to disturb the approximate equilibrium between the processes of erosion and resistance of the ground surface. On an extensive continental erosion surface the slope of the ground may be inadequate to provide sufficient potential energy for even a severe storm to disrupt the soil and vegetation cover. Extensive plains therefore require very large inputs of energy for climatic events to cause major modifications to the land surface. Consequently a steady state is likely to prevail for long periods, disruptions are few and, on sites of superior resistance, paleoforms are likely to survive for tens or hundreds of millions of years. The stability of a land surface is often evident from the depth of soils, weathering profiles, and cover beds upon it.

In hill and mountain country the available slope provides high potential energy for geomorphic processes, thresholds of resistance are low, and the climatic-biogeomorphic regime may permit rapid and large-scale responses to inputs of energy. Extreme events are consequently likely to have more effect upon the landscape than they are on a lowland (Fig. 20.7). Periods of steady state (Fig. 1.13) are likely to be brief, events may be extreme, recovery periods may be prolonged by interruptions from minor events, and states of form adjustment may dominate the long-term dynamic equilibrium. Paleoforms will seldom survive in a high energy environment.

Towards a philosophy of geomorphic change

Over a time-scale which is appropriate to geological thinking, there appears to be no simple and direct sequence of events of the kind envisaged by W. M. Davis in his 'cycle of erosion'. At this scale the pattern of continental arrangements and orogenies is determined by plate tectonic events and the marginal accretion of orogenic belts largely eliminates repeated orogenies on one site. Superimposed upon the plate tectonic regime is a climatic regime with many grades of intensity and periodicity.

Rates of erosion of the landscape depend largely upon the available slope for geomorphic purposes. Paleoforms, such as planation surfaces and inselbergs, which have survived with only minor modification for tens of millions of years are usually located far from the loci of late Cenozoic lateral or vertical erosion and are particularly important on the cratons with a long history of semi-arid climate. Steep slopes are the foci of change so that there is always a tendency for steep and irregular forms to be eliminated or to migrate laterally.

In studies of long-term geomorphic development of whole landscapes the researcher has to interpret ancient causes from their historical results. An understanding of the modern conditions under which soils and sediments may be formed has to be a guide to such interpretations, but a guide of sometimes questionable relevance for (1) the initial site conditions are never known; (2) the knowledge of past conditions is always poor; and (3) our simplifications may be inappro-

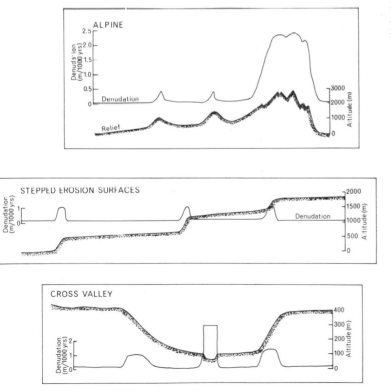

Fig. 20.7 A schematic representation of the variations in denudation rate across three types of landscape.

priate to explain past conditions of unknown complexity.

We can recognize areas of intense and extensive modification during late Cenozoic time. In geological terms transitions from glacial to interglacial climates were catastrophic and are certainly not reproduced in the time-span of human experience. A long-term catastrophism must be added to actualism in such studies, and leads to a recognition of the importance of interruptions to periods of relative stability. At the scale of cratons, long, stable periods may have durations of 10–100 My when deep chemical weathering prevails under a nearly continuous vegetation cover (known as *biostatic* conditions). Interruptions of short unstable (*rhexistatic*) phases of 1–10 My, or so, duration may occur as mean sea-levels fall, ground water tables fall, vegetation declines, climates become more semi-arid, and erosion lowers craton ground surfaces by several metres in a few million years. At the scale of individual valleys climatic oscillations of 10–100 ky will leave their imprint: individual hillslopes may be profoundly modified by landslides, or other erosional events which may occur with a frequency of 1–10 000 y.

In uplands and orogenic belts the short-term events become far more important modifiers of the landscape than they are on cratons. The land surface then evolves in a punctuated and episodic manner, even though the slow and nearly continuous processes of weathering, solution, and soil creep maintain their contribution to regolith development and preparation for more catastrophic changes which may occur with a periodicity of 1–1000 years.

The survival of young paleoforms is most obvious in temperate zones which were formerly glaciated, because the results of glacial erosion and deposition may be scarcely modified by later processes. Old paleoforms are most obvious in the tropics and subtropics where climatic oscillations over cratons have been from humid to sub-humid conditions and back again. Recognition of paleo-forms has importance in understanding evolution of whole landscapes.

The detailed study of processes, and the resulting minor landforms and sediments, tells us little about whole landscapes in any direct sense, but it has two values. (1) The study of sediments and results of single processes can aid our interpretation of deposits of the past; and (2) the study of modern processes permits geomorphologists to predict the development of small-scale landforms, and so use their scientific expertise to solve practical problems. While emphases upon contemporary process–landform relationships and systems should lead to greater understanding of modern and late Cenozoic landforms, the deficiencies of applying strict actualistic or uniformitarian concepts must be recognized. It may be prophesied that geomorphologists will forge closer links with pedologists and sedimentologists as they attempt to unravel the record of the past left in deposits, while maintaining their links with glaciologists, hydrologists, geologists, geophysicists, and engineers as they seek a better understanding of modern exogenetic and endogenetic processes and their results. In this aspect of their science geomorphologists will increasingly use experimental, empirical (inductive) methods, and so will have much in common with other natural scientists.

Further reading

A review of denudation rates is given by Selby (1982) and of rates of tectonic deformation by Ollier (1981). The reader who wishes to develop a philosophy of geomorphic change will find many stimulating ideas in Chorley and Kennedy (1971), in papers by Hack (1960), Schumm and Lichty (1965), Twidale (1976), and in the collection published in *Zeitschrift für Geomorphologie, Supplement*, 36, (1980). Evidence for variability in denudation rates in Australia is offered by Ollier (1982).

References

Adams, C. S. and Swinnerton, A. C. (1937). 'Solubility of limestone', *American Geophysical Union, Transactions*, 2, 504–8.

Alimen, M.-H. (1967). 'The Quaternary of France', pp. 89–238 in K. Rankama (ed.) *The Quaternary*, Vol. 2 (Interscience: New York).

Allègre, C. J. *et al.* (1984). 'Structure and evolution of the Himalaya-Tibet orogenic belt', *Nature*, 307, 17–22.

Allen, J. R. L. (1968). *Current Ripples* (North-Holland: Amsterdam).

Allen, J. R. L. (1970). *Physical Processes of Sedimentation* (Allen & Unwin: London).

Allen, J. R. L. (1971). 'Transverse erosional marks of mud and rock: their physical basis and geological significance', *Sedimentary Geology*, 5, 167–385.

Allen, P. (1975). 'Ordovician glacials of the central Sahara', pp. 275–85 in A. E. Wright and F. Moseley (eds.) *Ice Ages: Ancient and Modern* (Seel House Press: Liverpool).

Anderton, R., Bridges, P. H., Leeder, M. R., and Sellwood, B. W. (1979). *A Dynamic Stratigraphy of the British Isles* (Allen & Unwin: London).

Andrews, J. T. (1975). *Glacial Systems* (Duxbury: North Scituate, Mass.).

Andrews, J. T. (1982). 'On the reconstruction of Pleistocene ice sheets: a review', *Quaternary Science Reviews*, 1, 1–30.

Argand, E. (1934). *Guide Geologique de la Suisse*, 3, 149–90.

Association of American Petroleum Geologists (1982). *Plate Tectonic Map of the Circum-Pacific Region* (A. A. P. G.: Tulsa, Okla).

Axelrod, D. I. (1952). 'A theory of Angiosperm evolution', *Evolution*, 6, 29–60.

Bagnold, R. A. (1941). *The Physics of Blown Sand and Desert Dunes* (Chapman & Hall: London).

Baker, B. H. (1965). 'An outline of the geology of the Kenya rift valley', *Report on the Geology and Geophysics of the East African Rift System* (Nairobi).

Ball, J. W. (1976). 'Cavitation from surface irregularities in high velocity flows', *Journal of the Hydraulics Division, American Society of Civil Engineers*, 102 (HY9), 1283–97.

Barat, C. (1957). 'Pluviologie et Aquidimétrie dans la Zone Intertropicale', *Mémoires de l'Institut Francais d'Afrique Noire* (Ifan-Dakar), no. 49, 1–80.

Barnes, H. H. (1967). 'Roughness characteristics of natural channels', *US Geological Survey Water Supply Paper*, 1849.

Bascom, W. (1964). *Waves and Beaches* (Doubleday: Garden City, NY.).

Beckinsale, R. P. and Chorley, R. J. (1968). 'Geomorphology, history of', pp. 410–16 in R. W. Fairbridge (ed.) *Encyclopedia of Geomorphology* (Reinhold: New York).

Bigarella, J. J., Mousinho, M. M., and Da Silva, J. X. (1969). 'Processes and environments of the Brazilian Quaternary', pp. 417–87 in T. L. Péwé (ed.) *The Periglacial Environment* (McGill-Queen's University Press: Montreal).

Bird, E. C. F. (1976). *Coasts* (2nd ed.) (Australian National University Press: Canberra).

Bird, J. B. (1972). *The Natural Landscapes of Canada* (Wiley: Toronto).

Bird, J. M. (ed.) (1980). *Plate Tectonics* (2nd ed.) (American Geophysical Union: Washington, DC).

Bird, J. M. and Isacks, B. (eds.) (1972). *Plate Tectonics* (American Geophysical Union: Washington, DC).

Blake, D. H. and Ollier, C. D. (1971). 'Alluvial plains of the Fly River, Papua', *Zeitschrift für Geomorphologie, Supplementband*, 12, 1–17.

Blatt, H., Middleton, G., and Murray, R. (1972). *Origin of Sedimentary Rocks* (Prentice-Hall: Englewood Cliffs, NJ).

Blong, R. J. and Pain, C. F. (1976). 'The nature of highland valleys, central Papua New Guinea', *Erdkunde*, 30, 212–17.

Bloom, A. L. (1978). *Geomorphology: A Systematic Analysis of Late Cenozoic Landforms* (Prentice-Hall: Englewood Cliffs, NJ).

Bolt, B. A. (1976). *Nuclear Explosions and Earthquakes* (W. H. Freeman: San Francisco, Calif.).

Bolt, B. A., Horn, W. L., Macdonald, G. A., and Scott,

R. F. (1975). *Geological Hazards* (Springer-Verlag: Berlin).

Bott, M. H. P., Holder, A. P., Long, R E., and Lucas, A. L. (1970). 'Crustal structure beneath the granites of south-west England', in G. Newall and N. Rast (eds.), *Mechanisms of Igneous Intrusion* (Gallery Press: Liverpool).

Boulton, G. S. (1979). 'A model of Weichselian glacier variation in the North Atlantic region', *Boreas*, 8, 373–95.

Boulton, G. S., Jones, A. S., Clayton, K. M., and Kenning, M. J. (1977). 'A British ice-sheet model and patterns of glacial erosion and deposition in Britain', pp. 231–46 in F.W. Shotton (ed.), *British Quaternary Studies* (Clarendon Press: Oxford).

Boulton, G. S. *et al.* (1982). 'A glacio-isostatic facies model and amino acid stratigraphy for late Quaternary events in Spitsbergen and the Arctic', *Nature*, 298, 437–41.

Bowen, D. Q. (1978). *Quaternary Geology* (Pergamon: Oxford).

Bowen, D. Q. (1979). 'Glaciations past and future', *Geographical Magazine*, 52, 60–7.

Bowler, J. M. (1976). 'Aridity in Australia: age, origins and expression in aeolian landforms and sediments', *Earth-Science Reviews*, 12, 279–310.

Bowler, J. M., Hope, G. S., Jennings, J. N., Singh, G., and Walker, D. (1976). 'Late Quaternary climates of Australia and New Guinea', *Quaternary Research*, 6, 359–94.

Bowles, F. A. (1975). 'Paleoclimatic significance of quartz/illite variations in cores from the eastern equatorial North Atlantic', *Quaternary Research*, 5, 225–35.

Brakenridge, G. R. (1978). 'Evidence for a cold, dry full-glacial climate in the American southwest', *Quaternary Research*, 9, 22–40.

Brakenridge, G. R. (1980). 'Widespread episodes of stream erosion during the Holocene and their climatic cause', *Nature*, 283, 655–6.

Bremer, H. (1980). 'Landform development in the humid tropics, German geomorphological research', *Zeitschrift für Geomorphologie, Supplementband*, 36, 162–75.

Bretschneider, C. L. (1959). 'Wave variability and wave spectra for wind generated gravity waves', *US Army Corps of Engineers, Beach Erosion Board Technical Memorandum*, 118.

Brook, G. A. and Ford, D. C. (1978). 'The origin of labyrinth and tower karst and the climatic conditions necessary for their development', *Nature*, 275, 493–6.

Brooks, C. K. (1973). 'Rifting and doming in southern East Greenland', *Nature Physical Sciences*, 244, 23–5.

Brown, E. H. and Waters, R. S. (eds.) (1974). *Progress in Geomorphology* (Institute of British Geographers, Special Publication no. 7: London).

Brown, R. J. E. (1968). 'Occurrence of permafrost in Canadian peatlands', *3rd International Peat Congress, Proceedings, Quebec*, 174–81.

Brown, R. J. E. (1970). *Permafrost in Canada* (Toronto University Press: Toronto).

Bruun, P. (1962). 'Sea-level rise as a cause of shore erosion', *American Society of Civil Engineers Proceedings, Journal of the Waterways and Harbour Division*, 88, 117–30.

Büdel, J. (1977) (translated by L. Fischer and D. Busche, 1982). *Climatic Geomorphology* (Princeton University Press: Princeton, NJ).

Büdel, J. (1979). 'Reliefgenerationen und Klimageschichte in Mitteleuropa', *Zeitschrift für Geomorphologie, Supplementband*, 33, 1–15.

Bullard, E., Everett, J. E., and Smith, A. G. (1965). 'The fit of the continents around the Atlantic', *Philosophical Transactions of the Royal Society, London*, A258, 41–51.

Burggraf, D. R. and Vondra, C. F. (1982). 'Rift valley facies and paleoenvironments: an example from the East Africa Rift System of Kenya and southern Ethiopia', *Zeitschrift für Geomorphologie, Supplementband*, 42, 43–73.

Butzer, K. (1974). 'Climatic change', *Encyclopedia Britannica*, 4, 730–41.

Butzer, K. W. and Isaac, G. L. (eds.) (1975). *After the Australopithecines* (Mouton: The Hague).

Butzer, K. W., Isaac, G. L., Richardson, J. L., and Washbourn-Kamau, C. (1972). 'Radiocarbon dating of East African lake levels', *Science*, 175, 1069–76.

Calkin, P. E., Behling, R. E., and Bull, C. (1970). 'Glacial history of Wright Valley, southern Victoria Land, Antarctica', *Antarctic Journal of the United States*, 5, 22–7.

Carozzi, A. V. (translator) (1967). *Studies on Glaciers, by Louis Agassiz* (Hafner: New York).

Carson, M. A. and Kirkby, M. J. (1972). *Hillslope Form and Process* (Cambridge University Press: Cambridge, England).

Catt, J. A. (1977). 'Loess and coversands', pp. 221–9 in F.W. Shotton (ed.), *British Quaternary Studies* (Clarendon Press: Oxford).

Chorley, R. J. (1957). 'Climate and morphometry', *Journal of Geology*, 65, 628–68.

Chorley, R. J., Beckinsale, R. P., and Dunn, A. J. (1973). *The History of the Study of Landforms*, Vol. 2 (Methuen: London).

Chorley, R. J., Dunn, A. J., and Beckinsale, R. P. (1964). *The History of the Study of Landforms*, Vol. 1 (Methuen: London).

Chorley, R. J. and Kennedy, B.A. (1971). *Physical*

Geography: A Systems Approach (Prentice-Hall International: London).

Chorley, R. J., Malm, D. E. G., and Pogorzelski, H. A. (1957). 'A new standard for estimating basin shape', *American Journal of Science*, 255, 138–41.

Chorley, R. J. and Morgan, M. A. (1962). 'Comparison of morphometric features, Unaka Mountains Tennessee and North Carolina, and Dartmoor, England', *Geological Society of America Bulletin*, 73, 17–34.

Chow, V. T. (ed.) (1964). *Handbook of Applied Hydrology* (McGraw-Hill: New York).

Clague, D. A., Dalrymple, G. B., and Moberly, R. (1975). 'Petrology and K-Ar ages of dredged volcanic rocks from the western Hawaii Ridge and the southern Emperor Seamount Chain', *Geological Society of America Bulletin*, 86, 991–8.

Clark, J. A., Farrell, W. E., and Peltier, W. R. (1978). 'Global changes in postglacial sea level: a numerical calculation', *Quaternary Research*, 9, 265–87.

Clayton, K. M. (1974). 'Zones of glacial erosion', pp. 163–76 in E.H. Brown and R.S. Waters (eds.) *Progress in Geomorphology*, (Institute of British Geographers Special Publication, no. 7).

CLIMAP Project Members, (1976). 'The surface of the ice-age earth', *Science*, 191, 1131–7.

Coastal Engineering Research Center (1977). *Shore Protection Manual*, 3 vols., (US Army Corps of Engineers: Fort Belvoir, Va.).

Coates, D. R. (ed.) (1976). *Geomorphology and Engineering* (Dowden, Hutchinson & Ross: Stroudsburg, Penn.).

Coates, D. R. (1981). *Environmental Geology* (Wiley: New York).

Cook, F. A. *et al.* (1979). 'Thin-skinned tectonics in the crystalline southern Appalachians, *Geology*, 7, 563–7.

Cooke, R. U. and Doornkamp, J. C. (1974). *Geomorphology in Environmental Management* (Clarendon Press: Oxford).

Cooke, R. U. and Warren, A. (1973). *Geomorphology in Deserts* (Batsford: London).

Cole, J. W. (1970). 'Structure and eruptive history of the Tarawera volcanic complex', *New Zealand Journal of Geology and Geophysics*, 13, 879–902.

Cotton, C. A. (1944). *Volcanoes as Landscape Forms* (Whitcombe & Tombs: Christchurch NZ).

Cox, A. (ed.) (1973). *Plate Tectonics and Geomagnetic Reversals* (W.H. Freeman: San Francisco, Calif.).

Cox, K. G. (1970). 'Tectonics and volcanism of the Karroo period and their bearing on the postulated fragmentation of Gondwanaland', in T.N. Clifford and I.G. Glass (eds.), *African Magmatism and Tectonics* (Oliver & Boyd: Edinburgh).

Craig, R. G. and Craft, J. L. (eds.) (1980). *Applied Geomorphology* (Allen & Unwin: London).

Dalrymple, J. B., Blong, R. J., and Conacher, A. J. (1968). 'A hypothetical nine unit landsurface model', *Zeitschrift für Geomorphologie*, 12, 60–76.

Damuth, J. E. and Fairbridge, R. W. (1970). 'Equatorial Atlantic deep-sea arkosic sands and ice-age aridity in tropical South America', *Geological Society of America Bulletin*, 81, 189–206.

Dansgaard, W., Johnsen, S. J., Clausen, H. B., and Langway, C. C. (1971). 'Climatic record revealed by the Camp Century ice core, pp. 37–56 in K.K. Turekian (ed.), *The Late Cenozoic Glacial Ages* (Yale University Press: New Haven, Conn.).

Davies, G. L. (1969). *The Earth in Decay* (Macdonald Technical and Scientific: London).

Davies, J. L. (1980). *Geographical variation in Coastal Development* (2nd ed.) (Longman: London).

Davies, J. L. and Williams, M. A. J. (1978). *Landform Evolution in Australasia* (Australian National University Press: Canberra).

Davis, W. M. (1902). 'Base-level, grade, and peneplain', *Journal of Geology*, 10, 77–111.

Davis, W. M. (1912). *Die Erklärende Beschreibung der Landformen*, XVII (B.G. Teubner: Leipzig).

De Jong, K. A. (1973). 'Mountain building in the Mediterranean region', in K.A. De Jong and R. Scholten (eds.), *Gravity and Tectonics* (Wiley: New York).

Denton, G. H., Armstrong, R. L., and Stuiver, M. (1971). 'The Late Cenozoic glacial history of Antarctica', pp. 267–306 in K.K. Turekian (ed.), *The Late Cenozoic Glacial Ages* (Yale University Press: New Haven, Conn.

Denton, G. H. and Hughes, T. J. (eds.) (1981). *The Last Great Ice Sheets* (Wiley: New York).

Derbyshire, E. (ed.) (1973). *Climatic Geomorphology* (Macmillan: London).

Derbyshire, E. (1975). 'Distribution of glacial soils in Britain', in *The Engineering Behaviour of Glacial Materials* (Geo Abstracts: Norwich).

Detwyler, T. R. (ed.) (1971). *Man's Impact on Environment* (McGraw-Hill: New York).

Dewey, J. F. (1972). 'Plate tectonics', *Scientific American*, 226, 56–8.

Dewey, J. F. and Bird, J. M. (1970). 'Mountain belts and the new global tectonics', *Journal of Geophysical Research*, 75, 2625–47.

Dietz, R. S. (1972). 'Geosynclines, mountains and continent building', *Scientific American*, 226(3), 30–3.

Dietz, R. S. and Holden J. C. (1970). 'The breakup of Pangaea', *Scientific American*, 223, 30–41.

Donovan, D. T. and Jones, E. J. W. (1979). 'Causes of world-wide changes in sea level', *Journal of the Geological Society, London*, 136, 187–192.

Dott, R. H. and Batten, R. L. (1981). *Evolution of the Earth* (McGraw-Hill: New York).

Douglas, I. (1977). *Humid Landforms* (Australian National University Press: Canberra).

Douglas, I. (1978). 'Tropical geomorphology: present problems and future prospects', pp. 168–84 in C. Embleton, D. Brunsden, and D.K.C. Jones, *Geomorphology, Present Problems and Future Prospects* (Oxford University Press: Oxford).

Drake, C. L. and Girdler, R. W. (1963–4). 'A geophysical study of the Red Sea', *Journal of the Royal Astronomical Society*, 8, 473–95.

Dubois, R. N. (1977). 'Predicting beach erosion as a function of rising water level', *Journal of Geology*, 85, 470–6.

Durand, A. (1982). 'Oscillations of Lake Chad over the past 50 000 years: new data and a new hypothesis', *Palaeogeography, Palaeoclimatology, Palaeoecology*, 39, 37–53.

Dury, G. H. (ed.) (1970). *Rivers and River Terraces* (Macmillan: London).

Du Toit, A. L. (1937). *Our Wandering Continents* (Oliver & Boyd: Edinburgh).

Eissmann, L. (1981). 'Periglaziäre Prozesse und Permafroststrukturen aus sechs Kaltzeiten des Quartärs', *Altenburger Naturwissenschaftliche Forschungen*, 1, 1–171.

Embleton, C. and King, C. A. M. (1975). Glacial Geomorphology (Arnold: London).

Embleton, C. and King, C. A. M. (1975). *Periglacial Geomorphology* (Arnold: London).

Embleton, C. and Thornes, J. (1979). *Process in Geomorphology* (Arnold: London).

Emiliani, C. (1966). 'Paleotemperature analysis of Caribbean cores P 6304-8 and P 6304-9 and a generalized temperature curve for the last 425 000 years', *Journal of Geology*, 74, 109–26.

Ernst, W. G. (1969). *Earth Materials* (Prentice Hall: Englewood Cliffs, NJ.).

Eyles, V. A. (1970). *James Hutton's 'System of the Earth, 1785; Theory of the Earth, 1788; Observations on Granite, 1794.'* (Hafner: Darien, Conn.).

Fairbridge, R. W. (1960). 'The changing level of the sea', *Scientific American*, 202(5), 70–9.

Fairbridge, R. W. (1976). 'Effects of Holcene climatic change on some tropical geomorphic processes', *Quaternary Research*, 6, 529–56.

Feininger, T. (1971). 'Chemical weathering and glacial erosion of crystalline rocks and the origin of till', *US Geological Survey Professional Paper*, 750-C, c65-c81.

Finkl, C. W. (1982). 'On the geomorphic stability of cratonic planation surfaces', *Zeitschrift für Geomorphologie*, 26, 137–50.

Finlayson, B. and Statham, I. (1980). *Hillslope Analysis* (Butterworth: London).

Fisk, H. N. (1944). *Geological Investigation of the Alluvial Valley of the Lower Mississippi River*. (Mississippi River Commission: Vicksburg, Miss.).

Flint, R. F. (1971). *Glacial and Quaternary Geology* (Wiley: New York).

Fong, P. (1982). 'Latent heat of melting and its importance for glaciation cycles', *Climatic Change*, 4, 199–206.

Ford, L. (1978). 'Alternative means of combating beach erosion', *Soil and Water*, April, 12–14.

Fournier, F. (1960). *'Climat et Érosion: la relation entre l'érosion du sol par l'eau et les précipitations atmosphériques'*, (Presses Universitaire de France: Paris).

Frakes, L. A. (1979). *Climates Throughout Geologic Time* (Elsevier: Amsterdam).

Frakes, L. A. and Kemp, E. M. (1972). 'Influence of continental positions on early Tertiary climates', *Nature*, 240, 97–100.

Francis, P. W. (1976). *Volcanoes* (Penguin: Harmondsworth).

French, H. M. (1976). *The Periglacial Environment* (Longman: London).

Frenzel, B. (1967) (translated A.E.M. Nairn, 1973). *Climatic Fluctuations of the Ice Age* (Case Western Reserve University Press: Cleveland, Ohio).

Furon, R. (1963). *Geology of Africa* (Oliver & Boyd: Edinburgh).

Gams, I. (1969). 'Some morphological characteristics of the Dinaric karst', *Geographical Journal*, 135, 563–72.

Garner, H. F. (1974). *The Origin of Landscapes* (Oxford University Press: New York).

Gass, I. G. and Masson-Smith, D. (1963). 'The geology and gravity anomalies of the Troodos Massif, Cyprus', *Philosophical Transactions of the Royal Society, London*, A255, 417–67.

Gasse, F., Fontes, J. C., and Rognon, P. (1974). 'Variations hydrologiques et extension des lacs Holocenes du Desert Danakil', *Palaeogeography, Palaeoclimatology, Palaeoecology*, 15, 109–148.

Gates, W. L. (1976). 'Modelling the ice-age climate', *Science*, 191, 1138–44.

Gilbert, G. K. (1877). *Report on the Geology of the Henry Mountains* (US Geographical and Geological Survey of the Rocky Mountain Region: Washington, DC).

Gilbert, G. K. (1914). 'The transportation of debris by running water', *US Geological Survey Professional Paper*, 86.

Gile, L. H. (1975). 'Holocene soils and soil-geomorphic relations in an arid region of southern New Mexico, *Quaternary Research*, 5, 321–60.

Gillott, J. E. (1968). *Clay in Engineering Geology* (Elsevier: Amsterdam).

Girdler, R. W. (1964–5). 'Research note—How genuine is the Circum-Pacific belt?', *Royal Astronomical Society, Geophysical Journal*, 9, 537–40.

Gorbunov, A. P. (1978). 'Permafrost in the mountains of central Asia', *Proceedings 3rd International Conference on Permafrost, Edmonton*, Vol. 1, 372–77.

Goudie, A. (1973). *Duricrusts in Tropical and Subtropical Landscapes* (Clarendon Press: Oxford).

Goudie, A. S. (1977). *Environmental Change* (Clarendon Press: Oxford).

Goudie, A. (1981). *The Human Impact: Man's Role in Environmental Change* (Blackwell: Oxford).

Goudie, A. (ed.) (1981). *Geomorphological Techniques* (Allen & Unwin: London).

Green, J. and Short, N.M. (1971). *Volcanic Landforms and Surface Features* (Springer-Verlag: New York).

Gregory, K. J. (ed.) (1977). *River Channel Changes* (Wiley: Chichester).

Gregory, K. J. and Walling, D. E. (1973). *Drainage Basin Form and Process* (Arnold: London).

Gregory, K. J. and Walling, D. E. (eds.) (1979). *Man and Environmental Processes* (Dawson: Folkestone).

Gribbin, J. (ed.) (1978). *Climatic Change* (Cambridge University Press: Cambridge, England).

Grim, R. E. (1962). *Applied Clay Mineralogy* (McGraw-Hill: New York).

Grove, A. T. (1977). 'Desertification', *Progress in Physical Geography*, 1, 296–310.

Grove, A. T. and Pullan, R. A. (1964). 'Some aspects of the Pleistocene palaeogeography of the Chad Basin', pp. 230–45 in F.C. Howell and P. Bourlière (eds.), *African Ecology and Human Evolution* (Methuen: London).

Guilcher, A. (1958). *Coastal and Submarine Morphology* (Methuen: London).

Hack, J. T. (1960). 'Interpretation of erosional topography in humid temperate regions', *American Journal of Science*, 258A, 80–97.

Hails, J. R. (ed.) (1977). *Applied Geomorphology* (Elsevier: Amsterdam).

Hallam, A. (1973). *A Revolution in Earth Sciences: From Continental Drift to Plate Tectonics* (Clarendon Press: Oxford).

Hamelin, L-E. and Cook, F.A. (1967). *Illustrated Glossary of Periglacial Phenomena* (Les Presses de L'Université Laval: Quebec).

Hamilton, A. (1976). 'The significance of patterns of distribution shown by forest plants and animals in tropical Africa for the reconstruction of Upper Pleistocene palaeoenvironments: a review', pp. 63–97 in E.M. Van Zinderen Bakker (ed.) *Palaeoecology of Africa*, Vol. 9 (Balkema: Cape Town).

Hamilton, W. (1977). 'Subduction in the Indonesian region', pp. 15—31 in M. Talwani and W.C. Pitman (eds.), *Island Arcs, Deep Sea Trenches and Back-Arc Basins* (American Geophysical Union: Washington, DC).

Hammen, T. van der, Wijmstra, T. A., and Zagwijn, W. H. (1971). 'The floral record of the Late Cenozoic of Europe', pp. 391–424 in K.K. Turekian (ed.) *The Late Cenozoic Glacial Ages* (Yale University Press: New Haven, Conn.).

Hansen, W. R. (1965). 'Effects of the earthquake of March 27, 1964, at Anchorage, Alaska', *US Geological Survey Professional Paper*, 542-A, 1–68.

Harland, W. B. *et al.* (1982). *A Geologic Time Scale* (Cambridge University Press: Cambridge, England).

Harris, C. (1981). *Periglacial Mass-wasting: a Review of Research* (Geo Books: Norwich).

Harris, L. D. and Bayer, K. C. (1979). 'Sequential development of the Appalachian orogen above a master decollement—a hypothesis', *Geology*, 7, 568–72.

Harris, S. A. and Brown, R. J. E. (1981). 'Permafrost distribution along the Rocky Mountains in Alberta', *Proceedings Fourth Canadian Permafrost Conference, Calgary*, 59–67.

Haynes, S. J. and McQuillan, H. M. (1974). 'Evolution of the Zagros suture zone, southern Iran', *Geological Society of America Bulletin*, 85, 739–44.

Hills, E. S. (1972). *Elements of Structural Geology* (Chapman & Hall: London).

Hjulström, F. (1935). 'Studies of the morphological activity of rivers as illustrated by the River Fyris', *Bulletin of the Geological Institute of the University of Uppsala*, 25, 221–527.

Hodgson, J. M. (1967). 'Soils of the West Sussex Coastal Plain', *Bulletin of the Soil Survey of England and Wales*, 3.

Horton, R. E. (1932). 'Drainage basin characteristics', *Transactions American Geophysical Union*, 13, 350–61.

Horton, R. E. (1945). 'Erosional development of streams and their drainage basins: hydrophysical approach to quantitative morphology', *Geological Society of America Bulletin*, 56, 275–370.

Howard, A. D. (1967). 'Drainage analysis in geologic interpretation: a summation', *American Association of Petroleum Geologists Bulletin*, 51, 2246–59.

Hsü, K. J. (ed.) (1982). *Mountain Building Processes* (Academic Press: London).

Hsü, K. J. (1983). 'Actualistic catastrophism', *Sedimentology*, 30, 3–9.

Hutchinson, J. N. (1980). 'Possible late Quaternary pingo remnants in central London', *Nature*, 284, 253–5.

Illies, J. H. and Baumann, H. (1982). 'Crustal dynamics and morphodynamics of the western European

Rift system', *Zeitschrift für Geomorphologie, Supplementband*, 42, 135–65.

Illies, J. H. and Greiner, G. (1978). 'Rhinegraben and the Alpine system', *Geological Society of America Bulletin*, 89, 770–82.

Imbrie, J. and Imbrie, K. T. (1979). *Ice Ages: Solving the Mystery* (Macmillan: London).

Institute of Geological Sciences (1974). *Volcanoes* (HMSO: London).

Isaac, K. P. (1981). 'Tertiary weathering profiles in the plateau deposits of East Devon', *Proceedings of the Geologists' Association*, 92, 159–68.

Ives, J. D. and Barry, R.G. (eds.) (1974). *Arctic and Alpine Environments* (Methuen: London).

Jahn, A. (1971). *Problems of the Periglacial Zone* (PWN: Warsaw). (translated for National Science Foundation, Washington, DC as T.T.72-54011, 1975).

Jahn, A. (1980). 'Main features of the Tertiary relief of the Sudetes Mountains', *Geographia Polonica*, 43, 5–23.

James, D. E. (1973). 'The evolution of the Andes', *Scientific American*, 229, 60–9.

Jennings, J. N. (1971). *Karst* (Australian National University Press: Canberra).

Jennings, J. N. and Mabbutt, J.A. (1967). *Landform Studies from Australia and New Guinea* (Australian National University Press: Canberra).

John, B.S. (ed.) (1979). *The Winters of the World* (David & Charles: Newton Abbot).

John, B. S. and Sugden, D. E. (1975). 'Coastal geomorphology of high latitudes', *Progress in Physical Geography*, 7, 53–132.

Johnson, A. (1970). *Physical Processes in Geology* (Freeman Cooper: San Francisco, Calif).

Johnson, R. G. and McClure, B. T. (1976). 'A model for northern hemisphere continental ice sheet variation', *Quaternary Research*, 6, 325–53.

Jones, D. K. C. (ed.) (1980). *The Shaping of Southern England* (Academic Press: London).

Jones, D. K. C. (1981). *Southeast and Southern England* (Methuen: London).

Jumikis, A.R. (1956). The soil freezing experiment. *Highway Research Board Bulletin*, 135, 150–65.

Kalinin, G. P. and Szesztay, K. (1971). 'Surface waters as elements of the World water balance', *International Association for Scientific Hydrology*, 92, *Water Balance*, Vol. I. 102–15.

Karig, D. E. (1977). 'Growth patterns on the upper trench', pp. 175–85 in M. Talwani and W.C. Pitman (eds.), *Island Arcs, Deep Sea Trenches and Back-Arc Basins* (American Geophysical Union: Washington, DC).

Karlén, W. and Denton, G.H. (1976). 'Holocene glacial variations in Sarek National Park, northern Sweden', *Boreas*, 5, 25–56.

Karte, J. (translated by P. Hessel) (1981). *Development and Present State of German Periglacial Research in the Polar, Subpolar and Alpine Environment* (Canada Institute for Scientific and Technical Information (NRC/CNR T.T.-1983): Ottawa).

Keller, E.A. (1972). 'Development of alluvial stream channels: a five-stage model', *Geological Society of America Bulletin*, 83, 1531–6.

Kendall, R. L. (1969). 'An ecological history of the Lake Victoria Basin', *Ecological Monographs*, 39, 121–76.

King, C. A. M. (1972). *Beaches and Coasts* (2nd ed.) (Arnold: London).

King, L. C. (1957). 'The uniformitarian nature of hillslopes', *Transactions Edinburgh Geological Society*, 17, 81–102.

King, L. C. (1967). *The Morphology of the Earth* (Oliver & Boyd: Edinburgh).

King, L. C. (1976). 'Planation remnants upon high lands', *Zeitschrift für Geomorphologie*, 20, 133–48.

King, P. B. (1959). *The Evolution of North America* (Princeton University Press: Princeton, NJ).

King, P. B. and Schumm, S. A. (1980). *The Physical Geography (Geomorphology) of William Morris Davis* (Geo Books: Norwich).

Klammer, G. (1981). 'Landforms, cyclic erosion and deposition, and Late Cenozoic changes in climate in southern Brazil', *Zeitschrift für Geomorphologie*, 25, 146–65.

Knox, J. C. (1972). 'Valley alluvium in southwestern Wisconsin', *Annals of the Association of American Geographers*, 62, 401–10.

Komar, P. D. (1976). *Beach Processes and Sedimentation* (Prentice-Hall: Englewood Cliffs, NJ).

Kukla, G. J. (1975). 'Loess stratigraphy of Central Europe', pp. 99–188 in K.W. Butzer and G.L. Isaac (eds.) *After The Australopithecines* (Mouton: The Hague).

Lachenbruch, A. H. (1962). 'Mechanics of thermal contraction cracks and ice-wedge polygons in permafrost', *Geological Society of America Special Paper*, 70, 1–69.

Lambert, J. M., Jennings, J. N., Smith, C. T., Green, C., and Hutchinson, J. N. (1960). 'The Making of the Broads', *Royal Geographical Society Research Series*, no. 3.

Langbein, W. B. and Schumm, S. A. (1958). 'Yield of sediment in relation to mean annual precipitation', *Transactions of the American Geophysical Union*, 39, 1076–84.

Leopold, L. B., Wolman, G. M., and Miller, J. P. (1964). *Fluvial Processes in Geomorphology* (W.H. Freeman: San Francisco, Calif.).

Lewis, W. V. (1932). The formation of Dungeness foreland. *Geographical Journal*, 80, 309–24.

Linton, D. L. (1955). 'The problem of tors', *Geographical Journal*, 121, 470–86.

Lipman, P. W. and Mullineaux, D. R. (eds.) (1981). 'The 1980 eruptions of Mount St. Helens, Washington', *United States Geological Survey, Professional Paper*, 1250, 1–844.

Löffler, E. (1977a). *Geomorphology of Papua New Guinea* (Australian National University Press: Canberra).

Löffler, E. (1977b). 'Tropical rainforest and morphogenetic stabilty', *Zeitschrift für Geomorphologie*, 21, 251–61.

Lowman, P. D. (1981). 'A global tectonic activity map', *Bulletin of the Association of Engineering Geology*, 23, 37–49.

Lundqvist, J. (1962). 'Patterned ground and related frost phenomena in Sweden', *Sveriges Geologiska Undersökning Årsbok*, 55, 1–101.

L'vovitch, M. I. (1961). 'The water balance of the land', *Soviet Geography*, 2, 14–28.

L'vovitch, M. I. (1970). 'World water balance (general report)', *International Association for Scientific Hydrology Publication*, 93, 401–15.

Mabbutt, J. A. (1977). Desert Landforms (Australian National University Press: Canberra).

Mackenzie, R. O. and Mitchell, B. D. (1966). 'Clay mineralogy', *Earth Sciences Reviews*, 2, 47–91.

Mackay, J. R. (1972). 'The world of underground ice', *Annals of the Association of American Geographers*, 62, 1–22.

Mackin, J. H. (1948). 'Concept of the graded river', *Geological Society of America Bulletin*, 59, 463–512.

Mankinen, E. A. and Dalrymple, G. B. (1979). 'Revised geomagnetic polarity time scale for the interval 0-5 m.y. B.P.', *Journal of Geophysical Research*, 84, 615–25.

Manley, S. and Manley, R. (1968). *Beaches: Their Lives, Legends and Lore* (Chilton: Philadelphia, Pa).

Marcussen, I. (1977). 'Supposed area wasting of the Weichselian ice sheet in Denmark', **Boreas**, 6, 167–173.

Marsh, G. P. (1864). *Man and Nature; or, Physical Geography as Modified by Human Action* (Scribners: New York).

McElhinny, N. W. (1973). *Paleomagnetism and Plate Tectonics* (Cambridge University Press: Cambridge, England).

McIntyre, A., Kipp, N. G., Bé, A. W. H., Crowley, T., Kellogg, T., Gardner, J. V., Prell, W., and Ruddiman, W. F. (1976). 'Glacial North Atlantic 18 000 years ago: a CLIMAP reconstruction', *Geological Society of America Memoir*, 145, 43–76.

McKee, E. D. (ed.) (1979). 'A study of global sand seas', *US Geological Survey Professional Paper*, 1052, 1–429.

McKenzie, G. D. and Utgard, R. O. (eds.) (1972). *Man and His Physical Environment* (Burgess: Minneapolis, Minn.).

McKnight, D. M., Feder, G. L., and Stiles, E. A. (1981). 'Toxicity of Mount St. Helens ash leachate to a blue-green alga', *US Geological Survey Circular*, 850-F, 1–14.

McMillian, N. J. (1973). 'Shelves of Labrador Sea and Baffin Bay, Canada', pp. 473–517 in 'The future petroleum provinces of Canada—their geology and potential', *Canadian Society of Petroleum Geology, Memoir*, 1, 473–517.

Meckelein, W. (1959). *Forschungen in der zentalen Sahara, Klimageomorphologie* (Georg Westermann Verlag: Braunschweg).

Michael, H. N. and Ralph, E. R. (eds.) (1971). *Dating Techniques for the Archeologist* (MIT Press: Cambridge, Mass).

Miller, V. C. (1953). 'A quantitative geomorphic study of drainage basin characteristics in the Clinch Mountain area: Va. and Tenn.', *Office of Naval Research Project NR 389–042, Technical Report*, 3, Columbia University.

Millot, G. (translated W.R. Farrand and H. Paquet) (1970). *Geology of Clays* (Springer-Verlag: New York).

Mollard, J. D. (1973). 'Airphoto interpretation of fluvial features', *Fluvial Processes and Sedimentation*, Proceedings of the Hydrology Symposium, University of Alberta, 342–80.

Morgan, A. V. (1973). 'The Pleistocene geology of the area north and west of Wolverhampton, Staffordshire, England', *Philosophical Transactions of the Royal Society, London*, 265B, 233–97.

Morgan, J. P. (1970). 'Deltas—a résumé', *Journal of Geological Education*, 18, 107–17.

Mörner, N-A. (ed.) (1980), *Earth Rheology, Isostasy and Eustasy* (Wiley: Chichester).

Morrison, R. B. (1965). 'Quaternary geology of the Great Basin', pp. 265–85 in H.E. Wright and D.G. Frey (eds.) *The Quaternary of the United States* (Princeton University Press: Princeton, NJ).

Mottershead, D. N. (1977). 'The Quaternary evolution of the south coast of England', pp. 299–320 in *The Quaternary History of the Irish Sea*, Geological Journal Special Issue, 7, 299–320.

Nairn, I. A. and Self, S. (1978). 'Explosive eruptions and pyroclastic avalanches from Ngauruhoe in February, 1975', *Journal of Volcanology and Geothermal Research*, 3, 39–60.

Nye, J. K. (1965). 'The flow of a glacier in a channel of rectangular, elliptic or parabolic cross-section', *Journal of Glaciology*, 5, 661–90.

Oberlander, T. (1965). *The Zagros Streams* (Syracuse Geographical Series no. 1, Syracuse University Press: Syracuse, NY).

Oberlander, T. M. (1972). 'Morphogenesis of granitic

boulder slopes in the Mojave desert, California, *The Journal of Geology*, 80, 1–20.

Ollier, C.D. (1969a). *Weathering* (Oliver & Boyd: Edinburgh).

Ollier, C.D. (1969b). *Volcanoes* (Australian National University Press: Canberra).

Ollier, C. D. (1981). *Tectonics and Landforms* (Longman: London).

Ollier, C. D. (1982). 'The Great Escarpment of eastern Australia: tectonic and geomorphic significance, *Journal Geological Society of Australia*, 29, 13–23.

Paylore, P. and Hanley, R. A. (1976). *Desertification: Process, Problems, Perspectives* (University of Arizona Press: Tucson, Ariz.).

Peltier, L. (1950). 'The geographic cycle in periglacial regions as it is related to climatic geomorphology', *Annals of the Association of American Geographers*, 40, 214–236.

Peterson, D. W. (1967). 'Geologic Map of the Kilauea Crater Quadrangle Hawaii, GQ-667', US Geological Survey.

Péwé, T. L. (ed.) (1969). *The Periglacial Environment* McGill-Queen's University Press: Montreal).

Péwé, T.L. (ed.) (1981). 'Desert dust: origin, characteristics, and effect on man', *Geological Society of America, Special Paper*, 186, 1–303.

Péwé, T. L. (1983). 'Alpine permafrost in the contiguous United States: a review', *Arctic and Alpine Research*, 15, 145–56.

Pickford, M. (1982). 'The tectonics, volcanics and sediments of the Nyanza Rift Valley, Kenya', *Zeitschrift für Geomorphologie, Supplementband*, 42, 1–33.

Piechocki, A. (1977). 'The Late Pleistocene and Holocene Mollusca of the Kunów region (N-E margin of the Swietokrzyskie Mts.)', *Folia Quaternaria* (Kraków), 49, 23–36.

Pitman, W. C. and Heirtzler, J. R. (1966). 'Magnetic anomalies over the Pacific-Antarctic Ridge', *Science*, 154, 1164–71.

Pomerol, C. (1973). *Ère Cénozoique* (Doin: Paris).

Potter, P. E. (1976). 'Significance and origin of big rivers', *Journal of Geology*, 86, 13–33.

Press, F. and Siever, R. (1974). *Earth* (Freeman: San Francisco, Calif.).

Price, R. A. and Doeglas, R. J. W. (eds.) (1972). 'Variations in tectonic styles in Canada', *Geological Association of Canada, Special Paper*, no. 11.

Prinz, W. (1908). *Les Cristallisations des Grottes de Belgique* (Brussels).

Pyne, S. J. (1980). *Grove Karl Gilbert: A Great Engine of Research* (University of Texas Press: Austin).

Raff, A. D. and Mason, R. G. (1961). 'Magnetic Survey off the west coast of North America, 40 °N latitude to 50 °N latitude', *Geological Society of America Bulletin*, 72, 1267–70.

Raymond, C. F. 'Flow in a transverse section of Athabasca glacier, Alberta, Canada', *Journal of Glaciology*, 10, 55–84.

Research Institute of Glaciology, Cryopedology and Desert Research, Academia Sinica, Lanchou, China (translated) (1981). *Permafrost* (Canada Institute for Scientific and Technical Information (NRC/CNR T.T.-2006): Ottawa).

Richards, K. (1982). *Rivers: Form and Process in Alluvial Channels* (Methuen: London).

Rittman, A. and L. (1976). *Volcanoes* (Orbis: London).

Rognon, P. and Williams, M. A. J. (1977). 'Late Quaternary climatic changes in Australia and North Africa: a preliminary interpretation', *Palaeogeography, Palaeoclimatology, Palaeoecology*, 21, 285–327.

Ruddiman, W. F. and Glover, L. K. (1975). 'Subpolar North Atlantic circulation at 9300 yr B.P.: faunal evidence', *Quaternary Research*, 5, 361–89.

Rutten, M. G. (1969). *The Geology of Western Europe* (Elsevier: Amsterdam).

Ruxton, B. P. and Berry, L. (1957). 'The weathering of granite and associated erosional features in Hong Kong', *Geological Society of America Bulletin*, 68, 1263–92.

Sancetta, C., Imbrie, J. and Kipp, N. (1973). 'Climatic record of the past 130 000 years in the North Atlantic deep sea core V23-82: correlation with the terrestrial record', *Quaternary Research*, 3, 110–16.

Sarnthein, M. (1978). 'Sand deserts during glacial maximum and climatic optimum', *Nature*, 272, 43–46.

Savat, J. (1975). 'Some morphological and hydraulic characteristics of river-patterns in the Zaïre Basin', *Catena*, 2, 161–80.

Scheidegger, A. (1965). 'The algebra of stream-order numbers', *US Geological Survey Professional Paper*, 525B, 187–9.

Scheidegger, A. E. (1975). *Physical Aspects of Natural Catastrophes* (Elsevier: Amsterdam).

Schumm, S. A. (1956). 'The evolution of drainage systems and slopes, in badlands at Perth Amboy, New Jersey', *Geological Society of America Bulletin*, 67, 597–646.

Schumm, S. A. (1968). 'River adjustments to altered hydrologic regimen—Murrumbidgee River and paleochannels, Australia', *US Geological Survey Professional Paper*, 598, 1–65.

Schumm, S. A. (ed.) (1972). *River Morphology: Benchmark Papers in Geology* (Dowden, Hutchinson & Ross: Stroudsburg, Penn.).

Schumm, S. A. (1977). *The Fluvial System* (Wiley: New York).

Schumm, S. A. (1979). 'Geomorphic thresholds: the concept and its applications', *Institute of British Geographers, Transactions* (New Series), 4, 485–515.

Schumm, S. A. and Lichty, R. W. (1965). 'Time, space and causality in geomorphology', *American Journal of Science*, 263, 110–19.

Schumm, S. A. and Mosley, M. P. (1973). *Slope Morphology: Benchmark Papers in Geology* (Dowden, Hutchinson & Ross: Stroudsburg, Penn.).

Searle, E. J. (1964). *City of Volcanoes* (Paul's Book Arcade: Auckland).

Sedimentation Seminar, H.N. Fisk Laboratory of Sedimentology (1977). 'Magnitude and frequency of transport of solids by streams in the Mississippi basin', *American Journal of Science*, 277, 862–75.

Selby, M. J. (1970). *Slopes and Slope Processes* (Waikato Branch, New Zealand Geographical Society).

Selby, M. J. (1974). 'Dominant geomorphic events in landform evolution', *Bulletin of the International Association of Engineering Geology*, 9, 85–9.

Selby, M. J. (1976). 'Loess', *New Zealand Journal of Geography*, 61, 1–18.

Selby, M. J. (1977). 'Transverse erosional marks on ventifacts from Antarctica', *New Zealand Journal of Geology and Geophysics*, 20, 949–69.

Selby, M. J. (1980). 'A rock mass strength classification for geomorphic purposes: with tests from Antarctica and New Zealand', *Zeitschrift für Geomorphologie*, 24, 31–51.

Selby, M. J. (1982a). *Hillslope Materials and Processes* (Oxford University Press: Oxford).

Selby, M. J. (1982b). 'Controls on the stability and inclinations of hillslopes on hard rock', *Earth Surface Processes and Landforms*, 7, 449–67.

Selby, M. J., Hendy, C. H., and Seely, M. K. (1979). 'A late Quaternary lake in the central Namib Desert, southern Africa, and some implications', *Palaeogeography, Palaeoclimatology, Palaeoecology*, 26, 37–41.

Selby, M. J., Rains, R. B., and Palmer, R. W. P. (1974). 'Eolian deposits of the ice-free Victoria Valley, southern Victoria Land, Antarctica', *New Zealand Journal of Geology and Geophysics*, 17, 543–62.

Self, S. and Rampino, M. R. (1981). 'The 1883 eruption of Krakatau', *Nature*, 294, 699–704.

Self, S. and Sparks, R. S. J. (eds.) (1981). *Tephra studies as a tool in Quaternary Research* (Reidel: Dordrecht).

Servant, M. and Servant, S. (1970). 'Les formations lacustres et les diatomées du quaternaire récent du fond la cuvette tchadienne', *Revue de Geographie Physique et Geologie Dynamique*, 12, 63–76.

Shackleton, N. J. and Opdyke, N. D. (1976). 'Oxygen-isotope and paleomagnetic stratigraphy of Pacific core V28-239, Late Pliocene to Latest Pleistocene',

Geological Society of America Memoir, 145, 449–464.

Shen, H. W. (1971). *River Mechanics* (2 Vols), (H.W. Shen: Fort Collins, Colo).

Shepard, F. P. (1981). 'Submarine canyons: multiple causes and long-term persistence', *American Association of Petroleum Geologists Bulletin*, 65, 1062–77.

Shepard, F. P. and Wanless, H. R. (1971). *Our Changing Coastlines* (McGraw-Hill: New York).

Sheridan, M. F. (1979). 'Emplacement of pyroclastic flows: a review', *Geological Society of America, Special Paper*, 180, 125–36.

Shotton, F. W. (ed.) (1977). *British Quaternary Studies* (Clarendon Press: Oxford).

Shreve, R. L. (1967). 'Infinite topologically random channel networks', *Journal of Geology*, 75, 178–86.

Simons, D. B. and Richardson, E.V. (1961). 'Forms of bed roughness in alluvial channels', *American Society of Civil Engineers, Proceedings*, 87, no. HY3, 87–105.

Slaymaker, H. O. and McPherson, H. J. (eds.) (1972). *Mountain Geomorphology: Geomorphological Processes in the Canadian Cordillera* (Tantalus Research: Vancouver).

Small, R. J. (1980). 'The Tertiary geomorphological evolution of south-east England: an alternative interpretation, Ch. 3, pp. 49–70 in D.K.C. Jones *The Shaping of Southern England* (Academic Press: London).

Smalley, I. J. (1972). 'The interaction of great rivers and large deposits of primary loess', *Transactions of the New York Academy of Sciences*, Series II, 34, 534–42.

Smith, A. G. and Hallam, A. (1970). 'The fit of the southern continents', *Nature*, 255, 139–44.

Smith, D. I. and Atkinson, T. C. (1976). 'Process, landforms and climate in limestone regions, pp. 367–409 in E. Derbyshire (ed.), *Geomorphology and Climate* (Wiley: London).

Soons, J. M. (1968). 'Canterbury landscapes: a study in contrasts', *New Zealand Geographer*, 24, 115–32.

Soons, J. M. and Selby, M. J. (eds.) (1982). *Landforms of New Zealand* (Longman Paul: Auckland).

Sparks, R. S. J., Self, S., and Walker, G. P. L. (1973). 'Products of ignimbrite eruptions', *Geology*, 1, 115–18.

Starkel, L. (1977). 'The palaeogeography of mid- and east Europe during the last cold stage, with west European comparisons', *Philosophical Transactions of the Royal Society, London*, B280, 351–72.

Stearns, H. T. (1966). *Geology of the State of Hawaii* (Pacific Books: Palo Alto, Calif.).

Steers, J. A. (1948). *The Coastline of England and Wales* (Cambridge University Press: Cambridge, England).

Steers, J. A. (1962). *The Sea Coast* (3rd edn.) (Collins: London).

Stevens, G. (1980). *New Zealand Adrift* (A.H. and A.W. Reed: Wellington).

Strahler, A. N. (1952). 'Dynamic basis of geomorphology', *Bulletin of the Geological Society of America*, 63, 923–38.

Strahler, A. N. (1964). 'Quantitative geomorphology of drainage basins and channel networks', sect. 4–11 in V. T. Chow (ed.) *Handbook of Applied Hydrology* (McGraw-Hill: New York).

Strahler, A. N. (1981). *Physical Geology* (Harper & Row: New York).

Strakhov, N. M. (1967). *Principles of Lithogenesis*, Vol. 1 (Oliver & Boyd: Edinburgh).

Sugden, D. E. (1977). 'Reconstruction of the morphology, dynamics, and thermal characteristics of the Laurentide Ice Sheet at its maximum', *Arctic and Alpine Research*, 9, 21–47.

Sugden, D. E. (1978). 'Glacial erosion by the Laurentide ice sheet', *Journal of Glaciology*, 20, 367–91.

Sugden, D. E. and John, B. S. (1976). *Glaciers and Landscape* (Arnold: London).

Suggate, R. P. (1965). 'The definition of "interglacial" ', *Journal of Geology*, 73, 619–26.

Sundborg, A. (1956). 'The River Klarälven: a study of fluvial processes', *Geografiska Annaler*, 38, 127–316.

Swan, L. W. (1967). 'Alpine and aeolian regions of the world', pp. 29–54 in H. E. Wright and W. H. Osburn (eds.), *Arctic and Alpine Environments* (Indiana University Press: Bloomington, Ill).

Sweeting, M. M. (1972). *Karst Landforms* (Macmillan: London).

Synge, F. M. (1977). 'Records of sea level during the Late Devensian', *Philosophical Transactions of the Royal Society, London*, B280, 211–28.

Takeuchi, H., Uyeda, S. and Kanamori, H. (1970). *Debate About the Earth* (Freeman Cooper: San Francisco, Calif.).

Talwani, M. and Pitman, W. C. (eds.) (1977). *Island Arcs, Deep Sea Trenches, and Backarc Basins* (Maurice Ewing Series 1, American Geophysical Union: Washington DC).

Tapponier, P. *et al.* (1982). 'Propagating extrusion tectonics in Asia: new insights from simple experiments with plasticine', *Geology*, 10, 611–16.

Thomas, M. F. (1974). *Tropical Geomorphology: A study of Weathering and Landform Development in Warm Climates* (Macmillan: London).

Thomas, M. F. (1978). 'Denudation in the tropics and the interpretation of the tropical legacy in higher latitudes—a view of British experience', pp. 185–202 in C. Embleton, D. Brunsden, and D.K.C. Jones, *Geomorphology, Present Problems and Future Prospects* (Oxford University Press: Oxford).

Thomas, M. F. and Thorp, M. B. (1980). 'Some aspects of the geomorphological interpretation of Quaternary alluvial sediments in Sierra Leone', *Zeitschrift für Geomorphologie, Supplementband*, 36, 140–61.

Thomas, W. L. (ed.) (1956). *Man's Role in Changing the Face of the Earth* (University of Chicago Press: Chicago, Ill.).

Tricart, J. (1965) (translated by C.J.K. de Jonge, 1972). *The Landforms of the Humid Tropics, Forests and Savannas* (Longman: London).

Tricart, J. (1968) (translated by S.H. Beaver and E. Derbyshire, 1974). *Structural Geomorphology* (Longman: London).

Tricart, J. (1975). 'Influence des oscillations climatiques récentes sur le modelé en Amazonie Orientale (Region de Santarem) d'après les images radar latéral', *Zeitschrift für Geomorphologie*, 19, 140–63.

Tricart, J. and Cailleux, A. (1965) (translated by C.J.K. de Jonge, 1972). *Introduction to Climatic Geomorphology* (Longman: London).

Trimble, D. E. (1981). 'The geologic story of the Great Plains', *US Geological Survey Bulletin*, 1493, 1–55.

Twidale, C. R. (1971). *Structural Landforms* (Australian National University Press: Canberra).

Twidale, C. R. (1976). 'On the survival of paleoforms', *American Journal of Science*, 276, 77–95.

Umbgrove, J. H. F. (1950). *Symphony of the Earth* (Nijhoff: The Hague).

Uyeda, S. (1978). *The New View of the Earth* (W.H. Freeman: San Francisco, Calif.).

Vail, P. R. (1977). 'Sea level changes and global unconformities from seismic interpretation', (a report of the JOIDES Subcommittee on the Future of Scientific Ocean Drilling: Woods Hole, Mass.).

Varnes, D. J. (1975). 'Slope movements in the western United States', pp. 1-17 in *Mass Wasting* (Geo Abstracts: Norwich).

Vuilleumier, B. S. (1971). 'Pleistocene changes in the fauna and flora of South America', *Science*, 173, 771–80.

Walcott, R. I. (1972). 'Past sea levels, eustasy and deformation of the Earth', *Quaternary Research*, 2, 1–14.

Walcott, R. I. and Cresswell, M. M. (eds.) (1979). 'The Origin of the Southern Alps', *Royal Society of New Zealand, Bulletin*, 18.

Walker, G. P. L. (1980). 'The Taupo pumice: product of the most powerful known (Ultraplinian) eruption?', *Journal of Volcanology and Geothermal Research*, 8, 69–94.

Walker, G. P. L. (1981). 'Generation and dispersal of

fine ash and dust by volcanic eruptions', *Journal of Volcanology and Geothermal Research*, 11, 81–92.

Warren, A. (1979). 'Aeolian processes', Ch. 10 in C. Embleton and J. Thornes (eds.), *Process in Geomorphology* (Arnold: London).

Warren, A. and Maziels, J. K. (1976). *Ecological Change and Desertification* (University College Press: London).

Washburn, A. L. (1979). *Geocryology* (Arnold: London).

Watkins, J. S., Montadert, L., and Dickerson, P. W. (eds.) (1979). *Geological and Geophysical Investigations of Continental Margins* (Memoir 29, American Association of Petroleum Geologists: Tulsa, Okla.).

Watson, E. (1977). 'The periglacial environment of Great Britain during the Devensian', *Philosophical Transactions of the Royal Society, London*, B 280, 183–98.

Wayland, E. J. (1934). 'Peneplains and some other erosional platforms', *Bulletin of the Geological Survey of Uganda, Annual Report, Notes*, 1, 74, 366.

Wegener, A. (1966). *The Origin of Continents and Oceans* (Dover: New York).

West, R. G. (1968). *Pleistocene Geology and Biology* (Longman: London).

White, G. W. (1956). '*Illustrations of the Huttonian Theory of the Earth', by John Playfair* (University of Illinois Press: Urbana, Ill.).

Whittow, J. (1980). *Disasters: The Anatomy of Environmental Hazards* (Pelican: Harmondsworth).

Wijmstra, T. A. (1975). 'Palynology and palaeoclimatology of the last 100 000 years', *World Meteorological Organisation*, WMO-No. 421, 5–20.

Wilcoxson, K. (1966). *Volcanoes* (Cassell: London).

Williams, H. and McBirney, A. R. (1979). *Volcanology* (Freeman Cooper: San Francisco, Calif).

Williams, M. A. J. and Faure, H. (eds.) (1980). *The Sahara and the Nile* (Balkema: Rotterdam).

Williams, P. W. (1972). 'Morphometric analysis of polygonal karst in New Guinea', *Geological Society of America Bulletin*, 83, 761–96.

Williams, P. W. (1976). 'Impact of urbanisation on the hydrology of Waiau Creek, North Shore, Auckland', *Journal of Hydrology* (New Zealand), 15, 81–99.

Williams, R. B. G. (1975). 'The British climate during the Last Glaciation; an interpretation based on periglacial phenomena', *Geological Journal Special Issue*, 6, 95–120.

Wilson, J. T. (ed.) (1976). *Readings from Scientific American: Continents Adrift and Continents Aground* (W.H. Freeman: San Francisco, Calif.).

Woldstedt, P. (1950). *Norddeutschland und angrenzende Gebiete im Eiszeitalter* (K.F. Koehler Verlag: Stuttgart).

Wolman, M. G. (1967). 'A cycle of sedimentation and erosion in urban river channels', *Geografiska Annaler*, 49A, 385–95.

Wolman, M. G. and Gerson, R. (1978). 'Relative scales of time and effectiveness of climate in watershed geomorphology', *Earth Surface Processes*, 3, 189–208.

Wolman, M. G. and Miller, J. P. (1960). 'Magnitude and frequency of forces in gemorphic processes', *Journal of Geology*, 58, 54–74.

Wooldridge, S. W. and Linton, D. L. (1955). *Structure, Surface and Drainage in South-east England* (George Philip: London).

Woollard, G. P. (1966). *The Earth Beneath the Continents* (Monograph No. 10, American Geophysical Union: Washington, DC).

Wright, H. E. and Osburn, W. H. (eds.) (1967). *Arctic and Alpine Environments* (Indiana University Press: Bloomington, Ill).

Wyllie, P. J. (1971). *The Dynamic Earth: Textbook in Geosciences* (Wiley: New York).

Yochelson, E. L. (ed.) (1980). 'The Scientific Ideas of G.K. Gilbert', *Geological Society of America Special Paper*, 183, 1–148.

Yoshikawa, T., Kaizuka, S., and Ota, Y. (1981). *The Landforms of Japan* (University of Tokyo Press: Tokyo).

Young, A. (1972). *Slopes* (Oliver & Boyd: Edinburgh).

Zenkovitch, V. P. (1967). *Processes of Coastal Development* (translated by D. G. Fry and edited by J. A. Steers) (Oliver & Boyd: Edinburgh).

Zinderen Bakker, E.M. Van (ed.) (1978). *Antarctic Glacial History and World Palaeoenvironments* (Balkema: Rotterdam).

Zinderen Bakker, E.M. Van (ed.) (1980). *Palaeoecology of Africa and the Surrounding Islands*, Vol. 12 (Balkema: Rotterdam) (previous volumes in the series have been published at roughly one year intervals).

Zinderen Bakker, E.M. Van and Clark, J.D. (1962). 'Pleistocene climates and cultures in north-eastern Angola', *Nature*, 196, 639–42.

Appendices

Appendix 1: *The International System (SI) of Units*

(1) The SI is based on six primary units

Quantity	Unit	Symbol
length	metre	m
mass	kilogram	kg
time	second	s
electric current	ampere	A
temperature	degree Kelvin	°K
luminous intensity	candela	cd

(Note: the spellings 'gramme' and 'meter' are used in some countries.)

(2) The following are some of the supplementary and derived units used in the system:

Quantity	Unit	Symbol
plane angle	radian	rad
area	square metre	m^2
volume	cubic metre	m^3
velocity	metre per second	m/s
acceleration	metre per second squared	m/s^2
density	kilogram per cubic metre	kg/m^3
force	newton	$N = kg\ m/s^2$
moment of force	newton metre	N m
pressure, stress	$\begin{cases} \text{newton per square metre} \\ = \text{pascal} \end{cases}$	N/m^2 Pa
viscosity: kinematic	metre squared per second	m^2/s
dynamic	newton second per metre squared	$N\ s/m^2$
work, energy, quantity of heat	joule	$J = N\ m$
power, heat flow rate	watt	$W = J/s$
temperature (customary unit)	degree Celsius	°C
density of energy flow rate	watt per square metre	W/m^2

(3) Prefixes used to extend the magnitude of the basic units are in steps of one thousand, and are:

Prefix	Symbol	Magnitude
giga	G	10^9
mega	M	10^6

kilo	k	10^3
milli	m	10^{-3}
micro	μ	10^{-6}
nano	n	10^{-9}
pico	p	10^{-12}

(4) *Writing SI units*

(a) The abbreviations do not take an 's' in the plural: thus km (NOT kms). Symbols are NOT followed by a full stop.

(b) Only one multiplying prefix is applied at one time to a given unit: thus two mega-metres squared (2 Mm^2) (NOT two kilo-kilo-metres squared).

(c) Multiplying prefixes are printed immediately adjacent to the unit symbols with which they are associated: thus km, mW.

(d) The product of two or more units is indicated by a space between unit symbols: N m (NOT Nm).

(e) The quotient of two units may be expressed by use of the solidus or of negative exponents: m/s or m s^{-1}; kg m^2/(s^3 A) or kg m^2 s^{-3} A^{-1}. The solidus must not be repeated on the same line of one expression unless ambiguity is avoided by the use of parentheses: m/s^2 or (m/s)/s, but NOT m/s/s.

(f) SI symbols and numerals should be spaced from one another: thus 13 g (NOT 13g).

(g) To facilitate reading of numbers of more than four digits, the digits should be spaced in groups of three on either side of the decimal point: 10 479.014. The comma is NOT used for spacing because in many countries it is used for the decimal point.

(h) The use of the prefixes hecta (10^2), deca (10), deci (10^{-1}), and centi (10^{-2}) is discouraged.

(5) *Notes*

(a) There is a basic coherence in the SI, for example, 1 watt = 1 joule/second exactly = 1 newton × 1 metre/1 second exactly.

(b) The kilogram is a unit of mass (i.e. quantity of matter), not of weight. Weight varies slightly with gravity at different places on Earth (and more in space) and is the force applied by a unit of mass. A force is a product of mass (kg) and acceleration (m/s^2). Thus force is kg m/s^2 and 1 kg m/s^2 = 1 N. The acceleration due to gravity on Earth is approximately 9.81 m/s^2. Thus the weight of a half kilogram mass is 0.5 × 9.81 = 4.9 N.

(c) The tonne (t) is commonly used for one thousand kilograms (1000 kg = 1 t = 1 Mg).

(d) The term 'unit weight' expresses the gravitational force per unit volume (N/m^3). The term 'density' refers to mass per unit volume (kg/m^3).

(e) The litre per second is sometimes used for small discharges: 1 m^3/s = 1000 litres per second.

Appendix 2: *Gravity anomalies*

On a planet of perfectly uniform layered rock the value of gravity at any site would be related to the shape of that planet and its rotation. It would be everywhere the same on a sphere, but on the oblate spheroid of Earth it would be rather less at the equator than the poles, because the poles are a little nearer to the centre of Earth. A departure of gravity values from those at the theoretical surface of a uniform Earth, which would be the sea-level ellipsoid (or *geoid*), is known as a *gravity anomaly*. Gravity anomalies are of two main types: 'Bouguer anomalies' which are concerned with topography, and 'isostatic anomalies' which are concerned with rock density.

Bouguer anomalies take into account three corrections: the Bouguer correction, a free-air correction, and a topographic correction. (1) The Bouguer correction arises because the mass of high mountains and plateaux exerts an attraction towards the pendulum of a gravimeter, or any other object, so the value of gravity measured at the foot of high continental landforms is less than the value measured on an open plain at the ellipsoid surface. (2) A free-air correction is made to compensate for elevation of the point at which gravity measurements are made, because the higher the elevation of the point the lower is the value of gravity, as this is controlled by distance from the Earth's centre. (3) A topographic correction is made, in addition to the Bouguer correction, because the latter assumes that the land on which the gravimeter rests is a horizontal plain.

Isostatic anomalies arise because the Earth's crust is not uniformly layered, consequently corrections have to be made for variations in rock density and the variations in thickness and composition of crustal layers (as indicated by seismic

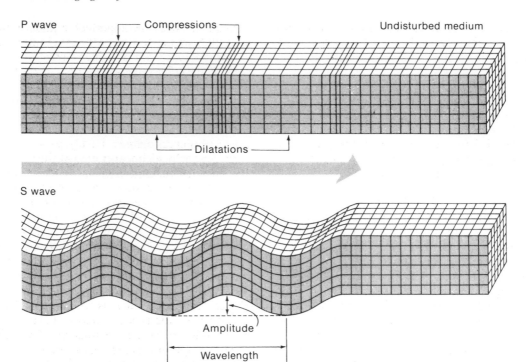

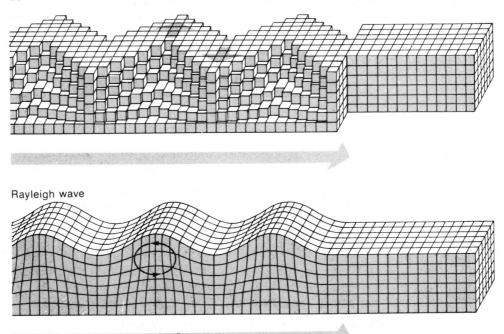

Fig. A1 A representation of the motion of seismic waves in an elastic medium (after Bolt, 1976 copyright W. H. Freeman & Co. All rights reserved).

evidence). Much of the crust is thought to be close to isostatic equilibrium but small anomalies are recognizable.

Appendix 3: *Seismic Waves*

Pressure applied to a rock surface first deforms the rock elastically. If the pressure is sustained, the rock may deform plastically or, if the pressure is great enough, it may break. Seismic waves at some distance from the focus of an earthquake are results of purely elastic behaviour.

Within the body of an elastic material only two types of seismic wave can propagate. The faster of these is a *P wave* (primary wave) which consists of compressions and dilations of the elastic material. This behaviour is analogous to that of sound waves or the backward and forward motion of a coil spring. The slower type of body wave is an *S wave* (secondary wave) which has motion vertically at right angles to the direction of travel of the energy, behaviour analogous to that of light waves or a vibrating guitar string (Fig. A1). The elastic body is sheared and twisted as the wave travels through it. Because a perfect liquid cannot be sheared S waves do not propagate in liquids.

Both P and S waves which arrive at the land surface are refracted and reflected like sound and light, with one type of wave generating the other type at boundaries of different materials. Multiple reflections of body waves between the surface of the Earth and some boundary below, where the elastic properties change sharply, produce *L waves* (Love waves) which travel only at the surface and with a vibratory motion horizontally at right angles to the direction of travel. A second type of surface wave is the *Rayleigh wave*. Its particle motion is like that of a sea wave with the surface rock moving in an elliptical path, but with an amplitude that decreases with depth until it ceases.

Surface waves, because of their large amplitude, produce the most violent earthquake shocks. The *Richter scale* is used to measure their magnitude. The scale of magnitudes is so arranged that each unit on the scale is equivalent to thirty times the energy released by the previous unit. Precisely, the Richter magnitude is the logarithm (to base 10) of the maximum seismic wave amplitude measured at

Table A1   The Modified Mercalli (M.M.) intensity scale of earthquakes (1931)

I   Not felt except by a very few under especially favourable circumstances.

II   Felt only by a few persons at rest, especially on upper floors of buildings. Delicately suspended objects may swing.

III   Felt quite noticeably indoors, especially on upper floors of buildings, but many people do not recognize it as an earthquake. Standing motor cars may rock slightly. Vibration like passing of truck. Duration estimated.

IV   During the day felt indoors by many, outdoors by few. At night some awakened. Dishes, windows, doors disturbed; walls make cracking sound. Sensation like heavy truck striking building. Standing motor cars rocked noticeably.

V   Felt by nearly everyone, many awakened. Some dishes, windows, etc., broken; a few instances of craked plaster; unstable objects overturned. Disturbances of trees, poles, and other tall objects sometimes noticed. Pendulum clocks may stop.

VI   Felt by all, many frightened and run outdoors. Some heavy furniture moved; a few instances of fallen plaster or damaged chimneys. Damage slight.

VII   Everybody runs outdoors. Damage negligible in buildings of good design and construction; slight to moderate in well-built ordinary structures; considerable in poorly built or badly designed structures; some chimneys broken. Noticed by persons driving motor cars.

VIII   Damage slight in specially designed structures; considerable in ordinary substantial buildings, with partial collapse; great in poorly built structures. Panel walls thrown out of frame structures. Fall of chimneys, factory stacks, columns, monuments, walls. Heavy furniture overturned. Sand and mud ejected in small amounts. Changes in well water. Persons driving motor cars disturbed.

IX   Damage considerable in specially designed structures; well-designed frame structures thrown out of plumb; great in substantial buildings, with partial collapse. Buildings shifted off foundations. Ground cracked conspicuously. Underground pipes broken.

X   Some well-built wooden structures destroyed; most masonry and frame structures destroyed with foundations; ground badly cracked. Rails bent. Landslides considerable from river banks and steep slopes. Shifted sand and mud. Water splashed (slopped) over banks.

XI   Few, if any, (masonry) structures remain standing. Bridges destroyed. Broad fissures in ground. Underground pipelines completely out of service. Earth slumps and land slips in soft ground. Rails bent greatly.

XII   Damage total. Practically all works of construction are damaged greatly or destroyed. Waves seen on ground surface. Lines of sight and level are distorted. Objects are thrown upward into the air.

a distance of 100 km from the epicentre. A magnitude of 2 is hardly felt, while the Richter magnitude 7 is the lower limit of an earthquake which has a devastating effect over a large area. The greatest magnitudes reached this century have been 8.9.

For many purposes a scale of felt intensity, or damage, is useful and the scale devised by Mercalli in 1902, and modified in 1931, is used for this. It is a 12-point scale and describes the conditions at a site.

Index